# International Scientific Conference of Communications, Information, Electronic and Energy Systems (CIEES 2020)

IOP Conference Series: Materials Science and Engineering
Volume 1032

Borovets, Bulgaria
26 - 29 November 2020

ISBN: 978-1-7138-4557-7
ISSN: 1757-8981

**Printed from e-media with permission by:**

Curran Associates, Inc.
57 Morehouse Lane
Red Hook, NY 12571

CURRAN ASSOCIATES INC. proceedings .com

**Some format issues inherent in the e-media version may also appear in this print version.**



Institute of Physics
Dirac House, Temple Back
Bristol BS1 6BE UK

Phone: 44 1 17 929 7481
Fax: 44 1 17 920 0979

techtracking@iop.org

**Additional copies of this publication are available from:**

Curran Associates, Inc.
57 Morehouse Lane
Red Hook, NY 12571 USA
Phone: 845-758-0400
Fax: 845-758-2633
Email: curran@proceedings.com
Web: www.proceedings.com

# TABLE OF CONTENTS

IOP Conf. Series: Materials Science and Engineering 1032 (2020) 011001 doi:10.1088/1757-899X/1032/1/011001

# The CIEES 2020 Conference Chair's Preface

Welcome to the International Scientific Conference of Communications, Information, Electronic and Energy Systems – CIEES 2020. The goal of the CIEES 2020 conference is to provide an international forum for researchers, academics, people in industry, and students to address recent research results and to present and discuss their ideas, theories, technologies, systems, tools, applications, work in progress and experiences on all theoretical and practical issues arising in telecommunications, electronics, signal processing and applications, information technologies and energy.

CIEES 2020 is scheduled to be held on 26 – 29 November, 2020 in Borovets, Bulgaria. The year of 2020 has been unprecedented with the COVID-19 pandemic disrupting our life. Due to the recent coronavirus outbreak, the Organizing Committee and the CIEES 2020 Program Committee, saw that it would not have been possible to hold CIEES 2020 in November in person, we decided to turn CIEES 2020 in a virtual format – online conference on the same period. The health and safety of the participants involved in CIEES 2020 is our priority. We all know that the value of CIEES is not only in listening to the papers, but in the networking, the face-to-face interactions, the exchange of ideas, also the gathering in the social events. Many things are changing and our perceptions are different these days. It may happen that we have to rethink the way we are used to organize the conferences.

The International Scientific Conference of Communications, Information, Electronic and Energy Systems – CIEES 2020 was held from November 26 to 28, 2020 virtually in Ruse, Bulgaria, organized by the Association CIEES. The partners of the conference were the University of Ruse (Bulgaria), Politehnica University of Bucharest (Romania), University "Vasile Alecsandri" of Bacau (Romania), Union of Scientists in Ruse (Bulgaria) and Encon Services Ltd (Bulgaria).

The technical program of this year's edition of the conference consists of two distinguished keynotes by world renowned experts with their lecture as following: Prof. Dr. Renato Filjar of Croatia, title of the lecture "*Resilient satellite navigation empowers modern science, economy, and society*", and Prof. Dr. Goran T. Djordjevic of Serbia, title of the lecture "*Performance of free-space optical systems in the presence of receiver imperfections*".

All authors prepared a video presentation up to 12 minutes that were played on stream with authors participating in the discussion during the session. All of the video presentation are available on the conference YouTube channel. The conference programme consists of 5 tracks: Telecommunications, Information Technologies, Electronics, Energy and Renewable Sources and Smart Farming. We are open 10 sessions, some of them in parallel. The time spent to deliver the video presentation by each speaker was 10 - 12 minutes. The maximum time slot for presentation and Q&A was 15 minutes. Questions were asked in the chat section of the zoom channel or by raised hand when the video presentation is finished. All Q&A were made at face-to-face in the Zoom channel, between the CIEES Host (technical moderator), Session Chair and participants. The Organizers provided to all participants a virtual room named CIEES 2020 Virtual Lounge Area, to continue the discussion outside the sessions.

The research track attracted submission from 187 distinct authors from 14 countries spanning four continents. Submissions were rigorously reviewed by an International Scientific Committee and external reviewers. A total of 190 reviews were provided, with each paper receiving at least two reviews. During the discussion phase, International Scientific Committee members and the Programme Committee members discussed the reviews in detail. The rebuttals from the authors were also seriously considered. For consistency across tracks, final decisions on papers were made at the face-to-face Programme Committee meeting between the Organizing Committee chairs and the Programme Committee chairs.

During the online conference, we used Zoom Meeting platform for opening ceremony, all conference sessions and meetings. In advance the Conference Organizing Committee of CIEES 2020 organized a test sessions for participants, Conference chairs and plenary lecturers. The Zoom Meeting platform permissions to 100 person in the one session, record all events, open more session in the same time, chatting, rise hand and the CIEES 2020 technical moderator (host), sessions chair (chair) and co-chair (co-host) have been manage it.

This year's CIEES 2020 conference has received a record high number of submissions 115 abstracts and 95 full papers submitted. After a single-blind review process were accepted 55 papers (57.9% acceptance rate) for presentation as full papers at the conference. The decision to accept a paper was entirely based on quality and not its length.

We are thankful to all the people that helped us for organizing the conference. In particular, special thanks to Prof. Dr. Iliya Iliev from Bulgaria, Prof. Dr. Renato Filjar from Croatia, the Conference Chairs, the Program Committee members, the authors and the reviewers for their hard work, efforts and expertise in making a successful technical program. Without them, this event would have never happened.

Finally, we grateful to all members of the International Scientific Committee, for providing insightful guidelines and support in preparing and organizing the conference, and for the highly professional work done on all aspects of the local organization of the Conference.

We also record our sincere gratitude to our publishing partner IOP Publishing, United Kingdom for their collaboration and full cooperation.

CIEES is an annual conference. For more information about this year conference and the upcoming conferences, please visit the conference website: http://ciees.eu.

We are looking forward to the next International Scientific Conference of Communications, Information, Electronic and Energy Systems – CIEES 2021 that will be held on November 2021 in Ruse, Bulgaria. We will enjoy together not only the conference but also the special "light" of city of Ruse and the wonderful view of the Danube River.

**Teodor ILIEV**
*General Conference Chair*
*tiliev@uni-ruse.bg*

**Ivaylo STOYANOV**
*General Conference Chair*
*stoyanov@uni-ruse.bg*

IOP Conf. Series: Materials Science and Engineering 1032 (2020) 011001 doi:10.1088/1757-899X/1032/1/011001

# CIEES 2020 ORGANIZING COMMITTEE

**General Conference Chairs:**

Teodor ILIEV (Bulgaria)

Ivaylo STOYANOV (Bulgaria)

**Honorary Chair:**

Hristo BELOEV (Bulgaria)

**Technical Programme Chair:**

Jacob FANTIDIS (Greece)

**Scientific Committee Chair:**

Bogdan POPA (Romania)

**Awards Committee Chair:**

Iliya ILIEV (Bulgaria)

**Guest Editors:**

Jordan HRISTOV (Bulgaria)

Renato FILJAR (Croatia)

**Programme Committee**

**Chair**: Filip TSVETANOV (Bulgaria)

Gani BALBAYEV (Kazakhstan)

Goran DJORDJEVIC (Serbia)

Dejan MILIC (Serbia)

Elena IVANOVA (Bulgaria)

Jacob FANTIDIS (Greece)

Eleni PAPADOPOULOU (Greece)

**Web Chair:**

Yordan DOYCHINOV (Bulgaria)

IOP Conf. Series: Materials Science and Engineering 1032 (2020) 011001 doi:10.1088/1757-899X/1032/1/011001

# CIEES 2020 INTERNATIONAL SCIENTIFIC COMMITTEE

**Chair**: Renato FILJAR (Croatia)

Carol SCHNAKOVSZKY (Romania)

Felipe PENARANDA-FOIX (Spain)

Valentin NEDEFF (Romania)

Sclahattin KOSUNALP (Turkey)

Filip TSVETANOV (Bulgaria)

Ivanka GEORGIEVA (Bulgaria)

George CULEA (Romania)

Panagiotis KOGIAS (Greece)

Svilen RACHEV (Bulgaria)

Bogdan POPA (Romania)

Iliya ILIEV (Bulgaria)

Frank GRÄBNER (Germany)

Dragos-Alexandru ANDRIOAIA (Romania)

Ozgur CULFA (USA)

Lawal Mohammed BELLO (Nigeria)

Gani BALBAYEV (Kazakhstan)

Goran DJORDJEVIC (Serbia)

Petru-Gabriel PUIU (Romania)

Dejan MILIC (Serbia)

Veneta ALEKSIEVA (Bulgaria)

# Peer review declaration

All papers published in this volume of *IOP Conference Series: Materials Science and Engineering* have been peer reviewed through processes administered by the Editors. Reviews were conducted by expert referees to the professional and scientific standards expected of a proceedings journal published by IOP Publishing.

- **Type of peer review: Single-blind**
- **Conference submission management system: Microsoft CMT - https://cmt3.research.microsoft.com/**
- **Number of submissions received: 95**
- **Number of submissions sent for review: 95**
- **Number of submissions accepted: 54**
- **Acceptance Rate (Number of Submissions Accepted / Number of Submissions Received X 100): 56.8%**
- **Average number of reviews per paper: 3.1**
- **Total number of reviewers involved: 61**
- **Any additional info on review process:** All manuscripts submitted to the CIEES 2020 conference were single-blind peer-reviewed and are carried out to meet the scientific criteria of novelty and academic excellence. Submissions were rigorously reviewed by International Scientific Committee and external reviewers. A total of 190 reviews were provided, with each paper receiving at least two reviews. During the discussion phase, International Scientific Committee members and the Programme Committee members discussed the reviews in detail. The rebuttals from the authors were also seriously considered. The final decision to accept a paper was entirely based on quality and not its length.
- **Contact person for queries:** Assoc. Prof. Teodor Iliev, PhD, Department of Telecommunications, University of Ruse, Bulgaria, E-mail: tiliev@uni-ruse.bg

# Resilient satellite navigation empowers modern science, economy, and society

**R Filjar**[1,2] **M C Damas**[3] **and T B Iliev**[4]

[1] Faculty of Engineering, University of Rijeka, Rijeka, Croatia
[2] Krapina University of Applied Sciences, Krapina, Croatia
[3] City University of New York (CUNY), New York, NY, USA
[4] Department of Telecommunications, University of Ruse, 8 Studentska Street, Ruse, 7017, Bulgaria

E-mail: renato.filjar@gmail.com

**Abstract.** Established on the exact mathematical principles, understanding of radio-wave propagation, and statistical signal processing and information theory, satellite navigation has become an essential cornerstone of modern civilisation, and an indispensable component of the national infrastructure. The increasing number of both navigation and non-navigation applications of the Global Navigations Satellite System (GNSS) utilise its Positioning, Navigation, and Timing (PNT) service for technology and business development, daily operation of technology and socio-economic systems, and improvement of the quality of life. The inherent shortcomings and limitations of GNSS ask for a transition towards the GNSS resilient to natural and artificial detrimental effects, which degrade the GNSS positioning performance. Here a systematic overview of the causes of the GNSS positioning performance degradation is outlined. Recent developments in mathematics and computer science are discussed as fundamental in the GNSS resilience development. Formulation of the Satellite-Positioning-as-a-Service (SPaaS) is outlined, as the new fundamental paradigm for resilient GNSS. Finally, the effects of the SpaaS on the wide range of GNSS applications in science, economy, and society are discussed. The contributions to SPaaS concept and the related developments through the application of statistical learning, mathematical methods and models development, and applications discussed result from the author's involvement in numerous international strategic, technology, regulatory, standardisation, business, and academic education development and collaboration activities.

## 1. Introduction

Navigation and orientation have remained the essential human virtues throughout the history of humanity. They have ensured scalable features, ranging from the individual survival and safety, to sustainable economic development and operations. The related human skills have been accompanied and augmented with various technology assistance. Satellite-based position determination (satellite positioning) is the most recent, and the most powerful one. A triumphant synergy of mathematics, science, and engineering, satellite positioning allows for estimation of the user's position, velocity, and time at the unprecedented accuracy and robustness levels, based on the sole reception of satellite-originated radio signals.

With non-navigation applications overtaking the navigation ones, satellite navigation has become a cornerstone of modern society, economy and everyday life [1, 2]. The Global Navigation Satellite

System (GNSS) empowers a growing number of technology and socio-economic systems, while enabling development and operations of a wide range of their applications [1, 2]. As with the other human-made technologies, satellite navigation is not faultless, since it inherits the intrinsic shortcomings and vulnerabilities of its core technologies [3, 4, 5, 6, 7]. Research efforts are now committed world-wide to the establishment of the GNSS resilient to natural and artificial sources of disruptions and operational positioning performance deterioration [4]. The ultimate goal is to secure satellite positioning as an essential component of the national infrastructure, thus allowing for sustainable socio-economic development based on both the navigation and non-navigation GNSS applications.

Here we propose a solution for the resilient GNSS in the form of the Satellite-Positioning-as-a-Service (SPaaS). Challenges of traditional concept of satellite positioning are addressed and discussed. Satellite navigation is considered from its core perspectives of mathematics and computer science. Modern design of GNSS user receivers using the Software-Defined Radio (SDR) approach allows for spreading and distributing the satellite positioning process to a computing service, rendering it optimised, and more robust and efficient [3, 8, 9]. The manuscript concludes with the benefits, opportunities and challenges of the proposed Satellite Positioning-as-a-Service (SPaaS) concept.

## 2. A Traditional Concept of Satellite Positioning

The satellite-based positioning is based on the accurate measurement of time a satellite signal needs to propagate between a satellite and a receiver aerials [6, 10, 11]. Satellite positioning process is conducted within the system architecture that comprises the four essential segments, as depicted in figure 1: (i) satellite (space) segment, (ii) control segment, (iii), user segment, and (iv) positioning environment (propagation media) [6].

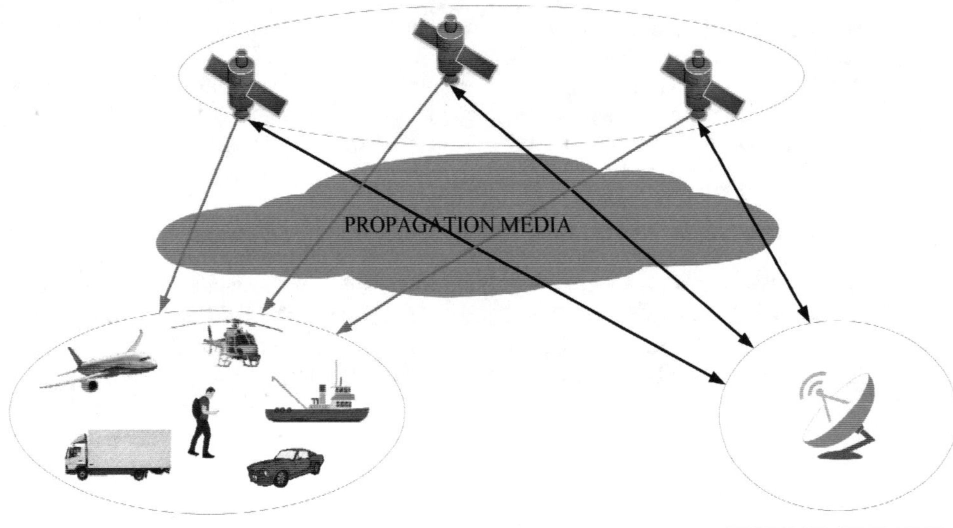


**Figure 1.** The four-segment architecture of a satellite positioning system.

The process requires the fulfilment of three essential presumptions [6, 12]: (i) utilisation of the common spatial reference co-ordinate system (World Geodetic System 84, WGS84, in case of the US-operated Global Positioning System, GPS), (ii) utilisation of the common time frame (Universal Time Co-ordinated, UTC), (iii) satellite signals propagation at the constant velocity equal to velocity of light in vacuum. The (i) presumptions ensures compliance in position expression throughout the architecture segments. The (ii) presumption allows for the precise satellite signal propagation time measurement. The utilisation of the (iii) presumption renders the measured propagation time equivalence with the propagated distance, the so-called pseudorange. The aim of the satellite positioning process is to determine values of the three components of the user position $x_u$, $y_u$, $z_u$ in the common reference system, and the user receiver clock error $b$, since it fails initially to synchronise with the common UTC time-frame [6, 10, 12, 13]. The unknowns may be expressed in the form of the user state vector $\vec{x}$:

$$\vec{x} = (x_u, y_u, z_u, b) \tag{1}$$

The pseudorange measurement from the i-th satellite $\rho_i$, elements of the user state vector (1), the estimated i-th satellite position co-ordinates $x_{si}$, $y_{si}$, and $z_{si}$, and velocity of the satellite signal propagation c form the satellite positioning mathematical model, as shown in (2) [6, 12].

$$\rho_i = \sqrt{(x_{si} + x_u)^2 + (y_{si} + y_u)^2 + (z_{si} + zx_u)^2} + c \cdot b + \varepsilon_i \tag{2}$$

The inevitable satellite-related pseudorange measurement errors are expressed through the introduction of the random variable $\varepsilon_i$. A numerical solution that yields (1) from (2) exists, for the system of equations of the type (2), with simultaneous pseudorange measurements from at least four satellites [6, 7, 12, 13].

The non-linear nature of the model, and the presence of the random variable in the model require utilisation of statistical modelling and advanced optimisation techniques to solve the system of equations for the user state vector, while at the same time mitigating the pseudorange measurement errors [13]. Using the common practice approach, the GNSS pseudoranges are corrected for the known systematic errors (ionospheric delay, tropospheric delay, and satellite clock error) using standard GNSS models [12]. Measured pseudoranges still remain contaminated with the random measurement errors [6, 7].

The satellite-based positioning process is traditionally user-centric [8, 11]. A GNSS receiver spreads the process in three processing domains, as depicted in figure 1: (i) Radio Frequency Domain (RFD), (ii) Base-Band Domain (BBD), and (iii) Navigation Domain (ND). A modern approach in the GNSS receiver design utilises the mathematical processing methods deployment with the so-called Software-Defined Radio (SDR) approach [3, 6]. The utilisation of the processing specific algorithmic realisation of mathematical methods on the general purpose computing devices, instead of approximate implementation as processing specific electronic circuits, has improved the satellite positioning process enormously, rendering it flexible, adaptable, re-configurable, and transparent for customised improvements and modifications [9, 14].

Traditionally, RFD serves for the initial received signals conditioning and A/D conversion with digitalisation. GNSS pseudoranges are measured in the satellite positioning process using the satellite-specific Pseudo-Random Noise (PRN) code sequences within the BBD, while the position estimation is performed in ND [6, 7, 11, 12]. Measured raw GNSS pseudoranges are treated in the ND with the standard correction models for ionospheric and tropospheric delays, and satellite clock errors, before initiating the GNSS position estimation process [6, 10]. Velocity estimation is conducted separately from the position estimation process, using Doppler effect measurements [12].

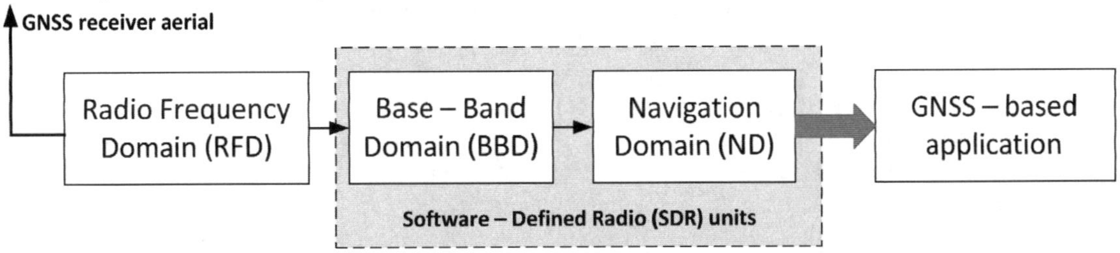


**Figure 2.** The GNSS receiver processing domains.

Commonly used algorithm for the GNSS position estimation is of iterative nature, and exploits a simple residual analysis, as evident from the equation (3), presented in the case of the four simultaneous GNSS pseudorange measurements ($\rho_1$, $\rho_2$, $\rho_3$, $\rho_4$). Additionally, the GNSS position estimation algorithm utilises the precise satellite positions ($x_i$, $y_i$, $z_i$), $i$=1…,4 at the time of the satellite signal broadcast, determined by a GNSS receiver using the satellite orbit parameters provided in the separate Navigation Message, and velocity of electromagnetic wave (light) in vacuum, $c$.

$$\begin{bmatrix} x_{k+1} \\ y_{k+1} \\ z_{k+1} \end{bmatrix} = \begin{bmatrix} x_k \\ y_k \\ z_k \end{bmatrix} + \begin{bmatrix} \Delta x \\ \Delta y \\ \Delta z \end{bmatrix}$$

$$\begin{bmatrix} \Delta x \\ \Delta y \\ \Delta z \\ c \cdot b \end{bmatrix} = \begin{bmatrix} \frac{x_k - x_1}{R_1,k} & \frac{y_k - y_1}{R_1,k} & \frac{z_k - z_1}{R_1,k} & 1 \\ \frac{x_k - x_2}{R_2,k} & \frac{y_k - y_2}{R_2,k} & \frac{z_k - z_2}{R_2,k} & 1 \\ \frac{x_k - x_3}{R_3,k} & \frac{y_k - y_3}{R_3,k} & \frac{z_k - z_3}{R_3,k} & 1 \\ \frac{x_k - x_4}{R_4,k} & \frac{y_k - y_4}{R_4,k} & \frac{z_k - z_4}{R_4,k} & 1 \end{bmatrix} + \begin{bmatrix} \rho_1 \\ \rho_2 \\ \rho_3 \\ \rho_4 \end{bmatrix} \tag{3}$$

$$R_i,k = \sqrt{(x_k - x_i)^2 + (y_k - y_i)^2 (z_k - z_i)^2}, i \text{...index of the satellite}, k - \text{th iteration}$$

While computationally efficient, the algorithm (3) is still a compromise between the achievable accuracy and computational complexity, the latter usually constrained with the battery drain on mobile devices [6, 12, 13].

The authors in [13] proposed an advanced approach that mitigates targeted natural sources of GNSS positioning performance degradation (random effects of ionospheric delay, tropospheric delay, and multipath effects, in particular) at the cost of the increased computational complexity.

The standard GNSS Position, Navigation, and Timing (PNT) service extends the position estimation with the accuracy of several metres, in positioning environment conditions that allow for good visibility of horizon (i. e. good availability of satellite signals), and common range of satellite signal delays [12]. The main causes of the GNSS positioning errors are unequal spatial distribution of satellites, as seen from the perspective of receiver, and the satellite signal delays caused by the physical effects of the positioning environment (ionospheric delay, as the most prominent, multipath effects, dangerous for their random nature, and tropospheric delay) [6, 12, 15]. The positioning environment conditions may deteriorate and even temporarily disrupt the GNSS PNT service, causing degradation of the GNSS positioning performance quality and system outages [4, 8, 15]. Development of a robust and resilient GNSS has become a mandatory accomplishment to sustain the GNSS-based modern civilisation.

## 3. GNSS Applications as the Foundation of Modern Science, Economy, and Society

Satellite navigation technology was developed with the concern of navigation applications. However, a wide proliferation of GPS/GNSS receivers in general public kicked off immediately the development of a range of GNSS-based applications (systems and services) that do not necessarily concern navigation tasks. Market acceptance of satellite navigation technology renders it an essential element of the national infrastructure. Science, economy, and society utilise GNSS as an enabler for sustainable development. Development of commercial applications of GNSS initiated formation of the GNSS business development environment (figure 3), since GNSS has been a frequent and unavoidable compound in multidisciplinary synergy of technologies and protocols.

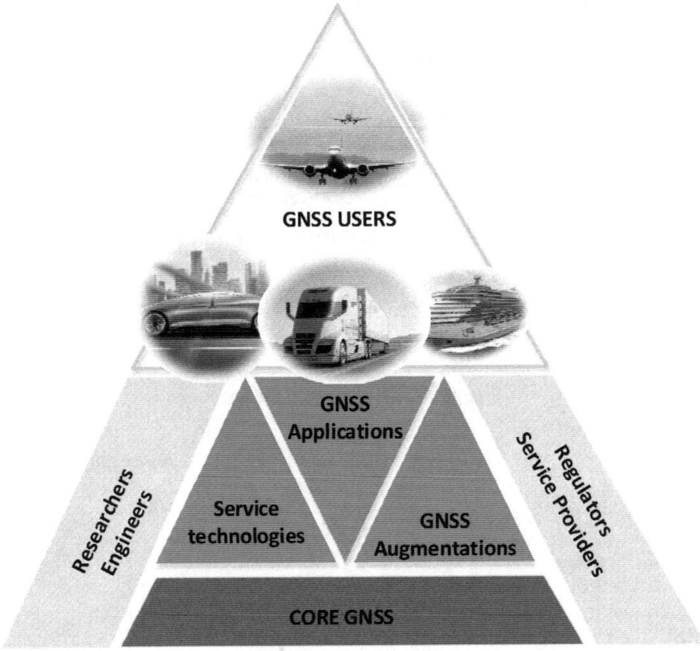


**Figure 3.** The GNSS applications business environment.

The core GNSS system provide PNT services with the quality of service that should be considered the best estimate of the user state vector, considering the current general condition of the positioning environment assuming the prefect conditions in the direct surroundings of a GNSS receiver, in order to render the GNSS positioning performance comparable across the range of potential applications. The GNSS system operators declare the core GNSS PNT performance in the specification that are available publicly. An example of the common practice is [12] for the US system GPS.

While core GNSS positioning performance describes the best available PNT service quality, the requirements for the GNSS positioning performance for particular GNSS applications should be addressed from a completely different perspective. Since the best GNSS positioning performance generally requires the additional effort and resources (sources of augmentation data, more dedicated position estimation algorithms etc.), it is essential not what a particular GNSS application would need for its flawless operation. A GNSS application needs neither the best, but the most suitable GNSS positioning performance, based on the conditions of positioning environment, as well as optimised utilisation of the communication and computational resources for GNSS position estimation. The establishment of a systematic classification of GNSS applications and their particular requirements for the GNSS PNT services quality determine a framework for systematic and sustainable GNSS application development.

Attempts were made to systematise GNSS applications per disciplines and market segments concerned. Telecommunications standardisation yield a set of requirements for the GNSS utilisation for Location-Based Services in mobile networks [2, 16, 17, 18] attempted at development of a cross-disciplinary classification of the GNSS-based services, but fails short from completing a detailed list of the individual GNSS applications, as well as in presenting their justified and quantitative requirements for the GNSS position estimation accuracy.

The European GNSS Agency conducted the most detailed and systematic study of the GNSS positioning performance requirements and the GNSS application classification across the wide range of disciplines [1]. Published on-line in a sense of the book series, the study addressed GNSS application classification and the requirements for GNSS position accuracy in the following disciplines, and market segments: Agriculture, Aviation, Location-Based Services, Maritime and Inland Waterways, Rail Users, Road User, Surveying User, and Time & Synchronisation User Needs. The study concerned both navigation and non-navigation GNSS applications.

A thorough and methodical classification and GNSS applications positioning performance requirements studies, such as the one performed by European GNSS Agency [1], emphasise the need for the provision of scalable and adaptable GNSS applications positioning performance for sustainable GNSS applications development with the robust quality of service.

## 4. Satellite Positioning-as-a-Service as an ICT Solution for Resilient GNSS

It was shown in previous Sections that the traditional satellite-based positioning approach assumes a stand-alone mobile unit that performs signal and information processing in all three domains of satellite positioning process (figure 1). The approach ensures the position estimate stays with the equipment owner, unless he or she agrees to share it through some form of a GNSS service. The concept stretches both energy and computational resources of a mobile unit. Furthermore, it limits the prospects of scaling levels of GNSS positioning performance, and utilisation of the high-accuracy and high-reliability positioning that requires additional infrastructure for provision of augmentation through third-parties information sources. Concerning the Navigation Domain (ND) in particular, the core GNSS receiver concept relies upon the measured GNSS pseudoranges, and the information distilled from the received Navigation Message to produce the GNSS position estimates. The task is conducted using the only position estimation algorithm deployed in a GNSS receiver, without opportunity to make any selection. Measured GNSS pseudoranges are corrected for known systematic errors using the standard correction models for ionospheric and tropospheric delay, and satellite clock errors, prior to be processed with the position estimation algorithm (figure 4.).

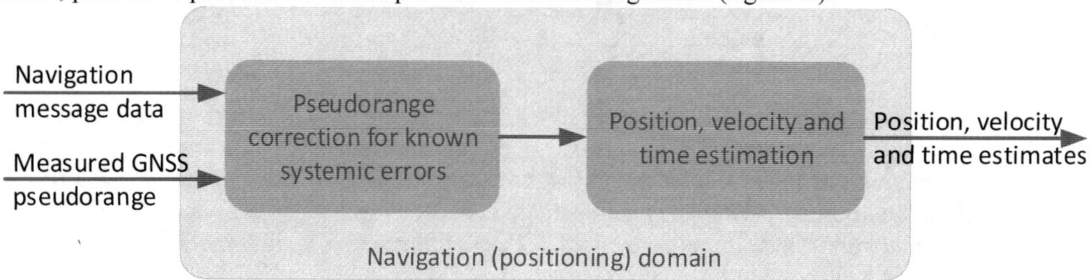


**Figure 4.** Traditional GNSS position estimation process in Navigation Domain.

Here we propose a redesigned approach, based on utilisation of contemporary ICT infrastructure and advances in signal processing, and computer science. Dubbed the Satellite-Positioning-as-a-Service, the proposed concept splits the responsibilities for signal and information processing between a mobile unit and a supporting cloud computing infrastructure, in a sense presented in figure 5.

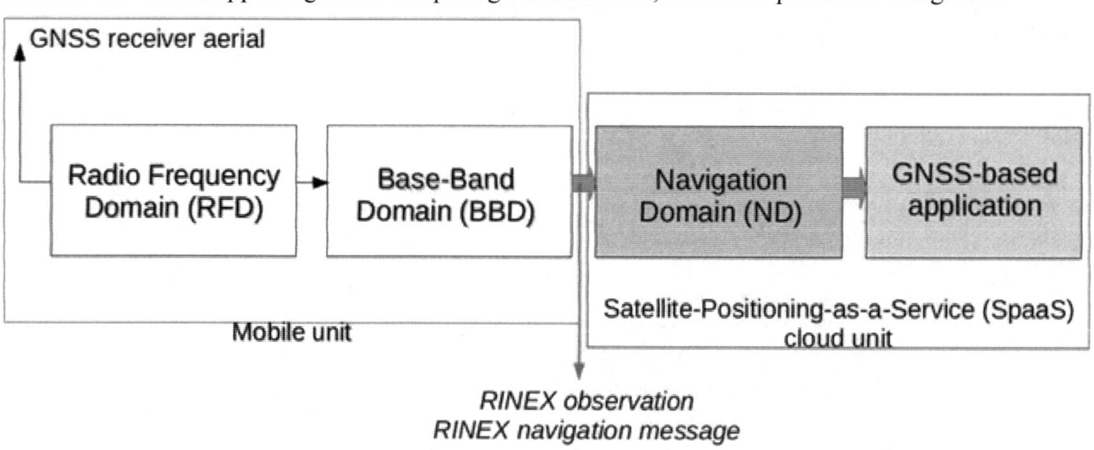


**Figure 5.** Architecture of Satellite-Positioning-as-a-Service.

In the proposed concept, a mobile unit (a user GNSS receiver) takes responsibility for satellite signals reception and conditioning, and GNSS pseudorange measurement. Measured GNSS pseudoranges and a received GNSS Navigation Message (NM) are then transferred to a cloud computing-based facility that performs the ND processing (figure 6).

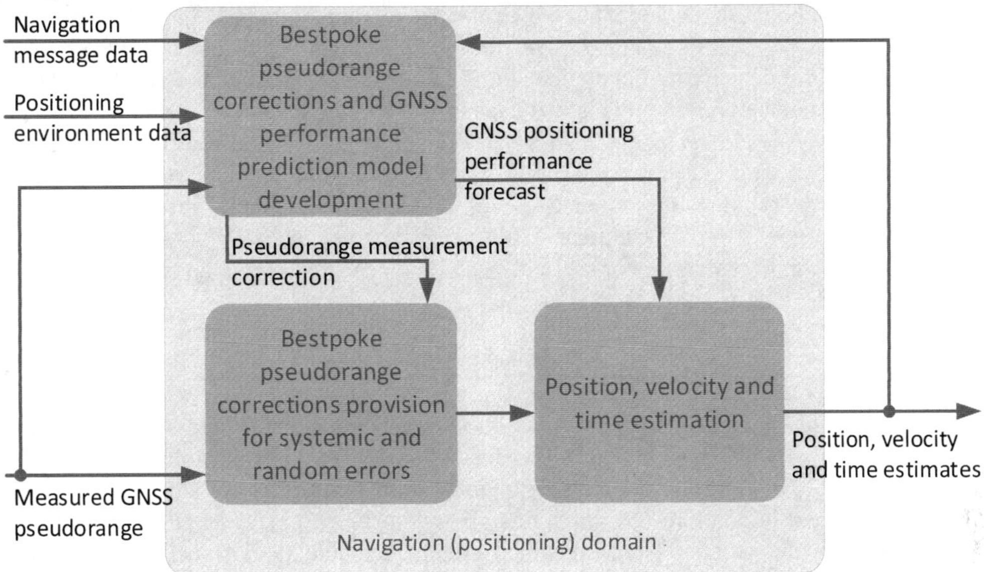


**Figure 6.** Adaptive GNSS position estimation process as the foundation of Satellite-Positioning-as-a-Service.

Tasks performed in the SpaaS unit may involve: (i) correction for ionospheric, tropospheric, satellite clock errors, and multipath effects, using more advanced and accurate correction model, and (ii) utilisation of the advanced computationally intensive GNSS position estimation algorithms (for instance, Weighted Least-Squares methods that target remaining random ionospheric delay effects). Observations (GNSS pseudoranges and NM) from mobile unit may be structured in a standardised manner, using formats common in the discipline, such as RINEX, as already deployed in the Android-based smartphone devices [19].

The Satellite-Positioning-as-a-Service (SpaaS) unit of the proposed system may take the advantage of the availability of more accurate and detailed information on the GNSS positioning environment conditions, and access to more accurate error compensation information and/or correction models, both provided either by third-parties (such as NASA, NOAA, ESA, national meteorological offices etc. that provide near real-time data openly available through) or using a private sensing network. Collection of both the broadcast and received NM may serve in the counter-spoofing process, thus rendering satellite positioning more resilient to adverse GNSS information security attacks [3].
The proposed SpaaS concept reveals the mobile unit from a demanding computational tasks in ND, thus saving energy, and rendering computational resources available for the other purposes.

## 5. Discussion

The proposed SpaaS concept offers numerous processing and computational efficiency advantages. In the Proof-of-Principle (PoP) development, we demonstrated the SPaaS unit deployed in the R environment for statistical computing. We utilised a simple interfaces to space weather condition data provided by NOAA, and meteorological data provided by Croatian Meteorological Office to serve the purpose of the accurate description of the positioning environment. For demonstration purposes, we deployed our own GNSS position estimation algorithm based on Weighted Least-Squares to compensate for the remaining ionospheric random error in GNSS pseudoranges. We also deplyoed our own local machine learning-based model of ionospheric delay, instead of the standard Klobuchar model usually deployed in the GPS position estimation. A rich fund of additional R libraries allows for implementation of: (i) different GNSS position estimation methods, and (ii) tailored GNSS pseudorange measurement correction models for mitigation of ionospheric and tropospheric effects. The PoP developed in R proves both the scalability of GNSS positioning performances for targeted GNSS applications, and adaption to changing GNSS positioning environment conditions, using advanced correction models, and embedding the GNSS pseudorange mitigation methods into machine learning-optimised GNSS position estimation methods [6, 13]. As we reach the PoP phase of our research, we intend to proceed with development and systematic examination of performance of: (i) local GNSS pseduorange correction models, (ii) advanced GNSS position estimation models, and (iii) computational efficiency advancement. A challenge of the near-real GNSS position estimation using a distributed system, involving telecommunications infrastructure, will be addressed in the scenario of the supporting 5G network utilization, to tackle potential delays in the GNSS observations delivery to SPaaS unit. With the use of NFV Dataplane [20], it can be provided a flexible and programmable infrastructure, which can host a variety of services, including near-real GNSS position estimation.

## 6. Conclusion

Satellite navigation systems have become a cornerstone of modern civilisation, empowering raising number of technology and socio-economic applications (systems and services). Global Navigation Satellite Systems (GNSS) has to provide uninterrupted, robust, reliable, and resilient Positioning, Navigation, and Timing (PNT) services to allow for sustained GNSS applications development and operation, based on GNSS as a component of the national infrastructure.

Here a novel concept of Satellite-Positioning-as-a-Service (SPaaS) is proposed for accomplishment of resilience against adverse natural and artificial effects, scalable GNSS positioning performance related to requirements of particular GNSS applications, and optimise resource utilisation for sustained GNSS positioning performance. SpaaS is founded on mathematical methods, statistical learning, computational advancement, and emerging telecommunications systems (5G).

A Proof-of-Principle (PoP) of the SPaaS was developed in the open-source R environment for statistical computing, demonstrating the SPaaS features, and rendering the foundation for further research in the field.

## References

[1] GSA 2018 *GNSS user needs and requirements library*. European GNSS Agency (GSA). (Prague, Czech Republic). Available at: https://www.gsa.europa.eu/gnss-applications/user-needs-and-requirements

[2] Thomas M *et al.* 2011 *Global navigation space systems: reliance and vulnerabilities*. Royal Academy of Engineering (London, UK) Available at: http://bit.ly/1vrIenu

[3] Filić M 2018 Foundations of GNSS spoofing detection and mitigation with distributed GNSS SDR receiver. *TransNav*, **12**(4), pp 649-656. doi: 10.12716/1001.12.04.01

[4] Filić M and Filjar R 2018 *Forecasting model of space weather-driven GNSS positioning performance.* (Lambert Academic Publishing. Riga, Latvia) ISBN 978-613-9-90118-0

[5] Filić M and Filjar R 2018 A south pacific cyclone-caused GPS positioning error and its effects on remote island communities. *TransNav*, **12(4)**, 663-670 doi: 10.12716/1001.12.04.03

[6] Filić M and Filjar R 2018 modelling the relation between GNSS positioning performance degradation, and space weather and ionospheric conditions using RReliefF features selection. *Proc of 31st International Technical Meeting ION GNSS+* 2018, 1999-2006. Miami, FL https://doi.org/10.33012/2018.16016

[7] Filjar R 2019 On suitability of massive GNSS pseudorange data for GNSS positioning performance studies. *Proc of 12th Baška GNSS Conference*, pp 7 - 27. Baška, Krk Island, Croatia. Available at: https://bit.ly/2T1pOeA

[8] Filic M, Filjar R and Ruotsalainen L 2016 An SDR-based study of multi-GNSS positioning performance during fast-developing space weather storm. *TransNav*, **10(3)**, pp 395-400 doi: 10.12716/1001.10.03.03

[9] Filjar R, Huljenić D and Lenac K 2013 Enhancing performance of GNSS position estimator by cloud-based GNSS SDR reciver architecture utilisation. *Proc ELMAR-2013*, pp 315-319. Zadar, Croatia

[10] Panny T 2010 *Navigation signal processing for GNSS software receivers* (Artech House. Norwood, MA, and London, UK) ISBN 978-1608070275

[11] Van Diggelen F 2009 *A-GPS: Assisted GPS, GNSS, and SBAS* (Artech House. Boston, MA, and London, UK) ISBN 1630812621

[12] US Department of Defense 2020 *Global positioning system standard positioning service performance standard* (5th ed.). (Office of the Department of Defense. Washington, DC). Available at: https://www.gps.gov/technical/ps/2020-SPS-performance-standard.pdf

[13] Filić M, Grubišić L and Filjar R 2018 Improvement of standard GPS position estimation algorithm through utilization of Weighted Least-Square approach. *Proc of 11th Annual Baška GNSS Conference*, pp 7-19 Baška, Krk Island, Croatia. Available at: https://bit.ly/3b5BnMo

[14] Filjar R, Huljenić D and Dešić S 2002 Distributed positioning: A network-supported method for satellite positioning performance. *Journal of Navigation*, **55**, pp 477-484 doi: 10.1017/S0373463302001996

[15] Filić M 2018 On Development of the forecasting model of GNSS positioning performance degradation due to space weather and ionospheric conditions *2nd URSI Atlantic Radio Science Meeting (AT-RASC)*, Meloneras, pp 1-4 doi: 10.23919/URSI-AT-RASC.2018.8471628.

[16] ETSI. 2015 *Satellite Earth stations and systems (SES); GNSS based location systems; Part 1: functional requirements*. ETSI TS 103 246-1 V1.1.1. (European Telecommunications Standardisation Institute. Sophia Antipolis, France). Available at: https://bit.ly/2YVooYP

[17] ETSI. 2015 *Satellite Earth stations and systems (SES); GNSS based location systems; Part 3: performance requirements*. ETSI TS 103 246-3 V1.1.1. (European Telecommunications Standardisation Institute. Sophia Antipolis, France) Available at: https://bit.ly/3jxJHYq

[18] ETSI. 2016 *Satellite Earth stations and systems (SES); GNSS based location systems; Part 5: performance test specification*. ETSI TS 103 246-5 V1.1.1. (European Telecommunications Standardisation Institute. Sophia Antipolis, France) Available at: https://bit.ly/2Dow2DK

[19] Filić M and Filjar R 2018 Smartphone GNSS positioning performance improvements through utilisation of Google location API. *Proc of 41st International Convention MIPRO/CTI*, pp 507-510. Opatija, Croatia doi: 10.23919/MIPRO.2018.8400087

[20] Iliev T B, Nikolaev N T, Mihaylov G Y, Ivanova E P, Stoyanov I S and Radev D I, 2018 A critical view of the NFV dataplane implementations for 5G networks *26th Telecommunications Forum* (TELFOR), pp 1-4 doi: 10.1109/TELFOR.2018.8611980

# Performance of free-space optical systems in the presence of receiver imperfections

**G T Djordjevic[1], D N Milic[1], and I D Radojkovic[2]**

[1] University of Nis, Faculty of Electronic Engineering, Department of Telecommunications, A. Medvedeva 14, 18000 Nis, Serbia
[2] Dunav osiguranje, 18000 Nis, Serbia

E-mail: goran.t.djordjevic@elfak.ni.ac.rs

**Abstract.** We present a review of papers considering the subcarrier intensity modulation free-space optical (FSO) communication systems in the presence of phase noise. This phase noise is a consequence of imperfect reference signal extraction in electrical domain after photodetector. We illustrate the effects of this phenomenon on the system symbol-error rate performance. The examples show that receiver imperfections, presented through phase noise, can degrade the FSO system performance to a large extent.

## 1. Introduction

In this Section, we describe different application areas of optical wireless communications (OWC). After that, we give an overview of the effects that are commonly present during signal transmission in free-space optical (FSO) system, with an emphasis on the impact of receiver imperfections. At the end of the Section, we provide an overview of the literature regarding the influence of the phase noise on the performance of the FSO system.

### *1.1. Application of OWC systems*

The OWC represents a promising technology for a variety of applications. This technology can be applied in different scenarios, like: 1) ultra-short range communications (interconnects within integrated circuits); 2) short range communications (wireless body area networks and device-to-device communications in dense Internet-of-Things cells); 3) medium range communications (visible light communications and infrared communications); 4) long range terrestrial communications; 5) ultra-long range communications (inter-satellite, deep-space and ground-to-satellite links) [1]-[6].

Nowadays, there are 8 billion mobile subscriptions, and this number is predicted to increase to 8.9 billion by the end of 2025, out of which 88 percent will be for mobile broadband. The number of smartphone subscriptions is forecast to reach 7.5 billion in 2025, which accounts for around 85 percent of all mobile subscriptions [7]. The demand for high data rates have been increasing dramatically. Emerging wireless communication networks should support high data rates, high reliability, security, low latency and high density of terminals.

In this work, we are focused to terrestrial links up to several kilometres enabling fibre-like data rates and being essential to backhaul and fronthaul needs. Wireless optical technology is a promising alternative scheme to radio frequency communication systems for addressing last mile bottleneck in broadband access networks. With increasing the number of nano and microcells in 5G systems, the demands will increase for backhaul connections between base stations, as well as base stations and mobile switching centres. The FSO technology is used for last mile access, i.e., for bridging the gap between backbone network and end-users. It can be used for providing back-up of the optical fibre links. The FSO is a favourite technology for establishing temporary links under disaster conditions, too.

### *1.2. Transmission effects in FSO links*

The optical signal propagation is influenced by three atmospheric factors: absorption, scattering and refractive-index fluctuations. Absorption and scattering are usually considered together like a deterministic process, which can be described jointly by an equation, while signal intensity fluctuations induced by turbulence are a consequence of the stochastic process. An optical signal propagating over an FSO channel experiences random intensity fluctuations due to turbulence, which is caused by the fluctuations of the refractive index of the medium [1]-[5]. These fluctuations of the received signal can be described by several statistical distributions, which are in agreement with measured results [3]. Some of these distributions are as follows: K distribution, lognormal, modified Ricean, Gamma-Gamma, Malaga, etc. Gamma-Gamma distribution is widely accepted and it can describe turbulence conditions from weak to strong ones [1]-[6].

The misalignment between transmitter and the receiver of the FSO link is known as pointing errors. This phenomenon is caused by building sway and vibrations, and is treated as a stochastic process that can be described by different distributions. In the case when horizontal and vertical displacements are mutually independent and when they can be described by normal distribution with zero mean value, the radial displacement of the photodetector can be described by Rayleigh distribution. The intensity of the pointing errors is fixed by the standard deviation of horizontal and vertical displacements [6].

On the basis of the detection type, there are two types of the FSO systems: intensity-modulation direct-detection (IM/DD) systems and coherent systems. In IM/DD approach, the information is conveyed by the light intensity, which is directly detected by a photodetector without applying local oscillator. Contrary, in the case of coherent systems, amplitude, frequency or phase modulation is applied and received optical signal is mixed with local laser signal before photo-detection. Due to simplicity and low cost, IM/DD systems are usually applied in practical terrestrial FSO systems. In order to improve performance of FSO systems, a subcarrier intensity modulation technique has been suggested to be an alternative to IM/DD systems [4], [5].

One of the negative effects in an FSO system is imperfection of the receiver in electrical domain. In the case of SIM-based systems, coherent modulations require carrier phase recovery for achieving the optimal performance. In practice, the carrier phase cannot be tracked with perfect accuracy. The difference between carrier phase and its estimate is called phase noise, and it is a stationary random process whose distribution conforms to a random variable following Tikhonov distribution [8]-[15].

Two main approaches were applied in previous works on this topic. In the first approach, loop SNR is specified by the phase noise standard deviation. This standard deviation is fixed over all values of channel SNR [8]-[11]. This approach corresponds to the case when there is a limiter in loop circuit. In the second approach, the SNR in the loop filter is dependent on the channel SNR, and loop SNR is linear function of channel SNR [13-16]. In both cases, we assume that intensity fluctuations bandwidth is much smaller than PLL loop filter and loop SNR remains constant over many symbol intervals.

### *1.3. Literature review*

The effect of imperfect reference signal recovery on BER performance of SIM-based BPSK system was firstly considered in [8]. Using the simulation approach, the results confirmed that error performance of an FSO system containing Gamma-Gamma channel can be strongly degraded by the phase noise. The authors of [9] presented the analysis of symbol error rate performance of SIM-based MPSK systems under strong atmospheric turbulence conditions. The Fourier series method [16]-[19] was applied in estimating the error performance. The Fourier series method was used in analyzing the performance of SIM-based M-ary differential PSK FSO system over Gamma-Gamma turbulence and pointing errors in [10]. In DPSK receiver, a local oscillator used for down-conversion generates a signal which is not perfect in the sense that phase of this signal is a random process fluctuating over a time. The aim of the paper was to estimate the effect of these fluctuations on error performance. The authors of [11] applied the Fourier series method in analyzing the effect of phase noise on SIM MPSK system when turbulence is modeled by Malaga distribution. However, numerical results presented in [11] are valid only for the case of Gamma-Gamma turbulence, and it is open question the effect of specific parameters of Malaga distribution together with phase noise on overall system performance.

Reference [12] studied the noisy reference loss of SIM-based BPSK and QPSK in lognormal fading. The analysis was extended to M-ary PSK in [13]. They assumed that signal-to-noise ratio in PLL loop filter is linearly dependent on the SNR over the FSO channel and the expressions were derived for estimating the asymptotic noisy reference loss. The analysis from these two references were further extended in [14] to the case of Gamma-Gamma channel with pointing errors, where the approximate expressions for error probability were derived. Reference [15] presented approximate expressions for error probability in detecting SIM-based MPSK signal transmitted over an FSO channel in the presence of diversity reception when there is a phase noise in diversity receiver.

## 2. Fourier series method in error rate estimation

In this Section, we introduce the system model that is used in illustrating the Fourier series method for estimating the effects of receiver imperfections on SIM FSO system performance. We present the basic steps in mathematical analysis and some numerical results with appropriate comments.

### 2.1. System model

Let us observe the system presented in figure 1. The SIM technique was borrowed from RF communications, and the signal is modulated using appropriate modulation technique in the electrical domain. The modulator output is denoted by $s(t)$. A dc bias is then added to keep the resulting modulated signal positive at all times. The signal thus obtained modulates directly and linearly the intensity of the laser. The optical transmitting power of the laser is denoted by $P_t$. The intensity of the optical signal at the output of the receiver has a form of $I_t(t) = P_t(1 + m \times s(t))$, where $m$ denotes the modulation index. The signal is then transmitted over an FSO channel. At the reception, direct detection is performed by a PIN photodiode mounted directly after the receiving optics, and also, the dc bias of the signal is removed. After that, the signal is demodulated in the electrical domain in classical way, well known and widely used in existing wireless devices. This electrical signal can be represented as $y(t) = RP_t mI \times s(t) + n(t)$.

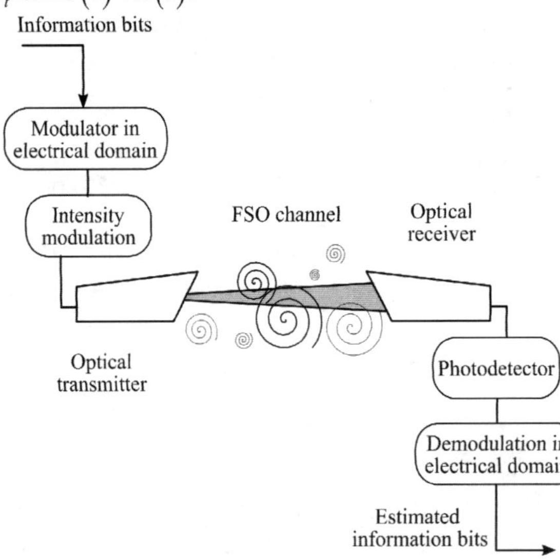


**Figure 1.** Model of the system.

The instantaneous signal-to-noise ratio is given by $\gamma = (RP_t mI)^2 / (2\sigma^2)$. The average SNR is then given by the mathematical expectation of this value, as $\bar{\gamma} = E\left\{\left(RP_t mI\right)^2 / \left(2\sigma^2\right)\right\} = E\left\{I^2\right\} \times \left(RP_t m\right)^2 / \left(2\sigma^2\right)$, while average electrical SNR is defined as $\mu = \left(E\{I\}\right)^2 \times \left(RP_t m\right)^2 / \left(2\sigma^2\right)$.

The analysis is provided for the example when turbulence induced signal intensity fluctuations over a channel can be described by a Gamma-Gamma probability density function, given by

$$p_I(I) = \frac{2(\alpha\beta)^{(\alpha+\beta)/2}}{\Gamma(\alpha)\Gamma(\beta)} I^{\frac{\alpha+\beta}{2}-1} K_{\alpha-\beta}\left(2\sqrt{\alpha\beta I}\right), \quad I > 0, \tag{1}$$

where $K_n(.)$ denotes the second kind modified Bessel function of order $n$ [22, (8.407)], while $\alpha$ and $\beta$ are parameters defining the turbulence strength. For the case of zero inner scale plane waves, these parameters can be expressed as

$$\alpha = \left(\exp\left(\frac{0.49\sigma_R^2}{\left(1+1.11\sigma_R^{12/5}\right)^{7/6}}\right)-1\right)^{-1}, \quad \beta = \left(\exp\left(\frac{0.51\sigma_R^2}{\left(1+0.69\sigma_R^{12/5}\right)^{5/6}}\right)-1\right)^{-1}, \tag{2}$$

where $\sigma_R^2$ is the Rytov variance given by $\sigma_R^2 = 1.23 C_n^2 k^{7/6} L^{11/6}$, with $L$ being the link length in km and $k$ being the wave number given by $k = 2\pi/\lambda$. We use the following numerical values: $\lambda = 1.55$ μm, $L$=2000 m. The turbulence strength is defined by:

$$C_n^2 = \begin{cases} 6\times10^{-15} \ \ \mathrm{m}^{-2/3}, & \text{weak turbulence} \\ 2\times10^{-14} \ \ \mathrm{m}^{-2/3}, & \text{moderate turbulence} \\ 5\times10^{-14} \ \ \mathrm{m}^{-2/3}, & \text{strong turbulence} \end{cases} \tag{3}$$

Figure 2 illustrates the turbulence parameter values in Gamma-Gamma model for the given link distance.

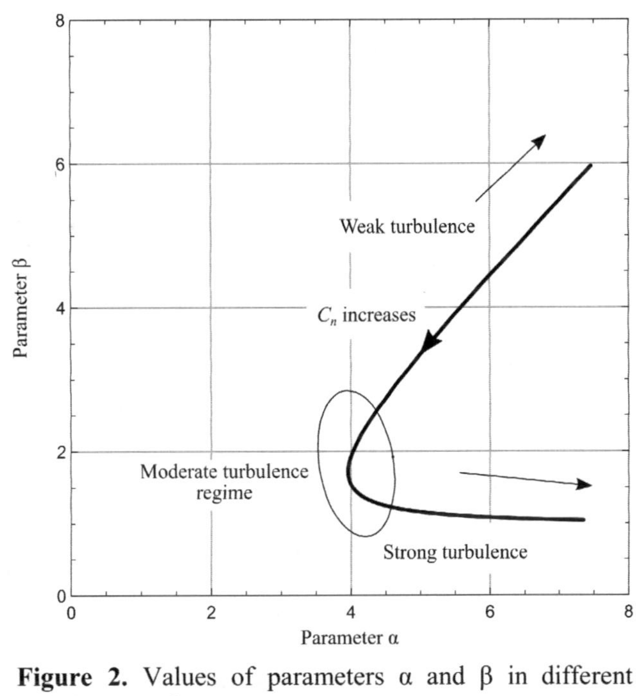


**Figure 2.** Values of parameters α and β in different turbulence regimes.

Weak turbulence is characterized by the relatively high values of α and β, which corresponds to relatively low values of atmospheric constant $C_n^2$. Limiting values of $\alpha, \beta \to \infty$ correspond to an ideal situation where there is no turbulence present. On the other hand, strong turbulence is represented by the low value of β that is close to one, while the other parameter α is large. In the limiting situation of infinite α, this leads to a simpler description of the turbulence channel corresponding to a Gamma distribution. Moderate turbulence regime is located between the two extremes and is represented by relatively low values of both parameters.

*2.2. Fourier series approach for error rate estimation*

The basic idea in determining the error performance lies in expressing the resulting received signal phase in the form of the Fourier series. The effects of intensity fluctuations and noise are incorporated in the Fourier series members. After the procedure described in [16]-[19] the phase of the received signal can be presented as

$$p_\psi(\psi) = \frac{1}{2\pi} + \sum_{n=1}^{\infty} b_n \cos(n\psi), \quad |\psi| \le \pi, \tag{4}$$

where coefficients $b_n$ are expressed as

$$b_n = \frac{n\Gamma(n+1)2^{\alpha+\beta-3}}{n!\pi^2\Gamma(\alpha)\Gamma(\beta)} G_{2,5}^{5,1}\left( \frac{\alpha^2\beta^2}{16\mu} \middle|_{\alpha/2,\ (\alpha+1)/2,\ \beta/2,\ (\beta+1)/2,\ 0}^{1-n/2,\ 1+n/2} \right), \tag{5}$$

where $G_{p,q}^{m,n}(.)$ is the Meijer's *G*-function [20, (9.301)]. The signal phase is estimated by a phase locked loop from the pilot signal. This estimation is not perfect and there is a difference between the actual signal phase and estimated phase. This difference as a random variable that can be described by Tikhonov distribution given by

$$p_\varphi(\varphi) = \frac{e^{\gamma_{PLL}\cos\varphi}}{2\pi I_0(\gamma_{PLL})}, \quad |\varphi| \le \pi. \tag{6}$$

This PDF can be expressed in the form of the Fourier series as

$$p_\varphi(\varphi) = \frac{1}{2\pi} + \sum_{n=1}^{\infty} c_n \cos(n\varphi), \quad |\varphi| \le \pi, \tag{7}$$

where coefficient $c_n$ is given by $c_n = I_n(1/\sigma_\varphi^2)/(\pi I_0(1/\sigma_\varphi^2))$. The estimation of truncation error in (7) can be found in [10].

After the quite similar procedure from [10], the symbol error probability in detecting SIM-based *M*-ary PSK signal can be derived in the form

$$P_e = 1 - \frac{1}{M} - \sum_{n=1}^{\infty} \frac{2\pi b_n c_n}{n} \sin\left(\frac{n\pi}{M}\right), \tag{8}$$

where coefficients $b_n$ and $c_n$ are previously derived, and *M* is the number of phase levels. In this way we expressed the symbol error probability in the form of the infinity series with known coefficients. The proof of convergence and estimation of the truncation error of similar series can be found in our previous works [10], [19]. This expression is convenient also for calculating numerical values in some boundary cases. In this paper we present some illustrative numerical examples.

*2.3. Numerical results*

In this paper, we have used a readily available method implemented in in *Mathematica* for numerical calculation of sums. In order to estimate the value of sum, Wynn-ε method uses additional number of terms in sum and then extrapolates them to a polynomial multiplied by decaying exponential function. We have used a minimum of 1000 terms in the sum, after which this error estimation is performed using built-in *Mathematica* summation algorithm. Since the series of terms is of alternating signs, working precision must also be increased in order not to lose significant digits because of subtraction operation. This also presents a difficulty in calculating Meijer's $G$ function values in (5) with required number of accurate digits, which consumes significant amount of CPU resources. Because of these difficulties, we have resorted to parallel computation using 24 kernels in a local network environment. 100 points in each of the symbol error curves (8) shown in figure 3 are computed in about 5 minutes using this setup.

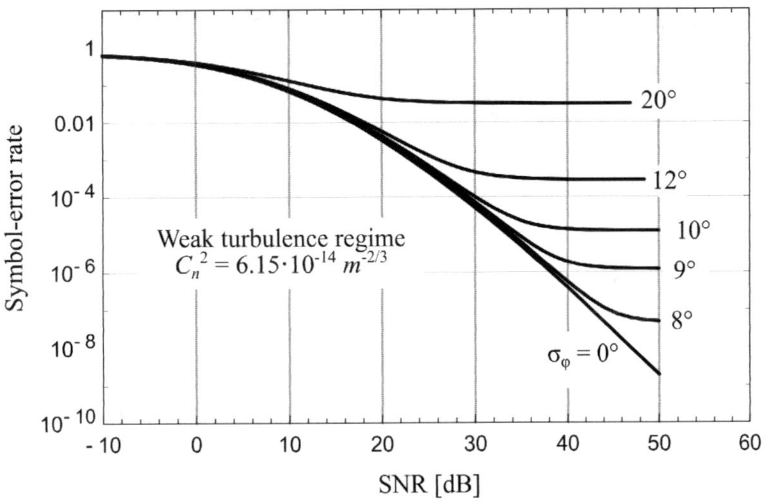


**Figure 3.** Symbol error rate of SIM QPSK system.

Figure 3 presents the symbol error probability dependence on electrical SNR for different values of phase noise standard deviation. The error probability decreases with increasing SNR until a given value. After that value, error probability stays constant and does not depend on SNR value. The value of this floor is dependent only on the phase noise standard deviation. The higher the value of this standard deviation, the lower the value is of error floor. That means that imperfect carrier phase recovery strongly influences the error performance. If the PLL loop circuit is poorly designed, the probability of error cannot fall below the floor despite the increase in signal strength.

Figure 4 presents the SER values for SNR of 50 dB for different values of phase noise standard deviation and under different turbulence conditions.

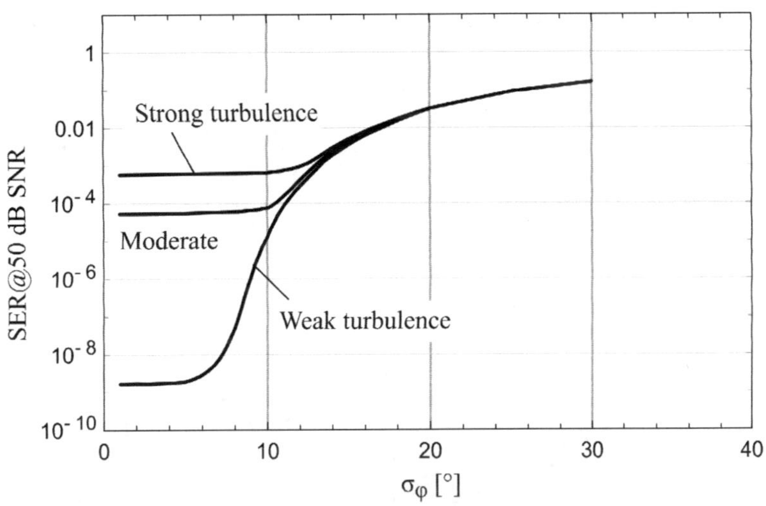


**Figure 4.** SER for fixed SNR of 50 dB in different turbulence strengths and phase noise standard deviations.

Two conclusions are visible. Firstly, if the system is designed to operate correctly in weak turbulence, the strong turbulence conditions could increase considerably the SER values. Secondly, the effect of phase noise is not primarily effect in strong turbulence condition up to values of phase noise standard deviation of 12 degrees. Contrary, in weak turbulence conditions, phase noise starts to have strong influence on SER values starting from phase noise standard deviation around 6 degrees.

## 3. Conclusion

In this Invited paper, we have highlighted the effect of imperfect reference signal recovery in electrical domain when SIM-based PSK is applied in FSO systems taking turbulence induced signal intensity fluctuations into account. We have presented the Fourier series method approach that was used by ourselves in some our previous works. The results have shown that phase noise can strongly degrade the SIM-based FSO systems, and the approach presented here has the ability to accurately predict these degradations. This approach can also be applied in designing of receivers for FSO systems.

## References

[1] Ghassemlooy Z, Popoola W and Rajbhandari S 2019 Optical wireless communications – System and channel modelling with MATLAB$^R$, *CRC Press*, Taylor&Fransis Group, Boca Raton

[2] Khalighi M A and Uysal M 2014 Survey on free space optical communication: a communication theory perspective, *IEEE Communications Surveys & Tutorials*, **16**, no. 4, pp 2231-2253

[3] Andrews L and Philips R 2005 Laser beam propagation through random media, *Bellingham, WA*, USA: SPIE

[4] Popoola W 2009 Subcarrier intensityy modulated free-space optical communication systems, *PhD thesis, University of Northumbria at Newcastle*

[5] Petkovic M 2015 Analysis and means to improve the performance of wireless optical telecommunication systems under conditions of atmospheric turbulence, *PhD thesis, University of Nis*

[6] Djordjevic G, Petkovic M, Cvetkovic A and Karagiannidis G 2015 Mixed RF/FSO relaying with outdated channel state information, *IEEE Journal on Selected Areas in Communications*, **33**, no. 9, pp 1935-1948

[7] Ericson Mobility Report, June 2020, Available on: https://www.ericsson.com/en/mobility-report/reports/june-2020/

[8] Djordjevic G 2013 Effect of phase noise on bit error rate performance of subcarrier intensity modulation binary phase shift keying wireless optical systems – simulation study, *Facta Universitatis, Series: Automatic Control and Robotics*, **12**, no. 3, pp 189-195

[9] Petkovic M and Djordjevic G 2015 SEP analysis of FSO system employing SIM-MPSK with noisy phase reference, *4th International Workshop on Optical Wireless Communications (IWOW)*, , Istanbul, Turkey, 7-8 September 2015, pp 46-50

[10] Petkovic M, Djordjevic G, Karagiannidis G and Milovanovic G 2017 Performance of SIM-MDPSK FSO systems with hardware imperfections, *IEEE Transactions on Wireless Communications*, **16**, no. 8, pp 5442-5451

[11] Chen D, Huang G and Wang C 2020 Influences of the phase noise on performance of subcarrier modulation system over Malaga turbulence channel, *Optical Engineering*, **59**(1), pp 016111-1 – 016111-13

[12] Song X, Yang F, Cheng J and Al-Dhahir N 2015 Asymptotic noisy reference Losses of subcarrier BPSK and QPSK systems in lognormal fading, *International Conference on Computing," Networking and Communications (ICNC), Workshop on Computing, Networking and Communications (CNC)*, Garden Grove, CA, USA, pp 352-356

[13] Song X, Yang F, Cheng J, Al-Dhahir N and Xu Z 2015 Subcarrier phase-shift keying systems with phase errors in lognormal turbulence channels, *Journal of Lightwave Technology*, **33**, no. 9, pp 1896-1904

[14] Gappmair W and Nistazakis H 2017 Subcarrier PSK performance in terrestrial FSO links impaired by Gamma-Gamma fading, pointing errors, and phase noise, *Journal of Lightwave Technology*, **35**, no. 9, pp 1624-1632
[15] Varrotsos G, Nistazakis H, Gappmair W, Sandalidis H and Tombras G 2019 SIMO subcarrier PSK FSO links with phase noise and non-zero boresight pointing errors over turbulence channels, *IET Communications*, **13**, no. 7, pp 831-836
[16] Prabhu V 1969 Error-rate considerations for digital phase-modulation systems, *IEEE Transactions on Communication Technology*, **COM-17**, no. 1, pp 33-42
[17] Tan C and Tjhung T 1983 On Fourier series expansion of the phase density function, *IEEE Transactions on Communications*, **COM-31**, no. 5, pp 736-738
[18] Kostić I 2007 Average SEP for M -ary CPSK with noisy phase reference in Nakagami fading and Gaussian noise, *European Transactions on Telecommunications*, **18**, no. 2, pp 109-113
[19] Marjanovic Z, Djordjevic G and Milovanovic G 2018 Truncation error analysis in computing of SEP and SEP floor for partially coherent receiver of MPSK signals over composite fading channels, *Journal of The Franklin Institute,* **355**, no. 2, pp 965-980
[20] Gradshteyn I and Ryzhik I 2000 *Table of integrals, series, and products*, NY, USA: Academic

**Acknowledgments**
This work was supported in part by the Ministry of Education, Science and Technological Development of Republic of Serbia, and in part by European Science Foundation under COST Action CA19111 - NEWFOCUS.

# Modelling and performance analysis of DWDM passive optical network

**S M Sadinov**[1], **K K Angelov**[2], **P Kogias**[3]

[1,2] Department of Communications Equipment and Technologies, Technical University of Gabrovo, 4 H. Dimitar str., 5300 Gabrovo, Bulgaria
[3] Department of Electrical Engineering, International Hellenic University, Ag. Loukas, 65404 Kavala, Greece

E-mail: murry@tugab.bg, kkangelov@tugab.bg, kogias@teikav.edu.gr

**Abstract.** Passive optical networks are emerging as the main technology for building access networks. The current metropolitan optical communication networks operate at 1Gbps, which appears to be insufficient for the future development and exponential growth of traffic from IP services. The existing various traffic-intensive services and the growing interest in 4K video content and streaming services require the introduction of new solutions within passive optical networks (PON). A promising option is the combination of 10 Gbps and above high-speed data transmission with spectral multiplexing. DWDM will provide the required high spectral efficiency, but in contrast will require the use of a proper modulation scheme to achieve a longer distance. In this paper a model for performance analysis and solving optimization problems for DWDM-PON networks is developed. Different signal modulations are considered and a comparative analysis is performed. The behaviour of the system at different input parameters is modelled. Performance analysis is performed by evaluating the spectral characteristics and values of BER and Q-factor.

## 1. Introduction

Passive optical networks (PON) offer a number of specific advantages compared to active optical networks [1]. They are characterized by higher reliability and significantly easier maintenance. Of interest is the entry of 10-Gbps PON networks. In order to ensure the development of PON networks in relation to the increased volume of traffic, one possible solution is to expand the opportunities by introducing DWDM technology.

This paper examines the development and research of a simulation model of a 10 Gbps DWDM-PON network. Of interest for research are the different effects of the introduction of DWDM [2]. The main goal is to increase the bandwidth capacity of classic PON networks by applying a combination of TDM and WDM multiplexing. Figure 1 shown the main principle of the WDM-PON. The downstream channel transmits signals with varied wavelengths intended for individual clusters of subscribers. The signals are multiplexed and demultiplexed by an AWG filter.

The use of channel multiplexing will lead to effects such as four wave mixing (FWM) at narrow interwave distances, chromatic dispersion, channel crosstalk and various non-linear effects [3]. These problems lead to signal degradation as it propagates into the optical fiber.

The application of simple optical modulation techniques in most cases provide sufficient measures for the problems [4]. However, to achieve high-speed transmission in the multichannel optical transmission network, it is necessary to use more efficient modulation formats. Duobinary modulation is a suitable option, as its wider optical signal spectrum allows to minimize the effect of chromatic dispersion and FWM [4].

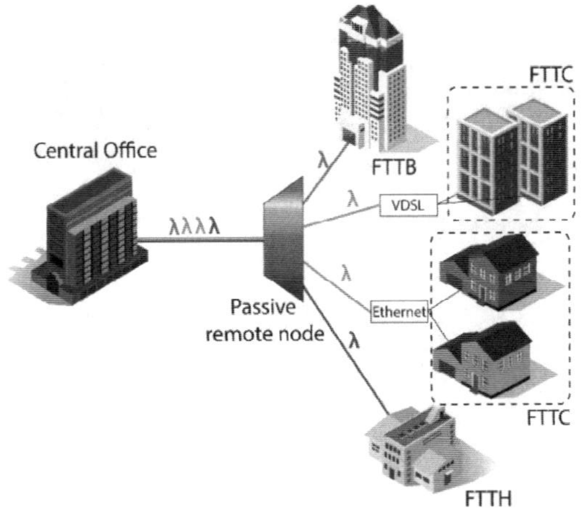


**Figure 1.** The principle of WDM-PON technique.

This paper will present an assessment and comparison of performance using the ordinary intensity modulation format. The performance of the modelled DWDM-PON optical network will be evaluated by the result values of Q-factor, BER, spectral characteristics and eye patterns of the optical signals.

## 2. Overview of the analysed modulation formats

The most common methods for optical signal modulation is external optical intensity modulation [5]. NRZ modulation is a popular solution in systems with intensity modulation [4]. It is characterized by simple implementation and easy signal detection. Figure 2 shows a Mach-Zender (MZM) modulator, which is used to obtain NRZ optical modulated signals.

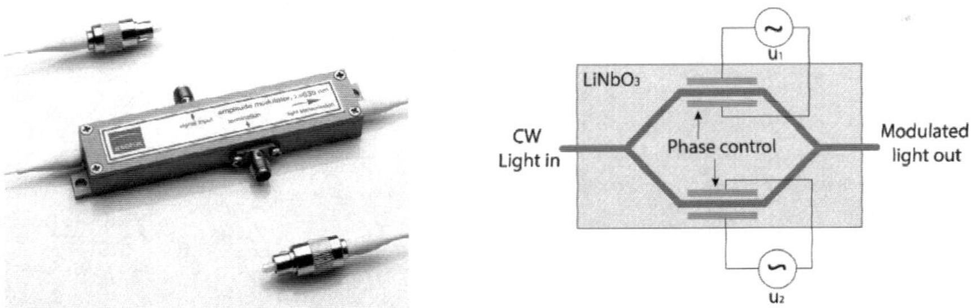


(a) (b)

**Figure 2.** Mach-Zender modulator (a) and its structure (б) for external optical modulation.

Duobinary modulation is an amplitude modulation, with significant resistance to interference in optical fibres and nonlinear effects (FWM, cross-phase and self-phase modulations) and is characterized by high spectral efficiency. The transmission of B bits per second requires the B/2 Hz bandwidth [4], therefore the bandwidth is reduced by half compared to NRZ modulation. The bitstream in duobinary modulation is manipulated in order to reduce the bandwidth and minimize the possibility of intersymbol interference (ISI) [3, 4].

One of the main difference in duobinary modulation compared to NRZ modulation is its ability to generate optical signals with narrower bandwidth. This allows minimizing the conditions for the occurrence of FWM in DWDM systems. However, the sensitivity of duobinary modulation to polarization mode dispersion (PMD) should be noted as a disadvantage.

IOP Conf. Series: Materials Science and Engineering 1032 (2020) 012003 doi:10.1088/1757-899X/1032/1/012003

## 3. Development of a model of DWDM-PON network

Figure 3 shows the proposed model of a DWDM-PON network for 10-Gbps transmission, which is simulated by the OptiWave software platform.

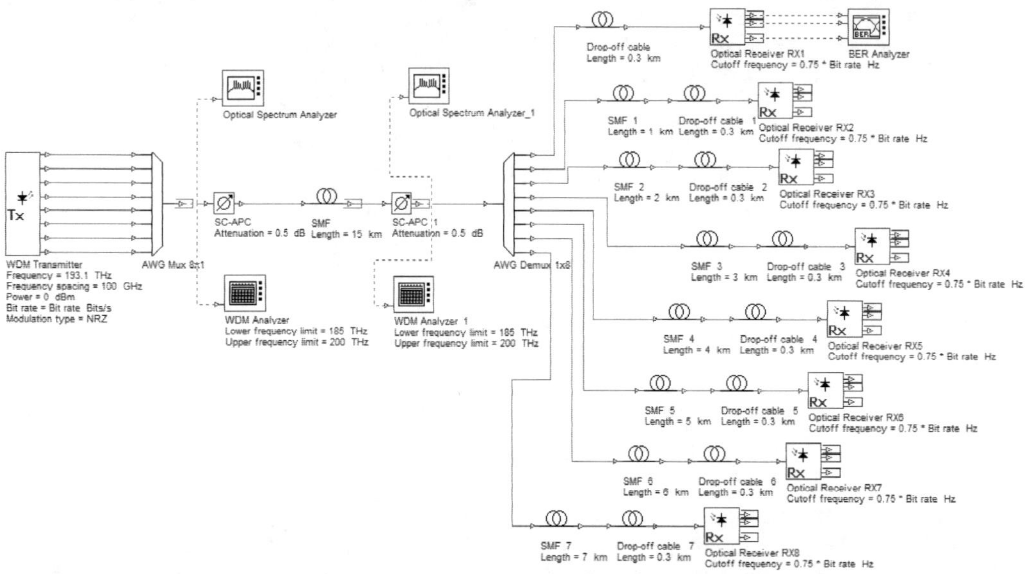


**Figure 3.** Modelling of 10-Gbps PON network with dense wavelength division multiplexing.

An 8-channel DWDM system with wavelength distribution in accordance with Recommendation ITU-T G.694.1 [6] was modelled. The channel distance between the individual wavelengths is 100 GHz. The model includes optical transmitters, AWG multiplexer, optical transmission line, AWG demultiplexer and optical receivers.

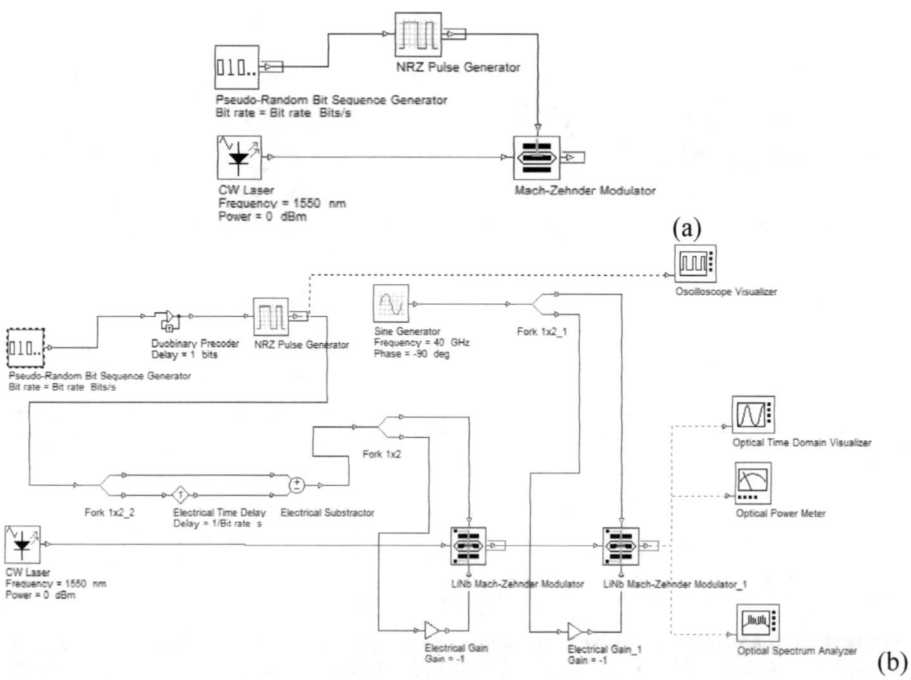


**Figure 4.** Optical transmitter model with (a) external NRZ modulation and (b) external duobinary modulation.

Either NRZ modulated optical transmitters or DM modulated optical transmitters are used as transmitters for the 8 optical channels. The model of the optical transmitter with NRZ modulation is shown in figure 4a, and with duobinary modulation – in figure 4b, respectively.

The optical transmission line includes a standard single mode optical fiber (SMF) at 15 km length with attenuation of 0.24 dB/km and a chromatic dispersion of 16.75 ps/(nm.km). Optical attenuators with 0.5 dB attenuation have also been added to model the presence of SC/APC connectors.

After the demultiplexer, 8 optical receivers are presented, each of which is 1 km away from the others. The proposed model can be applied in system with different optical line lengths to determine their performance and behaviour.

On figure 5 is showed the developed model of one of the receivers. The signal is initially fed to a band-pass Bessel optical filter, which removes the products from the nonlinear effects. The second block is PIN photodiode, which detects and transform the optical into electrical signal (with responsiveness of 1A/W). The second filter (low-pass Bessel electronic filter) applies a fine-tuning of the electrical signal.

The parameters and characteristics of the optical system are measured and displayed using different visualizers: optical spectrum analyser, optical power meter and BER analyser (figure 5).

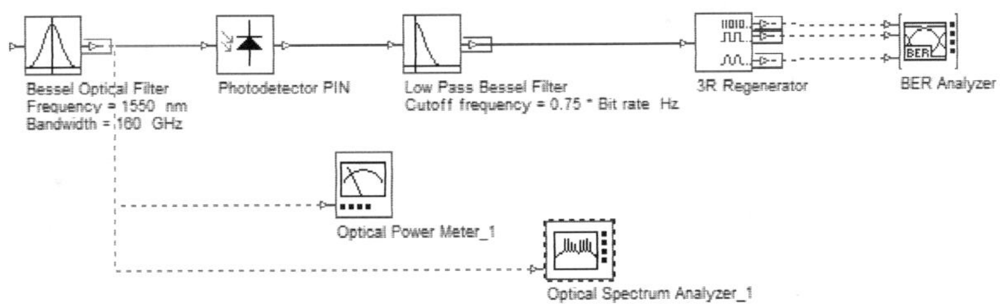


**Figure 5.** Optical receiver model.

## 4. Results

In figure 6 are shown the measured spectral characteristics for both types of modulated optical signals at the output of the AVG multiplexer.

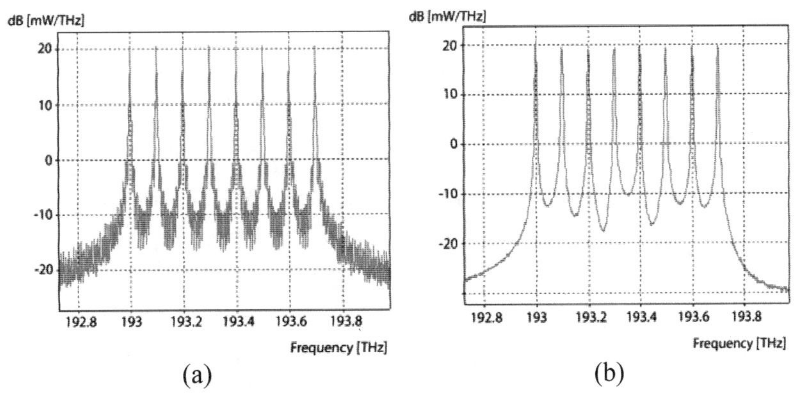


**Figure 6.** Spectrum of modulated optical signals at the output of the AWG multiplexer for (a) NRZ modulation and (b) duobinary modulation.

In figure 7 are shown the measured spectral characteristics for both types of modulated signals (optical) received at the input of the AWG demultiplexer. The formation of new parasitic wavelengths is clearly visible – they occur for both ends of the spectrum sidebands and in each cases of modulated signals. In case of NRZ modulation their amplitude is significantly larger – it is about 19 dB compared to 10 dB for duobinary modulation. In addition, the figure 7 shown the reduction of the optical signal in the case of NRZ modulation.

IOP Conf. Series: Materials Science and Engineering 1032 (2020) 012003 doi:10.1088/1757-899X/1032/1/012003

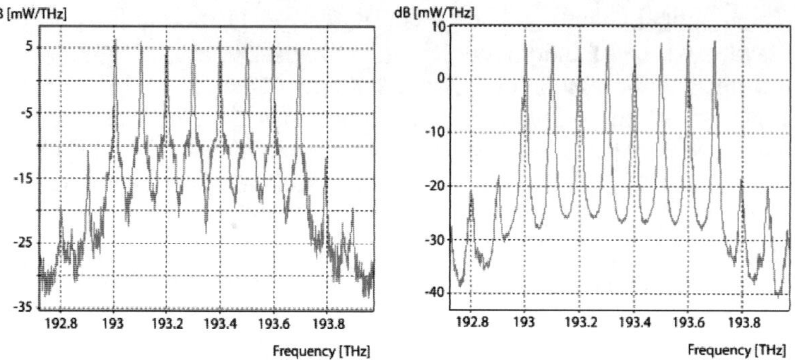


(a) (b)

**Figure 7.** Spectrum of modulated optical signals at the input of the AWG demultiplexer for (a) NRZ modulation and (b) duobinary modulation

In figures 8 and 9 are show the eye patterns of the received and demodulated signals obtained at the outputs of the nearest and farthest optical receivers, respectively.

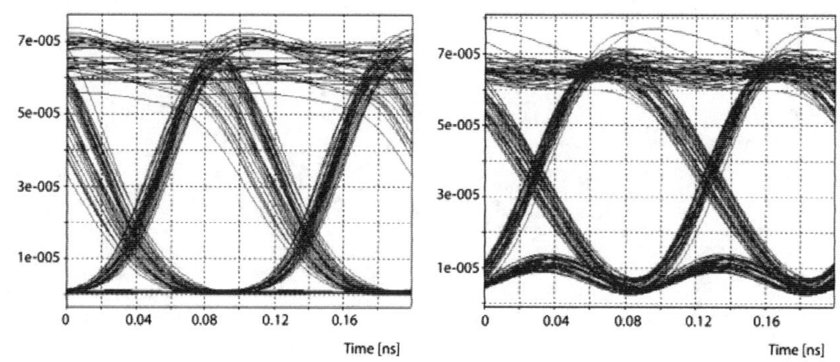


(a) (b)

**Figure 8.** Eye pattern of the received and demodulated signal at the output of an optical receiver RX1 for (a) NRZ modulation and (b) duobinary modulation.

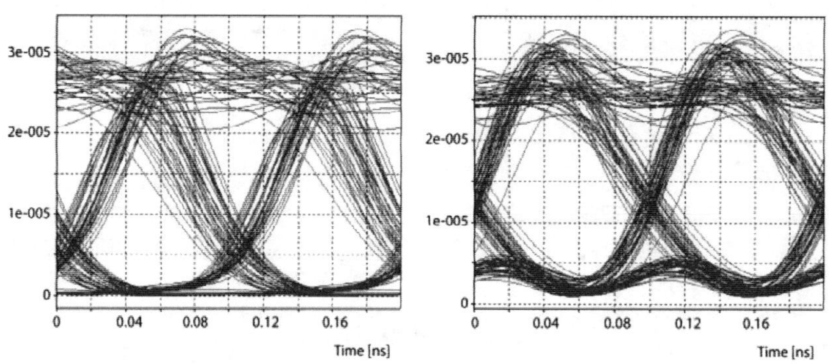


(a) (b)

**Figure 9.** Eye pattern of the received and demodulated signal at the output of an optical receiver RX7 for (a) NRZ modulation and (b) duobinary modulation.

Figures 7 and 9 clearly show the effect and advantage of duobinary modulation format. The BER analyzer determines the dimension of the BER / Q-factor of the signal for the two different modulation schemes and for different lengths of the fiber-optic (i.e. for different optical receivers 1 to 7). The results are shown in figure 10, where the value of the Q-factor of 16.94 dB corresponds to the maximum acceptable value of BER = $1.10^{-9}$.

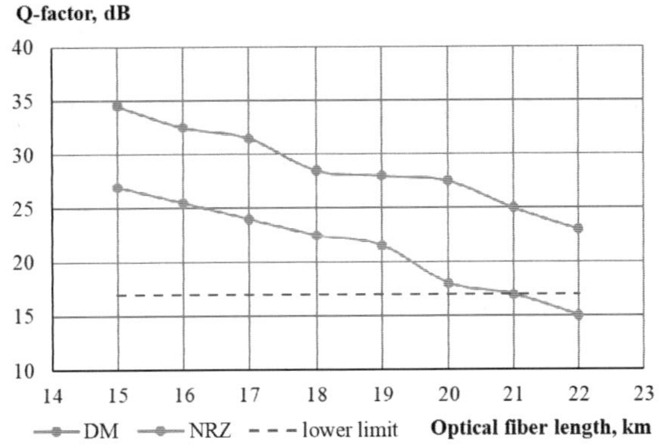


**Figure 10.** Dependence of the Q-factor on the optical fiber length for the two different types of modulation

## 5. Conclusion

The results of the simulation clearly show the advantages of the optical duobinary modulation scheme. The application of duobinary modulation format in DWDM-PON systems make available of a bigger number of communication channels per unit bandwidth, i.e. using more densely spaced wavelengths. The narrower frequency spectrum leads to greater signal stability in the event of ISI and FWM.

## 6. References

[1] Angelov K, S Sadinov and P Kogias, 2017 Practical Model for Management, Monitoring and Research of Passive Optical Network, *XXVI International Scientific Conference Electronics (ET)*, Sozopol, pp 1-4, doi: 10.1109/ET.2017.8124332.

[2] Horvath T, Cymorek P, Munster P, Oujezsky V and Vojtech J 2018 Simulations of grant allocation in NG-PON2 networks using OPNET modeler, *Journal of Communications Software and Systems*, **14**(4), pp 281 – 289, doi: 10.24138/jcomss.v14i4.525

[3] Angelov K, S Sadinov and P Kogias, 2019 *Modelling and optimization of multichannel optical communication lines*, International Scientific Conference UNITECH 2018, Gabrovo, **1**, pp. I-322-326

[4] Sadinov S, K Angelov and P Kogias, 2019 Modelling and Performance Analysis of Modulation Formats and Dispersion Compensation Schemes in a High-Speed Optical Communication Network, *Journal of Engineering Science and Technology Review (JESTR), Kavala Institute of Technology*, pp 22-25

[5] Ovseník L, Turán J, Ivaniga T and Ivaniga P, 2017 Deployment of the PON with an optical fibre G-652.B, *IEEE 14th International Scientific Conference on Informatics*, Poprad, pp 289-292, doi: 10.1109/INFORMATICS.2017.8327262

[6] ITU-T G.694.1. Series G: Transmission Systems and Media, Digital Systems and Networks: Transmission media and optical systems characteristics – Characteristics of optical systems. Spectral grids for WDM applications: DWDM frequency grid. 2012.

**Acknowledgments**

The studies were carried out with the financial support of the CoC "QUASAR" – project № BG05M2OP001-1.002-0006.

# Deployment of mesh network in an indoor scenario for application in IoT communications

**K K Angelov[1], S M Sadinov[2], P Kogias[3]**

[1,2] Department of Communications Equipment and Technologies, Technical University of Gabrovo, 4 H. Dimitar str., 5300 Gabrovo, Bulgaria
[3] Department of Electrical Engineering, International Hellenic University, Ag. Loukas, 65404 Kavala, Greece

E-mail: [1] kkangelov@tugab.bg, [2] murry@tugab.bg, [3] kogias@teikav.edu.gr

**Abstract.** The need for digital transformation in the industry and the related development of the Internet of Things (IoT) has led to the rapid development of various communication technologies and approaches for the implementation of flexible and intelligent communication networks. The main direction is wireless technologies, which make it possible to accelerate the wide range of new applications and services related to smart cities, industry, e-government and others. The possibilities are wide - use of different licensed or unlicensed frequency bands, channel bandwidths, modulation formats, power efficiency, reliability and security. Different network topologies could be applied for the access and collection of data from the sensor networks - star topology, mesh topology, tree topology, each of which has different advantages and disadvantages. This paper discusses the possibilities for deploying low-power mesh networks in an indoor scenario based on IQRF technology. Important communication parameters, approach advantages, and technological limitations will be highlighted. As an example, a study of network behaviour and efficiency in an indoor scenario will be considered.

## 1. Introduction

Wireless communication networks are widely used in various fields. Unlicensed bands, which for the European Union are 433 MHz and 868 MHz, are most often used for their application in Internet of Things (IoT) communications. These frequency bands are very often used for communication between sensors, actuators, equipment, controllers and remote controls, in home automation systems, in industrial automated systems, etc.

The IoT concept requires the implementation of a heterogeneous network environment in which the huge amount of data from multiple devices leads to some important issues: limited frequency spectrum resource, bandwidth efficiency, security requirements and more [1, 2]. Improving efficiency in IoT networks is most often achieved by aggregating traffic between end devices and applications. A huge number of IoT end devices share their data through a gateway and its management can be realized through various wireless technologies for data transmission over small and long range of distance. For this purpose, many communication protocols and standards have been developed (ZigBee, SigFox, LoRa, IQRF, etc.) [3, 4]. They use different approaches to design the network architecture, different transmission rates and maximum amount of data payload, different signal processing (modulation, error protection) and communication parameters (latency, transmitted power, transmission distance, power consumption, etc.) [5, 6, 7, 8].

One of the appropriate and promising technologies is the IQRF wireless communication platform. The IQRF platform was designed to address smaller segments of the wireless market – buildings automation, telemetry, utilization reporting for utilities, etc. The platform is distinguished by its specific methods for design of generic network communication platform, special signal coding scheme and direct peripheral addressing in a wireless network.

The IQRF technology/platform includes transceivers, gateways, development tools, communication protocols, and supporting services. The main parameters of IQRF are as follows [5]: sub-GHz ISM bands (433 MHz, 868 MHz, and 916 MHz); mesh or p2p network topology; reliable low-power communication (hundreds of nA in sleep mode and hundreds of μA in transmission mode); low data rate (about 19,2 kbps); low to medium coverage range (tens of meters in indoor scenario, hundreds of meters in outdoor environment); security level (the nodes are bound to the mesh network via password, AES-128 network encryption with self-distribution of the keys, AES-128 optional user encryption, CRC-16 message checksum, and block checksums); DPA (Device Peripheral Access) protocol for services and peripherals control. IQRF supports different MESH routing algorithms (full-mesh routing, reduced mesh routing, optimized mesh routing and others) and up to 240 devices in a single network with a maximum of 240 hops per packet (33 hops in 1 seconds with DPA method) [5].

Table 1 shows a comparison of IQRF with other short/medium and long communication range wireless technologies. The parameters presented in the comparison are derived from the technical documentation and various sources given in [8, 9, 10].

**Table 1.** Comparison between IQRF and other relevant wireless technologies.

| **Parameter** | **IQRF** | **LoRa** | **ZigBee** | **SigFox** |
|---|---|---|---|---|
| Frequency band | 433, 868, 916 (US) MHz | 433, 868, 780 (US), 915 (US) MHz | 2,4 GHz | 868, 902 MHz |
| Transmission rate | 20 kbps | 0,3 – 37,5 kbps | 250 kbps | 0,1 kbps |
| Maximum payload | 64 byte | 243 byte | 128 byte | 12 byte |
| Range (in free space) | 1 km | 15 km | 300 m | 50 km |
| Modulation format | GFSK | CSS | BPSK, OQPSK | DBPSK, GFSK |
| Network topology | mesh | star | star, mesh, tree | star |
| Latency | 400 ms | > 1000 ms | 16 ms | > 1000 ms |
| Maximum power consumption | 15 μA | 1 mA | 30 mA | 1 mA |
| Maximum transmitted power level | 11 dBm | 14 dBm | 20 dBm | 27 dBm |
| Maximum number of nodes per gateway | 240 (DPA), $65.10^3$ (OS) | $10^4$ | $65.10^3$ | $10^6$ |
| Maximum battery power time | 5 – 10 years | 5 – 10 yeras | 2 years | 5 – 10 yeras |

The mesh topology of the network suggests to create local-based sensor nests with a higher density of sensors instead of operator-based connectivity to single sensors or groups of sensors over large-scale geographical areas. For this reason, it is of interest to study the scenario of indoor measurements. As a particular case in this paper, the study is based on an indoor environment, which is a single floor of a university building (figure 2).

## 2. Design of mesh network for IoT application with IQRF

The typical design of IQRF network with mesh topology is shown in figure 1. This topology includes IQRF transceiver modules, gateway, cloud service (network server) and user application.

The mesh network topology is implemented by wirelessly interconnected IQRF transceivers, which use a specialized routing algorithm to transmit data packets. The IQRF transceiver uses an 8-bit microcontroller that performs the operations provided in the operating system.

Two different approaches are applied for realization of a certain functionality. In the first approach, the existing hardware profile (HWP) in the three-layer IQRF architecture can be used through application software plug-ins. It allows access to all services and resources. This is an important advantage, as it does not require any special programming, but only control of data flows. The second approach uses a user application layer and predefined operating system functions. By using the C programming language, the desired functionality can be programmed.

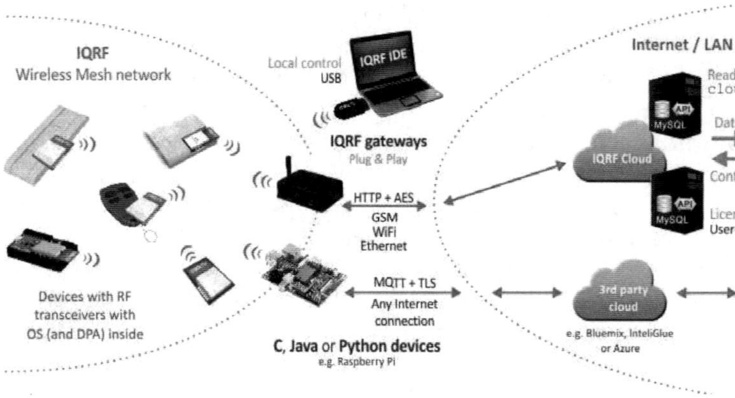


**Figure 1.** Typical design of mesh network for IoT application with IQRF [5].

In addition, there are two radio communication modes of operation. In the first mode, data packets are available to all devices in the range. Network features are not used here, i.e. this is a non-networking mode, which is suitable for peer-to-peer devices. In the second mode, one of the devices is assigned a coordinator role and manages the mesh network. A mesh network can contain up to 239 nodes, and data packets will be available only to the addressed nodes. The coordinator role can be dynamically assigned to each node in the network, and the other nodes are common nodes. This determines the maximum number of hops to be 240. Virtual routing of individual packets is applied, which allows to determine redundant paths in the network dynamically.

The IQRF gateway is used to provide an interface between the IQRF network and another type of network (Wi-Fi, Ethernet, 2/3/4G) to connect to the cloud server [11]. The gateways use a sensor node with the role of coordinator and it allows for data collection, accessing, and device controlling in the network.

The cloud server provides a service to end users and serves as an intermediary between them and the IQRF network. The gateway uploads the data received from the end devices directly to the server. Users can access the cloud server through a web and application user interface (API), which allows cloud monitoring.

## 3. Experimental setup for testing and measurements in an indoor scenario

Here will be considered an experimental setup for the implementation of IQRF network and scenarios for indoor measurement for a floor of the university building of Technical University of Gabrovo. In figure 2 is shown the floor architectural plan.

The following equipment was used for the experimental studies: two end devices (TX1 and TX2 – figure 2), a coordinator (C – figure 2), a Raspberry Pi based gateway (G – figure 2) and a cloud service. The network architecture for the considered indoor scenario is shown in figure 3.

The end devices (nodes) include a transceiver and a sensor module / relay module. The both transceivers are implemented with a DCTR-72DAT module operating at 868 MHz with 10 mW output power and –104 dBm sensitivity. The sensor module is connected to the first transceiver (TX1 – figure 2) and it is intended to measure temperature, light intensity (using a photoresistor) and voltage (using a potentiometer) (figure 4 (a)). The relay module is connected to the second transceiver (TX2 – figure 2) and it includes two bistable relays (figure 4 (b)). It enables to switch connected equipment on or off using two relays inside. Both sensor and relay modules are controlled by DPA commands implemented in custom DPA handlers delivered with the module. The handlers specify HWP identification (HWPID) to individual modules which allows their identification.

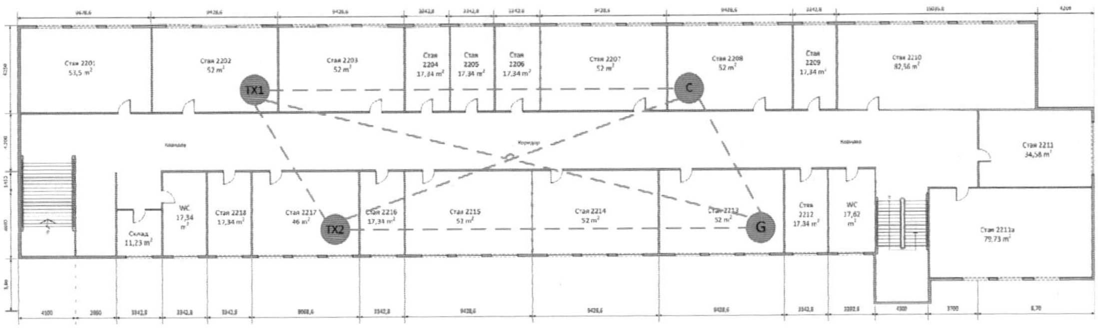


**Figure 2.** The architectural plan and experimental setup of IQRF network in indoor scenario (TX – transceiver, C – coordinator, G - gateway).

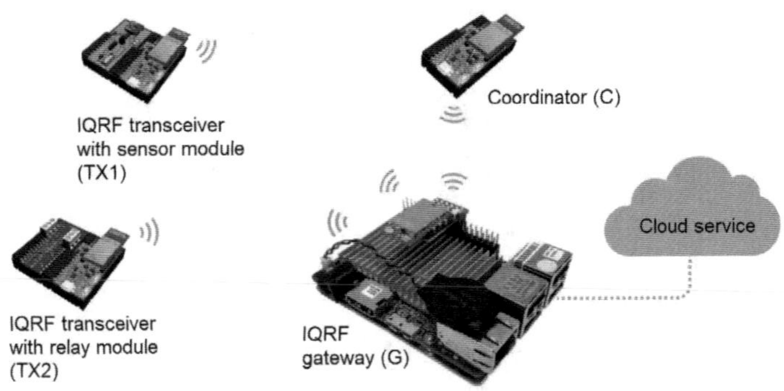


**Figure 3.** The network architecture of the deployed IQRF mesh network for indoor scenario.

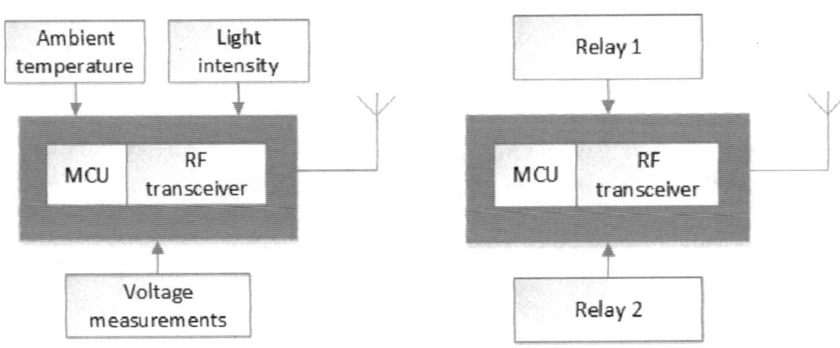


(a) (b)

**Figure 4.** Block diagram of the two IQRF end devices: (a) with sensor module, (b) with relay module.

The coordinator is implemented with a DCTR-72DAT transceiver (i.e. with the standard IQRF transceiver but configured as a coordinator). It manages the configuration of the IQRF mesh network and the communication in it, as well as provides the virtual routing of the data. The communication between the end devices and the coordinator is through IQRF technology.

The IQRF gateway is based on a Raspberry Pi single-board computer via a DCTR-72DAT transceiver connected to it using a KON-RASP-01 adapter with a SPI (Serial Peripheral Interface) interface.

IQRF gateway daemon software is used to connect to the cloud service. Different platforms can be used as a cloud service, for example Microsoft Azure, IBM Watson IoT, IQRF Cloud, Google Cloud, AWS and others. The IQRF gateway transmit data to the cloud by using MQTT protocol (Message Queuing Telemetry Transport). As this paper examines the physical deployment of the mesh network, the cloud service is not covered in detail here.

## 4. Results

The performed measurements aim to analyse the operability and stability of the mesh network. All nodes in the network are distributed according to figure 2 and communicate through the coordinator with the gateway. To assess the quality of the radio connection, the value of the Received Signal Strength Indicator (*RSSI*) parameter is reported at regular intervals of 3 minutes within half an hour.

The reported *RSSI* values are presented in table 2. Based on these reported values and the RF sensitivity level of the DCTR-72DAT transceiver, figure 5 presents a comparative analysis of the change in the *RSSI* level within the measurement duration for the three nodes (both terminal devices and the coordinator).

**Table 2.** Comparison between IQRF and other relevant wireless technologies.

| Timestamp, min | RSSI for TX1, dBm | RSSI for TX2, dBm | RSSI for C, dBm |
|---|---|---|---|
| 0 | -85 | -83 | -66 |
| 3 | -86 | -83 | -64 |
| 6 | -85 | -82 | -64 |
| 9 | -85 | -80 | -65 |
| 12 | -87 | -78 | -65 |
| 15 | -86 | -78 | -65 |
| 18 | -88 | -80 | -66 |
| 21 | -86 | -80 | -66 |
| 24 | -84 | -82 | -64 |
| 27 | -84 | -82 | -64 |
| 30 | -86 | -83 | -65 |
| **Average level of RSSI:** | -85,63 | -81,00 | -64,91 |

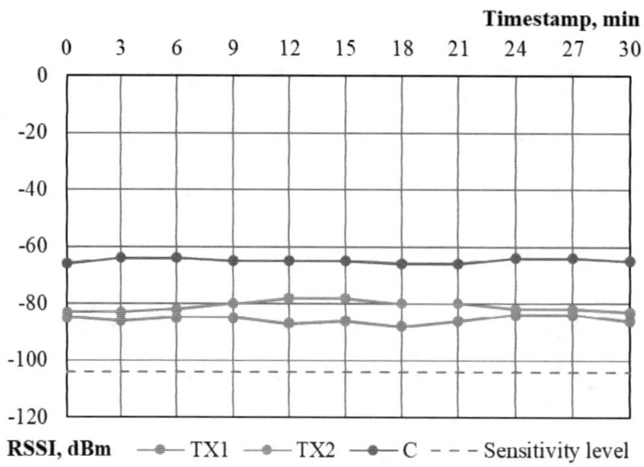


**Figure 5.** Time variation of the RSCI level for a measurement period of 30 minutes.

The RF sensitivity level of the transceivers is considered as the minimum allowable in assessing the quality of the transmission, i.e. the measured levels must be greater than it. The results clearly show that all measured values are greater than this limit value. This shows that the IQRF network is a convenient solution for the considered indoor scenario when deploying an IoT sensor network. It should be noted that the farthest node (TX1) is at a distance of about 45 m from the gateway, and communication with it takes place through multiple walls/obstacles (as well as communication with the node TX2). This further confirms the applicability, reliability and consistency of the implemented system.

As a future work it can be considered the application of IQRF technology in an outdoor environment with a much larger number of nodes, as well as to take into account the impact on network congestion and latency.

## 5. Conclusion

It is obvious that IQRF technology cannot be used for high data rate IoT applications. At the same time, the mesh topology, virtual routing and the specifics of data transmission ensure high network reliability. However, IQRF offers a longer range, higher sensitivity and better power-management (in exchange for lower throughput and higher latency). The presented results show good indoor and free-space signal spread, stable parameters for long-term measurements. However, there are some limitations such as the limited number of nodes in the network and average security without sufficient and clear key-management.

## References

[1] Hammi B, Khatoun R, Zeadally S, Fayad A, Khoukhi L 2018 IoT technologies for smart cities, *IET Networks*, **7**, no. 1, pp 1–13, https://doi.org/10.1049/iet-net.2017.0163

[2] Gilchrist A 2016 Industry 4.0: The Industrial Internet of Things, *Apress, Berkeley CA*, https://doi.org/10.1007/978-1-4842-2047-4

[3] Angelov K, Manchev N, Kogias P and Sadinov S 2019 Design and Development of a Platform for Test Applications in LoRa/LoRaWAN, *Journal of Engineering Science and Technology Review (JESTR)*, Kavala Institute of Technology, pp 17-21

[4] Manchev N., Angelov K, Kogias P and Sadinov S 2019 Development of Multichannel LoRaWAN Gateway for Educational Applications in Low-Power Wireless Communications, *IEEE XXVIII International Scientific Conference Electronics (ET)*, Sozopol, Bulgaria, pp 1-4

[5] IQRF. Technology for wireless. [Online]. Available: https://www.iqrf.org/technology/

[6] Iliev T, Bikov T, Mihaylov G, Ivanova E, Stoyanov I and Keseev V 2019 Algorithms for congestion control in LTE mobile networks, *Software Engineering and Algorithms in Intelligent Systems. CSOC2018 2018, Advances in Intelligent Systems and Computing*, **763**, ed R Silhavy, Springer, Cham, pp 474-484, https://doi.org/10.1007/978-3-319-91186-1_49

[7] Iliev T, Mihaylov G, Bikov T, Ivanova E, Stoyanov I and Radev D 2017, LTE eNB traffic analysis and key techniques towards 5G mobile networks, *40th International Convention on Information and Communication Technology, Electronics and Microelectronics (MIPRO)*, Opatija, Croatia, pp 497-500, doi: 10.23919/MIPRO.2017.7973476

[8] Centenaro M *et al* 2016 Long-range communications in unlicensed bands: the rising stars in the IoT and smart city scenarios, *IEEE Wireless Communications* **23** (5), pp 60–67

[9] Fujdiak R, Mlynek P, Malina L, Orgon M, Slacik J, Blazek P and Misurec J 2019 Development of IQRF technology: analysis, simulations and experimental measurements, *Elektronika ir Elektrotechnika*, **25**, no. 2, pp 72–79, http://dx.doi.org/10.5755/j01.eie.25.2.22739

[10] Krupka, L, Vojtech L and Neruda M 2016 The issue of LPWAN technology coexistence in IoT environment, *17th International Conference on Mechatronics - Mechatronika (ME)*, pp 1-8

[11] Calvo, I, Gil-García, J.M, Recio, I, López, A 2016 Quesada, J. Building IoT Applications with Raspberry Pi and Low Power IQRF Communication Modules, *Electronics* **5**, 54

**Acknowledgments**

The studies were carried out with the financial support of the CoC "QUASAR" – project № BG05M2OP001-1.002-0006.

# Security of the sensory data in the cloud

**F A Tsvetanov[1] and M N Pandurski[1]**

[1] Department of Communication and computer engineering and technology, South-West University, Blagoevgrad, 2700, Bulgaria

E-mail: ftsvetanov@gmail.com

**Abstract.** The integration between the clouds and the sensor networks makes it possible to solve the problem of the limited computing power and the limited capacity of the sensor networks, as well as for the storage and processing of the collected data, without unnecessarily increasing the cost of the sensor networks. A significant challenge is the security of the data transmission connection between the sensors and the cloud. An algorithm has been proposed that ensures security in the transmission of data from the sensor network to the cloud via the DLS security protocol. The algorithm is implemented through the ThingSpeak cloud, and the connection between the sensor network and the cloud is implemented through the MQTT protocol, creating the flow that will gather sensor data, using Node-RED and Generate a self-signed certificate with OpenSSL. The algorithm was tested with Wireshark software, which confirmed that the data packets sent from the sensor network to ThingSpeak use the MQTT data transfer protocol and connect to communication port 8883, which is protected/encrypted via TLS / SSL protocol.

## 1. Introduction

A typical wireless sensor network contains many spatially distributed, self-regulating sensors that jointly monitor and measure various environmental parameters. The sensors provide a variety of useful data that can be used to monitor and control the environment in which they are located. The amount of data generated by sensor networks is enormous, heterogeneous and multidimensional in nature. Significant hardware storage and computing power are required to store and process this data [1].

Unlike traditional networks, WSN has its limitations in design and resources. *Resource constraints* include limited power, short communication range, low bandwidth, limited processing, and storage in each sensor node. Design limitations depend on the application, which is based on the observed environment [2]. Designing an additional ability to process the collected data leads to an increase in the cost of the sensors. Any increase in the price of a single sensor has a multiplier effect on the total cost of the sensor network, as the sensor network can contain hundreds or even thousands of sensors.

Cloud computing provides tremendous computing power and storage space. Cloud computing allows systems and users to use the platform as a PaaS service, infrastructure as an IaaS service, and software as a SAAS service at a very low cost. These services are provided by various cloud providers most often for a fee. The integration between the clouds and the sensor networks is a great opportunity and solution to the problem of the limited computing power and the limited capacity of the sensor networks, for storage and processing of the collected data, without unnecessarily increasing the cost of the sensor networks. The implementation of this integration has formulated a new paradigm called Sensor Cloud Computing [3].

One of the most important barriers to cloud acceptance is security, which is closely linked to legislative and technical components in terms of ensuring data protection. Traditional security mechanisms such as identity, authentication and authorisation are no longer sufficient for clouds in their current form [4]. The challenges of security and privacy in the cloud require further research. Researchers from academia, industry and standards organisations continue to work and provide potential solutions to these challenges [5], [6], [7], [8].

In this paper, an algorithm is proposed to increase the security in the transmission of data from the sensor network to the cloud via the SSL security protocol. The operability of the algorithm is checked and proved.

## 2. Algorithm for increasing the security of transmitted sensor data in the cloud

The algorithm of figure 1 presents one of the possible variants for increasing the security of the transmitted sensor data via TSL protocol. This variant includes a ThingSpeak cloud, MQTT protocol. We create a stream that collects sensor data using Node-RED. The protection of the connection from the sensor network to ThingSpeak is realised by generating a self-signed certificate with OpenSSL. The arguments for choosing this algorithm are given below.

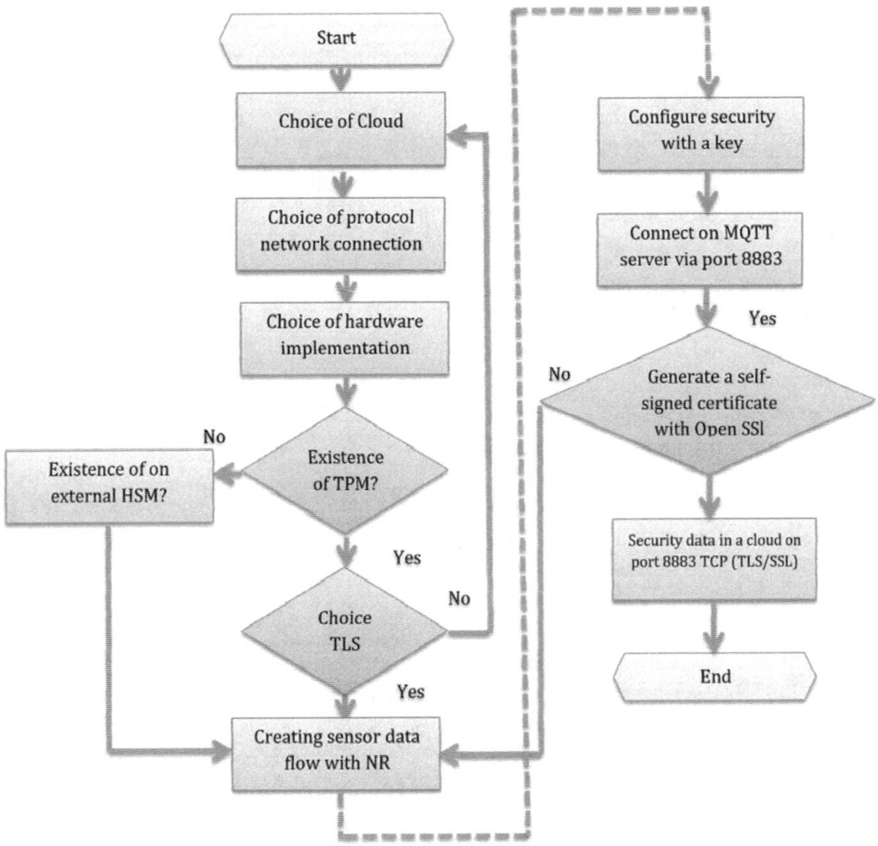


**Figure 1.** Algorithm for the security of sensor data via TLS.

### *2.1. Arguments for choosing the algorithm*

#### *2.1.1. Selecting Cloud*

To upload the sensor data, we select the ThingSpeak cloud [9], which is an open platform for IoT data based on public cloud technology. ThingSpeak enables real-time data collection, analysis and activation with the Open API (HTTP or MQTT). With applications and plugins, there is the ability to store data, visualise, monitor and integrate user data with various third-party platforms, including leading IoT platforms such as ioBridge, Arduino, Raspberry pi. Twilio, Twitter, ThingHTTP,

MATLAB. Sensor data is collected in each channel, which has eight fields that can contain any type of data, three location fields and one status field [10].

*2.1.2.* Select a protocol for connecting the network to the cloud and sending data
The MQTT protocol is selected, which is designed mainly for connecting devices with limited power over wireless networks and is used for publishing / subscribing. It is considered one of the main protocols in the implementation of IoT. MQTT is used to send short messages to and from a broker. ThingSpeak is used as an MQTT broker so that devices can send messages to ThingSpeak. Messages may contain the current office temperature collected by a sensor, etc. ThingSpeak accepts the message and stores its contents in a ThingSpeak channel. Once the data is in the channel, it can be easily visualised and analysed with MATLAB code. The protocol works with TCP / IP sockets or WebSockets. MQTT over WebSockets can be secured with SSL. The publish/subscribe architecture allows messages to be moved to client devices without the device constantly connecting to the server [11]. MQTT supports various mechanisms for data authentication and security. It is important to note that these security mechanisms configured on the MQTT broker, and the client must adhere to the existing mechanisms. There are three ways a broker can verify the identity of an MQTT client: client IDs, usernames and passwords, and client certificates. TLS (SSL) protocols can be used to protect the content of MQTT sensor data messages.

*2.1.3. Choosing a security protocol and hardware realisation*
The choice of this protocol determined by the capabilities of the selected hardware implementation. The result of a study by the company RF Desk security solution shows that there are two possibilities [12]:

- *First*, by selecting a computer motherboard with a chip that allows you to store cryptographic keys used for encryption. Many computers have Trusted Platform Modules hardware chip. Веднъж активиран, Trusted Platform Module предоставя пълни възможности за криптиране на диска. и поддържа защитен режим на свързване на SSL сокет за комуникация с брокера ThingSpeak MQTT.

- *Second, the* use of the external hardware security module (HSM). The module uses TCP / IP and is added to the management system, generating and securely storing cryptographic keys.

We choose the Transport Layer Security (TLS) protocol because it protects the data that is transmitted online between a web browser and a website via HTTPS and because it ensures that the data retains its original integrity and ensures confidentiality through encryption [24]. TLS security is part of the TCP / IP protocol, not MQTT. TLS operates over other transport protocols such as TCP, which means it is over the fourth layer of the OSI model. It is generally used as a transport layer, but in some cases may be part of the presentation or application layers. TLS security provides an encrypted channel using asymmetric cryptography for the personal message authentication code, which increases the reliability of the MQTT message—protecting all parts of the MQTT message, not just the payload of the message. At this stage, the TLS protocol is the most secure when transmitting data online [13].

*2.2. Security Configuration with certification (authority) keys*
*2.2.1. Create the flow that will gather sensor data, and pass it to Thing Speak cloud via MQTT protocol, using software Node-RED.*
Node-Red provides both MQTT nodes for a subscription (login) and publishing (login). We secure the connection via TLS / SSL encryption [14]. The nodes we use are DHT sensors, functional nodes (which use msg.payload to send data), MQTT output node, which connects to the server/cloud mqtt.thingspeak.com. It publishes in the topic about the channel temperature/humidity in our ThingSpeak accounts and accordingly channels fig.2.

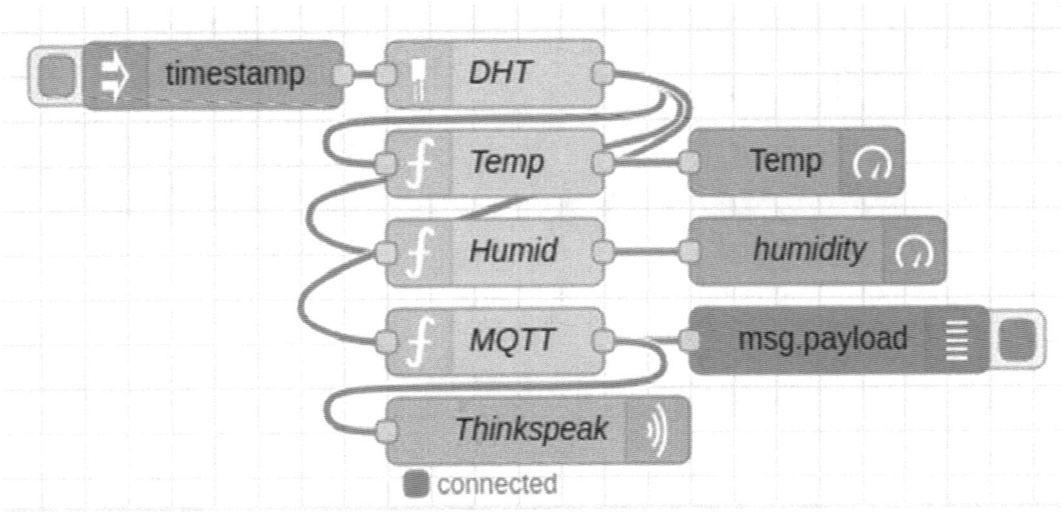


**Figure 2.** The flow using Node-Red.

*2.2.2. Connecting to an MQTT Broker/Server and Publish Using The Publish Node*
To connect to an MQTT broker/server, enter the IP address or name of the Broker and the port we are using (default 1883). The MQTT broker offers encryption and username/password authentication. When a client posts a message to the Broker, he sends the subject of the message and QoS.

*2.2.3. Security settings*

- *Generate a self-signed certificate*

When using our own MQTT broker, we use software such as OpenSSL (which is an open-source cryptographic tool for TLS / SSL protocols) to protect the connection. For this purpose, we create a self-signed certificate using open-SSL.

- *Cloud protection:*

An arbitrary username and MQTT API key provided to connect to ThingSpeak via MQTT. The MQTT URL broker used mqtt.thingspeak.com. To configure the MQTT client and communicate with the ThingSpeak MQTT broker, we must have one of the connection options shown in table 1:

**Table 1.** ThingSpeak MQTT broker connection options

| Port | Connection Type | Encryption |
|---|---|---|
| 1883 | TCP | None |
| 8883 | TCP | TLS/SSL |
| 80 | WebSocket | None |
| 443 | WebSocket | TLS/SSL |

By default, the MQTT broker does not use activated security but uses port 1883 for unencrypted communication. To use the ThingSpeak TCP security configuration, we change the default protected port for MQTT to 8883. The monitoring of the sensor data in ThingSpeak MQTT is shown in figure 3.

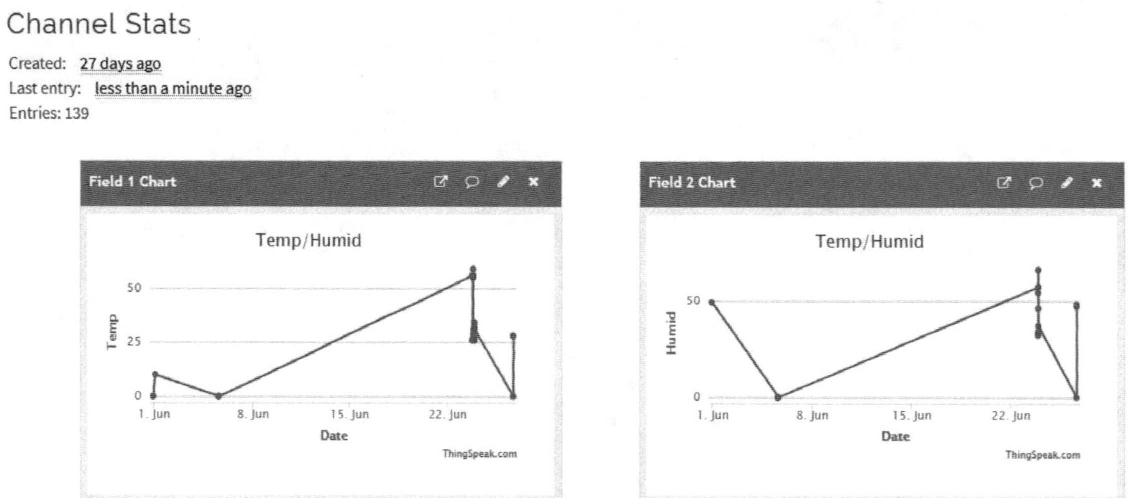


**Figure 3.** The results from the sensor data on the ThingSpeak.

## 3. Check the efficiency of the algorithm to transmit encrypted data

The validation of the algorithm for data encryption was performed using the Wireshark open-source software for monitoring and analysing sent/received data packets and analysis of network protocols. Wireshark is a tool for detecting (sniffer (a computer program that recognises and records a variety of restricted information, especially the secret passwords needed to gain access to files or networks)) and analysing packages. It captures network traffic on the local network and stores this data for offline analysis from Ethernet, Bluetooth, Wireless (IEEE.802.11), Token Ring, Frame Relay connections, etc. [15]. The result of the inspection is shown in figure 4.

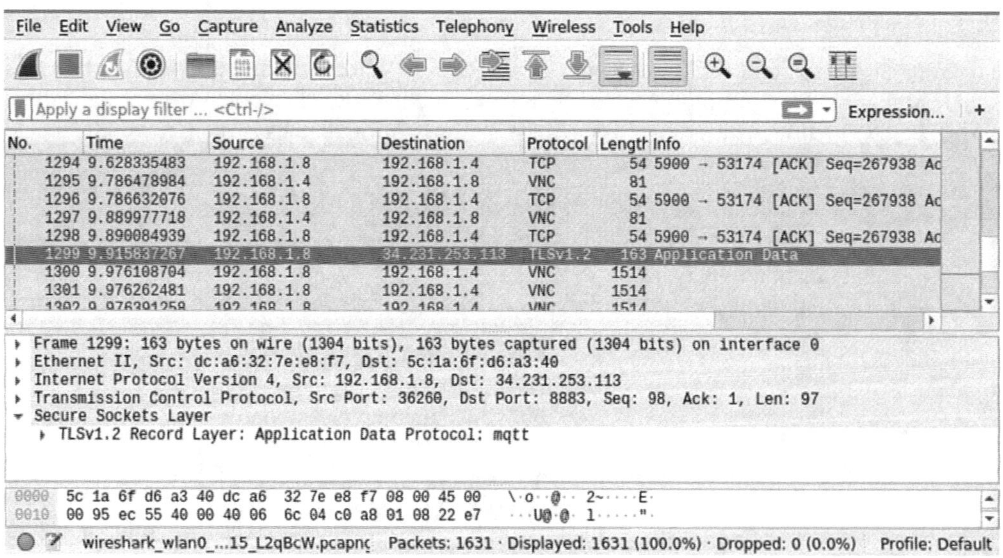


**Figure 4**. The results from the Wireshark packets capture.

The verification confirms that the data packets are sent from the sensor network to ThingSpeak, using MQTT protocol for data transmission, and connected to the communication port 8883 which is protected/encrypted by TLS / SSL protocol.

IOP Conf. Series: Materials Science and Engineering 1032 (2020) 012005 doi:10.1088/1757-899X/1032/1/012005

## 4. Conclusion

The integration between the clouds and the sensor networks makes it possible to solve the problem of storing and processing large volumes of data without unnecessarily increasing the cost of sensor networks. An important integration challenge is the security of the connection when transmitting data between sensors and the cloud. The researched literature sources show that the research problem is topical and the subject of research by both the academic and the industrial community. Different solutions are depending on the specific challenge to be solved by designing network controllers, establishing standards for security, optimisation number of sensors in the network, software countermeasures and implementation of specialised hardware security devices. In this paper, we propose an algorithm that ensures security in the transmission of data from the sensor network to the cloud via the DLS protocol. The algorithm is implemented through the ThingSpeak cloud, the connection between the sensor network and the cloud is through the MQTT protocol, creating the flow that will gather sensor data, using Node-RED and Generate a self-signed certificate with OpenSSL. The algorithm was tested with Wireshark software, which confirmed that the data packets sent from the sensor network to ThingSpeak use the MQTT data transfer protocol and connect to communication port 8883, which is protected/encrypted via TLS / SSL protocol.

## References


[1] Tsvetanov F and Pandurski M 2018 Some aspects for the integration of sensor networks in cloud structures *Proceedings of International Conference on High Technology for Sustainable Development* pp 249-253

[2] Hristov G Tsvetanov F and Georgieva I 2016 Cloudy Technologies in Industry, *Knowledge – International Journal* Vol 15.2 pp 891-897

[3] Mell P and Grance T 2011 NIST Definition of Cloud Computing, *Recommendations of the National Institute of Standards and Technology*

[4] Chadwick D Fan W Costantino G Lemos R Cerbo F Herwono I Manea M Mori P Sajjad A and Wang X 2020 A cloud-edge based data security architecture for sharing and analysing cyber threat information, *Future Generation Computer Systems* Volume 102 pp 710-722

[5] Mohammad S and Ebrahim Al 2019 A secure and energy-efficient platform for the integration of Wireless Sensor Networks and Mobile Cloud Computing *Computer Network* Volume 16

[6] Coppolino L, D'Antonio S Mazzeo G and Romano L 2019 A comprehensive survey of hardware-assisted security: From the edge to the cloud *Internet of Things* Volume 6 pp 1-17

[7] Intel Threat Detection Technology, 2018, Accessed June 23, 2020, https://www.intel.com

[8] Microchip, AN2559, Smart IoT Wireless Sensor Node using XMEGA® AU MCUand LoRa Technology, http://ww1.microchip.com/downloads/en/AppNotes/00002559A.pdf

[9] https://thingspeak.com

[10] Ray P 2017 A survey of IoT cloud platforms Future *Computing and Informatics Journal* Volume 1 Issues 1–2 pp 35-46 https://doi.org/10.1016/j.fcij.2017.02.001

[11] https://uk.mathworks.com/help/thingspeak/mqtt-basics.html

[12] https://www.rcdevs.com/products/openotp-hsm-hardware-security-modules/

[13] Lake J 2019 What is TLS, and how does it work? Available at: https://www.comparitech.com/blog/information-security/tls-encryption/, Accessed 23 June 2020

[14] https://nodered.org/

[15] Wireshark – Go Deep, Available at: https://www.wireshark.org

# Statistical techniques to determine of optimal and acceptable noise levels

**I S Balabanova[1], S S Kostadinova[2], V I Markova[2], S M Sadinov[1] and G I Georgiev[1]**

[1] Department of Communications Equipment and Technologies, Technical University of Gabrovo, 4 H. Dimitar, Gabrovo, 5300, Bulgaria
[2] Department of Communication Engineering and Technologies, Technical University of Varna, 1 Studentska, Varna, 9010, Bulgaria

E-mail: ivstoeva@abv.bg

**Abstract.** The paper presents an approach for an application of QoS procedures, on noise impacts in communications, respectively Uniform White Noise (UWN), Gaussian White Noise (GWN), Bernoulli Noise (BN) and Poisson Noise (PN). The approach consists in experimental establishment of recommended optimal and acceptable limit levels of the noise indicator Root Mean Square based (RMS) on the processing of registered information sets for each specific noise. A set of methodological statistical procedures are applied to the experimental data with respect to the complete and individual input sets (RMS levels for each individual noise). Types of family characteristics are analyzed and evaluated about different quality indicators as "Mean Plot of multiple variables", "Normal Probability Plots", "Individual Plots", "X-Bar and R char variable", "Capability Plots" and "Capability Histograms". The approach is also associated with the detection of RMS samples with deviations outside the defined statistical levels, as well as their exclusion in order to improve the quality of the processed information sample.

## 1. Introduction

The sources of noise in communication systems can be external - atmospheric, galactic, synthetic noise, etc., or internal for the systems. The statistical description and modelling of noise impacts are based on concepts such as "random variable", "Cumulative and Probability Density Functions" and "Statistical Averages" [1].

Algorithms for automatic detection of peak areas and interpolation process are often used in the practice of signal recovery at the appearance of pulse noise, in the basis of which is the principle of realization of an autoregressive process [2]. Other recovery approaches are associated with the use of statistical processing based on the inverse error functions, etc. [3].

The reduction of the effect of noise impacts is an important problem in a number of areas such as cellular mobile communications, speech recognition, image processing, medical, radar and other signals. That's why, is important to determine and maintain optimal noise levels in the communication channels, which would introduce minimal disturbances in the forms of signal and save the integrity of the transmitted information.

The paper proposes a statistical approach related to the assessment of experimental noise, establishment of recommended RMS optimal and permissible variance levels. The approach also involves the quality analysis and composition adjustment of the sample being processed for noise samples outside the statistically defined limits when such a procedure is required.

IOP Conf. Series: Materials Science and Engineering 1032 (2020) 012006 doi:10.1088/1757-899X/1032/1/012006

## 2. Exposition

Regarding the purpose of the study, four types of interference at fixed parameters were simulated for each specific noise, as follows; Uniform White Noise: Amplitude = 1; Gaussian White Noise: Standard deviation = 1; Bernoulli Noise: Ones probability = 0.5 and Poisson Noise: Mean = 1. The simulation of the signals was made trough of a virtual instrument developed on the basis of LabVIEW with included control elements for regulation of the experimental values and parameters of the noise, given in figure 1. The instrument also has digital and graphical indicator devices for complex and individual monitoring of the experimental signals. For each of them, a sub-virtual tool for statistical analysis and calculation of Root Mean Square parameter was used.

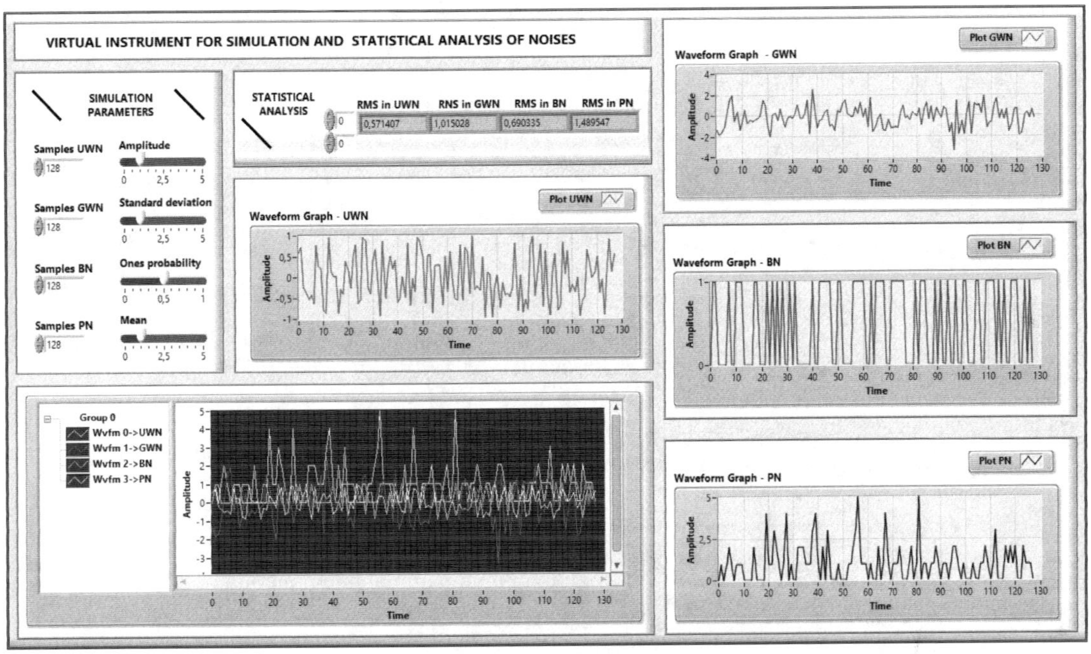


**Figure 1.** Front panel of virtual instrument for noise simulation and statistical analysis.

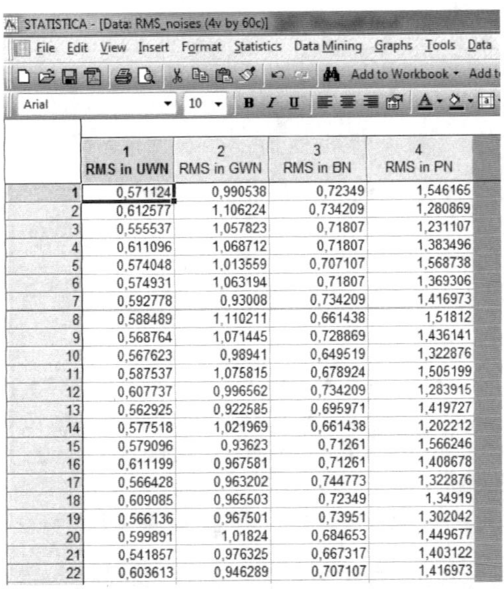


| | 1<br>RMS in UWN | 2<br>RMS in GWN | 3<br>RMS in BN | 4<br>RMS in PN |
|---|---|---|---|---|
| 1 | 0,571124 | 0,990538 | 0,72349 | 1,546165 |
| 2 | 0,612577 | 1,106224 | 0,734209 | 1,280869 |
| 3 | 0,555537 | 1,057823 | 0,71807 | 1,231107 |
| 4 | 0,611096 | 1,068712 | 0,71807 | 1,383496 |
| 5 | 0,574048 | 1,013559 | 0,707107 | 1,568738 |
| 6 | 0,574931 | 1,063194 | 0,71807 | 1,369306 |
| 7 | 0,592778 | 0,93008 | 0,734209 | 1,416973 |
| 8 | 0,588489 | 1,110211 | 0,661438 | 1,51812 |
| 9 | 0,568764 | 1,071445 | 0,728869 | 1,436141 |
| 10 | 0,567623 | 0,98941 | 0,649519 | 1,322876 |
| 11 | 0,587537 | 1,075815 | 0,678924 | 1,505199 |
| 12 | 0,607737 | 0,996562 | 0,734209 | 1,283915 |
| 13 | 0,562925 | 0,922585 | 0,695971 | 1,419727 |
| 14 | 0,577518 | 1,021969 | 0,661438 | 1,202212 |
| 15 | 0,579096 | 0,93623 | 0,71261 | 1,566246 |
| 16 | 0,611199 | 0,967581 | 0,71261 | 1,408678 |
| 17 | 0,566428 | 0,963202 | 0,744773 | 1,322876 |
| 18 | 0,609085 | 0,965503 | 0,72349 | 1,34919 |
| 19 | 0,566136 | 0,967501 | 0,73951 | 1,302042 |
| 20 | 0,599891 | 1,01824 | 0,684653 | 1,449677 |
| 21 | 0,541857 | 0,976325 | 0,667317 | 1,403122 |
| 22 | 0,603613 | 0,946289 | 0,707107 | 1,416973 |

**Figure 2.** RMS input dataset for QoS statistical analysis.

For each examined noise, a sample of 60 experimental observations was accumulated for the indicator, for subsequent processing and analysis with the tools of the STATISTICA software environment (figure 2) [4, 5]. A group analysis of the experimental data is applied, consisting in constructing a Histogram of the average values of RMS in a given confidence interval "0.95", shown in figure 3. No deviations from the defined confidence area have been identified.

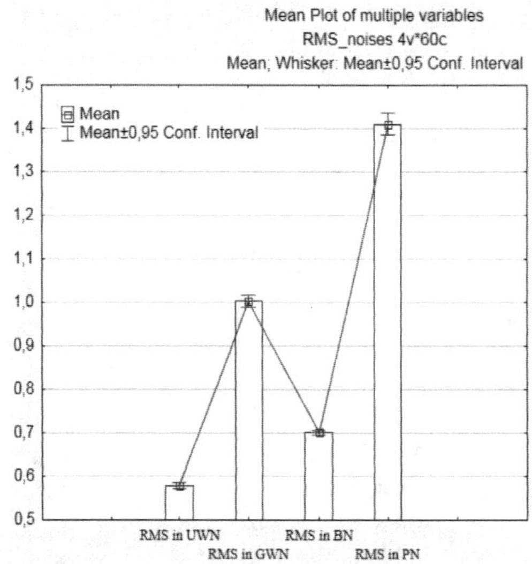


**Figure 3.** Mean plot about RMS in UWN, RMS in GWN, RMS in BN and RMS in PN..

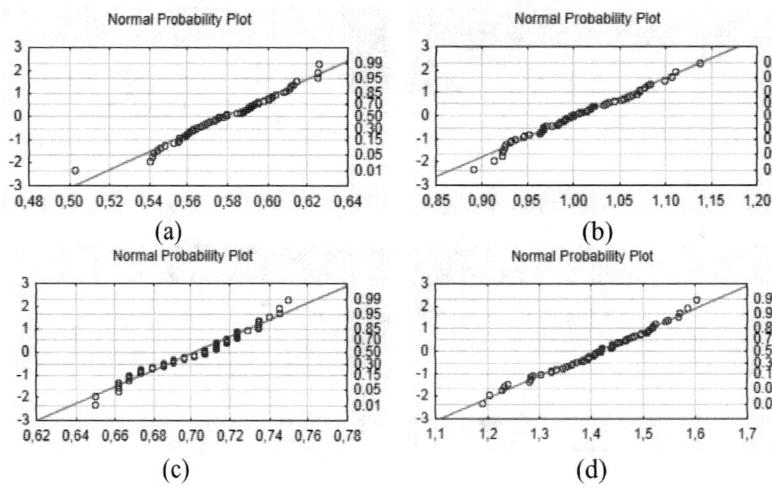


**Figure 4.** Normal probability plots of (a) RMS in UWN, (b) RMS in GWN, (c) RMS in BN and (d) RMS in PN.

An individual quality analysis was conducted for the indicated information groups, the first step from which it is aimed at assessing the nature of the random change of the data by means of Normal probability graphs in figure 4. Examining the location of the test points in the direction of the 45 ° line, no discrepancies regarding normal distribution are detected visually. Based on individual graphs of the observations in figure 5 the following optimal RMS values of noises and their variation levels of change are found, in which they are subject to parametric control: UWN: RMS = 0.57860 compared to the levels from 0.53787 to 0.61933; GWN: RMS = 1.0029 in the range from 0.91317 to 1.0926, BN: RMS = 0.70145 according to the range from 0.65054 to 0.75235 and PN: RMS = 1.4101, limited from 1.2258 to 1.5945.

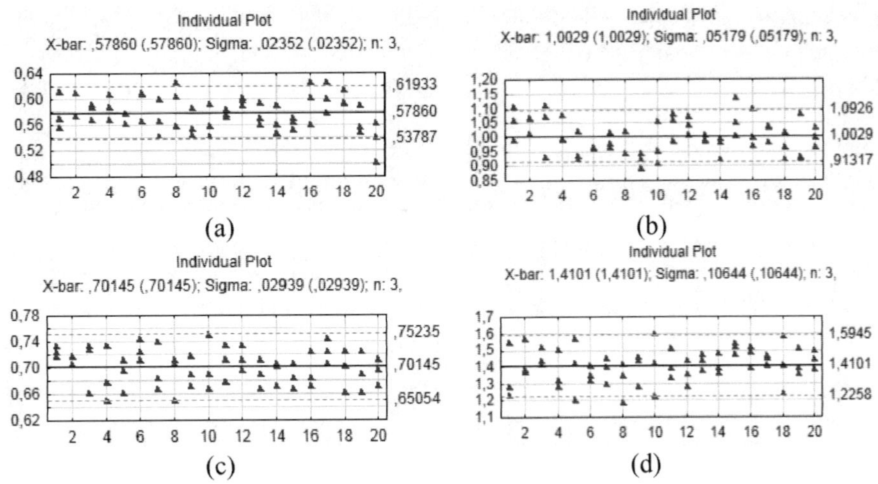


**Figure 5.** Individual plots for (a) RMS in UWN, (b) RMS in GWN, (c) RMS in BN and (d) RMS in PN.

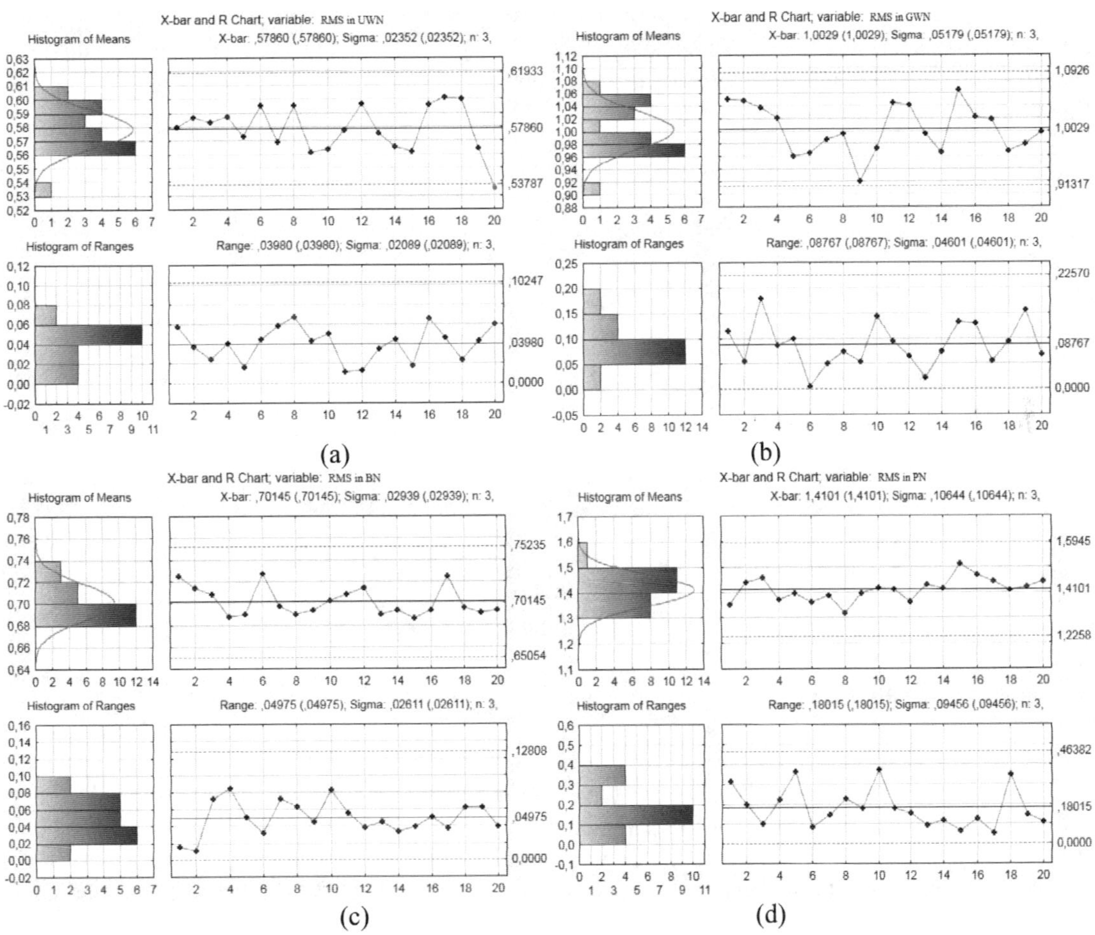


**Figure 6.** Individual plots for (a) RMS in UWN, (b) RMS in GWN, (c) RMS in BN and (d) RMS in PN.

In each test group, individual points are observed, located relative to the respective boundaries, or outside them. In order to limit the number of RMS observations that are not characterized by a normal distribution, it is necessary to group the data and establish a new information sample with a reduced dimension. In other words, each group of 60 observations is limited to a composition of 30 RMS samples. In this way, an increase of level of the data subject to regulation (falling into the established variation RMS levels) is achieved.

According to the presented X-Bar and R graphs in figure 6 only in the analysis regarding the RMS levels of Constant white noise a problematic sample was found below the lower optimal threshold (figure 6(a)). According to the diagrams, maximum permissible deviations of the RMS samples from the found optimal levels were determined, respectively for UWN: <0.10247; GWN: <0.22570; BN: <0.12808; PN: <0.46382. With regard to the registered object of non-parametric control, more complete detailed statistical information can be obtained and its constituent observations should be excluded from the sample for the specific signal (figure 6).

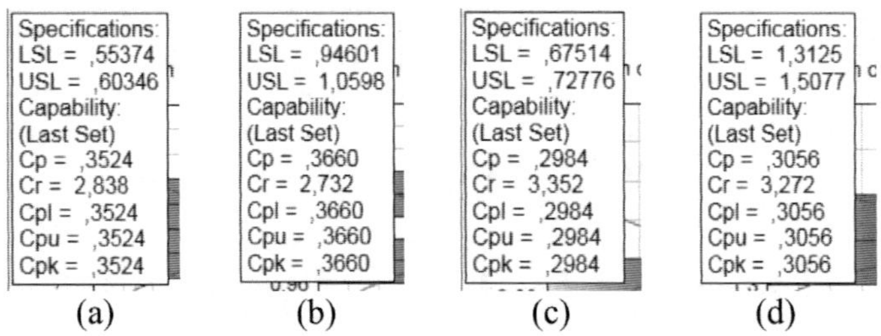


**Figure 7.** Capability indicators of (a) RMS in UWN, (b) RMS in GWN, (c) RMS in BN and (d) RMS in PN.

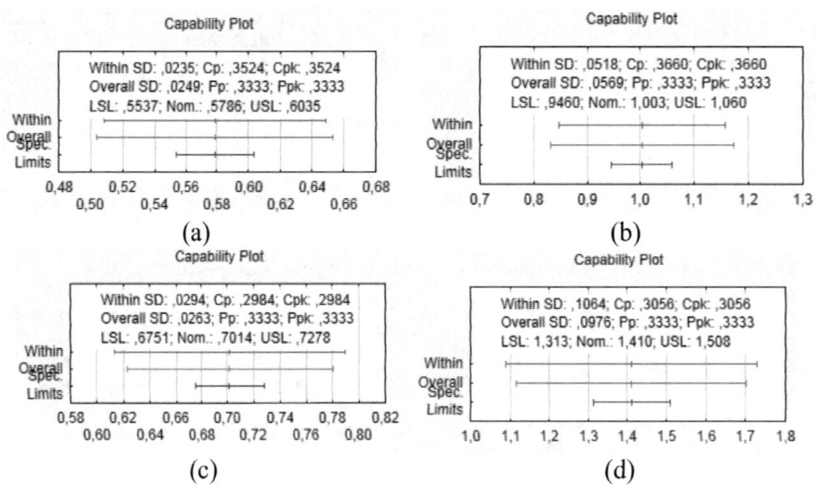


**Figure 8.** Capability plots for (a) RMS in UWN, (b) RMS in GWN, (c) RMS in BN and (d) RMS in PN.

Based on the processing and analysis of the experimental data, capability indicators are obtained. The entered indicators are presented in numerical and graphical form in figure 7 and figure 8. The Cr indicator can be accepted with the highest degree of significance, according to which the group "RMS at BN" is determined with the best quality, followed by "RMS at PN", "RMS at UWN" and "RMS at GWN".

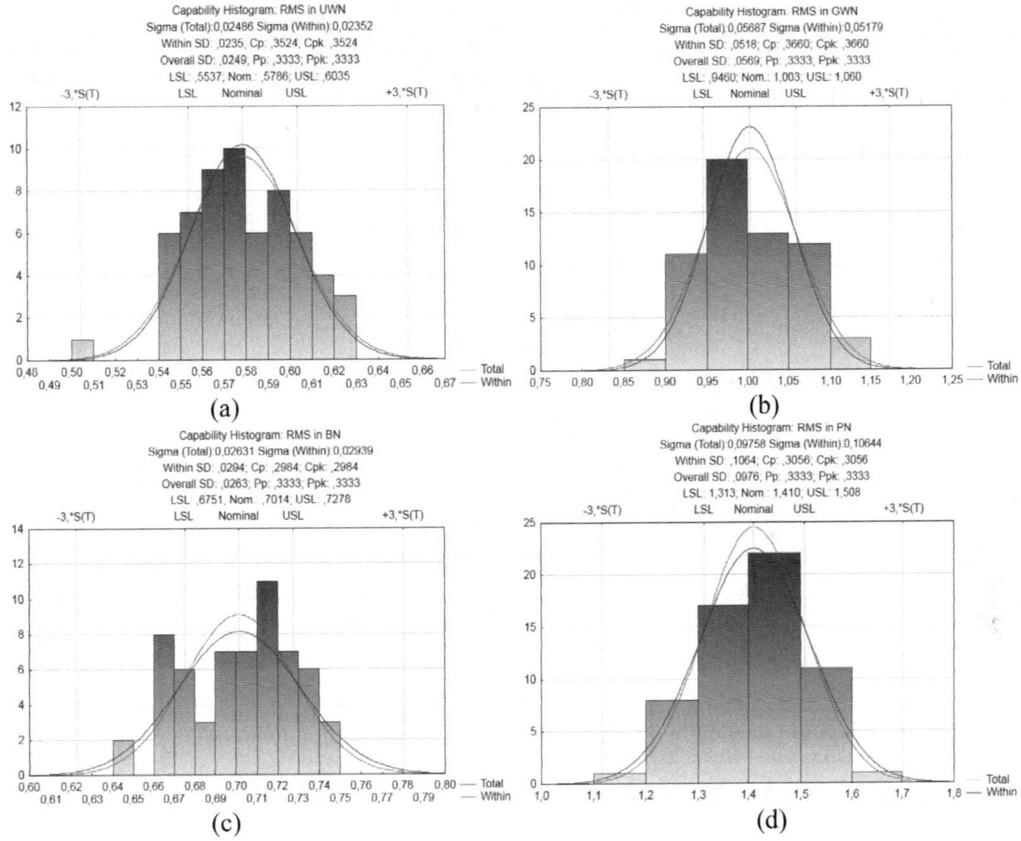


**Figure 9.** Capability histograms in analysis of RMS in (a) UWN, (b) GWN, (c) BN and (d) PN.

In figure 9 a probability density functions are built. Total and Within, referring to the analyzed information observations and samples. The given dependences describe to a good extent the experimental data, confirming their normal nature of distribution.

## 3. Conclusion

The proposed methodology can be successfully applied for statistical analysis of different types of information arrays, concerning embarrassing impacts on electrical signals, traffic input and output flows in communication systems.

## References

[1] Ward D 2004 *Ee2/ise2 Communications Ii Part I – Communications Principles* pp 1–61
[2] Oudre L 2015 Automatic Detection and Removal of Impulsive Noise in Audio Signals, *Image Processing On Line*, **5** , pp, 267–281, https://doi.org/10.5201/ipol.2015.64
[3] Mohonta Ch Sh, Ali F M and Sadeque G M 2015 Study of different types of noise and its effects in communication systems *International Journal and Engineering and Management Research* **5** pp 410-413
[4] Ivanova E, Iliev T, Mihaylov G, Keseev V and Stoyanov I 2019 Significant simulation parameters for RESTART/LRE method in teletraffic systems of SDN. *Lecture Notes in Networks and Systems,* **68** pp 265-273, https://doi.org/10.1007/978-3-030-12450-2_25
[5] Ivanova E, Iliev T, Mihaylov G, Keseev V and Stoyanov I 2019 Significant simulation parameters for RESTART/LRE method in teletraffic systems of ultra broadband convergence networks *Advances in Intelligent Systems and Computing Springer Cham* **860**, pp 69-77, https://doi.org/10.1007/978-3-030-00184-1_7

# Voice control and management in smart home system by artificial intelligence

**I S Balabanova[1], S S Kostadinova[2], V I Markova[2], S M Sadinov[1] and G I Georgiev[1]**

[1] Department of Communications Equipment and Technologies, Technical University of Gabrovo, 4 H. Dimitar, Gabrovo, 5300, Bulgaria
[2] Department of Communication Engineering and Technologies, Technical University of Varna, 1 Studentska, Varna, 9010, Bulgaria

E-mail: ivstoeva@abv.bg

**Abstract.** The paper provides a 3D architectural model of Smart Home system. An information data sets of parameters in sound analysis of test voice commands were collected. The following analyzed indicators are included, respectively LZE, LZeq, LZF, LZS, LZI, LAE, LAeq, LAF, LAS, LAI, LCE, LCeq, LCF, LCS, LCI and LEX8h. Backpropagation and Hybrid algorithms based Artificial intelligence (AI) and Adaptive neuro-fuzzy interface system (ANFIS) were synthesized. Selected architectures are integrated in intelligent automated voice control system for human access control, power switching and lighting, air conditioning systems and home appliances. In the process of synthesis, different criteria for network performance in the analysis of activation type in the output layers in AI and input layer in ANFIS are applied. About all considered voice categories for functional control an accuracy of 100.0% was established. Verification procedures concerning reliability of the achieved results were performed for correct confirmation.

## 1. Introduction

Modern intelligent vision and speech systems are based on deep neural network architectures and algorithms, among which can be mentioned Convolutional Neural Networks (CNN), Deep generative models (Deep Belief Networks - DBNs), Stacjed Auto-Encoders (SAEs), Variational Autoencoders (VAEs), Generative Adversarial Networks (GANs), Flow-Based Models, Recurrent Neural Networks (RNNs) and Long Short-Term Memory (LSTM) networks. Trends show that smart home technologies are gaining ever higher speed given the growing urbanization of the urban environment, resulting in the emergence of new products based on artificial intelligence (AI) [1,2,3]. According to the literature studies, AI integration is associated with five main functions: Device Management, Energy Management, Healthcare, Intelligent Interaction and Security [4,5]. The role of AI is to provide activities such as Data Processing, Voice Recognition, Activity Recognition, Decision Making and Prediction Making [6,7].

The paper shows a conceptual design of a Smart Home type system for building automation with integration of voice analysis and control based on Artificial Intelligence and Adaptive Neurally Fuzzy interface systems [3,4]. The results of research and synthesis of ANN with Levenberg-Marquardt training algorithm and ANFIS architectures, based on extracted sound indicators in the processing of specified sets of voice commands for access control, activation and control of internal systems and appliances are presented.

## 2. 3D architectural model of Start Home project and concept of voice control systems

Using the capabilities of the 3Ds MAX platform, an architectural three-dimensional model of "Smart Home" was created. 3D views of various aspects of the project are shown in figure 1. Training and selective procedures for selection of neural architectures for control of installations and household appliances through initiated "voice commands" are provided. For each pronounced command, sound analysis procedures are applied on the basis of which the data for a family of 16 indicators are registered, among which: 1) LZE (DB), 2) LZeq (dB), 3) LZF (dB), 4) LZS (dB), 5) LZI (dB), 6) LAE (dBA), 7) LAeq (dBA), 8) LAF (dBA), 9) LAS (dBA), 10) LAI (dBA), 11) LCE (dBC), 12) LCeq (dBC), 13) LCF (dBC), 14) LCS (dBC), 15) LCI (dBC) и 16) LEX8h (dBA).

(a)

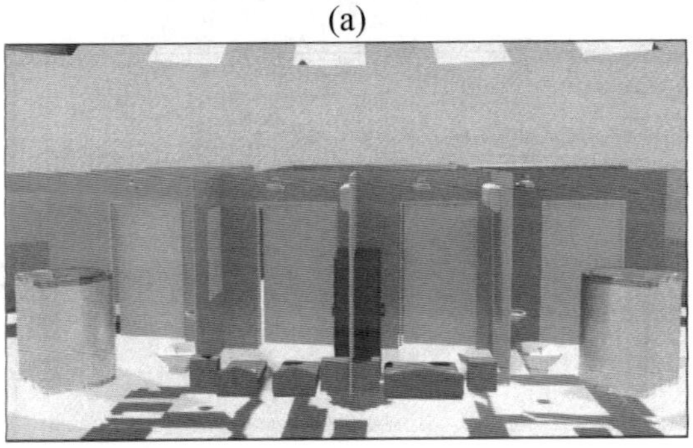

(b)

**Figure 1.** 3D views of Start Home – (a) south and (b) sides.

The recorded indicators are used as input variables in neuronal training as each voice command is coded with a separate output neuron and a corresponding and discrete code combination. For example, when identifying two commands, the following combinations are used: "1 0" for Command №1 and "0 1" for Command №2.

## 3. Analysis and synthesis of artificial neural networks for voice control systems

The research is focuses on activities for searching for optimal neural architectures, i.e. networks with an approximately balanced ratio between "maximum accuracy" and "minimum MSE", in a fixed range of variation of the hidden neurons, respectively from 2 to 15. By limiting the number of experimental hidden neurons, the goal is to increase the speed of the neural computational procedures.

### *3.1. ANN architecture for home access control and electrical power system activation*

A neural network for voice commands identification has been created for:

- switching on of the entire home electrical installation;
- establishing of the front door locking mechanism in an inactive state.

Table 1 illustrates the results of the training architectures within the accepted test limits concerning the hidden neurons. Correct voice identification was found at 2, 7, 8, 11 and 15 neural units, which suggests the network selection to be based relative to the change of the root mean square error. Comparing the tabular indications for MSE, the highest error 0.0567 was found at 7 and the lowest 0.0057 at 2 hidden neurons, determining the choice of the considered ANN (figure 2 (a)).

**Table 1.** Results in synthesis of ANNs for home access control and electrical power system activation.

| *Hidden Neurons* | Accuracy, % | *MSE* |
|---|---|---|
| 2 | 100.0 | 0.0057 |
| 3 | 81.8 | 0.0803 |
| 4 | 54.5 | 0.2449 |
| 5 | 100 | 0.0103 |
| 6 | 90.9 | 0.0701 |
| 7 | 100.0 | 0.0567 |
| 8 | 100.0 | 0.0063 |
| 9 | 90.9 | 0.2631 |
| 10 | 81.8 | 0.0901 |
| 11 | 100.0 | 0.0228 |
| 12 | 90.9 | 0.0339 |
| 13 | 100.0 | 0.0097 |
| 14 | 81.8 | 0.2136 |
| 15 | 100.0 | 0.0131 |

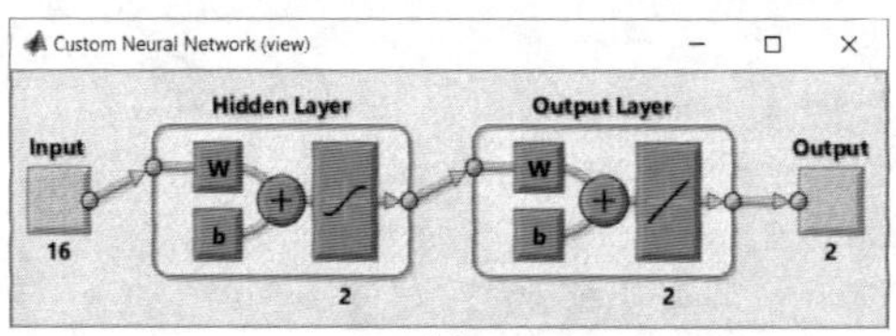


(a)

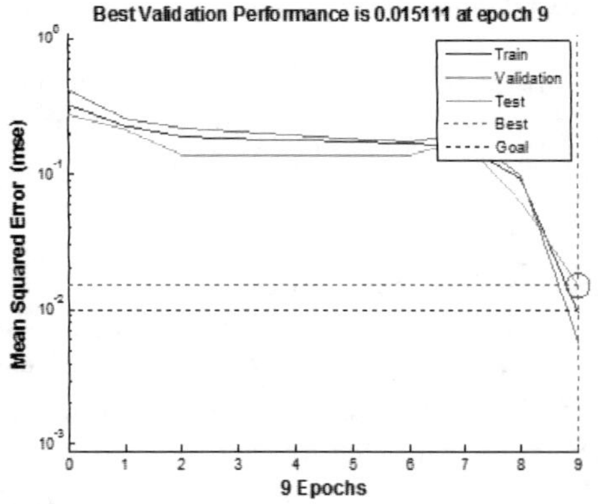


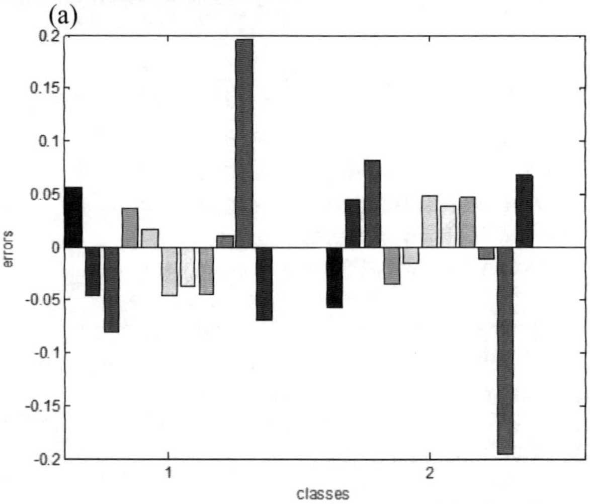


(b) (c)

**Figure 2.** Architecture (a), MSE (b) and Network errors (c) in ANN for home access control and electrical power system activation.

The training of the specific ANN takes place within 9 epochs for which the best achieved productivity is 0.015111 (figure 2 (b)).

With respect to the moment of the 9th iteration, the gradient and the variable Mu are equal to 0.10634 and 1e-05 when zero validation checks are performed. The final variation interval of the network errors is from - 0.2000 to +0.2000 – figure 2 (c).

### *3.2. ANN architecture for lighting and climate control system*

In this subsection the results of the set of procedures for voice control of the main and auxiliary lighting, as well as the ventilation air-conditioning modular subsystem for separate rooms of the home are presented, respectively:

- main lighting in the living room and kitchen;
- auxiliary lighting in the bedroom;
- air conditioning system in the living room and kitchen with a medium degree of heating/cooling;
- air conditioning system in the bedroom in a medium degree of heating/cooling;

**Table 2.** Results in synthesis of ANNs for lighting and climate control system.

| *Hidden Neurons* | Accuracy, % | *MSE* |
|---|---|---|
| 2 | 59.4 | 0.1327 |
| 3 | 65.6 | 0.1209 |
| 4 | 90.6 | 0.0831 |
| 7 | 81.3 | 0.0634 |
| 8 | 100.0 | 0.0237 |
| 9 | 93.8 | 0.0278 |
| 10 | 87.5 | 0.0582 |
| 11 | 93.8 | 0.0482 |
| 12 | 81.3 | 0.0591 |
| 13 | 93.8 | 0.0331 |
| 14 | 87.5 | 0.0643 |
| 15 | 96.9 | 0.0239 |
| 2 | 59.4 | 0.1327 |
| 3 | 65.6 | 0.1209 |

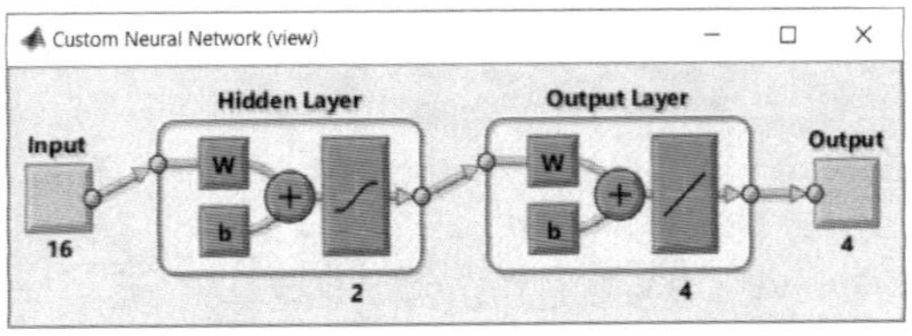


**Figure 3.** Architecture of ANN for lighting and climate control system.

The performance data from table 2 show a change in the accuracy criterion from 59.4% at 2 to 100.0% at 8 intermediate neurons. A variation MSE range from 0.0237 to 0.1327 was registered here, respectively, at 8 and 2 intermediate neurons. According to the analysis, the most suitable neural model regarding the specifics of the task is an ANN with 8 hidden neurons, shown in figure 3.

A relatively higher number of 61 training iterations were observed, with a maximum network performance of 0.10167 obtained within 55 epochs (figure 4 (a)). For the training period, 6 validation checks were performed. Similar error margins were found for the first and fourth groups, while the highest ones were observed in the second base group. In general, the obtained errors fall in the range of about -0.6000 to +0.5000, as shown in figure 4 (b).

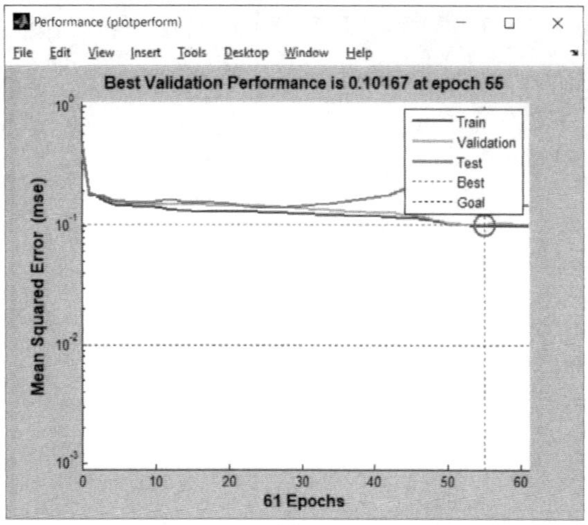


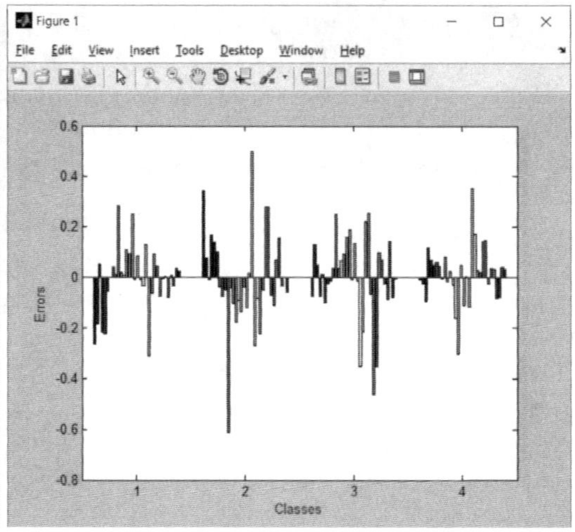


(a) (b)

**Figure 4.** MSE (a) and Network errors (b) in ANN for lighting and climate control system.

*3.3. ANFIS architecture for home access control and electrical power system activation*

Activities for training of neural-fuzzy architecture according to an algorithm combining the approaches of least squares and error backpropagation have been performed, in connection with voice control of "home unlocking" and "power-up" procedures. The procedures are targeted using a constant integer type output membership functions. In connection with the provided voice commands, a limit on the number of training sound indicators has been introduced to optimize the test ANFIS architecture given in figure 5 (a). Group of indicators 1) LZE (DB), 2) LZeq (dB), 3) LZF (dB), 4) LZS (dB), 5) LZI (dB) was used.

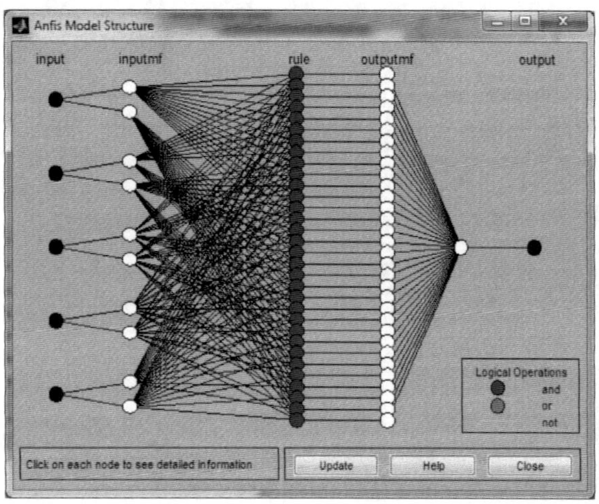


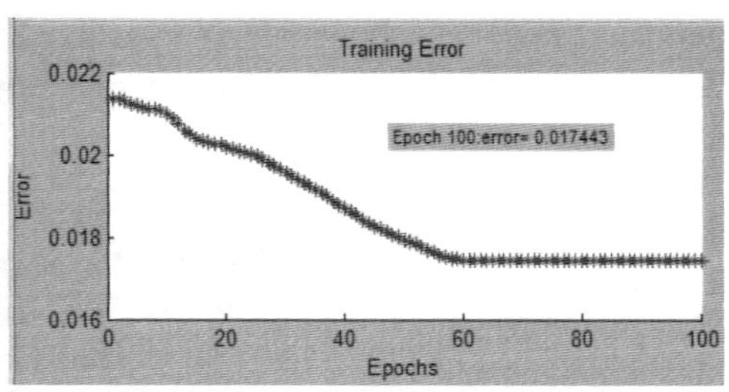


(a) (b)

**Figure 5.** Architecture (a) and RMSE in "trimf" input membership function of neuro-fuzzy classifier for home access control and electrical power system activation.

The activities for the synthesis of a neuron-fuzzy classifier consist in monitoring the "Root Mean Square Error (RMSE)" indicator at a different fixed type of membership function of the input variables. A feature of this type of systems is that the parameters of the input membership functions were adjusted automatically during the training, so as to create architecture with maximum efficiency. The selection of the most adequate ANFIS model is based on a minimum indication of the learning error, in this case 0.017443, registered with the "triangular" type of membership function shown in figure 5 (b).

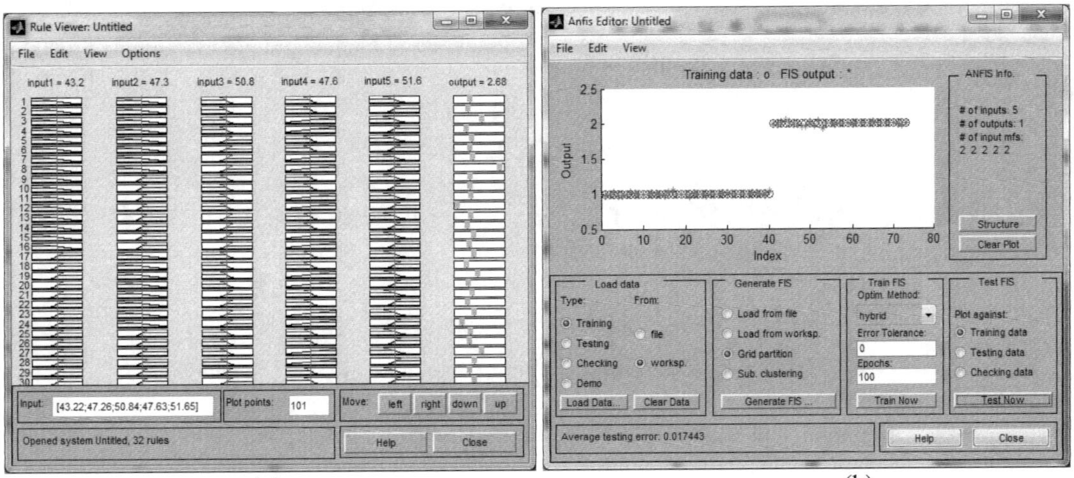


(a) (b)

**Figure 6.** Fuzzy rules (a) and testing of neuro-fuzzy classifier for home access control and electrical power system activation.

Figure 6 (a) is a graphical interpretation of the defined linguistic rules in the study of ANFIS. Complete correctness was found in the identification of user voice commands, defining the used tools as fully applicable in solving such recognition tasks, as confirmed by figure 6 (b).

## 4. Conclusion

Artificial intelligence finds more and more manifestations not only in the industry, but also in terms of increasing people's living comfort when building homes of a new modern type. The presented research demonstrates the strength of its computational advantages, as well as the possibility of combining with the mathematical apparatus of fuzzy logic, in the development of voice control systems with potential implementation in homes, offices and buildings.

## References

[1] Alam M, Samad M, Vidyaratne L, Glandon A and Iftekharuddin 2020 Survey of deep neural networks in speech and vision systems, *Neurocomputing* **417** pp 302-321, https://doi.org/10.1016/j.neucom.2020.07.053

[2] Ivanova E, Iliev T and Stoyanov I 2020 Modelling and simulation of big data networks, *7th International Conference on Energy Efficiency and Agricultural Engineering (EE&AE)*, Ruse, pp 1-4, doi: 10.1109/EEAE49144.2020.9278783

[3] Guo X, Shen Zh, Zhang Y and Wu T 2019 Review of the application of artificial intelligence in smart home, *Smart Cities* **2**(3) pp 402-420, https://doi.org/10.3390/smartcities2030025

[4] Kopytko V, Shevchuk L, Yankovska L and Semchuk Z 2018 Smart home and artificial intelligence as environment for the implementation of new technologies *Traektoriâ Nauki = Path of Science* **4**(9) pp 2007-2012, doi: 10.22178/pos.38-2

[5] Stoyanov I, Iliev T, Mihaylov G, Ivanova E and Kogias P 2018 Smart grid communication protocols in intelligent service for household energy use, *Advances in Intelligent Systems and Computing Springer* **661** pp 380-389, https://doi.org/10.1007/978-3-319-67618-0_35

[6] Kumar S and Qadeer M 2012 Application of AI home automation, *International Journal of Engineering and Technology* **4**(6) pp 803-807, doi: 10.7763/IJET.2012.V4.488

[7] Saberi O and Menes R 2020 Artificial intelligence and the future for the smart homes, *EM Compass* **78**, pp 1-8, Available at: http://hdl.handle.net/10986/33615

IOP Conf. Series: Materials Science and Engineering 1032 (2020) 012008 doi:10.1088/1757-899X/1032/1/012008

# Evolution of mobile networks and seamless transition to 5G

**E P Ivanova[1], T B Iliev[1], I S Stoyanov[2] and G Y Mihaylov[3]**

[1] Department of Telecommunications, University of Ruse, 7 Studentska Str., 7000, Ruse, Bulgaria
[2] Electrical and Power Engineering Department, University of Ruse, 7 Studentska Str., 7000, Ruse, Bulgaria
[3] Department of Telecommunications, University of Telecommunications and Post, 1 Akad. Stefan Mladenov Street, Sofia, 1700, Bulgaria

E-mail: epivanova@uni-ruse.bg

**Abstract.** The evolution in mobile networks continues with the next 5G technology, which is already released in 2020, and also has a distinct advantage from its predecessor. This difference also results in measurements of Quality of Service (QoS) and Quality of Experience (QoE) which are very diverse because there is no specific standard on 5G technology. This paper gives a review of an approach in the measurement as part of the process of QoS and QoE. In addition, this paper provides measurements on how to predict the availability of new 5G. The measurement process is related to the location and conditions, where 5G technology will be implemented.

## 1. Introduction

The first mobile cellular networks of the fifth generation 5G are expected to be available to the end user in 2020. The technology will be a combination of different networks, technologies and applications, and there will be no one universal 5G network. It is expected to have a set of different frequency spectra with different access options for different needs. Wider bandwidths will be used, the connection time is expected to be between 1 and 5 ms, ensuring full availability of radio coverage. 5G should provide real-time communication of time-dependent applications, transmission speeds in the DL direction up to 20 Gbps and up to 10 Gbps in the UL direction, complete reliable communication, improved energy efficiency, direct communication between mobile devices, etc. [1, 2,3]

The standardization of 5G technologies and solutions should be completed by 2021, so 5G currently means only fragmented solutions, which in the future will become part of the overall IMT2020 solution. Such solutions have already been implemented in various countries, but they are still local and test in nature and don't provide all the planned functionality of IMT2020 networks. [4,5]

5G networks significantly extend the limited functionality of previous generations of mobile networks (figure 1) [6]. The main functional characteristics of 5G networks can be summarized to:

- Enhanced Mobile Broadband (eMBB);
- Ultra-Low Latency Reliable Communication (ULLRC);
- Massive IoT/IIoT, mMTC (massive Machine Type Communication).

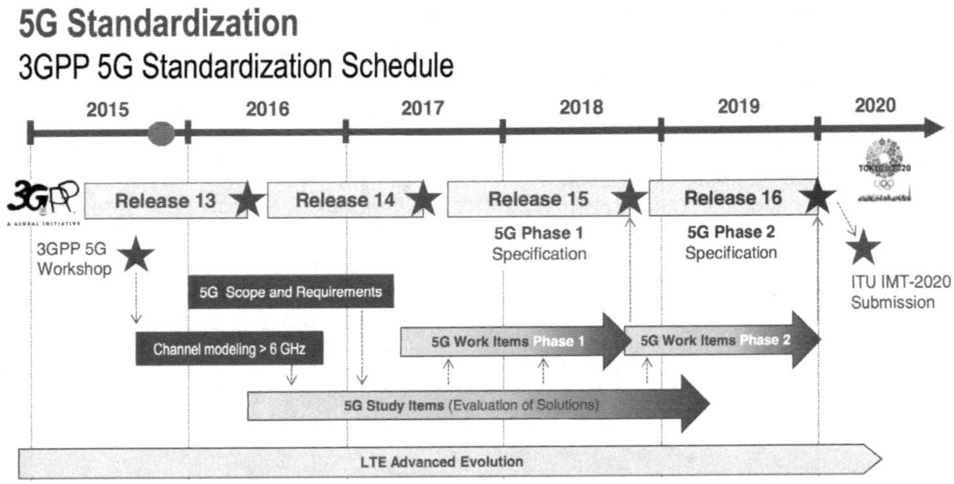


**Figure 1.** 5G evolution.

Technical characteristics and parallel between the generations of mobile cellular systems, mostly related to data transmission are presented in the article [3].

The paper describes characteristics of gNodeB and UE main parameters, measurements on how to predict the availability of new 5G in Sofia, measuring the important data [7, 8, 9].

## 2. 5G mobile networks

5G provides the ability to upgrade through cognitive radio technology, which includes various important features, such as the ability of devices to identify their geographical location, as well as time, temperature and more signals in its work environment. In addition, it timely distinguishes changes in its environment and responds accordingly to ensure continuous quality of service (QoS).[10, 11] Figure 2 shows the 5G system model, which is entirely IP based, designed for wireless and mobile network.[2]

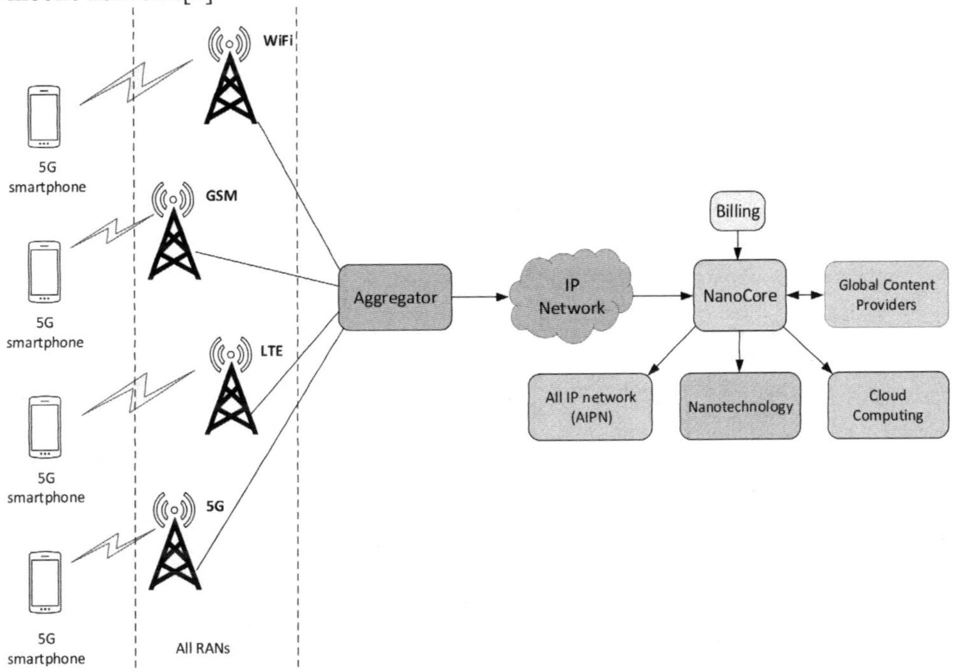


**Figure 2.** System model of 5G, entirely IP based

The 5G network architecture has been developed for the exchange of all types of data and the provision of various services using technologies such as network virtualization Network Function Virtualization (NVF) and software defined networks Software Defined Networking (SDN) [12, 13]. The new technology is mainly characterized by ultra-broadband mobile access improved mobile broadband, eMBB, URLLC, mass connection of various sensors and devices from the world of Internet of Things (IoT), mMTC [14, 15].

The system consists of a main user terminal and then a number of independent autonomous radio access technologies. Each of the radio technologies is considered as an IP connection to the outside world. IP technology is designed to provide sufficient control data for the appropriate routing of IP packets associated with certain application connections, i.e. sessions between client applications and servers somewhere on the Internet. In addition, in order to make packet routing available, it must be fixed in accordance with the given user policies (shown on figure 2).

The main idea of NanoCore is the combination of three innovation technologies:

- All IP Network (AIPN);
- Nanotechnology;
- Cloud Computing.

As shown in figure 2, the 5G NanoCore is a convergence point for other technologies that have their own impact on the existing wireless network. Interestingly, its design makes it easier for NanoCore to work in parallel multimode, including all IP network modes and 5G network mode. In this mode (as shown in the image below), all RAN network technologies and different access networks (DAT) are controlled. Because the technology is compatible and manages all new deployments (based on 5G), it is more efficient, less complex and more powerful. [2]

Each service mode can be opened under a 5G new deployment mode such as World Combination Service Mode (WCSM). WCSM is a wonderful feature of this technology; for example, if a professor writes on a whiteboard in a country, it can be shown on another whiteboard in any part of the world except a conversation and a video. In addition, a new service can be easily added via a parallel multimode service [4, 16].

## 3. Experimental results

Compared to the generation used for mobile coverage, different coverage is achieved, and with each more modern generation, the coverage area decreases, at the expense of the services used. The coverage is in accordance with a decision of the Communications Regulation Commission [8], as the admissible levels according to the decision of CRC are presented in table 1 and are shown on figure 3.

**Table 1.** Permissible coverage according to the generation of cellular networks used.

| Level of coverage | 2G network (GSM) *X-level of the signal* (RxLev) | 3G network (UMTS) *X-level of the signal* (CPICH RSCP) | 4G network (LTE) *X-level of the signal* (RSRP) | Color |
|---|---|---|---|---|
| Bad coverage | X ≤-104 dBm | X ≤ -105 dBm | X ≤ - 110 dBm | White |
| Not good enough | -85 dBm < X < -104 dBm | -85dBm < X < -105 dBm | - 110 dBm < X < - 100 dBm | Green |
| Good | -85 dBm ≤ X < -65 dBm | -85 dBm ≤ X < -65 dBm | - 100 dBm ≤ X < - 80 dBm | Yellow |
| Very good | X ≥-65 dBm | X ≥-65 dBm | X ≥- 80 dBm | Red |

The mobile coverage of 5G now (September 2020), is still a confidential project and there is no official information about the coverage and capacity of base stations in Bulgaria. In order to be able to present practical results, officially presented results are presented from city, where 5G technology for mobile coverage works in the same frequency range in which it will work in the future in Bulgaria - 3.3GHz - 3.8GHz.

The presented results show forecast data, which will most likely be valid for Bulgaria as well. The presented data show analysis and measurements of coverage in tests of 5G technology [7, 8].

IOP Conf. Series: Materials Science and Engineering 1032 (2020) 012008 doi:10.1088/1757-899X/1032/1/012008

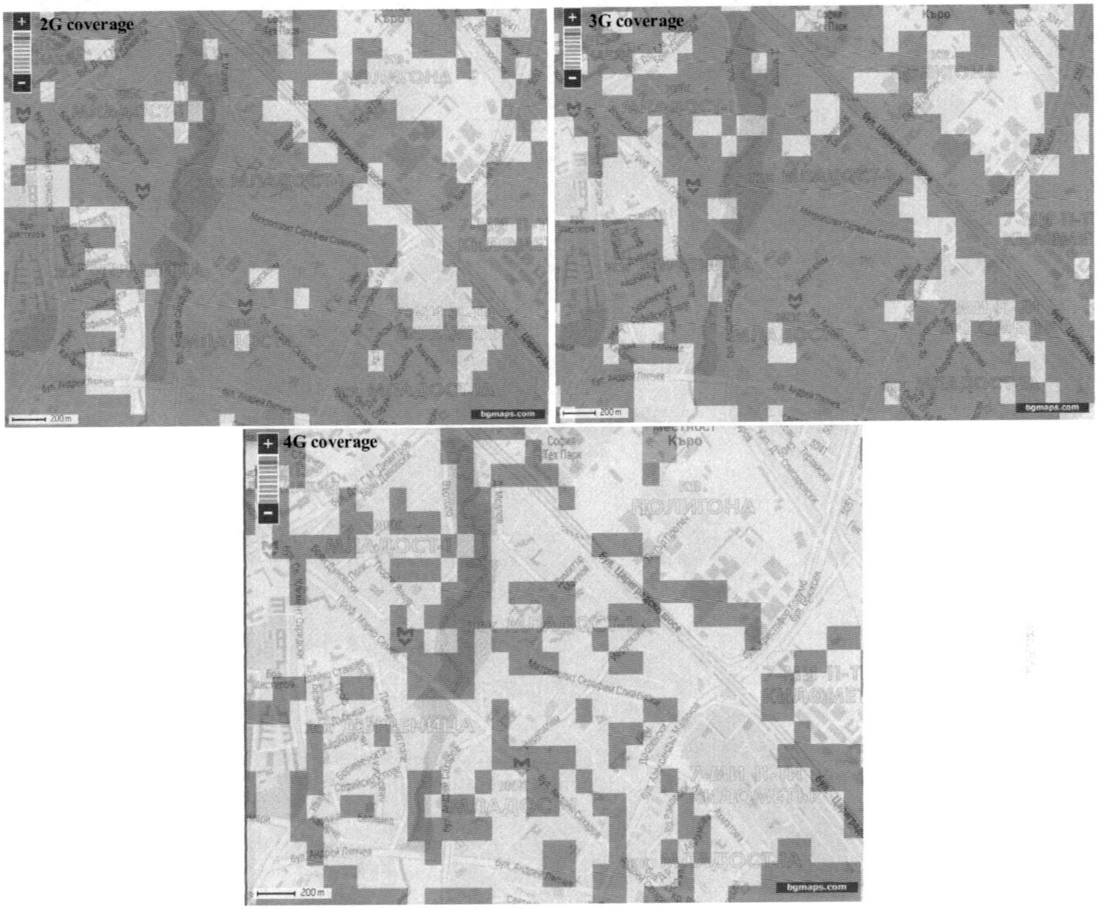


**Figure 3.** Coverage of 2G, 3G and 4G in Sofia

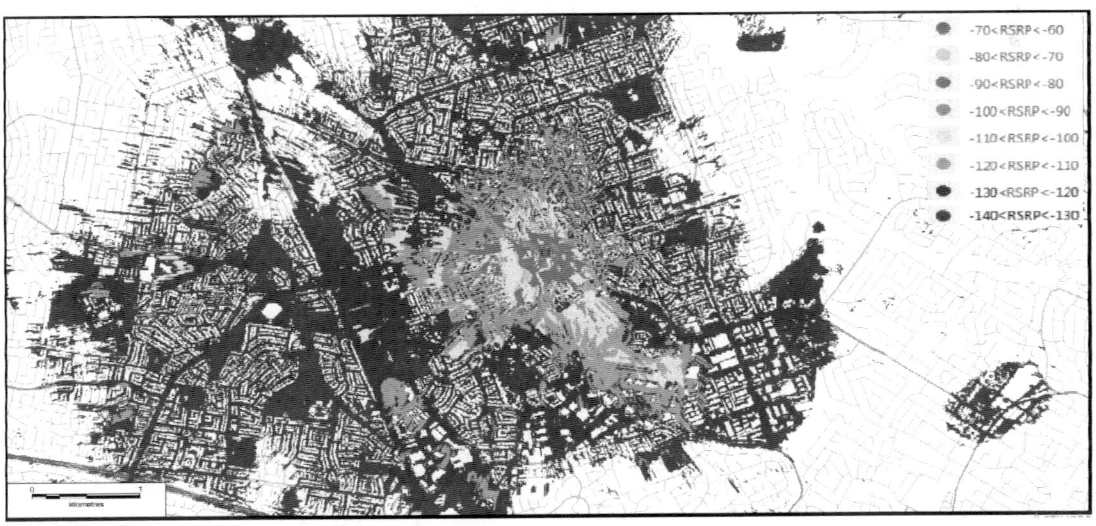


**Figure 4.** Prediction of NR coverage

The main parameters for main gNodeB are presented in table 2, there are presented Cell Edge Rate in Mbps. The thresholds of NRSRP are for area probability 90% and 75% for edge probability.

**Table 2.** 5G NR Link Budget.

| **Main gNodeB and UE parameters** | | |
|---|---|---|
| Data Channel Type | PUSCH (UL) | PDSCH (DL) |
| NR Duplex Mode | TDD | |
| System Frequency Band | n78 (3500MHz) | |
| System Bandwidth | 60 | |
| Min MCS requirements | 16 QAM | 64 QAM |
| **Transmitter gNodeB** | | |
| Max Tx power per port (dBm) | 23 | 43 |
| Tx Antenna Gain (dBi) | 2 | 24.7 |
| **Receiver (UE)** | | |
| Minimum SUNR (dB) | -0.54 | -2.29 |
| Receiver Sensitivity (dBm) | -107.23 | -91.97 |
| gNodeB Antenna Height (m) | 27.77 | |
| Area probability | 90% | |
| Edge Probability | 75% | |

A Rohde & Schwarz NR [9, 16] measuring instrument and a broadband measuring antenna were used to test the considered 5G coverage [4]. The device registers data from points in single input mode, single output (SISO) in more than 30 km in the surrounding streets. The second test is a stationary bandwidth test using a 5G UE (consumer equipment) placed through the window in the back of the car.

The UE used 2x2 Multiple Input, Multiple Output (MIMO) and speed tests were performed using a TCP based speed testing service on a smartphone connected to the UE via 5 GHz Wi-Fi network.

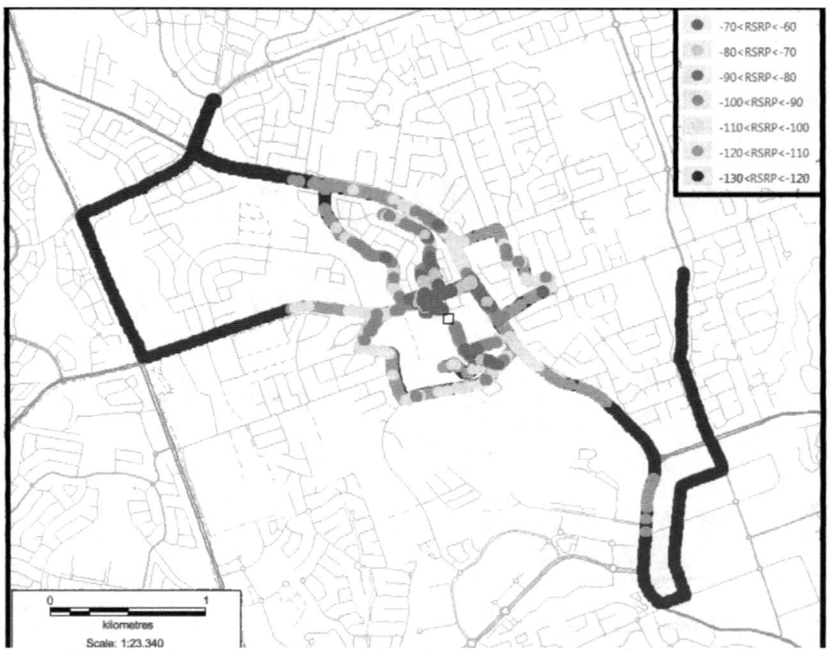


**Figure 5.** Coverage of 5G compared to the RSRP levels by numerical methods

Figure 5 shows the coverage of 5G compared to the RSRP levels by numerical levels. The figure 6 shows the same coverage but with applicable measurements. The results are in normal relative differences.

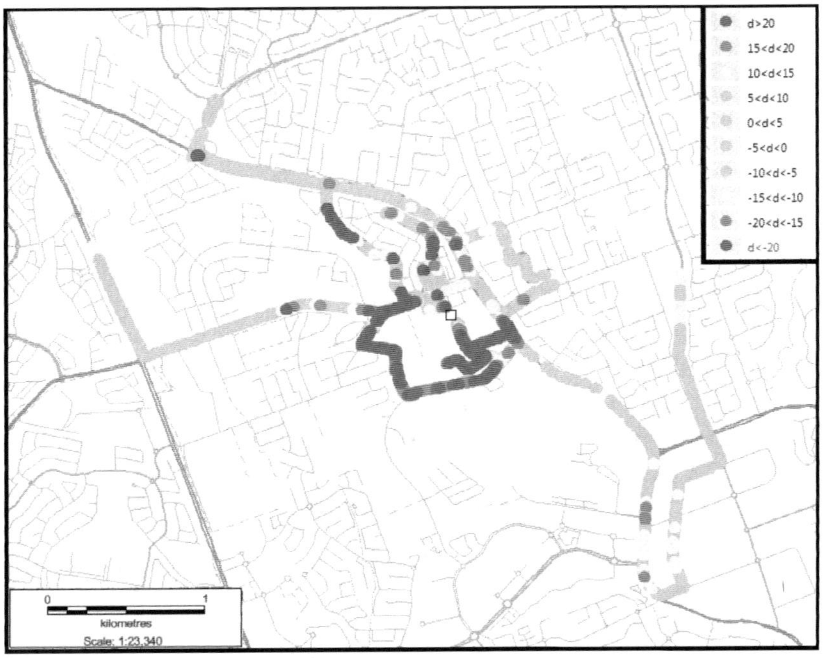


**Figure 6.** Coverage of 5G compared to the measurement of RSRP levels

The performed measurements of RSRP levels, compared to the needed values, allow a seamless introduction of the new technology. The 5th generation communications are compatible with the current location of base stations, and modern technologies allow the implementation of 5G modules in now existing ones.

## 4. Conclusion

The experimental results shown by the used equipment of different generations and analysis of existing coverage by mobile operators is possible to use a common collocation area for the equipment of different generations of cellular networks. The technology for mobile communications of the 5th generation are compatible with the current location of base stations, and modern technologies allow the implementation of 5G modules in existing ones.

From the presented solutions for coverage of different generations of mobile networks, currently operating on the territory of the Republic of Bulgaria, a complete idea can be obtained about the coverage provided by mobile operators.

## References

[1] Iliev T, Mihaylov G, Bikov T, Ivanova E, Stoyanov I and Radev D 2017 LTE eNB traffic analysis and key techniques towards 5G mobile networks," *40th International Convention on Information and Communication Technology, Electronics and Microelectronics* (MIPRO), Opatija, Croatia, 2017, pp 497-500, doi: 10.23919/MIPRO.2017.7973476

[2] Mihaylov G, Iliev T, Bikov T, Ivanova E, Stoyanov I, Keseev V and Dinov A 2018 Test cases and challenges for mobile network evolution from LTE to 5G, *41st International Convention on Information and Communication Technology, Electronics and Microelectronics* (MIPRO), Opatija, Croatia 2018, pp 449-452, doi: 10.23919/MIPRO.2018.8400085

[3] Lopa V 2015 Evolution of mobile generation technology: 1G to 5G and review of upcoming wireless technology 5G. *International journal of modern trends in engineering and research* **2**.10, pp 281-290

[4] Iliev TB, Mihaylov GY, Stoyanov IS and Ivanova EP 2020 LTE and 5G NR – Coexistence and Collaboration, *43rd International Convention on Information, Communication and Electronic Technology (MIPRO)*, Opatija, Croatia, pp 389-392, doi: 10.23919/MIPRO48935.2020.9245123
[5] Iliev T, Nikolaev N, Mihaylov G, Ivanova E, Stoyanov I and Radev D 2018 A Critical View of the NFV Dataplane Implementations for 5G Networks, *26th Telecommunications Forum* (TELFOR), Belgrade, 2018, pp 1-4, doi: 10.1109/TELFOR.2018.8611980
[6] https://www.rohde-schwarz.com/id/solutions/test-and-measurement/mobile-network-testing/stories-insights/article-real-5g-ue-measurements_252646.html
[7] Angelov K and Sadinov S 2019 Investigation of the Reliability of Service Equipment in Communication Networks, *27th National Conference with International Participation* (TELECOM 2019), Sofia, Bulgaria, pp 58-61, doi: 10.1109/TELECOM48729.2019.8994894
[8] https://crc.bg/files/_bg/resh_512-07_08_2014.pdf
[9] CURRY, Tristan; ABBAS, Robert. 5G Coverage, Prediction, and Trial Measurements. arXiv preprint arXiv:2003.09574, 2020
[10] Tikhvinskiy V, Bochechka G, Gryazev A and Aitmagambetov A 2018 Comparative Analysis of QoS Management and Technical Requirements in 3GPP Standards for Cellular IoT Technologies, *Journal of Telecommunications and Information Technology*, **2**, pp. 41-47
[11] Balabanova I, Kostadinova S, Markova V and Georgiev G 2019 Analysis and categorization of traffic streams by artificial intelligence. *International Conference on Biomedical Innovations and Applications (BIA)*, Varna, Bulgaria, pp 1-5, doi: 10.1109/BIA48344.2019.8967475
[12] Balabanova I, Kostadinova S, Georgiev G, 2019 Regression Analysis and Optimization of Teletraffic Models for Parameters Forecasting. *Journal of Engineering Science and Technology Review Special Issue on Telecommunications, Informatics, Energy and Management* 2019, Kavala, Greece, 2019, pp 9-13, ISSN: 1791-2377
[13] Ivanova E, Iliev T, Mihaylov G, Keseev V and Stoyanov I 2019 Significant simulation parameters for RESTART/LRE method in teletraffic systems of SDN, *Lecture Notes in Networks and Systems*, **68**. Springer, Cham. pp. 265-273, https://doi.org/10.1007/978-3-030-12450-2_25
[14] Ivanova E, Iliev T, Mihaylov G, Tsvetanov F, Otsetova E and Radev D 2015 Computer modeling and simulation of teletraffic models for 5G networks, *38th International Convention on Information and Communication Technology, Electronics and Microelectronics* (MIPRO), Opatija, Croatia, pp. 432-435, doi: 10.1109/MIPRO.2015.7160310
[15] Angelov K, Manchev N, Kogias P and Sadinov S 2019 Design and Development of a Platform for Test Applications in LoRa/LoRaWAN, *Journal of Engineering Science and Technology Review* (JESTR), Kavala Institute of Technology, pp. 17-21
[16] Koumaras V, Foteas A, Foteas A, Kapari M, Sakkas C and Koumaras H 2018 5G Performance Testing of Mobile Chatbot Applications, *IEEE 23rd International Workshop on Computer Aided Modeling and Design of Communication Links and Networks* (CAMAD), Barcelona, pp. 1-6, doi: 10.1109/CAMAD.2018.8515004

# Conceptual framework for quality assessment in human-centric 6G XR services

**V Stoynov**[1], **A Ivanov**[1] **and D Mihaylova**[1]

[1] Department of Communication Networks, Technical University of Sofia, bul. Kl. Ohridski 8, Sofia 1000, Bulgaria

E-mail: vstoynov@tu-sofia.bg

**Abstract.** The 6G communication networks and technologies have to enable the development of wide variety of low latency and ultra-reliable services. Furthermore, the 6G is foreseen to be the era of new devices which integrate our natural human senses in order to create a fully new concept of human-machine interactions. The eXtended Reality (XR) based wearables will surely improve the way we collaborate with modern technologies by merging the real and virtual world in one innovative continuum in order to expand the possibilities of using and interacting with different types of modern or already well-known IT services. In this context, a vital 6G research area is the design of different flexible quality assessment metrics, targets and measures that integrate human perceptions and physical factors from human physiology. This metric design should consider the multisensory nature of XR devices by incorporation of different objective network-related as well as strongly subjective human-related features like sight, touch, smell, taste, body physiology, psychological traits, gestures and even sixth sense. In this paper a conceptual framework incorporating a mixture of new and classical quality assessment constructs like Quality of Service (QoS) and Quality of Experience (QoE) is proposed for 6G networks from human-centric point of view. It considers the development of new types of intelligent XR technologies and give an insight in regard to overall human experience assessment, modelling and prediction.

## 1. Introduction

The 6G networks' paradigm will revolutionize our perception regarding the contemporary lifestyle, society, business and communication. The implementation of 6G systems will result in development of numerous new services and corresponding research areas. The main characteristics of next-generation networks - high level of node densification, higher throughputs, and large-scale antenna systems will be provided by new emerging trends that comprise new types of services and the current revolution in new categories of wireless devices - smart wearables, implants, XR devices, and artificial intelligence (AI), computing, and sensing [1].

The 6G networks will be able to establish a reliable and secure connection between a wide variety of smart devices. Furthermore, the communication network will be characterized by diverse types of coverage areas. In this context, it is not possible to ensure just a single global coverage to connect the whole world. For that reason, the communication networks will be divided into autonomous sub-networks. The latter will not be homogeneous but heterogenous – they will be composed by different types of intelligent nodes providing excellent Quality of Service (QoS) to the end-users. Furthermore, 6G networks will be characterised by integration of non-terrestrial and terrestrial communication access, while both are realised by heterogeneous networks by nature. Their different levels of heterogeneity must be taken into account in order to cost-effectively integrate them together. Consequently, 6G should be able to integrate various heterogenous components in one secure and reliable continuum which will be a big challenge [2], [3].

Nowadays, an exceptional creation of new types of Internet of Everything (IoE) services is ongoing. For example – the extended reality (XR) services, incorporate different augmented, virtual, and mixed reality (AR/VR/MR)) concepts, telemedicine, haptics, unmanned aerial vehicles, brain-computer interfaces, and connected autonomous systems. In this context, these various types of specific applications will disrupt the original goal of 5G networks to support short-packet, sensing-based Ultra Reliable Low Latency Communication (URLLC) services. Consequently, the wireless systems must simultaneously provide high reliability, low latency, and high data rates, for heterogeneous devices, across uplink and downlink. This will result in successful operation of diverse sets of specific services such as XR and autonomous systems. The development and operation of new IoE services will require an implementation of new intelligent and flexible architectures for reliable and secure end-to-end communication, control, and computing functionalities. In order to provide these modern services, exceptional challenges must be addressed. Firstly, it is extremely important to characterise the fundamental rate-reliability-latency trade-offs governing their performance to exploiting frequencies beyond sub-6 GHz. On the other hand, the transformation of wireless systems into a self-sustaining intelligent networks and the orchestration of communication-computing-control-localization-sensing resources personalised for various IoE scenarios are needed [4], [5].

The development of quality assessment metrics that integrate (i) physical factors from human physiology or from a management system for robotics or automation; (ii) psychophysical traits measurements and (iii) cognitive perceptions and behavioural consequences characterization is a vital area of 6G research, especially in the context of new types of intelligent wearables. This requires both real psychophysical experiments and new suitable and flexible mathematical apparatus determination that combine widely known QoS and QoE constructs. The theoretical development of such a complex metric can be achieved using methods from other disciplines, such as operations research (for example, multi-attribute utility theory and machine learning (ML)). Finally, 6G will be the first generation of communication services to use multiple human cognitive feelings which once again provokes the necessity to define new types of human-machine interaction metrics.

This paper provides a basic conceptual framework incorporating a combination of new and classical quality indicators like QoS and QoE in order to design a new construct called Quality of Interaction providing intelligent quality assessment in 6G XR networks from human-centric perspective.

## 2. Extended Reality concept for interconnected world

### *2.1. Introduction to XR*

New types of impressive technologies are related with diverse forms of realities such as Virtual Reality (VR), Augmented Reality (AR), Mixed Reality (MR), cross reality or eXtended Reality (XR). The expression "XR" is increasingly being used and represents an umbrella term for these technologies (figure 1).

XR-based technologies and services are able to improve the time-space flexibility of users. This can be achieved by avoiding their need to be at the same place at the same time, for example when working on a project. This results in the ability to create faster and more powerful decision-making processes for diverse sets of industries. The remote guidance systems for performing complex tasks such as maintenance and assembly is the main area of XR implementation and operation. Generally, XR will be the next-generation mobile computing platform that unites different types of realities together in order to create an entire new reality-virtuality continuum, depicted in figure 1, thus expanding the human experiences [6].

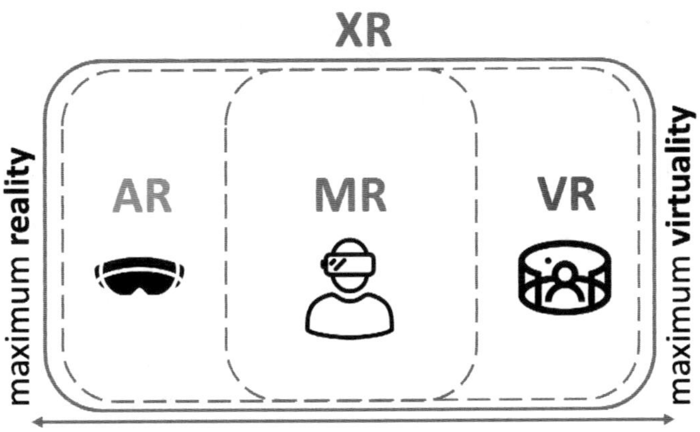


**Figure 1.** Representation of eXtended Reality concept.

Both endpoints of XR continuum are represented by AR and VR. The first one – AR, is created by presenting a virtually created information right on reality-based environment. In this context, the user can see through his eyes in real time a mixture of the world around him and virtual objects. This can be obtained only by using additional wearable devices. Obviously, the AR is able to provide the user with additional information that cannot be obtained by using only human senses. AR also has a big impact in education, and it is possible to change the way students will learn in the future. Also, the mobile devices which are used by many people nowadays are really powerful and can be seen as great possibility to use AR-based applications and services. On the other hand, VR bases virtual data as the main focus, having the user operating into the artificial reality virtual environment. The VR environment is created using computing technologies including mainly computer graphics and artificial intelligence. Generally, VR enables the user to perform navigation and interactivity with the virtual objects in a fully virtual situation in a real world. It has many applications, including military, healthcare, education, scientific visualization and entertainment.

The combination of two types of environments results in a mixed reality (MR) concept creation where AR meets VR, by merging the physical and virtual information. In this context, the MR allows users to be part of one environment combining physical objects and digital elements, for example, by the use of semitransparent displays. Furthermore, the operation of MR systems is based on the creation of an illusion that digital objects are in the same space as physical ones. This leads to the requirement each digital object to be precisely situated into the real environment and aligned with the real physical objects in real time. MR comprises systems in which the virtual features are of great importance as well as those in which the physical reality is dominant. Compared to AR, the MR has more virtual objects than physical elements.

Existing online social networking sites like Facebook and Twitter are precursors of what we will come to truly witness when social networking will encompass immersive virtual-reality technology. Actually, social virtual reality lets two people placed in different locations from geographical point of view, thus defining different forms of avatars, to transfer different type of data between them as if they were face-to-face. Also, they can make eye contact and can manipulate virtual objects that they both can see. A serious challenge to the current VR headsets is still the intelligent tracking of gaze direction. Nowadays, there are only few VR headsets able to perform eye tracking. Furthermore, current state-of-the-art VR technology is incapable to read detailed facial expressions and senses. Lastly, the most powerful VR prototypes are wired with cables because the wireless communication mediums created for examples by using 4G/LTE networks are not able to transmit huge amount of high-resolution video at high frame rates. Also, perfect user interface (VR equivalent of the mouse) is still needed [7], [8].

### *2.2. XR implementation challenges*

Application of systems based on XR technologies and services has been proven to be practical and effective way to create the sense of presence and engagement for users to explore inside a virtual-physical environment. One of the biggest challenges concerning the XR applications design is finding the balance of mixing two totally different concepts - the virtual and the physical (the real). Currently, the challenge of integration of virtual and physical worlds in one continuum has been addressed from different points of view and numerous disciplines, such as computer science, psychology, physiology, sociology and more other. Furthermore, several other important implementation challenges still exist within the application design and development phase of XR services and they can be summarised as follows:

- The VR experience can be accompanied by different symptoms like general discomfort, apathy, drowsiness, headache, disorientation or fatigue which are summarized by the term Motion Sickness (MS).
- The high battery consumption of AR services due to the intense use of sensors causes rapid battery drain on devices which results in limited mobility of the end-user.
- The greatest challenge for Internet of Things-based XR systems is achieving high levels of security and reliability. There are many studies which present a comprehensive surveys on security in IoT, highlighting an extensive list of issues from poor authentication controls to lack of encryption in communications.
- Definition and analysis of specific functional, performance and ergonomic requirements and characteristics for design of secure and reliable next generation XR systems.
- Definition, measurement and modelling of human perception-based QoE criteria for next-generation XR systems and devices using objective, subjective and hybrid approaches.
- Providing a high communication reliability and traffic safety for autonomous vehicles with XR functionalities.

## 3. Basic conceptual framework for quality assessment

The haptic XR applications being promising solution for 6G system implementation and development, there is still not clear determination of user experience when interacting with such new variety of devices and systems. Objective measurement or prediction of XR communication characteristics through a suitable evaluation model is highly needed. New specific QoE design for haptic virtual environments is obligatory in order to evaluate the user experience in the context of multi-modality applications. Usually, the traditional QoE metric focuses on the quality of involvement and interaction between the user and the specific application, which is the main difference compared to QoS typically indicating the functionality and the quality of the particular technology used in the application [9]. On the other hand, QoE is not able to determine the whole process of human-machine interaction when XR services are used and the reaction of human body and brain to this reality-virtuality continuum. In this context, it is essential to design a new type of quality assessment metric which considers the variety of different types of devices characterizing the next-generation communication technologies. Furthermore, the measurement process of such metric will require measuring the user's perception of a plenty of VR applications. In this context, we require not only improved performance measures over the well-established QoS measures but also new flexible modelling approaches in order to deal with one of the most important features of the human-machine interaction - the subjectivity of the user.

In figure 2 the structure of conceptual framework based on new complex metric called Quality of Interaction is depicted. It consists of four basic elements – Quality of Service – QoS, Quality of Psychophysical Experience (QoPsE), Quality of Physical Experience (QoPhE) and Quality of Perception. For each basic element the most important parameters are listed. Each of them has impact on overall QoI determination for the specific XR-based communication environment.

QoI is designed as a complex metric which focuses on human-machine interaction with respect of specific human body and perception experiences when operating in XR systems. In this context, the QoI refers to the reaction of the system as a result of user needs and reaction of the user from physiological, psychophysical and perceptive point of view when interacting with diverse set of applications or services based on usage of intelligent devices. As can be seen in figure 2 QoI design is based on differentiation of two different types of reaction resulting from the human-machine interaction described by *network reaction layer* and *human reaction layer*. Therefore, the QoI is principally subjective metric and can be flexibly modified according to the specific application, context or because of the specifics of user expectation and involvement. Additionally, the QoI indicator is characterised by high level of reliability for XR service providers to convey overall end-to-end system functioning (client, terminal, network, services infrastructure, etc.). Furthermore, QoI design is based on multidisciplinary approach involving user psychology, engineering science, economics, etc. As we already mentioned, the QoI depends on four different sub-elements that directly or indirectly affect the user's perception towards the XR services. These elements should perform to their best to provide high overall quality of interaction. However, the diversity in these elements makes the QoI estimation rather complex and unpredictable. For that reason, the implementation of different approaches for human reaction prediction based on machine learning (ML) and artificial intelligence is highly desirable. As is widely known, the intelligent ML algorithms are able to automatically learn from the past observations of QoI components' values to make accurate predictions in the future. Some of the most popular ML algorithms which can be used in QoI prediction are decision tree, neural networks, fuzzy expert system, etc. The QoI estimation system may be composed of ML algorithms in order to learn the correlation between QoS parameters and the subjective QoPsE, QoPhE and QoP. The learning process comprises an important phase of training the algorithm with subjective data set. Thus, it is able to predict the QoI based on any combination of input QoS parameters. It is important to mention that an effective QoI prediction model must automatically re-organise to each new dataset without much complexity and time. Finally, it is still necessary to design modern innovative mechanisms to efficiently correlate user performance parameters from QoS, in real time. A potent basis for them, which is also increasing in popularity is the reinforcement learning of deep neural networks [10]. Such algorithms can use the collected QoS parameters to learn the optimal QoI under particular QoPsE, QoPhE and QoP constraints and thus extend the cognitive capabilities of the system.

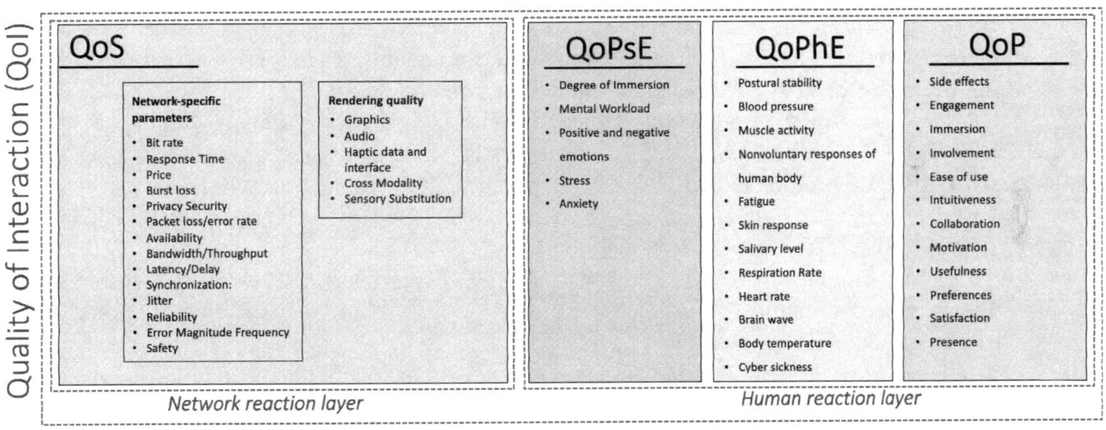


**Figure 2.** Basic conceptual framework for quality assessment

## 4. Conclusions

This paper presents a basic conceptual framework based on original Quality of Interaction determination characterizing the human-machine relationship in 6G XR environments. QoI is divided into four different quality assessment constructs defining at the same time the reaction of the specific communication environment (network) – QoS, and the reaction of the human body and brain, described by the QoPsE, QoPhE, QoP. This conceptual framework is a suitable basis for implementation and evaluation of different mathematical models describing human-machine communication in virtual environments. Future work may focus on implementation of different models for QoI estimation and prediction concerning machine learning and artificial intelligence methods [11]. Additionally, research concerning evaluation of correlation between the different QoI layers and basic elements is foreseen.

## References

[1] Chen S, Liang Y, Sun S, Kang S, Cheng W and Peng M 2020 Vision, requirements, and technology trend of 6G: How to tackle the challenges of system coverage, capacity, user data-rate and movement speed, *IEEE Wireless Communications*, **27** (2), pp 218-228, doi: 10.1109/MWC.001.1900333

[2] Tomkos I, Klonidis D and Pikasis E, Theodoridis S 2020 Toward the 6G network era: opportunities and challenges *IT Professional* **22** (1) pp 34–38, doi: 10.1109/MITP.2019.2963491

[3] Ziegler V and Yrjola S 2020 6G indicators of value and performance, *2nd 6G Wireless Summit (6G SUMMIT),* Levi, Finland, pp 1-5, doi: 10.1109/6GSUMMIT49458.2020.9083885

[4] Andrade T and Bastos D 2019 Extended reality in IoT scenarios: concepts, applications and future trends, *5th Experiment International Conference (exp.at'19)*, Funchal, Portugal, pp 107-112, doi: 10.1109/EXPAT.2019.8876559

[5] Srivastava S 2019 MobileXR: The future of extended reality (AR/VR/MR) - counterpoint research *Counterpoint Research,* Available at: https://www.counterpointresearch.com/future-xtended-reality-arvrmr/

[6] Chuah, S H W 2019 Wearable XR-technology: literature review, conceptual framework and future research directions *In International Journal of Technology Marketing* **13** (3/4), pp 205-259, doi: 10.1504/IJTMKT.2019.104586

[7] Bonetti F, Warnaby G and Quinn L 2018 Augmented reality and virtual reality in physical and online retailing: a review, synthesis and research agenda, ed T Jung and tom M Dieck *Augmented Reality and Virtual Reality. Progress in IS. Springer, Cham.* https://doi.org/10.1007/978-3-319-64027-3_9

[8] Yang H, Yu J, Zo H and Choi M 2016 User acceptance of wearable devices: an extended perspective of perceived value, *Telematics and Informatics* **33** (2) pp 256–269, https://doi.org/10.1016/j.tele.2015.08.007

[9] Hamam A and El Saddik A 2012 Evaluating the quality of experience of haptic-based applications through mathematical modelling, *IEEE International Workshop on Haptic Audio Visual Environments and Games (HAVE 2012) Proceedings*, Munich, pp 56-61, doi: 10.1109/HAVE.2012.6374439

[10] Zhang C, Patras P and Haddadi H 2019 Deep learning in mobile and wireless networking: A survey, *IEEE Communications Surveys & Tutorials,* **21** (3), pp 2224-2287, doi: 10.1109/COMST.2019.2904897

[11] Lopez-Martin M, Carro B, Lloret J, Egea S, Sanchez-Esguevillas A 2018 Deep learning model for multimedia quality of experience prediction based on network flow packets, *IEEE Communications Magazine* **56**(9) pp 110–117, doi: 10.1109/MCOM.2018.1701156

## Acknowledgments

This work was supported by the National Program “Young Scientists and Postdoctoral Students”, PMC No577, 17.08.2018, of Ministry of Education and Science of Bulgaria.

# Energy detectors performance evaluation for interweave cognitive radio network scenario in 5G

**A Ivanov[1], V Stoynov[1], D Mihaylova[1], V Poulkov[1]**

[1] Department of Communication Networks, Technical University of Sofia, bul. Kl. Ohridski 8, Sofia 1000, Bulgaria

E-mail: astivanov@tu-sofia.bg

**Abstract.** With the recent advances in resource allocation methods for 5G heterogeneous networks and dynamic spectrum access, the convergence between them becomes a more prevalent topic in scientific research. By implementing cognitive functionalities in the 5G network's nodes, its efficiency can be improved and, in addition, the integration with existing wireless standards will be smoother. To enable this, the intelligent sensors are required to achieve rigid detection accuracy. Thus, the purpose of this paper is to examine the ability of popular energy detection spectrum sensing techniques to identify the primary user's signal in an interweave (autonomous operation of the secondary users) 5G deployment scenario. This evaluation is performed through comprehensive simulations for differing number of primary and secondary users, and for different data stream types transmitted by the primary base station.

## 1. Introduction

The usefulness of Cognitive Radio (CR) enabled dynamic spectrum access (DSA) has been very well established over the course of the recent years. Its aspects of accurate spectrum characterization (either from the point of view of a single user equipment, UE or of the whole CR network) and available spectrum distribution among the cognitive nodes (secondary users), are still objects of intense scientific examination. This field is further expanded through the application scenarios of heterogeneous, ultra-dense and Internet of Things (IoT) networks within the scope of 5G [1], [2], [3]. Additionally, implementing device-to-device (D2D) communications in mobile networks can increase the utilization of the same through adaptive resource distribution [4]. The mobile network's performance indicators should also be carefully examined in order to suggest further improvement through general frameworks and within operator-specific scenarios [5], [6]. Furthermore, CR-aided non-orthogonal multiple access (NOMA) is envisioned as a promising technology for spectral efficiency improvement in 5G systems [7], [8]. All of these technologies can greatly benefit from the DSA capabilities of CR. Central place in the integration of CR capabilities in modern and future networks is given to the vital function of spectrum sensing which is required to provide accurate assessment of spectrum occupancy by the incumbent (primary) users so as to avoid unwanted interference in their communications. Energy detection has been established as the most prominent sensing method due to its speed which allows for decision on spectrum vacancy to be made within a short time so that the available frequency channel can be used before the incumbent users resume its utilization [3], [9], [10].

This paper evaluates the detection accuracy of two popular types of energy detectors (ED) for spectrum sensing within a 5G heterogeneous network scenario through the use of the novel 5G-air-simulator [11]. The effect of different data stream types and number of primary and secondary users on detection performance is studied via the simulation results.

## 2. System model and simulation details

The examined system model is a modification of the "Single Cell with Femto-Cells" model of the 5G-air-simulator. It incorporates a micro-cell (mC) with radius of 350 m within the scope of which operate the primary (micro-cell UEs, mUEs) and the secondary users (SUs), all of which are stationary. The SUs are based on the femto-cell UEs implemented in the simulator, and are randomly distributed within two single-floor buildings with random locations inside the micro-cell's coverage. The CR spectrum sensing function is implemented within the SUs to determine whether the frequency channel is free from mC and mUE transmissions or not. Within the simulator environment's core, functionality for idle periods for the mC PU is implemented, in order to simulate the PU's absence from the spectrum at random times during the simulation. The mC continuously transmits data streams throughout the simulation's runtime (600 s).

Two common types of ED methods are implemented - Traditional ED (TED) and Rayleigh-fading ED (RFED) [9], [10]. For each of them, the signal-to-noise ratio $\gamma$ or SNR (estimated using the Mutual Information Effective SNR Mapping [12] method implemented by default in the 5G-air-simulator) is compared to the respective decision threshold $\lambda$. For the TED and RFED the threshold values are defined respectively in [9] and [10]. It is assumed that uplink (UL) transmissions with overhead information from both mUEs and SUs are present for communication establishment. Thus, the SNR is defined as:

$$SNR = \frac{P_{mC} + \sum_{i=1}^{M} P_{mUE_i}}{P_N + \sum_{j=1}^{K} P_{SU_j}}, \tag{1}$$

where the UL transmissions from SUs (i.e. the sum of their power at the receiver - $\sum_{j=1}^{K} P_{SU_j}$) are interfering because from the detector's point of view, it is only required to detect the MC (represented as the received power $P_{mC}$) and mUEs ($\sum_{i=1}^{M} P_{mUE_i}$) signals. $P_N$ is the noise power. The detector's performance is characterized using the probabilities of detection ($P_D$) and of false alarm ($P_{FA}$) which are calculated from the respective equations in [9], [10].

## 3. Simulation results

The simulation is performed for two cases for the number of SUs (6 and 10) and for each of them, the detectors' performance is evaluated for 4 different types of data streams – Video (with quality of 440 kbps), Voice over Internet Protocol (VoIP), Constant Bit-Rate (CBR) data transfer and a combination of those three ("All types"), and for the cases of 5 and 10 mUEs. The relevant parameters for calculation are the following – $P_N$ and $N = \tau f_s$. The overall bandwidth is 1.4 MHz, with sub-channels 180 kHz wide. Therefore, the sampling frequency $f_s$ is assumed as 360 kSps, whereas $\tau$ is the reception time for each simulation instance. The transmission power of the mC is 43 dBm while that of all mUEs and SUs is 23 dBm.

The simulation results are shown on figure 1 to figure 4. For the case of 5 mUEs being present, it is obvious that for both detector types, the results' curves follow similar patterns, with the general difference that the TED shows slightly higher overall performance. Lowest detection accuracy is exhibited for the CBR traffic ($P_D$ of 90% and $P_{FA}$ of 10% at -8 dB SNR for 5 mUEs) whereas the highest is noted for the VoIP type ($P_D$ of 90% and $P_{FA}$ of 10% at -10 dB SNR for 5 mUEs, i.e. a 2 dB gain). This can be attributed to the traffic pattern of CBR which due to its non-real time requirements for communication, leads to more retransmissions in the examined low-SNR scenario. Video traffic just like the VoIP type has more stringent latency (and respectively, tolerating loss) requirements and thus less retransmissions and higher detection probability. It is also observed that there are some discrepancies in the results for the VoIP traffic which lead to sudden degradations in the $P_D$ and $P_{FA}$ (more specifically, at -17, -14 and -11 dB of SNR). They can be attributed to multiple retransmissions within shorter reception time and smaller number of samples and thus, lower performance for the same SNR values. For the case of "All types", the video stream being the most bandwidth intensive (440 kbps) takes precedence over the others and for this reason, its detection accuracy is nearly the same as that of the Video stream type. Thus, the curves for these two cases ("All types" and Video) overlap almost completely and they have about 1 dB difference in comparison to the results for VoIP and CBR. As for the $P_{FA}$ of the TED, the results show similar patterns but in the opposite direction. In regards to the RFED, the results are very similar though showing slightly lower performance for SNR levels above -10 dB. There is a bigger gap between the results for 5 and 10 mUEs of about 3 – 4 dB. Also, for more interfering mUEs, the difference between the result curves for the different traffic types is bigger. Thus, the REFD improves the detection performance for the different traffic types slightly more than the TED for the lower SNR ranges (below SNR of -10 dB) but the TED still has better $P_D$ at higher SNR but it has variations in its $P_{FA}$, whereas the same is always zero for the RFED which is why it is not plotted in figure 2 and figure 4. This advantage is due to the RFED accounting for the fading in its formulation.

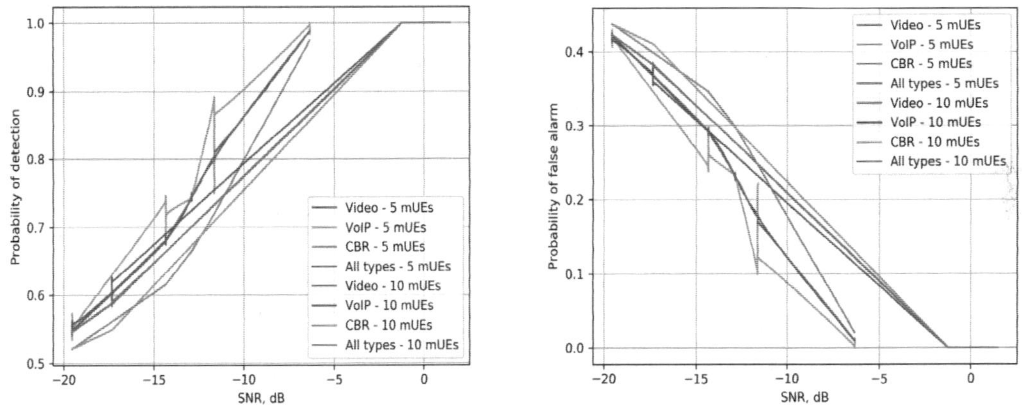


**Figure 1.** Probabilities of detection and of false alarm for TED for 5 and 10 mUEs, different types of traffic and 6 SUs.

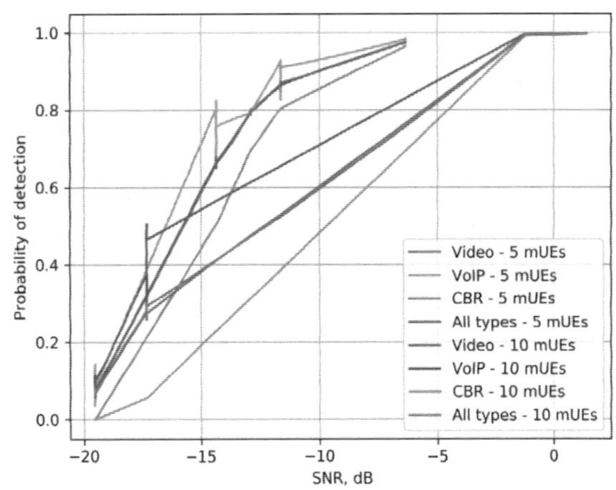


**Figure 2.** Probabilities of detection for RFED for 5 and 10 mUEs, different types of traffic and 6 SUs.

The results' patterns for the 10 SUs case follow very similar distributions to those of the alternative. A significant change, however, is exhibited in the results for 5 mUEs in that they are very close to those for 10 mUEs (the difference between them is less than 1 dB). The reason for this development is the interference created by the additional SUs for their UL overhead communications. It is also observed that nearly perfect performance ($P_D$ close to 100% and $P_{FA}$ near 0%, respectively) is reached for lower SNR of -8 dB, whereas the same is achieved at SNR of -2 dB in the alternative case.

Thus, increasing the number of SUs leads to higher detection. However, the increased interference levels as a consequence will deteriorate the SUs' communication exchange.

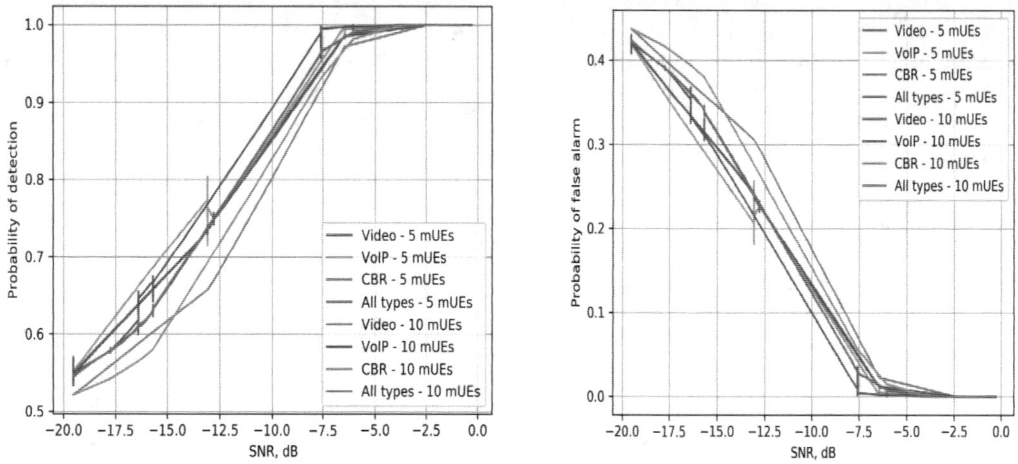


**Figure 3.** Probabilities of detection and of false alarm for TED for 5 and 10 mUEs, , different types of traffic and 10 SUs.

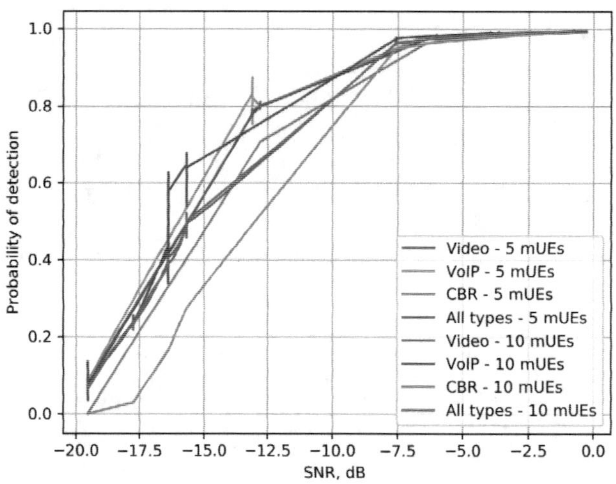


**Figure 4.** Probabilities of detection for TED for 5 and 10 mUEs, different types of traffic and 10 SUs.

## 4. Conclusions

This paper presents a performance study for two popular types of ED for spectrum sensing of the within a 5G heterogeneous network scenario through the use of the 5G-air-simulator. The simulation results show the effect of different data stream types and number of primary and secondary users on detection performance. Consequently, it has been found that primary user detection is more potent when real-time data streams (such as VoIP or video) are prevalent in modern mobile communications. In addition, increasing the number of SUs in the interweave CR network, has much greater effect on detection accuracy degradation for low SNR levels than the differing number of mUEs. Thus, adaptive interference cancellation is necessary for communications in dense and IoT networks for 5G.

## References

[1] Ivanov A, Tonchev K, Poulkov V and Manolova A Framework for implementation of cognitive radio based ultra-dense networks, *42nd International Conference on Telecommunications and Signal Processing (TSP),* Budapest, Hungary, pp 481-486, doi: 10.1109/TSP.2019.8769067

[2] Tarek D, Benslimane A, Darwish M and Kotb AM 2020 Survey on spectrum sharing/allocation for cognitive radio networks Internet of Things. *Egyptian Informatics Journal*, https://doi.org/10.1016/j.eij.2020.02.003

[3] Luo S, Xiao Y, Lin R, Xie X, Bi G, Zhao Y, Huang J 2019 Opportunistic spectrum access for UAV communications towards ultra dense networks, *IEEE Access,* **7**, pp 175021-175032, doi: 10.1109/ACCESS.2019.2956958

[4] Iliev T B, Mihaylov G Y, Ivanova E P, and Stoyanov I S 2017 Power control schemes for device-to-device communications in 5G mobile network, *40th International Convention on Information and Communication Technology, Electronics and Microelectronics (MIPRO)*, Opatija, pp 416-419, doi: 10.23919/MIPRO.2017.7973460

[5] Iliev T B, Mihaylov G Y, Bikov T D, Ivanova E P, Stoyanov I S and Radev D I 2017 LTE eNB traffic analysis and key techniques towards 5G mobile networks, *40th International Convention on Information and Communication Technology, Electronics and Microelectronics (MIPRO)*, Opatija, pp 497-500, doi: 10.23919/MIPRO.2017.7973476.

[6] Mihaylov G Y, Iliev T B, Bikov T D, Ivanova E P, Stoyanov I S, Keseev V P and Dinov A R 2018 Test cases and challenges for mobile network evolution from LTE to 5G, *41st International Convention on Information and Communication Technology, Electronics and Microelectronics (MIPRO)*, Opatija, pp 449-452, doi: 10.23919/MIPRO.2018.8400085

[7] Shi Z, Gao W, Zhang S, Liu J and Kato N 2020 AI-enhanced cooperative spectrum sensing for non-orthogonal multiple access, *IEEE Wireless Communications,* **27**(2), pp 173-179, doi: 10.1109/MNET.001.1900305

[8] Játiva PP, Cañizares MR, Arancibia CS and Freire JJ 2019 Non-orthogonal multiple access for cognitive mobile radio networks in 5G communications. *22nd Conference on Innovation in Clouds, Internet and Networks and Workshops (ICIN),* Paris, France, 2019, pp 344-350, doi: 10.1109/ICIN.2019.8685840

[9] Kumar A, Thakur P, Pandit S and Singh G 2020 Threshold selection and cooperation in fading environment of cognitive radio network: consequences on spectrum sensing and throughput. *AEU-International Journal of Electronics and Communications*, **117**, 153101, https://doi.org/10.1016/j.aeue.2020.153101

[10] Huang H, Zhu J and Mu J 2020 A novel sensing strategy based on energy detector for spectrum sensing. *Applied Sciences*, **9**(21), 4634, https://doi.org/10.3390/app9214634

[11] Martiradonna S, Grassi A, Piro G and Boggia G 2020 5G-air-simulator: An open-source tool modeling the 5G air interface. *Computer Networks*, **173**, 107151, https://doi.org/10.1016/j.comnet.2020.107151

[12] He X, Niu K, He Z and Lin J 2007 Link layer abstraction in MIMO-OFDM system. *International Workshop on Cross Layer Design,* Jinan, pp 41-44, doi: 10.1109/IWCLD.2007.4379036

**Acknowledgments**
This work was supported by the National Program "Young Scientists and Postdoctoral Students", PMC №577, 17.08.2018, of Ministry of Education and Science of Bulgaria.

# An overview of methods to combat eavesdropping in NOMA-based networks through physical layer security

**D Mihaylova[1], V Stoynov[1], A Ivanov[1], Z Valkova-Jarvis[1]**

[1] Department of Communication Networks, Technical University of Sofia, bul. Kl. Ohridski 8, Sofia 1000, Bulgaria

E-mail: dam@tu-sofia.bg

**Abstract.** Thanks to their potentials in providing reliable communications for a large number of devices, 5G non-orthogonal multiple access (NOMA) techniques have recently been the subject of considerable research interest. Accordingly, research of Physical Layer Security approaches in NOMA is essential. A primary purpose and challenge for security at the physical level of NOMA-based systems is to protect the information transmitted against both external and internal eavesdropping attacks. To this end, it is necessary to develop PLS schemes which allow each user to decode the signals transmitted by other users, thus guaranteeing normal operation of the successive interference cancellation, while also providing privacy of the information exchange. In this paper, different types of eavesdropping are studied and a summary of existing approaches to improve security at the physical layer in 5G NOMA networks is provided.

## 1. Introduction

One way to answer the need for high data rate and low latency communications in fifth-generation wireless networks is by applying new multiplexing techniques for channel access. Since NOMA provides efficient spectrum utilization, it is envisaged as a key enabling technology for future radio access. Depending on the multiplexing employed, NOMA can be realized in either the code or power domain. Code-domain NOMA provides the same transmit resources to several users through the assignment of spreading sequences with low mutual correlation [1]. Power-domain NOMA, as an alternative, shares the entire radio resource by allocating different transmit powers to users conforming to the estimated channel gain [2]. This paper aims attention at power-domain NOMA, for brevity called NOMA.

Conventionally-used orthogonal-multiple access (OMA) schemes share the available resources in time, frequency, code or space domain in such a way that the signals received at different users are orthogonal to each other. Although orthogonality ensures interference mitigation, the limited transmission resources make hard the support of high data rates and low latency. To improve system throughput and user fairness, NOMA can be employed instead, as an effective multiple access that permits multiple users to transmit simultaneously on the same frequency band by means of superposition coding (SC) at the transmitter and successive interference cancelation (SIC) at the receiver side [2], [3]. Furthermore, different novel types of waveforms can be applied collaboratively with NOMA, e.g. filter bank multicarrier (FBMC) and universal filtered multicarrier (UFMC), in order to improve signal performance and spectral efficiency of 5G mobile communication systems. Since the carriers are not yet orthogonal to each other, the innovative waveform design could lead to more flexible control of the bandwidth of each subcarrier and the degree of overlap between them. It is also to a certain extent capable of avoiding adjacent subcarrier interference, thereby leading to improved usage of spectrum resources [4]. However the operation principles of NOMA impose new challenges for the privacy enhancement of information exchange. Attractive solutions to protect the NOMA network against eavesdropping by unauthorized and/or untrusted users are the physical layer security (PLS) methods, which profits by the channel randomness to establish security instead of relying on complex computational algorithms. Moreover, PLS significantly reduces the complexity of cryptography approaches and obviates the need for secret key management, which makes it applicable in resource-constrained NOMA networks. In this paper, the PLS approaches that can be used to improve the security of downlink NOMA transmissions are investigated.

## 2. Main principles of NOMA

In NOMA, user ordering is established concerning their channel gains. In general, NOMA is investigated in a paired-users scenario, since increasing the number of users sharing a single resource block enlarges the complexity of SIC algorithm and increases the interference between users. According to the estimated CSI, the paired users are referred to as near user and far user, as shown in figure 1. The transmit power is shared between paired users in accordance with the following requirements: $\alpha_f > \alpha_n$ and $\alpha_f + \alpha_n = 1$, where $\alpha_f$ is the transmit power, assigned to the far user ($U_f$), and $\alpha_n$ is the power factor of the near user ($U_n$). During downlink transmission the base station (BS) performs SC of the signals $x_f$ and $x_n$ intended to $U_f$ and $U_n$ respectively and broadcasts the resultant composite signal $x$, determined by (1), to both users in the pair [2], [3]. $P$ stands for the total transmit power, intended for the NOMA pair:

$$x = \sqrt{P\alpha_n}x_n + \sqrt{P\alpha_f}x_f . \tag{1}$$

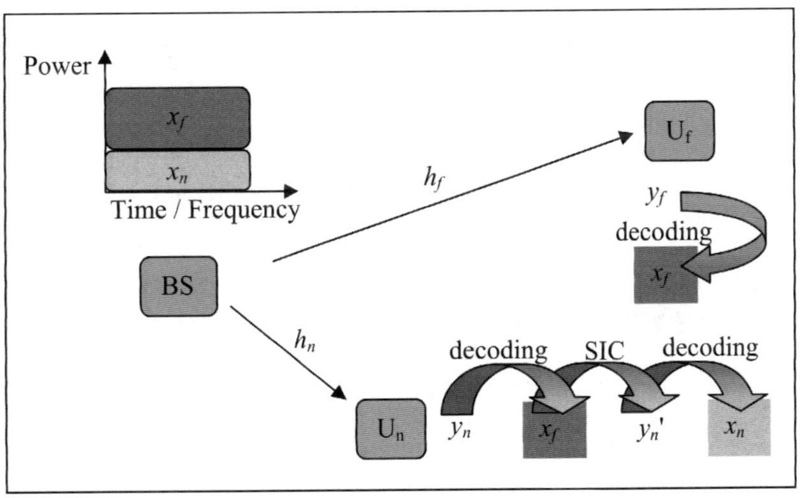


**Figure 1.** A model of downlink NOMA principles.

The signals received at $U_f$ and $U_n$ – $y_f$ and $y_n$ respectively are given in equation (2), where $h_f$ stands for the channel between the BS and $U_f$, $h_n$ is the downlink channel from the BS to $U_n$, $w_f$ and $w_n$ are noise variables [2], [3].

$$\begin{aligned} y_f &= h_f x + w_f , \\ y_n &= h_n x + w_n . \end{aligned} \tag{2}$$

As higher power weight is allocated to the $x_f$ component of $x$, at the far user $x_f$ is obtained by directly decoding the received signal $y_f$. $U_f$ treats the $x_n$ component of $x$ as interference and ignores its presence. On the $U_n$ side, in order to decode its own signal, the near user has first to decode $x_f$, considered as interference, and subtract it via SIC from the received signal $y_n$, following equation (3):

$$y_n^{'} = y_n - h_n\sqrt{P\alpha_f}x_f \,. \tag{3}$$

The residual signal after SIC, $y_n'$, is then decoded to obtain $x_n$.

## 3. Security concerns in NOMA-based networks

The SIC algorithm as applied in NOMA makes the concept of PLS different from that in OMA networks. On the one hand, $U_n$, whose channel with the BS outperforms the downlink channel of $U_f$, successfully decodes $U_f$'s signal, $x_f$, which is required for proper SIC implementation. Moreover, $U_f$ is capable of partly decoding $U_n$'s signal, $x_n$. On the other hand, the air interface of the downlink NOMA channel makes it vulnerable to eavesdropping outside the network. Therefore, NOMA downlink transmission is exposed to both external and internal eavesdropping attacks and its security enhancement can be studied separately depending on in which of the subsequent three groups it belongs [5]:

### *3.1. PLS to counteract external eavesdropping*

When the eavesdropper (ED) is not from the users of the NOMA system, the BS is not familiar with his location and channel state information (CSI). Depending on ED's location in the cell and the transmit power allocated to both users in the NOMA pair, different amounts of their private data can be eavesdropped by the unauthorized user. The most common PLS techniques to counteract external eavesdropping in NOMA-based networks include [5]:

- Power allocation corresponding to the estimated channel gains of $U_f$ and $U_n$;
- Beamforming aimed at $U_f$ and $U_n$;
- Artificial noise-aided (AN) beamforming to degrade the quality of the signals being eavesdropped;
- Distributed beamforming by cooperative relays with or without cooperative jamming.

The problem of eavesdropping from outside the single-input single-output (SISO) NOMA network was investigated in [6], where sharing of the total transmit power between users in the pair, so that to achieve maximal secrecy sum rate (SSR), is studied. The improved efficiency of NOMA compared to OMA is numerically proven. In [7] the SSR optimization problem under SIC and transmission power constraints of multiple-input multiple-output (MIMO) NOMA is solved by the alternating optimization method. Different strategies for choosing optimal transmit antenna in both SISO-NOMA and multiple-input single-output (MISO) NOMA scenarios are proposed in [8], where the relation between secrecy outage probability (SOP) and CSI availability is shown.

In [9] the eavesdropping of large-scale NOMA systems with one or multiple antenna elements and random user pairing is investigated. In both scenarios, it is assumed that a region exists around the BS where ED is not present, while in the multiple-antenna system, eavesdropping is additionally suppressed by generating AN. A novel beamforming scheme which exploits AN for safeguarding the downlink transmission of MISO-NOMA system is suggested in [10]. Unlike [9] and [10], in [11] AN is not induced in the system and security enhancement is accomplished by employing only the inter-user interference. Two different approaches for null-steering jamming beamforming are proposed in [12]. In the first scheme, called self-cooperative jamming, the BS sends the jamming signal during downlink transmission phase together with the information signal of legitimate users. The concept of the other scheme, nonself-cooperative jamming, includes two-stage communication of idle legitimate users for sending interfering signals towards the eavesdropper.

In [13] the privacy of unicasting data is achieved by specific beamforming and power allocation while multicasting is also accomplished. In [14] the NOMA network works cooperatively with a multi-antenna decode-and-forward relay node and a two-stage full-duplex jamming scheme with relay beamforming is proposed to improve secrecy performance. When it comes to practical cognitive radio network implementations the issue of PLS is substantial, due to their dynamic spectrum access (DSA) mode of operation. Its complexity depends on the degree of cooperation between the incumbent and cognitive networks, where the most sophisticated PLS will be necessary for interweave deployments. If an external malicious eavesdropper is not accurately detected, this can lead to unauthorized communications or throughput degradation for the cognitive and/or incumbent users [15]. MISO-NOMA in cognitive radio scenario with simultaneous wireless information and power transfer is examined in [16], where security enhancement against multiple eavesdroppers is realized by cooperative jamming. In [17], millimetre wave NOMA is considered with imperfect SIC, where security enhancement is proposed by two maximum ratio transmission beamforming strategies. A user pairing scheme based on minimal angle difference is also introduced to protect the system against external eavesdroppers.

### *3.2. PLS to counteract internal eavesdropping*

Since the internal ED is a user from the NOMA network, the BS is familiar with his CSI. However, the main challenge in this scenario is to establish privacy without preventing the proper accomplishment of SIC algorithm. Depending on whether the $U_f$ or $U_n$ is untrusted, internal NOMA eavesdropping is subdivided into the following types:

#### *3.2.1. Eavesdropping on $U_n$ by $U_f$*

According to the NOMA principle, $U_f$ decodes $x_f$ from the received superposed signal $x$, treating the information of the near user as interference. Even with less power, the component of $x_n$ still exists in the composite received signal which enables the far user to partially decode it. To alleviate this security issue, methods from the PLS can be employed such as power allocation, beamforming and AN aided beamforming. However, proper dimensioning of beamforming matrices is of significant importance, so as to guarantee the privacy of $U_n$'s signal, while achieving the basic rate of $U_f$.

In [18] the authors propose combining power allocation with zero-forcing beamforming to avoid information leakage to the far untrusted user of a MISO-NOMA network. An optimization approach is studied that maximizes the SSR of $U_n$ subject to a transmit power constraint at BS and transmission rate requirements for $U_f$. The same optimization problem for MIMO-NOMA is studied in [19]. Closed form expressions of the pair outage probability and the SOP of $U_n$ are derived in [20]. Two optimal schemes for relay selection in cooperative NOMA scenarios are proposed in [21], where closed form expressions of SOP and positive secrecy capacity probability are provided.

#### *3.2.2. Eavesdropping on $U_f$ by $U_n$*

As explained in *Section 2*, before fulfilling SIC operations, the near user $U_n$ has first to decode the signal of $U_f$. Moreover, NOMA principles for power allocation (namely, $\alpha_f > \alpha_n$) additionally facilitate detection of $x_f$ by $U_n$. Therefore, securing the far user's information from untrusted near user eavesdropping presents a main security concern in NOMA-based networks. Dealing with this security issue is a very challenging task, since information leakage of data from the far user must be prevented, while ensuring the normal operation of SIC decoding.

In [22], a combination of beamforming and power allocation strategies is proposed, whose successful resistance to internal eavesdropping by the near user is evaluated through SOP analysis. In the case of an untrusted near user, the authors of [23] propose an angle conversion method to transform the original signals of users before transmission. Here, the constellation of the original signal differs from that of the transmitted signal, which guarantees successful SIC decoding of the transmitted signals, while maintaining the privacy of the original signals. In order to be secure, the transformations for each user follow different principles, which can be learned by the intended user via the auxiliary mechanism proposed in [23]. However, as stated in [24], this auxiliary mechanism for sharing the transformation sequences between users suffers from key-management issues. A key-less PLS technique against eavesdropping on the far user by the near user of a MISO-NOMA using directional modulation is proposed in [24]. The method simultaneously ensures the privacy of the far user data and the reliability of SIC by distorting the transmitted symbols at the near user to be from a lower order symbol alphabet. Closed-form expressions are obtained for designing the optimal directional modulation vectors and bit-to-symbol mapping is suggested that improves the secrecy rates.

### *3.3. PLS to counteract both external and internal eavesdropping*

The case of both internal and external eavesdropping represents the most severe security issue of NOMA downlink transmission, as it involves the issues of both *Section 3.1* and *Section 3.2*. The signal transformation method, proposed in [23] against internal eavesdropping, can be applied to secure the system in such a scenario [5]. Another solution, that can be used to minimize the SOP of a system with both internal and external eavesdroppers, is the one suggested in [22].

## 4. Conclusion

Despite the numerous advantages provided by non-orthogonal multiple access, the essence of the applied SIC algorithm makes the NOMA system vulnerable to different types of eavesdropping attacks. In this paper, three varied types of NOMA eavesdropping are examined, together with an overview of methods to combat them. As far as the authors know, there is only a small number of works focused on security against internal NOMA eavesdropping on the far user by the near user. The limited number of studies on this severe problem for the security of NOMA, makes it an open research area that needs further comprehensive studies.

## References

[1] Duchemin D, Gorce J-M, Goursaud C 2018 Code Domain Non Orthogonal Multiple Access versus ALOHA: a simulation based study *ICT 2018 - 25th International Conference on Telecommunications* (*Saint-Malo, France*) *2018* pp 445-450

[2] Islam S R, Avazov N, Dobre O A, Kwak K-S 2017 Power-domain non-orthogonalmultiple access (NOMA) in 5G systems: potentials and challenges *IEEE Communications Surveys & Tutorials* vol. 19, no. 2 pp. 721-742

[3] Saito Y, Kishiyama Y, Benjebbour A, Nakamura T, Li A, Higuchi K 2013 Non-orthogonal multiple access (NOMA) for cellular future radio access *Proceedings of the 77th IEEE Vehicular Technology Conference (VTC Spring) 2013* pp. 1–5

[4] Cai Y, Qin Z, Cui F, Li G Y, McCann, J A 2018 Modulation and Multiple Access for 5G Networks *IEEE Communications Surveys & Tutorials* vol. 20 no.1 pp. 629–646

[5] Furqan H M, Hamamreh J, Arslan H 2019 Physical Layer Security for NOMA: Requirements, Merits, Challenges, and Recommendations *arXiv preprint arXiv:1905.05064*

[6] Zhang Y, Wang H, Yang Q, Ding Z May 2016 Secrecy Sum Rate Maximization in Non-orthogonal Multiple Access *IEEE Communications Letters* vol. 20 no. 5 pp. 930-933

[7] Tian M, Zhang Q, Zhao S, Li Q, Qin J Aug. 2017 Secrecy Sum Rate Optimization for Downlink MIMO Nonorthogonal Multiple Access Systems *IEEE Signal Processing Letters* vol. 24 no. 8 pp. 1113-1117

[8] Lei H *et al.* 2017 On Secure NOMA Systems With Transmit Antenna Selection Schemes *IEEE Access* vol. 5 pp. 17450-17464

[9] Liu Y, Qin Z, Elkashlan M, Gao Y, Hanzo L March 2017 Enhancing the Physical Layer Security of Non-Orthogonal Multiple Access in Large-Scale Networks *IEEE Transactions on Wireless Communications* vol. 16 no. 3 pp. 1656-1672

[10] Lv L, Ding Z, Ni Q, Chen J July 2018 Secure MISO-NOMA Transmission With Artificial Noise *IEEE Transactions on Vehicular Technology* vol. 67 no. 7 pp. 6700-6705

[11] He B, Liu A, Yang N, Lau V K N Oct. 2017 On the Design of Secure Non-Orthogonal Multiple Access Systems *IEEE Journal on Selected Areas in Communications* vol. 35 no. 10 pp. 2196-2206

[12] Alsaba Y, Leow C Y, Abdul Rahim S K 2019 Null-Steering Beamforming for Enhancing the Physical Layer Security of Non-Orthogonal Multiple Access System *IEEE Access* vol. 7 pp. 11397-11409

[13] Ding Z, Zhao, Peng M, Poor H V July 2017 On the Spectral Efficiency and Security Enhancements of NOMA Assisted Multicast-Unicast Streaming *IEEE Transactions on Communications* vol. 65 no. 7 pp. 3151-3163
[14] Cao Y *et al.* Aug. 2019 Secrecy Analysis for Cooperative NOMA Networks With Multi-Antenna Full-Duplex Relay *IEEE Transactions on Communications* vol. 67 no. 8 pp. 5574-5587
[15] Singh P, Pawar P, Trivedi A Dec. 2018 Physical layer security approaches in 5G wireless communication networks *2018 First International Conference on Secure Cyber Computing and Communication (ICSCCC) (Jalandhar, India)* pp. 477-482
[16] Zhou F, Chu Z, Sun H, Hu R Q, Hanzo L April 2018 Artificial Noise Aided Secure Cognitive Beamforming for Cooperative MISO-NOMA Using SWIPT *IEEE Journal on Selected Areas in Communications* vol. 36 no. 4 pp. 918-931
[17] Huang S, Xiao M, Poor V August 2020 On the Physical Layer Security of Millimeter Wave NOMA Networks *IEEE Transactions on Vehicular Technology*
[18] Li Y, Jiang M, Zhang Q, Li Q, Qin J Aug. 2017 Secure Beamforming in Downlink MISO Nonorthogonal Multiple Access Systems *IEEE Transactions on Vehicular Technology* vol. 66 no. 8 pp. 7563-7567
[19] Jiang M, Li Y, Zhang Q, Li Q, Qin J Dec. 2017 Secure Beamforming in Downlink MIMO Nonorthogonal Multiple Access Networks *IEEE Signal Processing Letters* vol. 24 no. 12 pp. 1852-1856
[20] ElHalawany B M, Wu K Physical-Layer Security of NOMA 2018 Systems Under Untrusted Users *2018 IEEE Global Communications Conference (GLOBECOM), (Abu Dhabi, United Arab Emirates)* pp. 1-6
[21] Cao K, Wang B, Ding H, Li T, Gong F Feb. 2020 Optimal Relay Selection for Secure NOMA Systems Under Untrusted Users *IEEE Transactions on Vehicular Technology* vol. 69 no. 2 pp. 1942-1955
[22] Cao K, Wang B, Ding H, Li T, Tian J, Gong F 2020 Secure Transmission Designs for NOMA Systems Against Internal and External Eavesdropping *IEEE Transactions on Information Forensics and Security* vol. 15 pp. 2930-2943
[23] Xu D, Ren P, Du Q, Sun L, Wang Y 2017 Combat eavesdropping by full-duplex technology and signal transformation in non-orthogonal multiple access transmission *2017 IEEE International Conference on Communications (ICC), (Paris)* pp. 1-6
[24] Christopher R M, Borah D K May 2020 Physical Layer Security for Weak User in MISO NOMA Using Directional Modulation (NOMAD) *IEEE Communications Letters* vol. 24 no. 5 pp. 956-960

**Acknowledgments**
This work was supported by the National Program "Young Scientists and Postdoctoral Students", PMC No577, 17.08.2018, of Ministry of Education and Science of Bulgaria.

# Storing data from sensors networks

**F A Tsvetanov**

Department of Communication and computer engineering and technology, South-West University, Blagoevgrad, 2700, Bulgaria

E-mail: ftsvetanov@gmail.com

**Abstract.** Sensor networks generate a massive amount of heterogeneous and multidimensional data with specific characteristics, for the storage, processing and visualisation of generated sensor data most often used specially designed databases for sensory data, which can integrate into cloud structures. This paper presents a diagram of the integration of data from sensor networks in the cloud, which solves essential issues about the cost of sensor networks and the possibilities for storing sensor data in specialised cloud databases. The peculiarities of the sensory data and the specific requirements for their storage and processing are analysed. Through which approaches and models in the storage are identified, as well as the possibilities for their application. TSBD time-series database and cloud databases suitable for storage of sensor data are systematised, and recommendations for the selection made.

## 1. Introduction

We live in an era of IoT, where billions of devices are connected and generate data with increasing speed. Obtaining information from these data gives a competitive advantage on people, businesses, organisations and governments. Networks made of intelligent IoT, also called sensor nodes in wireless sensor networks (WSN), have been a rapidly evolving field of research in recent years. These networks widely used to collect sensor data, which supports environmental monitoring, detection of anomalies in the health of patients of industrial applications, border security, etc. The amount of data generated by sensor networks is enormous, heterogeneous and multidimensional. Significant hardware resources of the sensor nodes (storage and computing) required for the storage and processing of this data. It known that the sensor nodes have limited computational and energy capacity [1],[2]. Integration between clouds and sensor networks is an excellent solution to the problem of the limited computing power of sensors, increasing the life of sensor networks, storing and processing the collected data. Storage of data is performed most frequently in various databases, which may be the integrin in the cloud structures. We often faced with the dilemma of what a database is suitable for the storage of sensor data.

This paper presents a block diagram of the integration of sensor data, solving the problem of storing large volumes of sensor data and the cost of sensor networks analysis of the features of the sensory data and the specific requirements for their storage and processing. Systematised are sensory databases and cloud database suitable for storing sensor data. Recommendations have made for the selection of an appropriate database for the storage and processing of sensor data.

## 2. Related work

In work [3] the issues related to data security in different cloud models are analysed. The authors compare the protection provided by the different cloud models and conclude that in all three models, data security depends on a secure and reliable network and a secure web browser. A distinction has been made between security in the clouds, depending on vulnerabilities and threats. The authors [4] present various energy-efficient techniques for data collection and storage concerning their energy efficiency. To reduce network traffic when there are strong Spatio-temporal correlations between sensor data is proposed, coding and classification of the sensor data category of IoT nodes. In [5] is presented an experimental study of the effectiveness of sensor systems for data storage, implemented based on file system Linux, depending on the number of sources data. The authors conclude that fewer hardware resources are needed when a log system used to build sensor storage systems. This system saves a small number of log files and uses a sensor-optimised mechanism for indexing (extracting time) and searching for data in log files. The authors [6] present an analytical study of trends in the field of storage and processing of sensory data. There is a tendency not to throw away anything when storing data, save all received data in the hope that existing or new technologies will use to allow the analysis of all collected data. The authors believe that not only secure data storage but also the ability to access them on demand effectively is essential. In [7] and [8], adaptive data downloading proposed to reduce the number of data samples, and the sensor data are redirected to the coordinator only when a significant change in the behaviour of the sensor nodes is detected.

## 3. Integration of sensors data in the cloud

### *3.1. A.Block diagram of the integration*

As we explained above, connecting the sensor network to a cloud structure solves the problem of storing and processing large volumes of data generated by the sensor networks. The communication between the WSN and the cloud computing system, according to figure 1. The data collected and recorded by the individual sensors is transferred to the cloud, processed and can be viewed, manipulated and modified via client devices displaying an application, web page or application program API [9]. Figure 2, a block diagram illustrating a communication method. Data stored in a cloud computing system can be accessed by multiple and different client devices, including mobile devices such as laptops and cell phones.

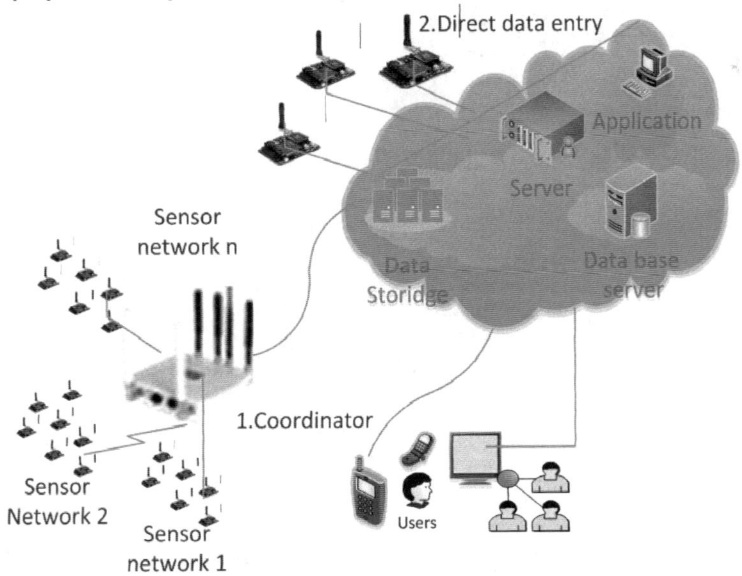


**Figure 1.** Integration of a sensor network with a cloud structure.

Cloud computing systems are more reliable than special server configurations, so measurements taken from sensor networks are more likely to be recorded and/or processed by such cloud systems. For example, data stored on a dedicated server it becomes inaccessible for some reason, the data generated and recorded by the sensor network during server failure is lost. In the case of medical sensors that record vital patient data, lost data can lead to incorrect treatment decisions. In case of malfunctioning temperature sensors of computer equipment, the lost data can lead to damaged computer equipment, downtime and unavailability in other network services. Cloud computing systems for recording measurements from sensor networks can also be more economical than dedicated servers, as cloud computing systems are often paid for use and do not require the purchase of large amounts of computer equipment.

### *3.2. Data transmission via the coordinator*

Different scenarios for data transmission are possible when integrating sensor networks with cloud structures. For example, the gateway can calculate the average measured values of the sensors, perform compression or other data encoding, transmit raw data or pre-processed data directly to the cloud structure. The data transmitted in the cloud structure encrypted. The coordinator may have an algorithm for detecting certain events and send a signal to the cloud computing system and/or send only the captured data. For example, the coordinator gateway can be configured to detect peaks in the measured data received from the sensor network or configured so that the sensor transmits only those data that have exceeded the configured threshold.

The cloud computing system receives the data from the coordinator and records it. The data most often is stored in appropriate databases. The relocation of processing functions to the cloud computing system allows the development of sensors and coordinators with proper components, which reduces the cost of sensor networks in general. The importance is, especially considering that there can hundreds or even thousands of sensors in sensor networks. Processing in the cloud computing system may include data analysis, data validation, data clean up and/or data transformation.

### *3.3. .Data transmission directly*

In this embodiment, the coordinator may miss or gateway to perform another role. In this case, the coordinate gateway or cloud computing system can activate the sensor via an RF module to read the measurement result.

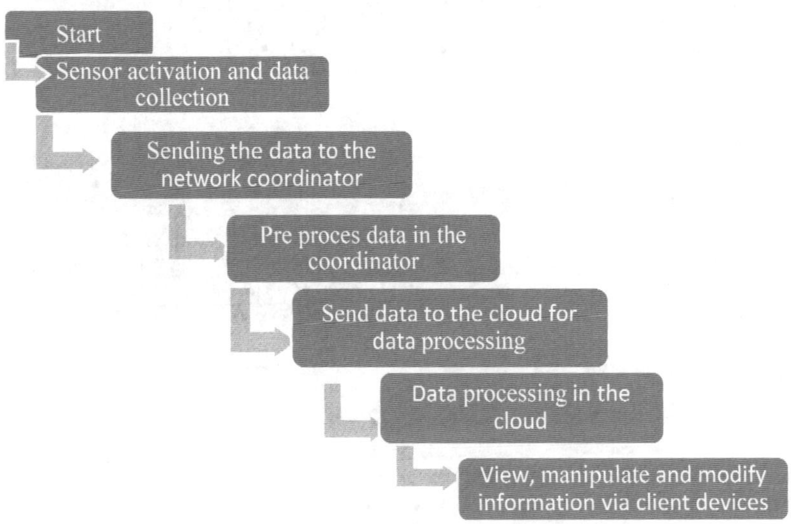


**Figure 2.** Block diagram of the method of communication between the sensor network and the cloud.

### *3.4. Functions of clients devices*

Client devices, such as a computer, tablet, or mobile phone, may be connected to the cloud computing system and have access. Access to the display, modify, delete, or manipulate data in the cloud computing system, including data received from the coordinate gateway sensor network, and other data stored in the cloud. Client devices can also direct measurement requests to each or only a specific sensor on the network. Besides, they can configure the sensors. For example, they were configuring the coordinator for pretreatment, configure the sensor to perform measurements at regular intervals such as every hour or every day. Also, specific client devices can view data or process data.

### *3.5. Communication between sensors and the cloud*

Block diagram of the method of communication between the sensor network and the cloud showing in figure 2. After activating the sensor, generated data transmitted to the coordinator. The coordinator can pre-process or transfer raw data to different databases in the cloud.

## 4. Databases for storing and managing data from sensors network

### *4.1. Characteristic of sensors data*

Sensor data from a single sensor are simple time series. Often for an extended period, the data from the sensor may not change its values. There is a case where the volume of data grows rapidly; several reports per second for one or more sensors from one network. The analysis of modern solutions for storage and processing of sensor data aims to identify approaches and opportunities for their application.

### *4.2. Basic models for storage and processing of sensory data*

At the moment, the following models distinguished:

- Storage and processing of data outside the sensor network;
- Distributed storage and processing on the network;
- Combined solutions;

*The first model* treats sensor data as a continuous stream that accumulates without loss in the sensor network, and then the data is transmitted and archived outside the sensor network. The collected data can be stored in different storages and can be made requests to access them using standard methods. This model implemented by a large number of already implemented networks designed to monitor all phenomena. Such networks are easy to implement but short-lived, especially when using high-speed sensors such as cameras, acoustic or vibration sensors, etc., as data transmission requirements often exceed the available energy resources.

*The second model* for working with sensor data considers the sensor network as a distributed database that supports requests for data provision and processing. In this model, a request directed to the sensor network, possibly to remote sensor nodes that store their data locally. This architecture is potentially more energy-efficient because the request processed on a smart sensor node, and the result of the request transmitted as processed data, rather than raw data. The energy costs of network transmission of large amounts of data exceed the costs of their local processing [6].

### *4.3. External storage and processing of sensory data*

In centralised processing, the storage of data from sensor nodes performed on a server outside the sensor network. A popular way to organise a network is as a *relational database.* Although general-purpose relational databases can store time-series data, they do not allow efficient work with an ordered set of time-series elements due to a lack of time optimisations for storing and retrieving data at time intervals.

### *4.4. TSBD time-series database*

TSBD time-series databases specially designed to store sensor data. They belong to the group of non-relational databases (Not only Structured Query Language / NoSQL) because they use search languages different but similar to SQL. NoSQL mostly used in real-time web and big data applications. A time-series database is a software system that optimised for handling time-series data, arrays of numbers indexed by time - DateTime or a DateTime range [10]. The timestamp usually has an accuracy of seconds and milliseconds, but with remote data collection, the accuracy can reach nanoseconds. The timestamp supports automatic adjustment of time zones. The frequency of data reporting can be periodic or when an event occurs. TSDB is optimised to measure change over time. Properties that make time series data very different from other data loads that are the generated data.

And these are data lifecycle management, aggregation and scanning of a large range of many records TSBD provide users with a service for storing, deleting, updating and modifying time series. All sol is supporting some time-series calculations (e.g. multiplication, addition and other manipulations to convert one or more lines to a new line). There are tools to perform filtering on random samples, while the values of one row can be a filter for another.

### *4.5. Type of TSBD time-series database*

According to the type of licenses, the databases classified as open sources, commercial, academic databases and industrial databases (figure3).

*InfluxDB* is a specially created, open-source time-series database. InfluxData is available in two versions InfluxDB Enterprise and InfluxDB Cloud. The InfluxDB data model has a linear protocol for sending data from time series, which takes the following form: measurement name, the tag-set field with a set timestamp. The measurement name is a string, the tag set is a collection of key/value pairs where all values are strings, and the fieldset is a collection of key/value pairs where the values can be int64, float64, bool, or string. Most other solutions for time series only support float64 values, which means that the user is unable to encode additional metadata along with timelines. Due to all these factors, InfluxDB is the best solution for working with time-series data lines [11].

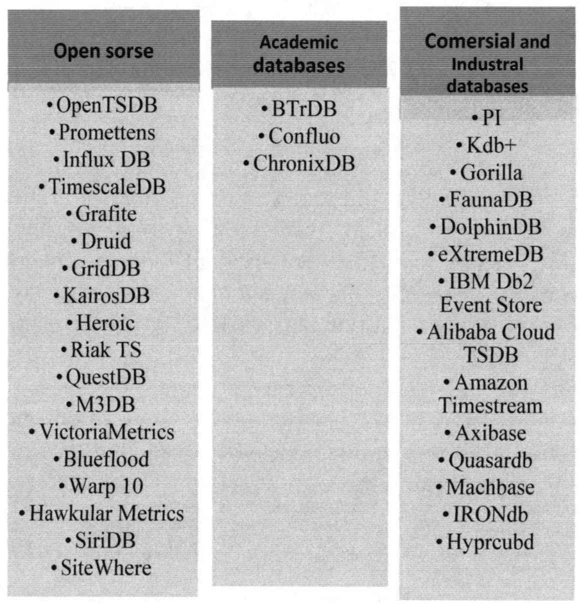


**Figure 3.** Type of TSBD time-series database.

Open Data Series Data Base (OpenTSDB) [12] is a time-series database built on *HBase*. It has excellent reading and writing performance, scales very well and is distributed free of charge. OpenTSDB can collect, store and service billions of data points without loss of accuracy, which makes it an ideal solution for the monitoring system. It consists of a *Time Series Daemon (TSD)* as well as a set of command-line utilities. Interaction with OpenTSDB is achieved primarily by starting one or more of the *TSDs*. Communication with TSD done with a simple protocol in *telnet* style, HTTP API or a simple built-in GUI. All communications are on the same port (TSD calculates the client protocol by looking at the first few bytes it receives). Data stored as timelines. Each time series is a sum of data points. *Prometheus* e TimeSeries DBMS and monitoring system with Open-source [12]. *Grafite*, Data logging and graphing tool for time series data Open-source [14]. *RRDtool,* is an industry-standard, with high-performance data logging and a graphical timeline data system. RRDtool easily integrated into scripts *Perl, python, ruby, Lua или Tcl* application [15]. *TimescaleDB* [16] offered as a cloud implementation *Timescale Cloud*, which can integrate into the cloud of a selected provider such as AWS, Azure or Google Cloud. *GridDB* optimised for IoT and big data. The basis of GridDB's principles based on offering a universal data cloud that is optimised for IoT, providing high scalability, performance and ensures high reliability. GridDB supports many time series functions, such as data compression for ever-increasing timeline data. APIs and other access methods Graphite protocol, HTTP REST, Telnet API [17].

The common characteristics of the different TSDBs are:

- ✓ storing timelines series without loss of resolution;
- ✓ high recording frequency (up to milliseconds);
- ✓ scaling to millions of records per second;
- ✓ horizontal scaling (increasing storage capacity by adding nodes);
- ✓ providing a graphical interface for graphics;
- ✓ read/write via HTTP.

Databases timelines series are the fastest-growing segment in the industry databases. Their popularity, according to an independent website, DB-Engines (figure 4).

| Rank | | | DBMS | Database Model | Score | | |
|---|---|---|---|---|---|---|---|
| Sep 2020 | Aug 2020 | Sep 2019 | | | Sep 2020 | Aug 2020 | Sep 2019 |
| 1. | 1. | 1. | InfluxDB | Time Series | 23.34 | +0.46 | +4.69 |
| 2. | 2. | 2. | Kdb+ | Time Series, Multi-model | 7.43 | +0.63 | +1.93 |
| 3. | 3. | 3. | Prometheus | Time Series | 5.69 | +0.33 | +2.24 |
| 4. | 4. | 4. | Graphite | Time Series | 4.31 | +0.20 | +0.99 |
| 5. | 5. | 5. | RRDtool | Time Series | 3.05 | +0.03 | +0.48 |
| 6. | 6. | ↑ 8. | TimescaleDB | Time Series, Multi-model | 2.73 | +0.19 | +1.33 |
| 7. | ↑ 8. | ↓ 6. | OpenTSDB | Time Series | 2.30 | +0.17 | +0.45 |
| 8. | ↓ 7. | ↓ 7. | Druid | Multi-model | 2.29 | +0.11 | +0.56 |
| 9. | 9. | ↑ 11. | FaunaDB | Multi-model | 1.86 | +0.15 | +1.43 |
| 10. | ↑ 11. | 10. | GridDB | Time Series, Multi-model | 0.75 | +0.09 | +0.26 |
| 11. | ↓ 10. | ↓ 9. | KairosDB | Time Series | 0.73 | +0.05 | +0.19 |
| 12. | 12. | ↑ 18. | DolphinDB | Time Series | 0.69 | +0.09 | +0.54 |

**Figure 4.** Applicability of TSBD [18].

The choice of a specific database may depend on the specifics of the implementation of the application that uses sensor data.

### *4.6. Online time series storage services*

The cloud database is a type of database service. Cloud Database typically works as a standard database solution, which typically implemented by installing database software on a cloud infrastructure. Internet services for storage and processing of sensor data are usually a cloud service that is built, implemented and delivered through a cloud platform. This platform allows the user to send their data to the cloud directly from the sources, storing and managing the life cycle of this data. One of them is a cloud service delivery platform (PaaS) model that allows organisations, end-users, and their applications to store, manage, and retrieve data from the cloud. Access can be made directly through a web browser or applied by the provider API for integration of applications and services.

Current trends in the storage and processing of sensor data require requests to the service, use the provided API for different types of data processing by displaying them in graphical form. The service works in multi-user mode, allowing each user to work safely with their data set. Different services have different policies regarding service charges. The amount of free data stored is limited, or the user released from service charges provided they make their data publicly available, or the service is provided free of charge without a guarantee of the level of service, etc. The most famous services are Xively, TempoIQ, Nimbits Public Cloud [6], which specially designed for storing and analysing data from time series of sensors, smart meters, servers and more.

## 5. Conclusion

The paper presents and analyses a block diagram for the integration of sensor data in a cloud structure, which allows solving important issues about the cost of the sensor network and the possibilities for storage and analysis of sensor data in specialised databases. The paper analyses the modern solutions for storage and processing of sensor data, identifying the models for storage of sensor data and the approaches and possibilities for their application.

The known TSBD time-series databases, specially designed for storing data from sensors, are systematised. Databases designed for time series data use 1/2 to 1/3 more efficient storage and process data by providing high performance in recording and requests. One of the most preferred databases is InfluxDB, as it provides plug-in capabilities, multi-cloud support, data analysis solution, query builder, powerful client libraries and 24-hour support for open source users. InfluxDB processes data quickly on the order of milliseconds, microseconds or nanoseconds, making the database preferred for financial and scientific purposes, and the excellent compression mechanism and storage architecture making it suitable for surveillance systems.

## References

[1] Tsvetanov F and Pandurski M 2018 Some aspects for the integration of sensor networks in cloud structures, *Proceedings of International Conference on High Technology for Sustainable Development,* pp 249-253

[2] Tsvetanov F, Tsvetanov K, Ivanova E and Georgieva I 2014 Modelling and Simulation of Communication Efficiency in Low-Speed Networks, *22nd Telecommunications Forum TELFOR,* pp 178-182

[3] Hashizume K, Rosado D and Fernández-Medina E 2013 An analysis of security issues for cloud computing, *Journal of Internet Services and Applications* **4-5**, pp 1-13

[4] Xinxin D, Zhangbing Z, Yuqing Z and Rahman T 2020 Energy-efficient sensory data gathering based on compressed sensing in IoT networks, *Journal of Cloud Computing: Advances, Systems and Applications*.

[5] Zhivchikova N and Shevchuk Y 2018 Experiments with sensor data storage, performance, *Program Systems: Theory and Applications,* **9:4** (39), pp 597–627

[6] Shevchuk E and Shevchuk Y 2015 Modern trends in the field of sensory data storage and processing, *Software Systems: Theory and Applications,* **6: 4** (27), pp 157–176

[7] Endo P T, Rodrigues M, Gonçalves G E et al. 2016 High availability in clouds: systematic review and research challenges, *J Cloud Comp* **5**, 16

[8] Al-Hoqani N and Yang S-H 2015 Adaptive sampling for wireless household water consumption monitoring, *Procedia Eng,* **119**, pp 1356–1365

[9] Dwivedi R, K Singh S and Kumar R 2019 Integration of Wireless Sensor Networks with Cloud: A Review, *9th International Conference on Cloud Computing, Data Science & Engineering (Confluence), Noida, India*, pp 114-119, doi: 10.1109/CONFLUENCE.2019.8776968.

[10] Dunning T and Friedman E 2014 Time series databases: new ways to store and access data, *O'Reilly Media Incorporated,* p 386

[11] Ploetz A, Kandhare D, Kadambi S and WuSeven X 2018 Get up and running with the fundamentals and functionalities of seven of the most popular NoSQL databases Paperback, *Packt Publishing* (Birmingham UK)

[12] Struckova A, Yufaa S, Visheratina A and Nasonova D 2019 Evaluation of modern tools and techniques for storing time-series data, *Procedia Computer Science,* **156**, pp 19-28

[13] Brazil B 2018 Prometheus: Up & Running: Infrastructure and Application Performance Monitoring, *O'Reilly Media Inc*, p 386.

[14] Bell P and Beer B 2015 Introducing HitHub *O'Reilly Media Inc.*

[15] Russell J and Cohn R 2012 Rrdtool *California: Book onDemand Ltd*, p 92

[16] Freedman M 2020 TimescaleDB vs InfluxDB: Purpose-built differently for time-series data https://blog.timescale.com/blog/

[17] Building IoT Applications with GridDB, available at: https://griddb.net

[18] B-Engines Ranking, https://db-engines.com/en/ranking

# Study the influence of intermodulation products on navigation signals

**T B Iliev[1], I S Stoyanov[2], G Y Mihaylov[3] and E P Ivanova[1]**

[1] Department of Telecommunications, University of Ruse, 7 Studentska Str., 7000, Ruse, Bulgaria
[2] Electrical and Power Engineering Department, University of Ruse, 7 Studentska Str., 7000, Ruse, Bulgaria
[3] Department of Telecommunications, University of Telecommunications and Post, 1 Akad. Stefan Mladenov Street, Sofia, 1700, Bulgaria

E-mail: tiliev@uni-ruse.bg

**Abstract.** The work examines terrestrial radio broadcasting using frequency modulation (FM), which interferes with the ILS (Instrument Landing System), VOR (VHF Omnial Radio range) and VHF communications systems, which is a widespread problem among aircraft users. In the case of ground/air communication receivers, this interference problem involves disturbing of audio communications and distorted reception of signals from the RV range. In aircraft ILS locators and VOR receivers, the interference problem is associated with interference of audio communications and errors related to course deviation and flag modes of control. Interference in these navigation receivers is considered as a more serious problem than errors in course deviation or impaired communication, especially in the approach and landing phases.

## 1. Introduction

GNSS multipath propagation phenomena [1] and intermodulation distortion are unwanted combination of multiple signals into a nonlinear device that can cause interference to adjacent receivers located in radio relay objects. Most part of the intermodulation occurs in the transmitter's nonlinear power amplifier. The front end of the receiver is the next most common mixing point is.[2] It usually occurs in the unprotected first mixer of older radio receivers or in some cases in overloading the rear front of RF amplifiers. There are many methods for reduction, identification and filtration the noise signals. [3,4] Intermodulation may also be caused by corroded tower connections, safety cables, screw clamps and coupling rods or any metal objects located nearby which may serve as a non-linear "mixer/rectifier" [5].

Of interest to the practice is the case of "transmission" intermodulation, in which disturbances occur only when the two transmitters are simultaneously on air [6]. This intermodulation shall be observed in the event when transmitter "A" and transmitter "B" are located close enough to each other, at the same time are on air at frequencies of 150.00 MHz and 151.00 MHz respectively. Under these conditions, it is possible to be received the signal from one transmitter from the antenna to the other and cause interference with the operation of the final power amplifier, i.e. a mixing may occur in one or both of the transmitter's power amplifiers, creating several new frequencies at intervals of 1.0 MHz. (deviation). Parasitic signals will appear at frequency intervals of ±1.0 MHz to the side of each transmission frequency (figure 1). For example, for frequencies 146.00 MHz, 147.00 MHz, 148.00

MHz, 149.00 MHz, 150.00 MHz, 151.00 MHz, 152.00 MHz, 153.00 MHz, 154.00 MHz, 155.00 MHz.[7] The main frequencies are 150.00 MHz and 151.00 MHz and the intermodulation products (IMP) are as follows:

- for the third order are 149.00 MHz and 152.00 MHz;
- for the fifth order is 148.00 MHz and 153.00 MHz;
- for the seventh order are 147.00 MHz and 154.00 MHz;
- for the ninth order are 146.00 MHz and 155.00 MHz.

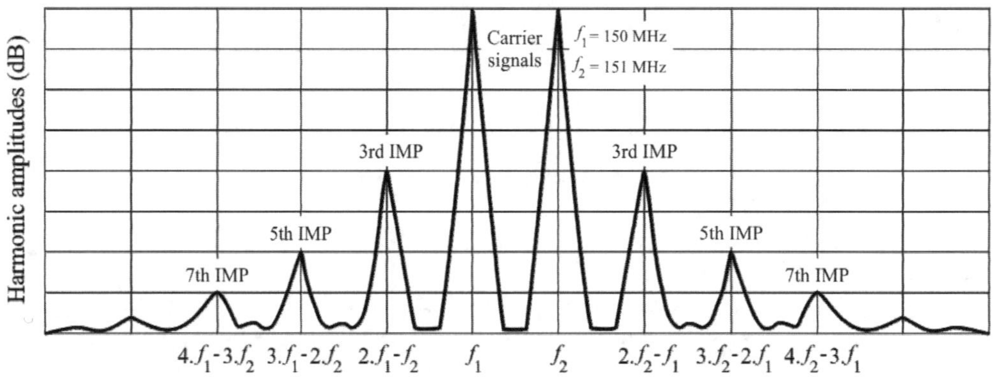


**Figure 1.** Spectral presentation of intermodulation products of the 3th, 5th, 7th and 9th order.

The common equation for the intermodulation product:

$$IM = n_1 A \pm n_2 B \pm n_3 B \pm \ldots, \tag{1}$$

where A, B, C, etc. are the mixing frequencies and $n_1$, $n_2$, $n_3$, etc. are harmonic, or multiples, of the mixing frequencies. For example, A+2B + 2C generates an intermodulation period of the fifth order (1 + 2 + 2).[8]

The greatest influence have the odd intermodulation disturbances of the 3[rd] and 5[th] order [9]. Higher order intermodal products have lower power levels compared to third- and fifth-order products.

## 2. Methodology for measuring the intermodulation distortion

A single linear system can't produce intermodulation if the input signal of a time-invariant system is a single-frequency signal, then the output signal has the same frequency, only the amplitude and phase can differ from the input signal. [10].

Nonlinear systems generate harmonious constituents. Therefore, if the input signal from the non-linear system has an $f_a$ frequency, then the output signal includes multiples of the input frequency. ($f_a$, $2f_a$, $3f_a$, $4f_a$, ……).

Intermodulation occurs when the input signal in a non-linear system is composed of two or more frequencies. For example, for an input signal containing three frequency components of $f_a$, $f_b$ and $f_c$ can be described with the expression [8]:

$$x(t) = M_a \sin(2\pi f_a t + \varphi_a) + M_b \sin(2\pi f_b t + \varphi_b) + M_c \sin(2\pi f_c t + \varphi_c), \tag{2}$$

where $M$ and $\varphi$ are the amplitudes and phases of the three components.

The output signal $y(t)$ is received after transmission of the signal through a non-linear function

$$y(t) = G(x(t)), \tag{3}$$

where y(t) contains the three frequencies of the input signal $f_a$, $f_b$ and $f_c$, as well as a number of linear combinations of the main frequencies, represented by (4):

$$k_a f_a + k_b f_b + k_c f_c \tag{4}$$

where $k_a$, $k_b$ and $k_c$ are random integers.

Each of mentioned frequency components has a different amplitude and phase, which depend on the nonlinear function. Therefore, when an input signal is set containing a random number $N$ from frequency components $f_a, f_b, \ldots f_N$, the output signal will contain a number of frequency:

$$k_a f_a + k_b f_b + \ldots + k_N f_N, \tag{5}$$

where $k_a$, $k_b$, …, $k_N$ are random integers.

The order O of an IMP is the sum of the absolute values of the coefficients [8,10]:

$$O = |k_a| + |k_b| + \ldots + |k_N| \tag{6}$$

i.e. third-order intermodulation products appear, if the condition is met:

$$|k_a| + |k_b| + |k_c| = 3 \tag{7}$$

$$(f_a + f_b - f_c),\ (f_a + f_c - f_b),\ (f_b + f_c - f_a) \tag{8}$$

$$(2f_a - f_b),\ (2f_a - f_c),\ (2f_b - f_a),\ (2f_b - f_c),\ (2f_c - f_a),\ (2f_c - f_b) \tag{9}$$

On figure 2 is shown the distribution of third-order intermodulation products. In blue are the main carriers, in red are the positions of the dominant intermodulation products, and in green - the positions of precisely defined intermodulation products.

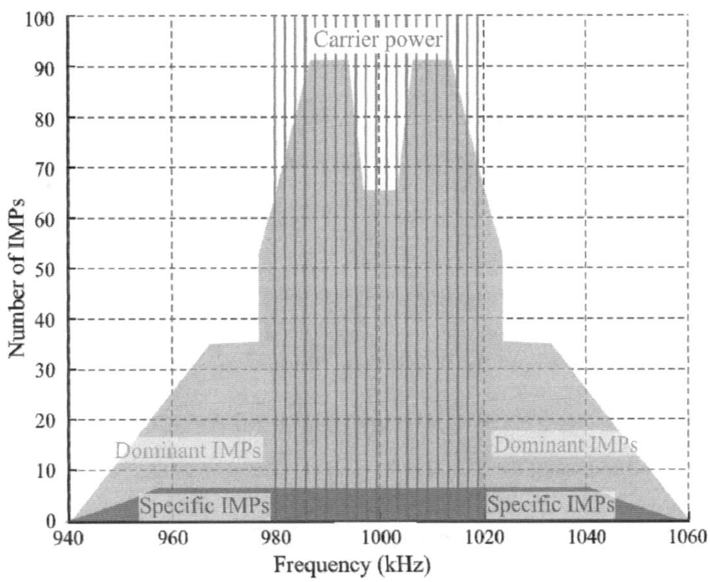


**Figure 2.** Distribution of third-order intermodulation products.

Analysis of the distribution of intermodulation products shows that the main carriers have the greatest amplitude, and the intermodulation products have a lower amplitude level relative to the main frequencies, but occupy a wider frequency spectrum. The odd intermodulation disturbances of the 3rd and 5th order have the greatest influence, because the observed maximum levels are closer to the main frequencies.

## 3. Common assessment method

The objective of the CAM (Common Assessment Method) is to calculate all potential incompatibilities in the aircraft volume at multiple points. The maximum potential incompatibility, associated with the specific RV service, is defined as a protective distance. The CAM is based on the need to protect civil aviation radio services within the established minimum distances to the antenna systems of repeaters, etc., depending on the air service (ILS or VOR) and its activities. [11, 12]

### *3.1. Test points in ILS*

Assessing compatibility with an ILS localizer, the CAM is based on a set of fixed test points, complemented by more test points for each designated operational coverage (DOC) transmission station of the ILS.

DOCs used in VOR services are big and therefore there are usually a lot of broadcast stations. The CAM assesses the compatibility regarding VOR, with test points selected above each broadcast station in the DOC and have in mind all stations outside the DOC.

Figure 3 shows the distribution of the fixed test points, and in table 1 - the minimum height, the distance from the location of the localizer and the navigation deviation dependent on the elongated axis line of the track. Fixed test points A, E, F, G have minimum heights of (0, 150, 300 and 450)m respectively, above the vertical section of the location where the ILS localizer is located. These values describe the descent trajectory with a slope of 3 $\pi$. All other fixed test points are at a minimum height of 600m.

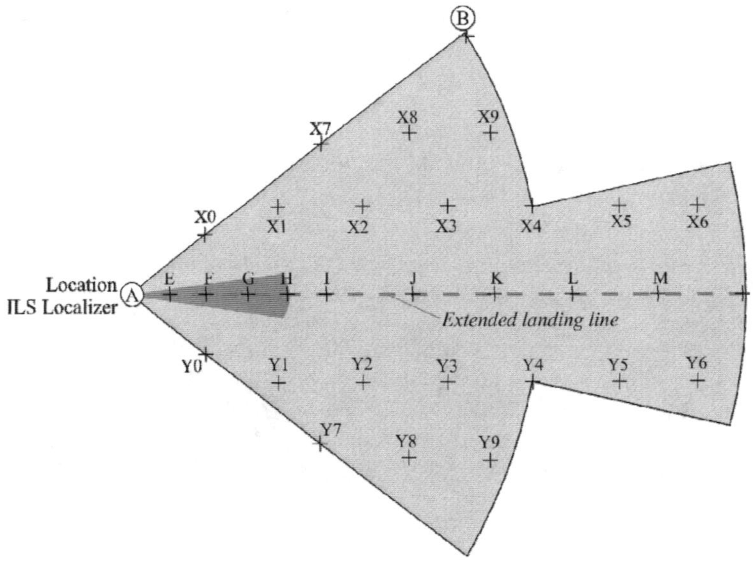


**Figure 3.** Location of fixed test points in the DOC zone of ILS.

**Table 1.** Distances of test points.

| Points on or above the Extended Centre Line | | | Points outside the Extended Centre Line (all at a height of 600m) | | |
|---|---|---|---|---|---|
| ID | Distance (km) | Minimum height | ID | Distance (km) | Navigational deviation, according to the Extended Centre Line |
| A | 0 | 0 | B,C | 31.5 | -35, 35 |
| E | 3 | 0 | X0,Y0 | 7.7 | -35, 35 |
| F | 6 | 150 | X1,Y1 | 12.9 | -25.5, 25.5 |
| G | 9 | 300 | X2,Y2 | 18.8 | -17.25, 17.2 |
| H | 12 | 450 | X3,Y3 | 24.9 | -12.9, 12.9 |
| I | 15 | 600 | X4,Y4 | 31.5 | -10, 10 |
| J | 21.25 | 600 | X5,Y5 | 37.3 | -8,6, 8.6 |
| K | 27.5 | 600 | X6,Y6 | 43.5 | -7.3, 7,3 |
| L | 33.75 | 600 | X7,Y7 | 18.5 | -35, 35 |
| M | 40 | 600 | X8,Y8 | 24.0 | -27.6, 27.6 |
| D | 46.3 | 600 | X9,Y9 | 29.6 | -22.1, 22.1 |

If the emitting station is inside the area shown in figure 3, an additional test point shall be created when the geographical coordinates of the emitting station are known and located at the same height as the emitting antenna. If the emitting station is within or below the ILS DOC, but outside the hatched area in figure 3, an additional test point shall be created having the geographical coordinates of the emitting station. The height of the test point is at least 600m above the ILS zone; or 150m above the emitting antenna.

*3.2. Test points for VOR*

The VOR test point is located at geographical coordinates of a emmiting station, at a minimum height, which is at least 600m above the terrain (approximated as 600m above the location of the broadcasting station), or 300m above the broadcasting antenna, or at height to which the height of the VOR plot has been added. Additional test points in the DOC can be used to ensure the specific operation of the VOR. For example, when used as a landing assistant, or when service is required at an elevation angle less than 0°.

*3.3. Selection of test points*

Broadcasting stations shall be included in analyses at a test point if there is a direct line of sight from the broadcasting antenna to the test point. If the calculated signal level is greater than the B1 cut-off value threshold or the field intensity in a free space is at least equal to the value that may cause TYPE Al, A2 or B2 incompatibility and are not more than 125km away for cases A1 and B2.

In order to assess the compatibility of a set of broadcast stations at each selected test point, it is necessary to:

- calculate the field intensity in the free space for each of the broadcasting stations at the test point, taking into account the slope distance, the maximum effective radiated power (EIRP) and the characteristics of the antenna;
- calculate the signal level of the ILS or VOR;
- calculate the incoming power at the inlet of the aircraft receiver.

The impact of frequency and type (ILS or VOR) of the RV compatibility assessment office for each interference type A1, A2, B1 and B2 must be taken into account.

## 4. Measurement results for the presence of IMP type A1

A measurement was carried out on the radio relay and TV station "Frangata" Varna to establish the presence of IMP type A1, created by mutual penetration of signals, when working on more than one transmitter on one object or in close proximity to different objects. The main objective is to establish whether VHF-FM radio stations, operating in the 87.5÷108 MHz range, located on a single site or on nearby sites can be a source of mutual intermodulation products. They may pose a threat to the security of the operation of civil aviation radio services within the scope of 108÷137 MHz. [10, 11, 12, 13, 14]

In table 2 are shown the basic technical parameters and characteristics of the measurements:

- Intensity of aircraft systems ILS, VOR, COM
  - ILS intensity - 32 dB(μV/m);
  - VOR intensity - 39 dB(μV/m);
  - COM intensity - 38 dB(μV/m).
- Altitude - 306.0 m;
- Frequency of radio stations;
- Maximum EIRP;
- Height of the antenna.

IMP in the frequency band above 108MHz are examined, since only they fall within the 108÷137 MHz frequency band assigned to civil aviation radio services:

- $2f_1 - f_2$ double-signal IMP; $f_1 > f_2$
- $f_1 + f_2 - f_3$ three-signal IMP; $f_1 > f_2 > f_3$

**Table 2.** Basic technical parameters.

| Radio station | Frequency (*MHz*) | Max. EIRP (*dBW*) | Height (*m*) |
|---|---|---|---|
| VESELINA | 106.30 | 28.0 | 38.0 |
| HR. BOTEV | 104.80 | 37.8 | 62.0 |
| BNR | 103.40 | 37.8 | 59.0 |
| RADIO 1 ROCK | 101.50 | 34.8 | 72.0 |
| HORIZONT | 100.90 | 44.6 | 62.0 |
| DARIK | 99.30 | 37.6 | 62.0 |
| CITY | 98.60 | 34.8 | 72.0 |
| BG RADIO | 97.80 | 34.8 | 78.0 |
| VERONIKA | 97.30 | 34.8 | 72.0 |
| RADIO 1 | 93.80 | 34.8 | 78.0 |
| RADIO ENERGY | 91.20 | 34.8 | 72.0 |
| FOCUS | 89.50 | 32.1 | 40.0 |

In the calculations are included:

- ILS systems located up to 125 km from the research site: Varna airport – 109.9 MHz; Burgas airport - 110.3 MHz. The requirement for electromagnetic compatibility (EMC) is ±200 kHz of frequency.
- Intermodulation products within the ILS range (108MHz ÷ 112MHz) using channels 108.10, 108.15, 108.30, 108.35MHz, etc. up to 111.70, 111.75, 111.90 and 111.95MHz do not comply with the EMC requirements:
- Intermodulation products within the VOR range (108MHz ÷ 112MHz) using channels 108.05, 108.20, 108.25, 108.40, 108.45 MHz, etc. up to 111.60, 111.65, 111.80, 111.85 MHz, also VOR frequencies occupying channels located at intervals of 50 kHz in the 112 ÷ 118MHz frequency band, as follows: 112.00, 112.05 ... 117.95 MHz do not comply with the EMC requirements.
- VOR systems located up to 278 km from the research VHF site - Varna airport - 109.9MHz, Burgas airport - 110.3MHz, Gorna Oryahovitsa airport - 113.6MHz, Ruse airport - 115.55MHz.

The measurement are carried out in accordance with the test staging presented in figure 4. IMP are measured after the output of the summing device (filter combiner) and before the feeder and antenna system.

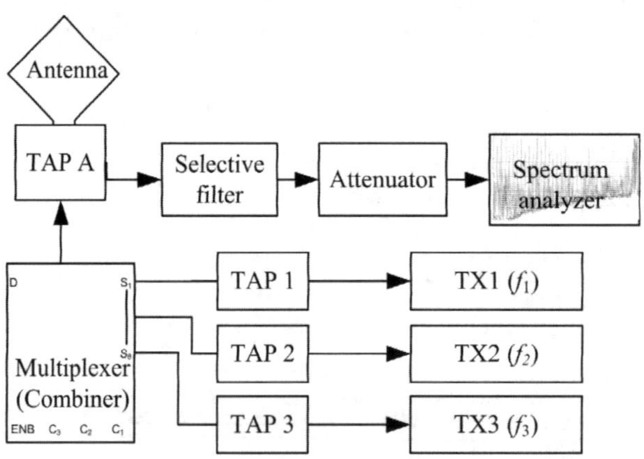


**Figure 4.** Test staging

The scheme consists of three transmitters, but the number may vary (*n*=2 or *n*=3), depending on whether a two or three signals A1 IMP are measured and perhaps other transmitters included in the same filter combiner, which are not involved in the product measured and may not be taken into account. An IMP can be measured from the interaction of few radio stations operating on two or three different filter combiners. In such as, the measurement of one A1 IMP shall be carried out sequentially in the outputs of the two or three filter combiners.

For the purposes of the study, a directional deflector, a selective band filter and a standard spectrum analyzer Keysight model "E4407B", with a frequency range from 9 KHz to 26.5 GHz, a power sensor Keysight type "U2000H", with a frequency range from 10 MHz to 18 GHz and equivalent loads are used.[15, 16]

The directional deflector emits a proportional portion of the energy supplied to the antenna. The band filter is set to the frequency of the A1 IMP, defined as incompatible, corresponding to the frequency of a particular VOR, ILS or COM for which the measurement is carried out. The filter achieves suppression of the main radiation of the transmitters to achieve the necessary measurement dynamics. Suppression of A1 IMP shall be determined by adding to the variance recorded by the spectral analyzer screen between the levels of a radio transmitter and the A1 IMP and the correction factors for the filter decay and the directional deflector "A" for the operating frequencies of the concerned transmitters.

For each IMP established, an amplitude frequency response (AFR) of the pass filter shall be made, including the transmitters that cause it. When setting the filter, the aim is to ensure minimal IMP suppression and maximum transmitted power.

Some of obtained results are presented in figures 5, 6, 7 and 8. On figure 5 is presented AFR for IMP frequency 112.40 MHz (marker 1), a combination of frequencies: marker 2 – 106.30 MHz, marker 3 – 103.4 MHz, marker 4 – 97.30 MHz, antenna 1 and antenna 2

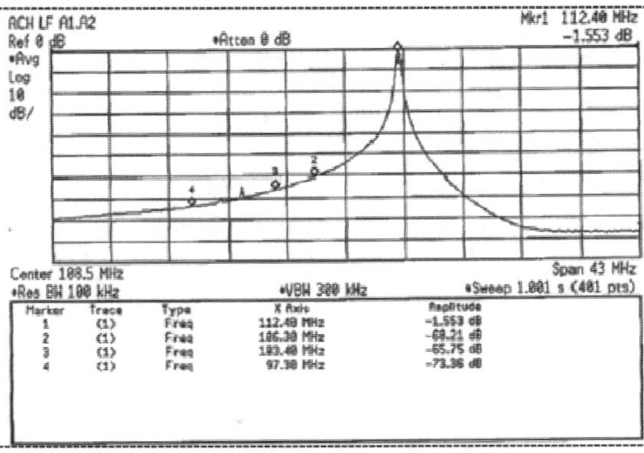


**Figure 5.** AFR for IMP frequency 112.40 MHz.

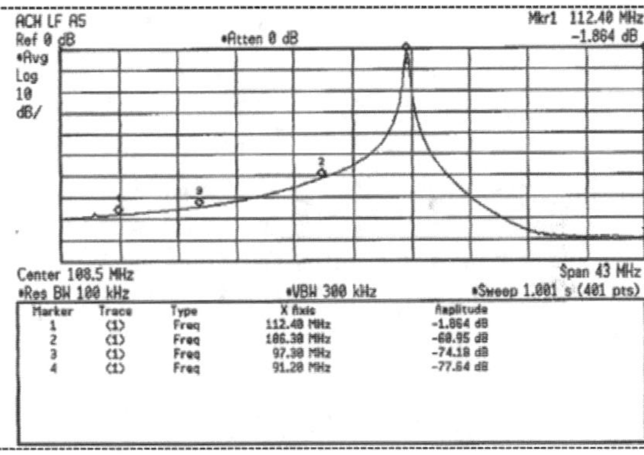


**Figure 6.** AFR for IMP frequency 112.40 MHz.

AFR for IMP frequency 112.40 MHz (marker 1), combination of frequencies: marker 2 – 106.30 MHz, marker 3 – 97.30 MHz, marker 4 – 91.20 MHz, antenna 5

Measurement for IMP frequency 112.40 MHz (marker 1), combination of frequencies: marker 2 – 106.30 MHz, marker 3 – 103.4 MHz, marker 4 – 97.30 MHz, antenna 2.

AFR for IMP frequency 123.10 MHz (marker 1), combination of frequencies: marker 2 – 106.30 MHz, marker 3 – 89.50 MHz, antenna 4 and antenna 5.

The carried out measurements show that there is no reported presence of type A1 IMP for the calculated frequencies from 109.9MHz to 120.2MHz. The presence of IMP type A1 for a frequency of 123.1 MHz belonging to the COM range has been reported. The measured 123.1 MHz IMP levels meet the requirements for permissible levels of interference signals (figure 8).

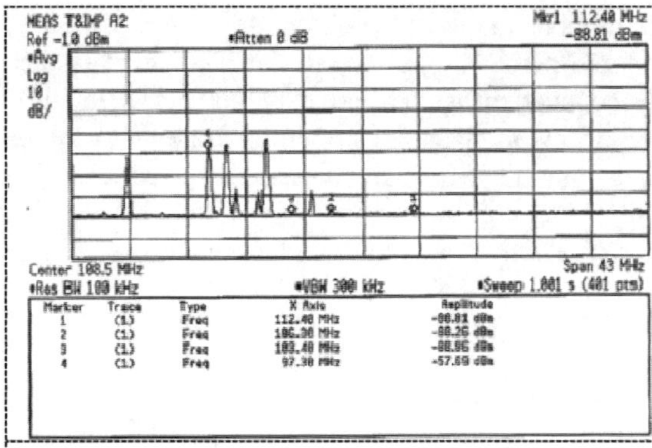


Figure 7. IMP frequency 112.40 MHz.

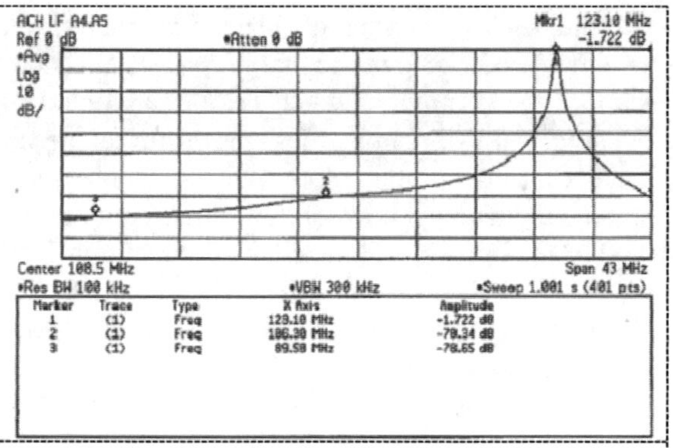


**Figure 8.** AFR for IMP frequency 123.10 MHz.

## Conclusion

IMP have an influence on the work of Instrument Landing System. A methodology has been proposed and a testbed for measuring IMP of type A1 has been proposed and studied. The experimental studies, carried out on-site, and the 3 antennas indicate that, there is no reported presence of IMP type A1 for the frequency range from 109.9MHz to 120.2MHz. The presence of type A1 IMP was recorded for only the 123.1 MHz frequency belonging to the COM range. The IMP levels for frequency 123.1 MHz meet the criteria of the EMC standards.

## References

[1] Iliev TB, Stoyanov IS, Sokolov SA and Beloev IH 2020 The influence of multipath propagation of the signal on the accuracy of the GNSS receiver, *43rd International Convention on Information, Communication and Electronic Technology (MIPRO),* Opatija, Croatia, pp 508-511, doi: 10.23919/MIPRO48935.2020.9245409

[2] Hegarty CJ, Bobyn D, Grabowski J and Van Dierendonck AJ 2020 An overview of the effects of out-of-band interference on GNSS receivers. *NAVIGATION*, 2020, 67, pp 143–161. https://doi.org/10.1002/navi.345

[3] Balabanova I, Georgiev G and Kostadinova S 2018 Artificial neural network for identification of signals with superposed noises, *Advances in Intelligent Systems and Computing*, **679** Springer, Cham., pp 488-495, https://doi.org/10.1007/978-3-319-68321-8_50

[4] Kogias P, Balabanova I, Malamatoudis M, Georgiev G and Sadinov S 2019 Reduction and identification of noise signals using artificial neural networks with various activation functions. *Journal of Engineering Science and Technology Review Special Issue on Telecommunications, Informatics, Energy and Management*, Kavala, Greece, 2019, pp. 89-93, ISSN: 1791-2377

[5] Wang Q, Su D, Jiang M and Xie S, 2006 A study on RF frequency optimization design system related to intermodulation interference, *4th Asia-Pacific Conference on Environmental Electromagnetics*, Dalian, pp 616-619, doi: 10.1109/CEEM.2006.258030.

[6] Handbook on radio frequency spectrum requirements for civil aviation, 2018 *ICAO spectrum strategy, policy statements and related information* I, 2nd edition, International Civil Aviation Organization

[7] Radio frequency interference best practices guidebook 2020, Cybersecurity and infrastructure security agency

[8] Maury microwave corporation, *Theory of intermodulation distortion measurement (IMD*), 1999, pp 1-3, Available at: https://www.maurymw.com/pdf/datasheets/5C-043.pdf

[9] Handbook on radio frequency requirements for civil aviation, Fifth Edition, 2009 (Doc 9718-AN/975), Available at: https://www.icao.int/safety/acp/ACPWGF/ACP-WG-F-22/ACP-WGF22-IP11-9718_5ed_unedited_version_en.pdf

[10] Methodology for measuring intermodulation products of type "A1", occurring during the operation of closely situated VHF-FM radio broadcasting stations, Available at: https://crc.bg/files/_en/Metodika_IMP1.pdf

[11] ITU-R SM.1009-1, *Compatibility between the sound-broadcasting service in the band of about 87-108 MHz and the aeronautical services in the band 108-137 MHz*, 10/1995, Available at: https://www.itu.int/dms_pubrec/itu-r/rec/sm/R-REC-SM.1009-1-199510-I!!PDF-E.pdf

[12] Rohde & Schwarz, Aeronautical radio navigation measurement solutions Application Note, Available at: http://www.rohde-schwarz.com/appnote/1MA193

[13] Acakpovi A, Tefutor I, Quist-Aphetsi K, Nwulu N, Sowah R and Abubakar R, 2019 Impact analysis of induced FM radio interferences on aeronautical radio navigation systems: case study of Kotoka international airport, Accra-Ghana, *International Conference on Computing, Computational Modelling and Applications* (ICCMA), Cape Coast, Ghana, 2019, pp 19-197, doi: 10.1109/ICCMA.2019.00011

[14] Skrypnik O 2019 *Radio Navigation Systems for Airports and Airways*, Springer Aerospace Technology

[15] https://www.keysight.com/us/en/assets/7018-01953/data-sheets/5989-9815.pdf

[16] https://www.keysight.com/us/en/assets/7018-01515/data-sheets/5989-6278.pdf

# An industrial IoT edge node for buffer level detection in a cardboard production line

**G Avramidis[1] and D Karampatzakis[1]**

[1] IEES Laboratory, Dept. of Computer Science, International Hellenic University, Kavala, 65403, Greece

E-mail: dkara@cs.ihu.gr

**Abstract.** Computer Science and Internet have evolved rapidly the last decades and equally impressive is the evolution of the Industrial Internet of Things technology into factories' shop floors. Among other technologies: modern CPU Architectures, Edge Computing, Deep Learning, Computer Vision, and Low Power Wide Area Networks, are playing a key role in this new competitive environment. In this paper, we present an Industrial IoT Edge Node for level detection on an overhead bridge conveyor (buffer) which is part of a 5-ply corrugated cardboard production line. We focused on the Edge Node and the development of the system was accomplished by using state of the art technologies from disciplines of computer vision and deep learning. We present two implementations using contour detection and CNN techniques. Finally, we implemented a LoRaWAN solution in the IIoT node to send alert messages to the control room. Experimental results are presented for the proposed system implementations.

## 1. Introduction

A new disruptive technology, the Industrial Internet of Things (IIoT), grows rapidly and its adoption is a challenge for industries worldwide. Based on claims of the research firm Meticulous (2020), the industrial IoT market is expected to grow at a CAGR of 16.7% from 2019 to 2027 to reach $263.4 billion by 2027. The IIoT applications requires to address new concerns about scalability, edge computing, communications, cyber security, and artificial intelligence.

Edge computing is an upcoming set of enabling technologies offering computational and storage resources at the edge of the network [1]. The increased number of IoT devices and applications produces huge amounts of data that needs processing closer to the data sources than to cloud infrastructures. In the IIoT environment the systems are time-sensitive and industrial applications generates enormous raw data which increases the network traffic causing high network latency and packet losses. So, an IIoT system is important to process data closer to the edge of the network performing complex computational tasks in limited hardware resources. Offloading a part of the workload to the edge in IIoT applications is an open research issue and several technological solutions are in progress. In the Industrial IoT because of the complex system characteristics the edge node is not only the producer of the data, but it is critical to be also the consumer of this data. In this context, Deep Learning [2] is a promising approach for data processing in proximity to the data source and succeeding computational offloading to the edge. Deep Learning uses big data to study complex data and in an IoT application could be more efficient compared to machine learning because it is able to exploit large data produced by IoT devices. Another important feature of Deep Learning is that intermediate data extracted from inner neural network layers is reduced compared to input data size which reduces network traffic to cloud servers and has different semantics which ensures privacy preservation. So, implementing Deep Learning techniques to the IoT edge layer (End Node or Gateways) is critical for industrial IoT applications and various solutions are demonstrated in recent literature. In the field of IoT communications one on the most widespread Low Power Wide Area Networks (LPWAN) is LoRaWAN [3]. LoRaWAN is an open MAC specification which uses the LoRa PHY. LoRa PHY is a proprietary radio technology of Semtech implementing a Chirp Spread Spectrum (CSS) modulation. In a typical start topology, a LoRaWAN consists of one or more LoRa End Node(s), one or more LoRaWAN Concentrator/Gateway(s), a LoRaWAN Network Server (NS) and Application Server(s) (AS). Depending on the application End Nodes and Gateways are at the edge of the network and NS and AS at the cloud. Also, LoRaWAN is a low power radio technology which extends End Nodes' battery life and provides two layers of security (network and application) using AES encryption.

In the industrial domain except for the big and multinational companies, in developing countries, many small to medium factories will implement IIoT solutions to improve their productivity and competence. In this paper, we present a low-cost IoT solution at the edge of the network which implements deep learning techniques to solve a specific industrial case. In more detail, we observe the level of flowing paper on the overhead bridge conveyor (buffer) which is part of a 5-ply corrugated cardboard production line. Our approach implements an edge device equipped with a camera that performs video sensing of a standard frame of buffer, implements a deep learning CNN to detect alerting states, and uses LoRaWAN to transmit alerts to the control room.

The remainder of this paper is organized as follows. In the next section we present related work. In Section 3 we present the architecture of our system and in Section 4 we provide more details about the two different implemented methods. In Section 5 we present experimental results and finally in the last section we conclude this article.

## 2. Related Work

Edge-based Deep Learning in Industrial Internet of Things is a hot issue in research and many approaches for edge offloading are presented in literature. The classical IoT applications with cloud offloading are not ideal for critical and time-sensitive industrial applications because data produced from end nodes needs massive data exchange between servers and IoT devices [4]. Also, because of the resource-constrained end nodes novel frameworks are presented for cooperative edge computing [5]. In the proposed approaches running deep learning models at the edge is challenging because of the limited computing capabilities of edge nodes. Nguyen Gia et al. [6] proposed a 5-layer network framework for smart farming applications with the following layers: Sensor, Edge, Fog, Cloud, and Terminal. In this paper, the authors are using CNN techniques at the edge gateway (Raspberry Pi3) to compress data produced at the end nodes. This approach reduces data transfer to cloud servers and make a LoRaWAN implementation feasible. A. Seferagic et al. in their review paper [7] use International Society of Automation (ISA) industrial systems classification and after their research conclude that LoRa networks, as a single-hop long range network, could be applied to non-critical IIoT systems of 4-6 classes. Based on ISA classification such kind of systems are: Open loop systems, alerting systems and information gathering systems. Additionally, Filho et al. [8] present the performance of LoRaWAN for co-existing telemetry and alarm messages in industrial applications. As LoRaWAN is a pure ALOHA network and packet collisions is one of the biggest issues, the authors presented three configurations of alarm topology using their Spreading Factor (SF) allocation schemes and test them using system level simulation (ns3). Simulation results show that it is possible to achieve 100% delivery of alarm messages in industrial environments but need to design a LoRa network with custom made configurations to have increased network performance. In applied domain, Bhattacherjee et al. [9] implemented an emotion detection IoT enabled edge-node for citizen security in a smart city environment. In this work, the authors present an edge-node running a trained CNN algorithm for face recognition on a Raspberry Pi3, detects facial emotions, and in case of distress situations sends an alarm signal via LoRaWAN. Finally, D. Ziouzios and M. Dasygenis [10], proposed a smart recycling bin which detects recycled materials captured by a camera module attached to the entry area of the bin and communicates with LoRaWAN. In this work the authors implemented the Convolutional Neural Network (resnet34 CNN) for image classification using a Single Board Computer (SBC) with an extra

FPGA SoC hardware (Xilinx Pynq-Z1). FPGAs is a promising hardware technology for executing computing tasks closer to the edge.

## 3. System Architecture

Our system architecture is presented in figure 1. The proposed IIoT system is attached to the frame of overhead bridge conveyor (buffer). The Edge node is composed of a Single Board Computer (Raspberry Pi3 Model B - ARM Cortex A53 – 1GB SDRAM) to run the trained CNN model (LeNet-5), a USB web camera (Microsoft LifeCam HD-3000) for video sensing, and also the SBC has an attached module of a LoRa Radio (Dragino LoRa/GPS for Raspberry) for wireless connectivity. In our experiments we used an outdoor RAK Wireless 7258 – EU868 LoRaWAN Gateway which communicates via internet with The Things Network (TTN) LoRaWAN Server. Finally, the Edge node system has a wall mount power supply.

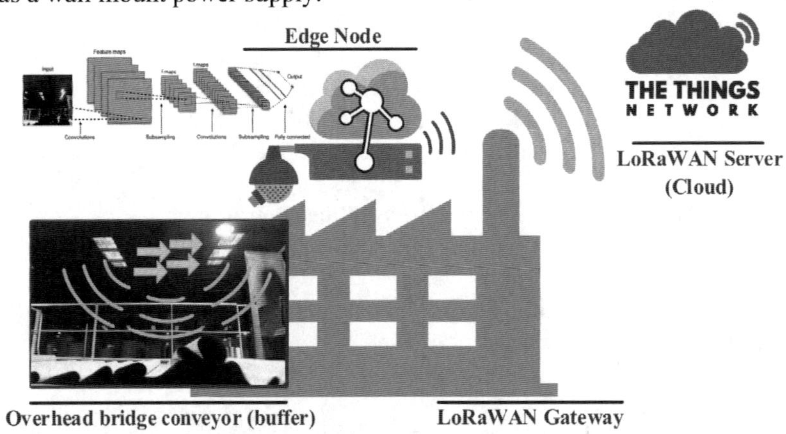


**Figure 1.** The overall architecture of the proposed Industrial IoT system.

## 4. Implementation

In our IIoT system, warning state detection was implemented with two different techniques: a) Contour detection using OpenCV and b) Deep learning using LeNet-5 CNN. In more detail, the two implementations are described in the following subsections.

### *4.1. Method A – Contour Detection*

The Method A is based on contour detection using OpenCV library. The basic concept is that, when there is enough paper on the overhead bridge conveyor (buffer), the level of paper mass is upper the pre-defined limit and when paper mass is lower from the limit, we have a warning state. This can be verified because as more paper is accumulated on the buffer, it will produce bigger layers of paper mass, that will sit one on top of the other. Responsible for the height detection is a USB camera that is connected to a Raspberry Pi3 computer. The USB camera is mounted in way that it captures the flow of paper.

Our system, in order to detect if there is paper in a specific area, tries to detect the colour of the paper in the specific area. So, at the start, the system asks the user to click on a pixel of the camera frame that is presented to the user using the RGB colour space. The next step is to increase the effectiveness of the system, regarding the colour detection, against the variations of the light during video capture from the camera. So, the system "moves" the frame from RGB colour space to HSV colour space, and as the user clicks the pixel with the wanted RGB values, it creates a range of HSV values that correspond to the wanted colour of the pixel. In other words, the system does not try to detect one specific only colour but a family of different shades of the first colour. At this point, the system can detect the wanted colour more easily, even with variations of light during capture. This colour detection is based on the use of a mask on the origin frame to produce a binary image. In order to create this mask, the system needs the upper and the lower HSV values from the colour range that was created in the previous step. Also, it must be noticed, that before the use of the mask the system performs a smoothing operation on the frame to eliminate some strong edges of the frame. As the mask has been used, and the binary image is created, the system performs morphological operations of erosions and dilations on the binary image of the frame. As a result, a modified binary image is created. In this modified binary image, the system will try to detect contours, that eventually correspond to the wanted colour.

*4.2. Method B – CNN Deep Learning*

The Method B is based on frame classification using the Lenet-5 Convolutional Neural Network. More specifically, this approach consists of the following parts.

**CNN model:** The Lenet-5 model is considered as one of the very first convolutional neural networks for deep learning as it has six layers. Among these layers there are 3 convolution layers, 2 sub-sampling layers and 1 fully connected layer. Finally, there is one output layer. The model was first used for handwritten digits classification. It takes a picture of small dimensions (e.g. 32 x 32 or 28 x 28) and it provides a classification label for this input image as an output. Deep learning models usually need large amount of data for better training.

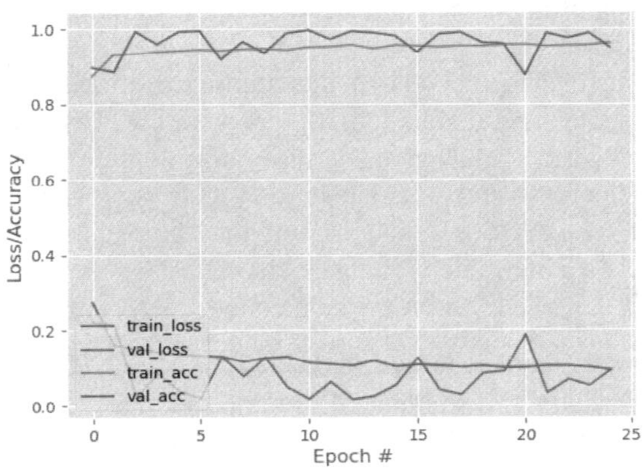


**Figure 2.** Training Loss and Accuracy on Warning State / Normal State.

**Supervised training:** As was mentioned above, the frames that fed our model were taken using a USB camera that was connected to a Raspberry Pi computer. The position of the camera was fixed but there was no need for a specific view angle. For the training process there were 4.840 frames captured, from actual paper flow on the moving belt. The operator of the machine labeled each frame as warning frame or as normal frame according to his specialized decision. The training process took approximately 15 minutes using an i7 CPU with 16GB of RAM. For comparison purposes the training was also executed on a Raspberry Pi 3 Model B and was completed after 105 minutes.

**Figure 3.** Warning and Normal states with high confidence.

**Learning Rate / Epochs / Batch size:** Regarding the training procedure, we should refer that we used 25 epochs to train our model, meaning that the whole training data set was loaded into our model

25 times. The batch size, which is the segment of the training set that is loaded into the model at each step, was set to 32. The learning rate was set to 0,001, which means that the weights of the model are being updated by 0,001 during the training procedure.

For network model implementation, training, and usage, the Keras and Tensorflow libraries are needed to be used in Python environment. In both methods, when the system detects 500 consequent warning frames the machine status is characterized as warning status (figure 3) and the system informs the operator to check the paper flow to prevent pausing of production machinery. In general, both methods succeed in detecting warning states.

## 5. Results

As it is shown in table 1, Method A seems to have difficulties in detecting frames of normal state, in comparison with Method B. This can be justified considering the difficulty of the detection of the contour above the line limit. Even if the HSV values of the colour that the user has selected, are expanded, it is still insufficient for the perfect detection of the contours inside the wanted area of the frame. On the other hand, Method B seems to analyse patterns from the frames and jumps to semantic conclusions during the training of the model. The process of the training procedure is shown in figure 2. It should be noticed that during the early epochs of the training the train loss is high and the train accuracy is low while during the last epochs of the training the model has achieved higher train accuracy and lower train loss, meaning that its training procedure is normal and efficient. This pattern analysis has as a result, Method B, to be independent of lighting conditions, color selection and the use of a limit line. Regarding the process time, it is shown that there is no significant difference for small duration videos (Video 1 and 2 have 60sec length) between the two Methods. However, for bigger videos (e.g. 240 sec) there is a noticeable difference between 10% and 17%. This can be justified because Method A uses procedures (Gauss, binary conversion, color space change etc.).

**Table 1.** Comparison of proposed status detection methods on i7 / 16GB.

| | | **Method A : Contour** | | | | **Method B : LeNet-5** | | | |
|---|---|---|---|---|---|---|---|---|---|
| | **Total Frames** | Normal State Frames | Warning State Frames | Time (sec) | Warnings | Normal State Frames | Warning State Frames | Time (sec) | Warnings |
| **Video 1** | **1.800** | 420 | 1.380 | 64 | 0 | 933 | 867 | 60 | 1 |
| **Video 2** | **1.800** | 682 | 1.118 | 63 | 0 | 1.711 | 89 | 62 | 0 |
| **Video 3** | **7.200** | 2.221 | 4.979 | 270 | 4 | 4.184 | 3.016 | 230 | 5 |
| **Video 4** | **7.200** | 2.816 | 4.384 | 255 | 2 | 4.829 | 2.371 | 229 | 4 |

In table 2 it is shown the noticeable difference regarding the execution time of the video on the Raspberry Pi3 SBC, with and without, the display of the frames, after they were classified by the model of Method B. The display of the frames causes an increment to the execution time of 50 to 70%. This shows that a CNN model can be executed on SBC and classify its input successfully, but when it comes to images, it is better not to force their display as this procedure has a high computational cost.

**Table 2.** Results of running trained CNN model on SBC.

| | | **Displaying Output Frames** | | | | **Not Displaying Output Frames** | | | |
|---|---|---|---|---|---|---|---|---|---|
| | **Total Frames** | Normal State Frames | Warning State Frames | Time (sec) | Warnings | Normal State Frames | Warning State Frames | Time (sec) | Warnings |
| **Video 1** | **1.800** | 936 | 864 | 163 | 1 | 936 | 864 | 96 | 1 |
| **Video 2** | **1.800** | 1.710 | 90 | 162 | 0 | 1.710 | 90 | 95 | 0 |
| **Video 3** | **7.200** | 4.190 | 3.010 | 609 | 5 | 4.190 | 3.010 | 385 | 5 |
| **Video 4** | **7.200** | 4.838 | 2.362 | 621 | 4 | 4.838 | 2.362 | 418 | 4 |

## 6. Conclusion and future work

This work presents an Industrial IoT system for buffer level detection in a cardboard production line. The system combines a low-cost SBC equipped with a USB camera, using Computer Vision methods and a LeNet-5 CNN model for warning state detection and communicates via a LoRaWAN network. Our experience from the implementation and the experimental results approves that both Methods A and B can solve the detection problem by analysing the frames from the USB camera. Method A requires the use of several procedures that have to do with image processing (filters, color space, etc.) and OpenCV library makes this task easier, while Method B uses a trained CNN model. Its supervised training needs time and processing power. In comparison with Method A, Method B gives better results, as during the training of the model, the pattern of the frame is analyzed, and semantic conclusions are made. It should be noticed that Method B is more robust as there are no tight restrictions regarding the camera capture position. The main requirement is the training procedure of the model with a training data set taken from real time data. Because of Covid-19 restrictions it was not feasible to test the system in real conditions, so for further work we will plan to test our system in the industrial shop floor and also to test more CNN models and SBCs.

## References

[1] Shi W, Cao J, Zhang Q, Li Y and Xu L 2016 Edge computing: vision and challenges. *IEEE Internet of Things Journal*, **3** (5), pp 637-646, doi: 10.1109/JIOT.2016.2579198

[2] Li H, Ota K and Dong M 2018 Learning IoT in edge: deep learning for the Internet of Things with edge computing. *IEEE Network*, **32** (1), pp 96-101, doi: 10.1109/MNET.2018.1700202

[3] Haxhibeqiri J, De Poorter E, Moerman I and Hoebeke J 2018 A survey of LoRaWAN for IoT: from technology to application, *Sensors*, **18** (11), 3995, https://doi.org/10.3390/s18113995

[4] Liang F, Yu W, Liu X, Griffith D and Golmie N 2020 Toward edge-based deep learning in industrial Internet of Things. *IEEE Internet of Things Journal*, **7** (5), pp 4329-4341, doi: 10.1109/JIOT.2019.2963635

[5] Galanopoulos A, Iosifidis G and Salonidis T 2019 Cooperative analytics for the Internet of Things *Twentieth ACM International Symposium on Mobile Ad Hoc Networking and Computing (Mobihoc'19)*, New York, NY, USA, pp 395–396. https://doi.org/10.1145/3323679.3326631

[6] Gia T N, Qingqing L, Queralta J P, Zou Z, Tenhunen H and Westerlund T 2019 Edge AI in smart farming IoT: CNNs at the edge and fog computing with LoRa. *IEEE AFRICON*, Accra, Ghana, pp 1-6, doi: 10.1109/AFRICON46755.2019.9134049

[7] Seferagić A, Famaey J, De Poorter E and Hoebeke J 2020 Survey on wireless technology trade-offs for the industrial internet of things, *Sensors* **20** (2) 488

[8] Santos Filho F H C, Dester P S, Stancanelli E M G, Cardieri P, Nardelli P H J, Carrillo D, Alves H 2020 Performance of LoRaWAN for handling telemetry and alarm messages in industrial applications, *Sensors*, **20** (11) 3061

[9] Bhattacherjee S S, Kumar N T S and Rajalakshmi P 2019 Emotion detection IoT enabled edge-node for citizen security, *IEEE 5th World Forum on Internet of Things (WF-IoT)*, Limerick, Ireland, pp 925-930, doi: 10.1109/WF-IoT.2019.8767173

[10] Ziouzios D and Dasygenis M 2019 A smart recycling bin for waste classification, *Panhellenic Conference on Electronics & Telecommunications (PACET)*, Volos, Greece, pp 1-4, doi: 10.1109/PACET48583.2019.8956270

## Acknowledgments

This work was supported by the MPhil program “Advanced Technologies in Informatics and Computers”, hosted by the Department of Computer Science, International Hellenic University, Greece.

# 6LoWPAN network analysis using simulations and experiments

**A M Haka[1], V P Aleksieva[1], H G Valchanov[1]**

[1] Faculty of Computer Sciences and Automation, Technical University of Varna, Varna 9010, Bulgaria

E-mail: aydin.mehmed@tu-varna.bg

**Abstract.** The constant expansion of Internet consumption increases the communication traffic on the network both locally and globally. To meet the needs of such consumption, technologies such as 4G, 5G and the Internet of Things (IoT) are increasingly being developed. Remote monitoring of environmental parameters or the condition of patients in real time can be organised using the capabilities of IoT networks. This paper presents a study of end-to-end delay and throughput in a 6LoWPAN IoT network through real network construction and simulation.

## 1. Introduction

According to Ericsson mobility report from June 2020 [1], the coronavirus crisis has caused an unprecedented number of people to continue their work remotely via the Internet. Thus, the levels of traffic on the Internet have increased. New generation communications such as 4G, 5G [2, 3, 4] and the Internet of Things (IoT) can be relied upon to successfully maintain such levels of traffic on the global network [5]. Global pandemics such as coronavirus may require remote control and monitoring of various indicators of both patients and the environment. This can be achieved through new IoT technologies such as 6LoWPAN, which allows communication over an IP network to transmit data from sensor nodes such as temperature, atmospheric pressure, humidity, ambient light and more. Continuous monitoring of various parameters of sensor networks requires the provision of good values for indicators such as throughput, end-to-end delay, packet loss, latency, etc. This can be achieved by studying a real network infrastructure or by simulating to study the impact of environmental interference, implemented algorithms or mechanisms on the operation of the network.

This paper presents the physical deployment of 6LoWPAN network and the study of throughput and end-to-end delay indicators, which compared with the results obtained through the 6LoWPAN simulation product presented in [6].

## 2. Related works

The study at [7] examined the neighbor discovery model and IPv6 frame in 6LoWPAN using the Contiki (Cooja) simulator. The experiments performed with 3, 5 and 7 sensor nodes. The studied parameters are throughput and packet loss. The aim of the study is to predict performance and characterise the proposed real network topology for 6LoWPAN.

In [8] the aim is to form an open, interoperable, scalable, reliable and low power wireless sensor network stack for 6LoWPAN. This stack is simulated through Contiki's Cooja simulator to studying performance under various conditions such as noise, topology, traffic, etc. For validation the simulated stack, a physical 6LoWPAN network was built using the Raspberry Pi as a 6LoWPAN border router with T-mote sky sensor nodes. The performance study bases on the values for throughput, latency and packet loss. The studies were performed for a 6LoWPAN network with 1 coordinator and 5 sensor nodes.

In [9] evaluates the performance of the communication stack containing CoAP and 6LoWPAN over the IEEE 802.15.4 radio link. The research realised through Contiki's Cooja simulator for 6LoWPAN. The parameters throughput, packet loss and end-to-end delay studied. The simulations realised for a 6LoWPAN network with 1 coordinator and 5 sensor nodes.

Similar to the researches considered, this paper presents studies of the 6LoWPAN sensor network through simulation and real deployment. The simulations realised through the product proposed in [6]. To validate and compare the simulation results, the same tests were performed over a real 6LoWPAN network using a BeagleBone Black board for border router, Texas Instruments (TI) CC2531EMK transceiver and TI CC2650STK sensor nodes.

## 3. 6LoWPAN simulation scenario and results

The tests for throughput and end-to-end delay from the simulator were obtained after building a 6LoWPAN network by one coordinator and 6 6LoWPAN sensor nodes connected in a star topology (figure 1). The coordinator configured to work on channel 25. Up to 6 end 6LoWPAN sensor nodes can be connected to the coordinator. All end nodes are static, perform the same type of application and located at the same distance from the coordinator (from 1m to 5m). The tests for reporting the values for throughput and end-to-end delay were made with 2, 4 and 6 sensor nodes connected to the 6LoWPAN coordinator. Once the coordinator and end node information added, a simulation performed to send a certain number of packets (figure 2-1). After adding the packets to the send queue (figure 2-2), the calculated values (figure 2-3) for end-to-end delay and throughput displayed.

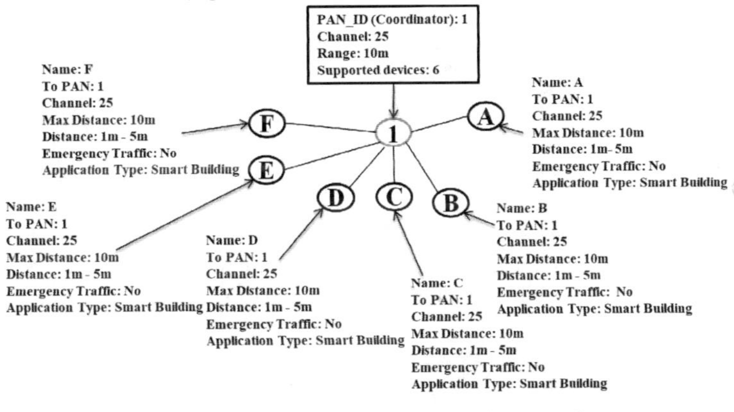


**Figure 1.** Used 6LoWPAN simulation topology.

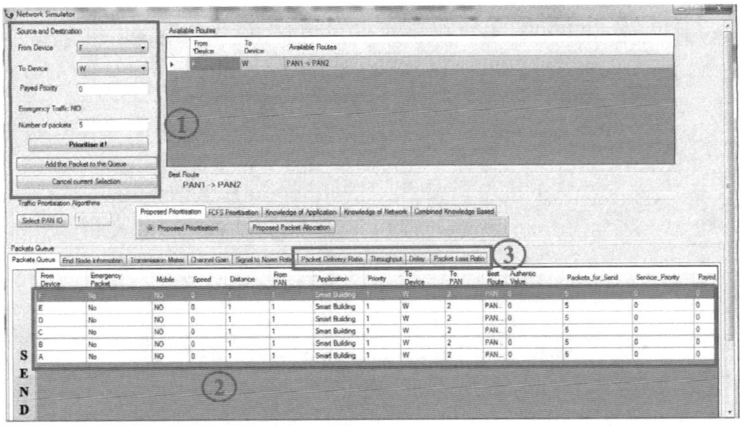

**Figure 2.** Simulation of 6LoWPAN communication.

The results of the conducted experimental studies are in large numbers, so they are summarised and presented in table 1. Since the simulation product considers tests in ideal conditions, at different distances the values obtained are identical. The difference in the experiments performed is manifested in relation to the different number of sent packets.

**Table 1.** Results for end-to-end delay and throughput according test by the 6LoWPAN simulator.

Results for End-to-End Delay

2 End Nodes

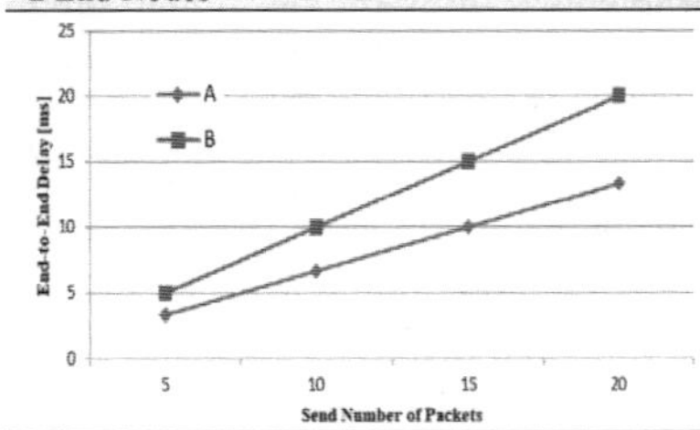


4 End Nodes

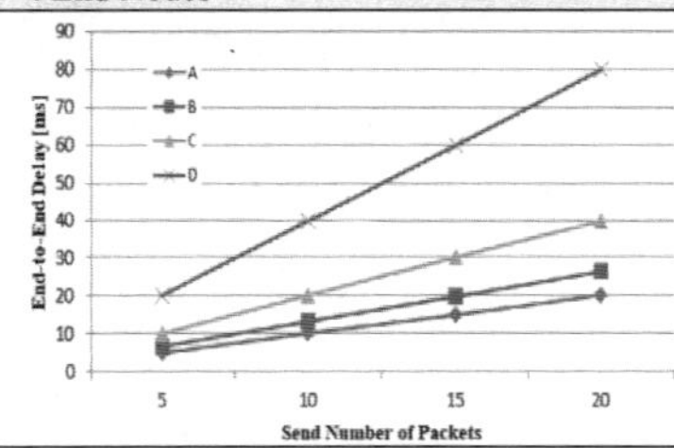


6 End Nodes

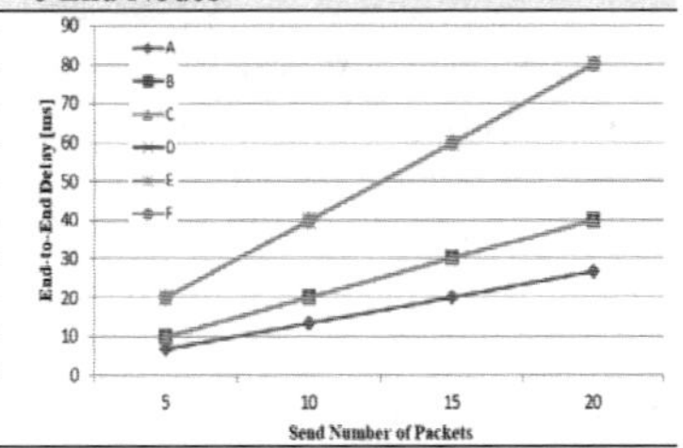


Results for Throughput

2 End Nodes

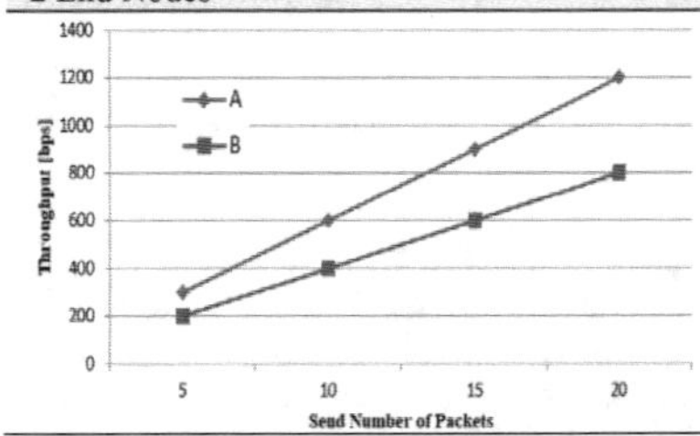


4 End Nodes

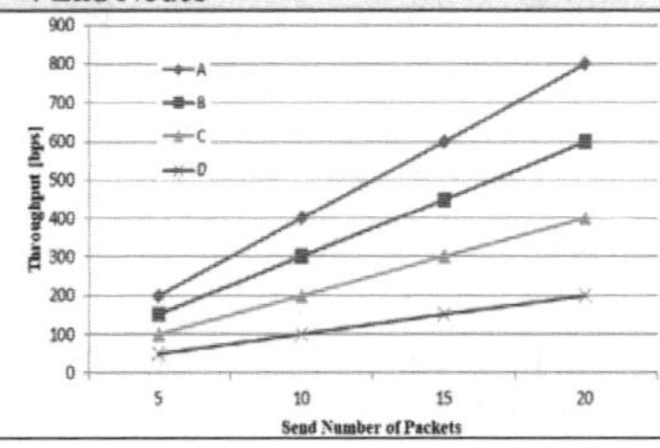


6 End Nodes

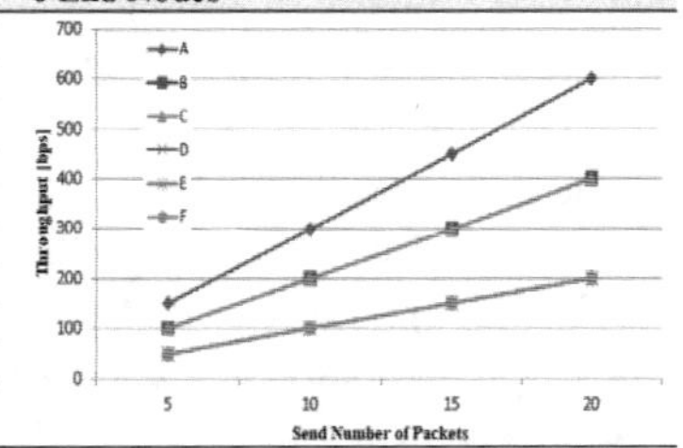


## 4. 6LoWPAN physical deployment, experimental scenario and results

The physical building of the 6LoWPAN network is done with BeagleBone Black – BBB01-SC-505 [10] board with Bone-Debian-9.9 operating system working as 6LoWPAN Gateway, TI transceiver - CC2531EMK [11] and TI multi-standard sensor nodes – CC2650STK [12]. The 6LoWPAN Gateway configured using Contiki Cetic-6lbr version 1.4.0 [13, 14]. The CC2531EMK board configured to operate as a 6LoWPAN transceiver and sensor nodes to operate in 6LoWPAN network using CC-DEVPACK-DEBUG of TI. The data transfer and the receipt number of bits from the end sensor nodes in the already built 6LoWPAN network can be tracked, when a second CC2531EMK transceiver configured to work as a 6LoWPAN sniffer. This can be done on a Linux machine using Sensniff program for 6LoWPAN [15]. The program is downloaded and installed, and at startup a traffic flow is created that can be visualised through Wireshark (figure 3). Packet source, destination and number of bits send performed by the IEEE address of the source and destination from the Wireshark output.

To ensure comparability of the results obtained by the simulator with those of the real network, the tests performed with the same topologies. According to figure 4, physical star topology realised, where **1** is 6LoWPAN coordinator and transceiver, and **2** are 6LoWPAN sensor nodes. Table 2 shows the results for end-to-end delay and throughput obtained through realised 6LoWPAN network. The results calculated based on the size of the sent and received packets between the sensor nodes and the coordinator in number of bits, as well as the delays received when transmitting 5, 10, 15 and 20 ICMPv6 packets. Situations with 2, 4 and 6 simultaneously connected in a 6LoWPAN network end sensor nodes were studied. The tests were performed by simultaneously sending ICMP packets to the end sensors. The values for end-to-end delay (1) and throughput (2) calculated similarly to [8].

$$\text{End\_to\_End Delay [ms]} = P + 2 * (L_{rq} + L_{rs}) \quad (1)$$

$$\text{Throughput [Kbps]} = \frac{N * (L_{rq} + L_{rs})}{T} \quad (2)$$

IOP Conf. Series: Materials Science and Engineering 1032 (2020) 012015 doi:10.1088/1757-899X/1032/1/012015

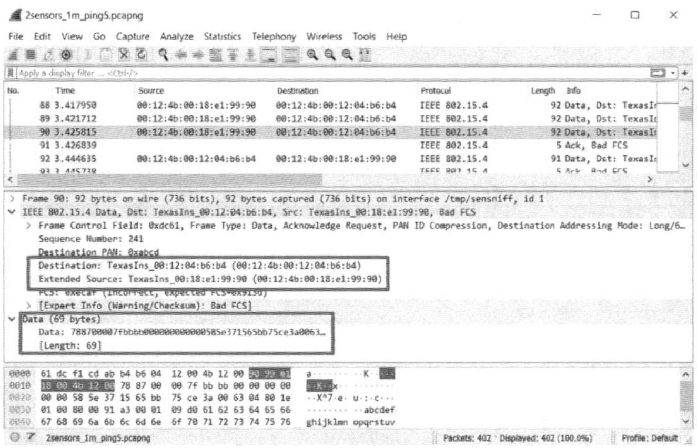


**Figure 3.** Captured 6LoWPAN packets through Wireshark.

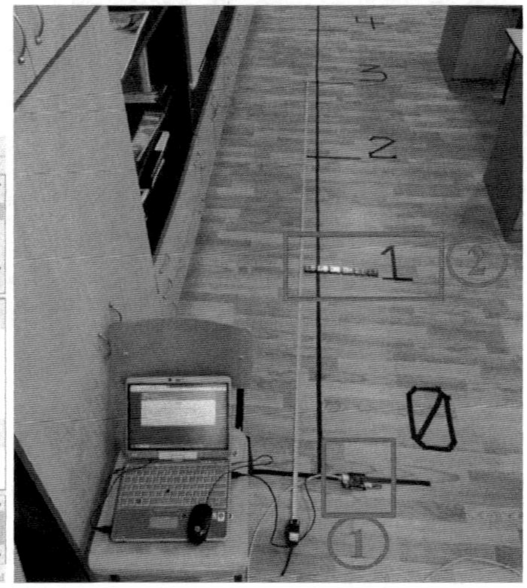

**Figure 4.** 6LoWPAN physical topology.

Where P is all processing delay, Lrq is length of request, Lrs is length of response, N is number of successful ICMP request/response pairs and T is total time of simulation. Throughput values converted from Kbps to bps.

According to the results of simulation with 2 sensors, the values for end-to-end delay and throughput are the best, as the resources from one IEEE 802.15.4 frame are distributed only between two devices. As the number of sent packets increases, there is a deterioration of the studied parameters. This is because more resources are needed to send more packets. The deviation of the results for end-to-end delay in simulation from the real ones is average 99% for 5, 10, 15 and 20 sent packets, and in the values for throughput for 5 sent packets is 97%, for 10 packets is 90%, for 15 packages is 80% and 20 packages is 71%.

The results for 4 sensor devices in the simulation show better values for the devices whose requests have a higher priority. This is due to the algorithm used to prioritise traffic and the mechanism for allocating resources to service requests. The tendency for deterioration of the studied parameters with increasing number of request is maintained. The deviation of the simulation results for end-to-end delay from the real ones at 4 end devices is average 99% for 5 and 10 sent packets and 98% for 10 and 15 packets, and in the values for throughput for 5 packets it is 98%, at 10 packages is 93%, 15 packages is 88% and 20 packages is 80%.

The results for 6 sensors in simulation show the highest values for the devices with the highest priority requests, and lower for the others. Some of the devices with lower priority requests have the same values for the studied parameters. As the number of sent packets increases, the values for the examined parameters deteriorate. The deviation in the simulated results with 6 sensors from the real ones for end-to-end delay is average 99% for 5, 10, 15 and 20 sent packets, and for throughput is 98% for 5 packets, 94% for 10 packets, 86% for 15 packets and 79% at 20 packets.

The results for end-to-end delay from the simulation differ by a very large percentage from the actual ones, regardless of the number of packets sent. In the results for throughput it is observed that with the increase of the number of sent packets the results obtained from the simulation are getting closer to the real ones, but nevertheless there is a large difference. This is because the simulation product does not take into account the presence of different noise effects in the communication environment and interference between nodes in the network.

The trend in the results of simulation and tests in a real network show that when the network is loaded, the studied parameters deteriorate.

**Table 2.** Results for end-to-end delay and throughput according test by the 6LoWPAN network.

| Node № | Results for End-to-End Delay | | | |
|---|---|---|---|---|
| | 5 ICMPv6 Packets | 10 ICMPv6 Packets | 15 ICMPv6 Packets | 20 ICMPv6 Packets |
| 2 | 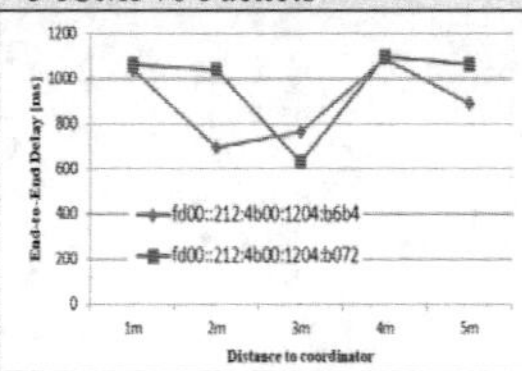 | 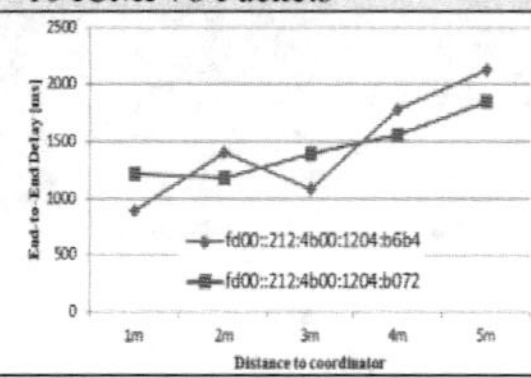 | 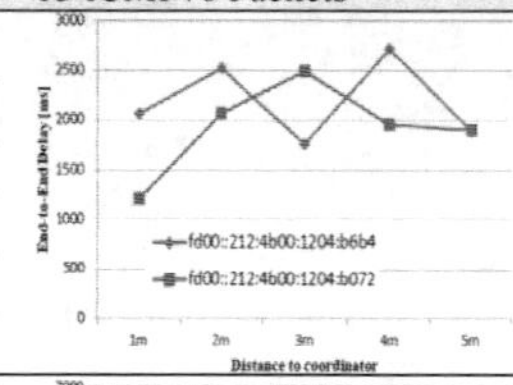 | 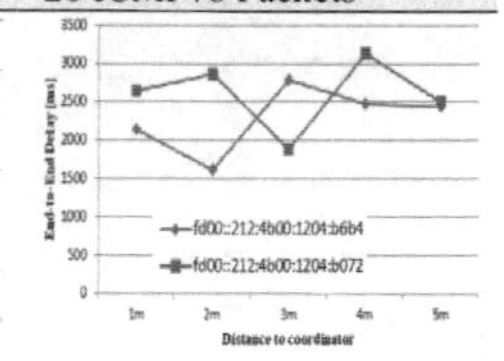 |
| 4 | 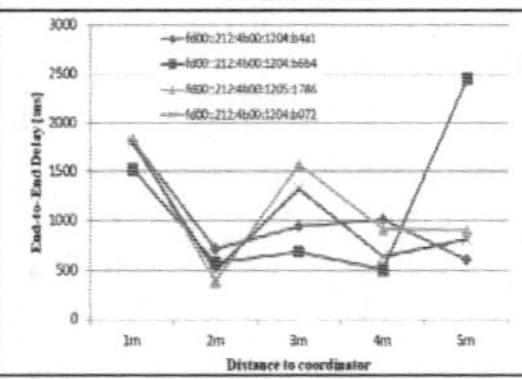 | 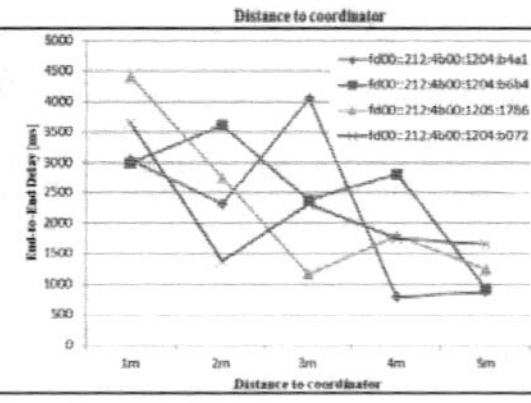 | 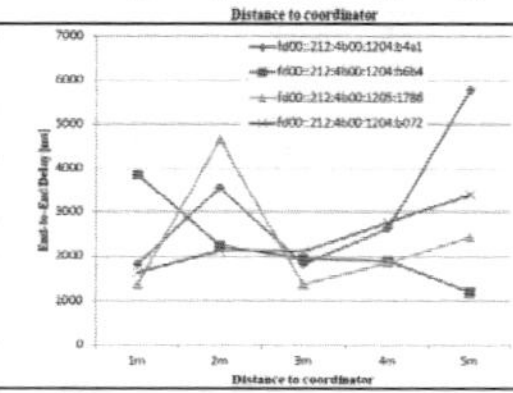 | 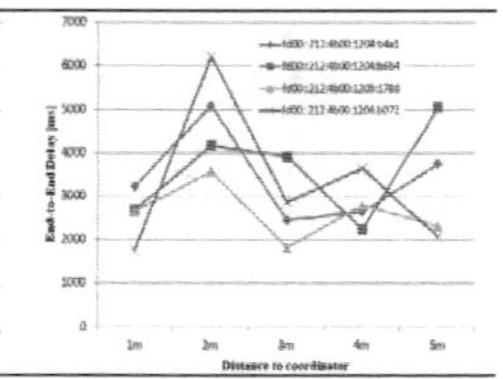 |
| 6 | 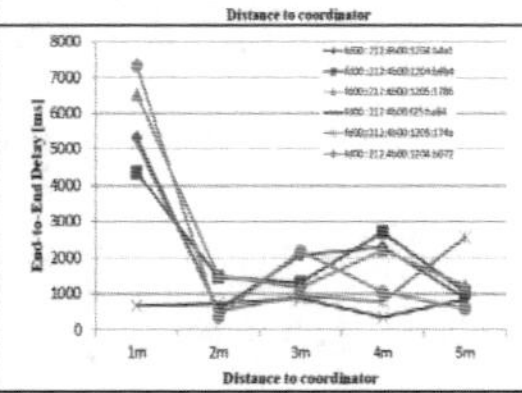 | 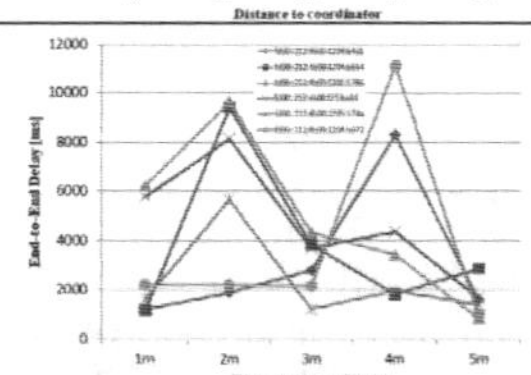 | 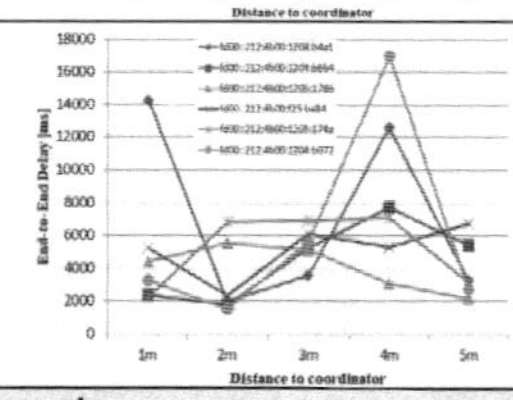 | 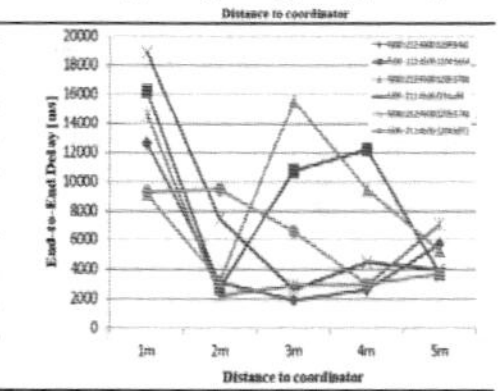 |
| Node № | Results for Throughput | | | |
| | 5 ICMPv6 Packets | 10 ICMPv6 Packets | 15 ICMPv6 Packets | 20 ICMPv6 Packets |
| 2 | 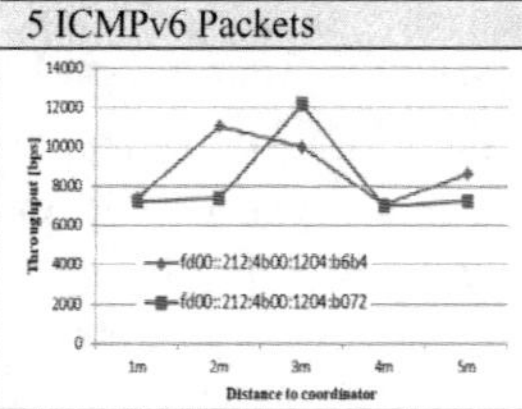 | 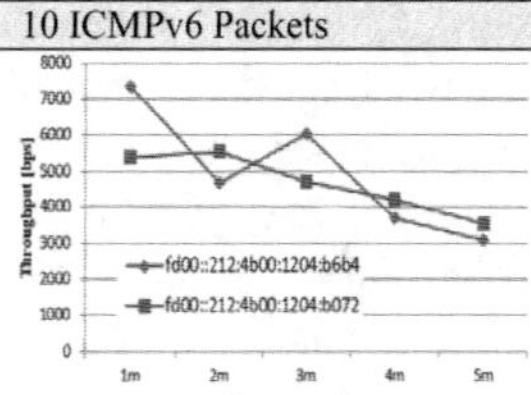 | 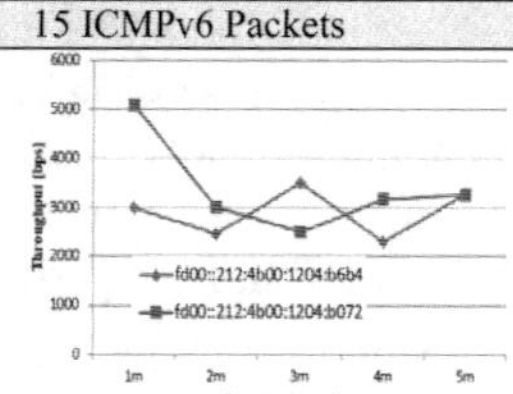 | 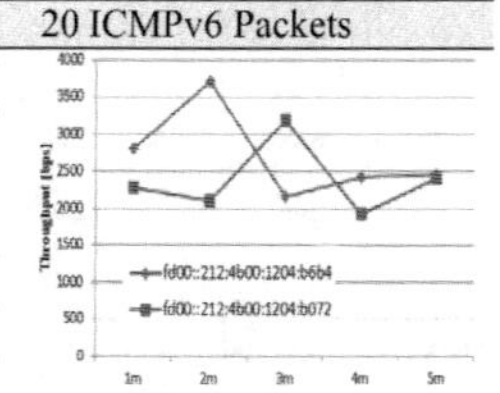 |
| 4 | 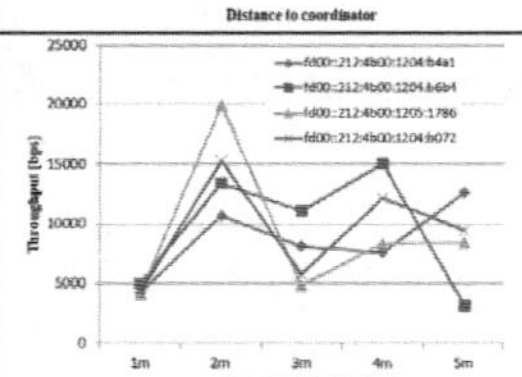 | 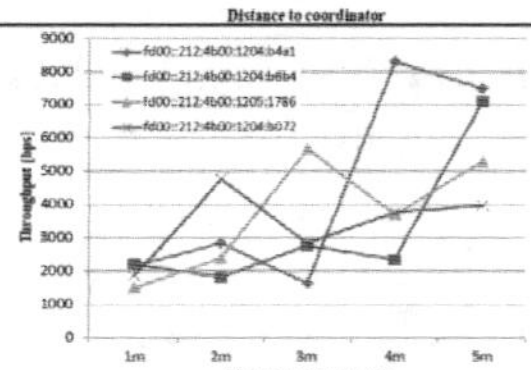 | 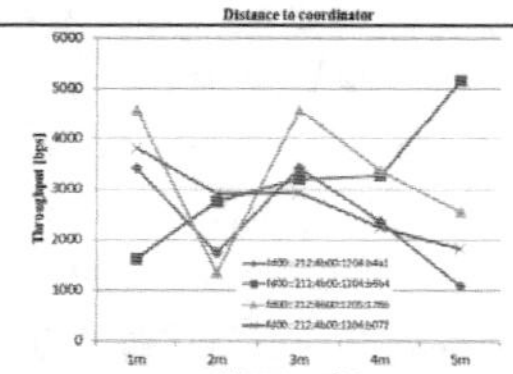 | 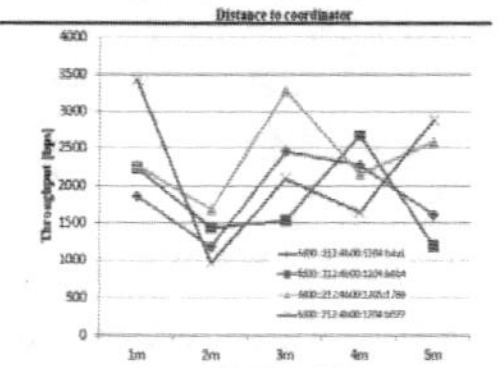 |
| 6 | 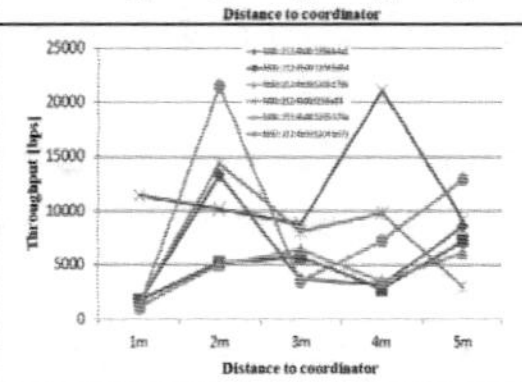 | 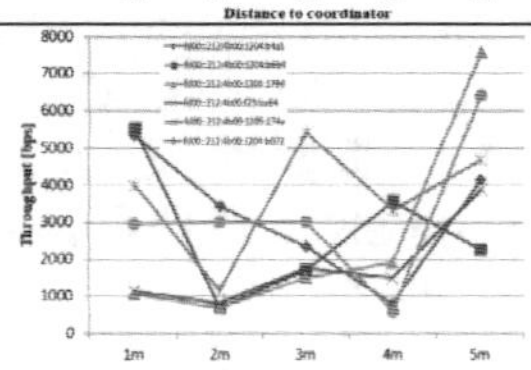 | 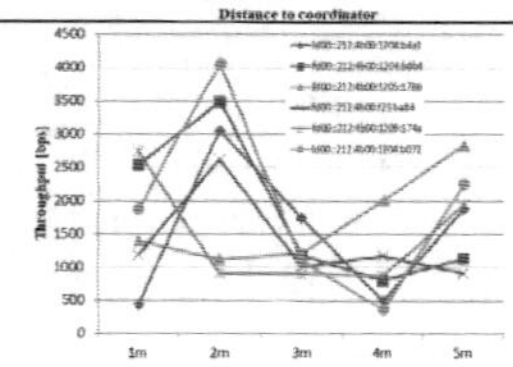 | 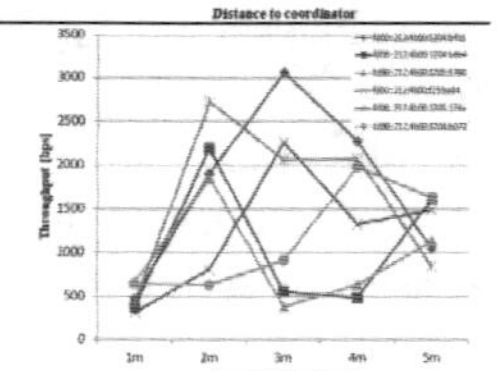 |

## 5. Conclusions

This paper compares end-to-end delay and throughput results reported in the simulation and the actual 6LoWPAN sensor network. The experimental results obtained from the simulation are low compared to the real ones. At different simulation distances, identical results are obtained, as the operation is performed under ideal conditions. The difference in the simulated results is manifested in the different number of packets sent in the simulation and in real conditions. The results of real network tests are variable because the communication between the sensors and the coordinator is influenced by environmental factors such as electromagnetic interference, radio interference, packet transmission errors, other sources operating on the same frequency, interference between sensors, etc. The obtained trend in the simulation results and real conditions are approaching, which gives reason to allege that the simulation product is suitable for purposes of education.

## References

[1] Ericsson Mobility Report, June 2020. Available at: https://www.ericsson.com/en/mobility-report/reports/june-2020, (Last visit on 05.07.2020)

[2] Iliev T, Mihaylov G, Ivanova E and Stoyanov I 2017 Power control schemes for device-to-device communications in 5G mobile network, *40th International Convention on Information and Communication Technology, Electronics and Microelectronics (MIPRO)*, Opatija, Croatia, pp 416-419, doi: 10.23919/MIPRO.2017.7973460

[3] Iliev T, Mihaylov G, Bikov T, Ivanova E, Stoyanov I and Radev D 2017 LTE eNB traffic analysis and key techniques towards 5G mobile networks, *40th International Convention on Information and Communication Technology, Electronics and Microelectronics (MIPRO)*, Opatija, Croatia, pp 497-500, doi: 10.23919/MIPRO.2017.7973476.

[4] Mihaylov G, Iliev T, Bikov T, Ivanova E, Stoyanov I, Keseev V and Dinov A 2018 Test cases and challenges for mobile network evolution from LTE to 5G, *41st International Convention on Information and Communication Technology, Electronics and Microelectronics (MIPRO)*, Opatija, Croatia, pp 449-452, doi:10.23919/MIPRO.2018.8400085

[5] Dinev D Z 2019 Simulation framework for realization of horizontal handover in Li-Fi indoor network, *IEEE XXVIII International Scientific Conference Electronics (ET)*, Sozopol, Bulgaria, pp 1-4, doi: 10.1109/ET.2019.8878509

[6] Haka A, Vasilev R, Aleksieva V and Valchanov H 2019 Simulation framework for building of 6LoWPAN network, *16th Conference on Electrical Machines, Drives and Power Systems (ELMA)*, Varna, Bulgaria, 2019, pp 1-5, doi: 10.1109/ELMA.2019.8771633.

[7] Agajo J, Kolo J G, Adegboye A, Nuhu B, Ajao L and Aliyu I 2017 Experimental performance evaluation and feasibility study of 6lowpan based internet of things, *Acta Electrotechnica et Informatica*, **17** (2) pp 16–22, doi: 10.15546/aeei-2017-0011.

[8] Thombre S, Islam R U, Andersson K and Hossain M S 2016 IP based wireless sensor networks: performance analysis using simulations and experiments, *J. Wirel. Mob. Networks Ubiquitous Comput. Dependable Appl.* **7** (3) pp 53-76

[9] Thombre S, Islam R U, Andersson K and Hossain M S 2016 Performance analysis of an IP based protocol stack for WSNs, *IEEE Conference on Computer Communications Workshops*, San Francisco, CA, pp 360-365, doi: 10.1109/INFCOMW.2016.7562102.

[10] BeagleBone Black. Available at: https://beagleboard.org/black. (Last visit on 05.07.2020)

[11] CC2531 USB EMK. Available at: http://www.ti.com/tool/CC2531EMK. (Last visit on 05.07.2020)

[12] SimpleLink Multi-standard SensorTag. Available at: http://www.ti.com/tool/CC2650STK?keyMatch= CC2650STK&tisearch=Search-EN-everything&usecase=GPN. (Last visit on 05.07.2020)

[13] 6LoWPAN Border Router. Available at: https://github.com/cetic/6lbr/wiki/Releases. (Last visit on 05.07.2020)

[14] A deployment-ready 6LoWPAN Border Router solution based on Contiki. Available at: https://github.com/cetic/6lbr. (Last visit on 05.07.2020)

[15] Sensniff Live Traffic Capture and Sniffer for IEEE 802.15.4 networks. Available at: https://github.com/g-oikonomou/sensniff. (Last visit on 05.07.2020)

## Acknowledgments

The research, the results of which are presented in the present paper, have been carried out under TU-Varna Scientific Project "Research of the possibilities of integrating machine learning and Blockchain technologies for the Internet of Things", which is financed by the state budget.

# Deployment and analysis of Bluetooth low energy network

**A M Haka[1], V P Aleksieva[1], H G Valchanov[1]**

[1] Faculty of Computer Sciences and Automation, Technical University of Varna, Varna 9010, Bulgaria

E-mail: aydin.mehmed@tu-varna.bg

**Abstract.** Along with the development of modern communication technologies such as 4G and 5G, Internet of Thing (IoT) technologies are being implemented more widely. The need for remote monitoring of environmental and patient parameters can be realised by means that represent IoT technologies. The localisation of mobile sensor nodes for low power IoT can be realised by tracking the Received Signal Strength Indicator (RSSI) values. This paper examines the RSSI values of sensor nodes in the Bluetooth Low Energy network for IoT.

## 1. Introduction

Together with the active proliferation and improvement of 4G and 5G networks, those for the Internet of Things (IoT) are also being developed. One of the designations of IoT networks is to supervise environmental parameters such as temperature, pressure, humidity, lighting and more [1]. Distant surveillance of diverse parameters is becoming busier in modern life, and the necessity for this is especially manifest during emergencies [2].

The monitoring of the condition of the environment or patients can be realised with the help of contemporary IoT sensor networks, which allow intermittent surveillance of parameters from both static and movable sensor nodes. In definite occasion, it may be requisite to locate the sensor nodes in the network. One of the ways to accomplish this with a network of low-power sensor units is to track the values for the Received Signal Strength Indicator (RSSI).

This paper presents the deployment of a physical Bluetooth Low Energy (BLE) sensor network for IoT and a study of the RSSI values derived from the end sensor units in the network.

## 2. Related works

The study in [3] presents a new method of localisation in a vehicle with high-precision, based on a functional study and full use of multi-channel RSSI in BLE technology. The objective of the research is to accurately localise the occupant in the vehicle. A hierarchical computational algorithm has been suggested to accomplish the desired precision. The conclusions show that the suggested method distinguishes the location of the driver or occupants with an accuracy ranging from 86.80% up to 92.02% for each seat for the Nexus phone with a standard deflection of 2.64% and from 85.43% up to 93.33% for Huawei phone with a standard deflection of 3.07%.

In [4] an exhaustive experimental investigation of RSSI values in a dense indoor BLE network presented. The main purpose of the study is to assign the deflections in RSSI bred by the noise effects in the environment and to elaborate a model for dissemination in the radio surrounding to minimise the evaluation of distance and localisation error. There is an average error in estimating the position of 1.32m with a deviation of 10dBm in the RSSI values, according to the experimental analysis. Also the environmentally specific radio constants strongly influence the accuracy of estimating the distance and location of the BLE units.

In [5] examined the positioning accuracy of various commercial BLE devices. The RSSI values from sensor knots are used to assign distances without using of a filter, to represent the raw productivity of the devices used for the purposes of the study. The results show a deviation in accuracy from 0.07m to 7.81m.

In [6] proposed an RSSI Gaussian mixture model (GMM) for performing indoor localisation using BLE sensors. GMM used to overcome the drastic fluctuations in RSSI and to present the distributed values more accurately. Significant improvements in terms of localisation accuracy the proposed model ensure, according to the experimental results.

In [7], a solution for an indoor localisation system based on the BLE standard investigated using RSSI values. The research considers RSSI based BLE indoor positioning. There are implemented five filtering algorithms: median, mode, single direction outlier removal (SDOR), feedback and shifting. The experiments show that indoor BLE-based localisation depends on different conditions, and improvement of the localisation system can be achieved by implementing more accurate filters.

In [8] presented an inverse model of the distance between two BLE devices depending on the measured RSSI values and Packet Error Rate (PER). A model of the accuracy of forecasting the distances calculated by RSSI and PER measurements presented.

The considered researches investigate the BLE standard, and examine the values of RSSI obtained from the end sensor nodes. The RSSI values in these studies are considered in order to be used to determine the location of the end nodes in the network. Alike to the research works considered, current paper examines the received RSSI values from end sensor nodes, based on the BLE standard. Tracing the variation of the RSSI values for static sensor nodes located at different distances from the master device and for mobile nodes that move at different speeds in a centralised BLE sensor network, is the major purpose of the study.

## 3. Bluetooth Low Energy (BLE) network deployment, experimental scenario and results

The physical building of the BLE network done with RaspberryPi 4 Model B [9] board with Raspbian operating system working as BLE master device, with built-in BLE transceiver and Texas Instruments multi-standard sensor nodes – CC2650STK [10].

The Bluepy program can be used to start the RaspberryPi to work as a master BLE device. Bluepy requires the installation of the Python programming language for its operation. Likewise the installation of libglib2.0-dev library required. Reading data from the end sensor nodes and listen to the communication medium on which the data transmitted, can realised after configuring the master BLE device. Thus, information can be obtained about the address of the device (figure 1-1), the measured RSSI values for the nodes in the network (figure 1-2), the name of the sensor node (figure 1-3), the transmitting power (figure 1-4) and manufacturer information (figure 1-5).

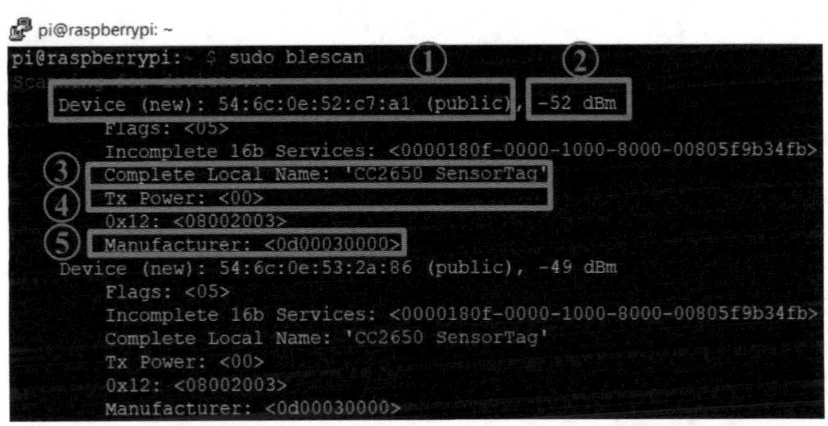


**Figure 1.** Captured BLE packets on the communication medium.

The master BLE device can receive information from the CC2650STK sensor nodes after they are configured to work with the BLE standard.

The star topology by connecting the end sensor nodes and the master one was realised to examination the alteration in RSSI values. According to figure 2, with 1 is marked the master device and with 2 the sensors that are used for experiments with static sensor nodes. Different experiments with 1, 2, 3, 4, 5 and 6 static nodes performed, where for every one the nodes are located at distances from 1m to 10m from the master device. The outcomes from the experiments performed, for static nodes, are in table 1.

**Figure 2.** Used physical topology for experiments with static nodes.

**Table 1.** Results for RSSI values from experiments with static sensor nodes.

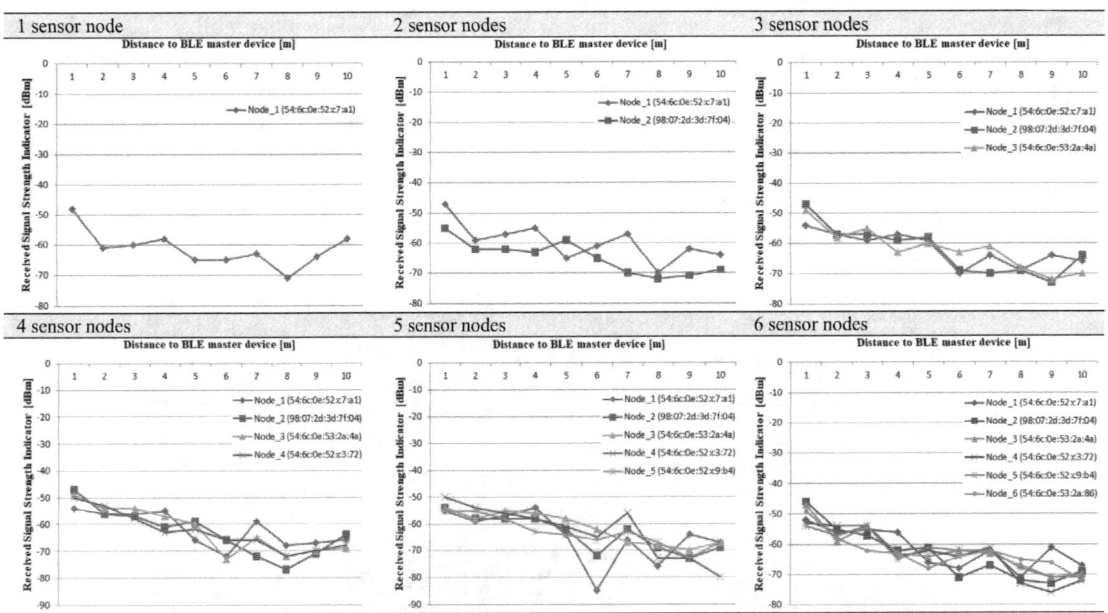


The experimental results presented in table 1 for all tests performed manifested that the RSSI values are not constant. For 1 node the results show that as the distance of the sensor from the master device increases, the received RSSI values deteriorate. However the value at 10 meters is significantly better than the previous ones. Although only one device is transmitting on the communication environment which is not loaded, the decline in the previous values may be due to external sources of interference. The measured RSSI values for one of the nodes are quite close to the values in the tests with one device, according the results for 2 sensor nodes. For the second node it is seen that the derived values are considerably lower. As the distance from the master device increases, the resulting RSSI values worsen. The trend that at closer distance to the service device the obtained RSSI values are better is confirmed from the other tests with 3, 4, 5 and 6 sensors. The measured values for RSSI decline more and more when the distance from the master device and the end nodes number in the network increase. Because of the load in communication medium and the simultaneous transmission of the end nodes in it there is caused interference and this decrease the quality and strength of the signal. A significant difference between the initially reported results and those reported at the end observed when the distance between the static sensors and the master device increases.

Similar experiments were performed with mobile nodes. According to figure 3, with 1 is marked the master device and with 2 the mobile nodes. The movement of the sensors is done as the nodes are arranged in a cart that moves. Different experiments with 1, 2, 3, 4, 5 and 6 mobile nodes performed, where for every one the nodes move at a speed of 0.5m/s to 3m/s. The outcomes from the experiments conducted, for mobile nodes, are presented in table 2.

**Figure 3.** Used physical topology for experiments with mobile nodes.

To examine the effect of interference between devices on the RSSI values gained by the coordinator for both static and mobile devices, the sensors are placed next to each other.

The results from the experiments for all tests with mobile sensor nodes shown in table 2 are also inconsistent in regard to the obtained RSSI values. At lower speeds is visible that the obtained RSSI values are better according to the results with 1 mobile sensor. As the speed amplifies, aggravation of the obtained values reported. The RSSI values for one of the nodes are close to the values reported for the first node, according the results for 2 mobile sensor nodes. For the second node it is seen that the values are considerably lower. This is bred by the load on the communication environment and the emerged interferences. The trend when the sensors moving at a lower speed, the received RSSI values are better is confirmed from the other tests with 3, 4, 5 and 6 sensors.

**Table 2.** Results for RSSI values from experiments with mobile sensor nodes.

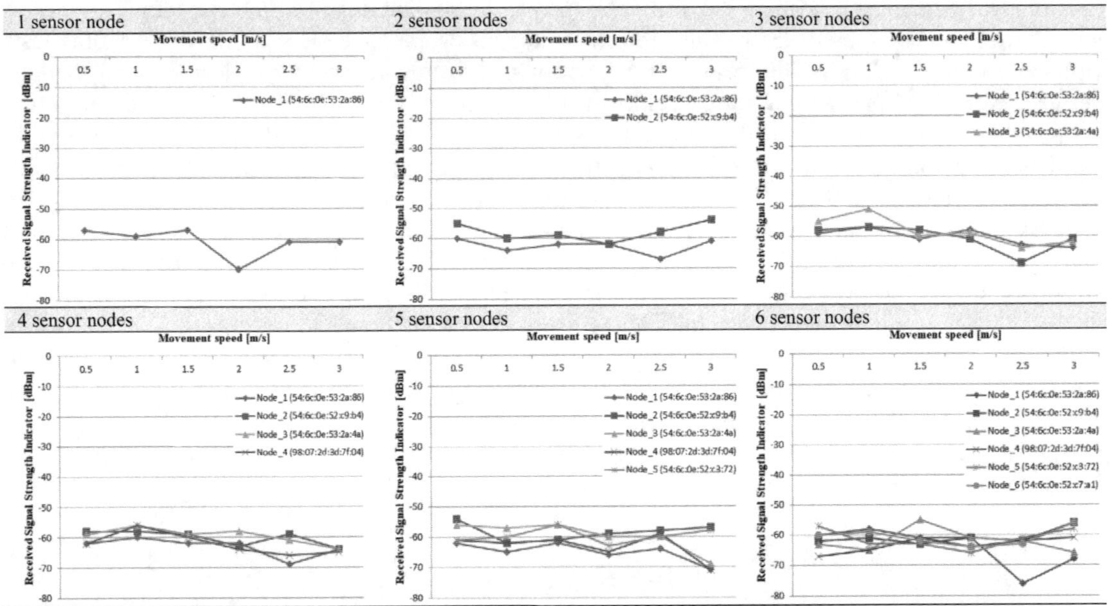


The experiments show deterioration of the reported RSSI values with increasing the number of sensors and the load on the communication environment. The received RSSI values decrease smoothly with increasing of the speed of the nodes. This is caused by the receiving and transmitting delay of RSSI values from the mobile nodes to the main device while movement.

## 4. Conclusion

This paper represents the deployment of Bluetooth Low Energy sensor network. The examination of alterations in the received RSSI values for static sensor nodes located at different distances from the master device and for mobile nodes moving at different speeds done. Experimental results for the RSSI with static sensor nodes show that with increasing distance between the end nodes and the master device, the received values aggravates with considerable changes. Experimental results for the RSSI with mobile sensor nodes show that with increasing the speed of end nodes, the received values aggravates, but the change in the results is smoother. For both static and mobile nodes sustained the tendency of aggravation of RSSI values with enlarging number of end sensor nodes in the network.

## References

[1] Dinev D Z 2019 Simulation Framework for Realization of Horizontal Handover in Li-Fi Indoor Network, *IEEE XXVIII International Scientific Conference Electronics (ET)*, Sozopol, Bulgaria, 2019, pp 1-4, doi: 10.1109/ET.2019.8878509

[2] Ericsson Mobility Report, June 2020. Available at: https://www.ericsson.com/en/mobility-report/reports/ june-2020, Last visit on 28.07.2020

[3] Yuan G, Ze Z, Changcheng H, Chuanqi H and Li C 2020 In-vehicle localization based on multi-channel Bluetooth Low Energy received signal strength indicator, *International Journal of Distributed Sensor Networks*, **16**(1), doi: 10.1177/1550147719900093 journals.sagepub.com/home/dsn

[4] Subhan F, Khan A, Saleem S, Ahmed S, Imran M, Asghar Z and Bangash J I 2019 Experimental analysis of received signals strength in Bluetooth Low Energy (BLE) and its effect on distance and position estimation. *Trans Emerging Tel Tech.* e3793. https://doi.org/10.1002/ett.3793

[5] Fachri M and Khumaidi A 2019 Positioning accuracy of commercial Bluetooth low energy beacon. *IOP Conf. Ser.: Mater. Sci. Eng.* **662** 052018, https://doi.org/10.1088/1757-899X/662/5/052018

[6] Malekzadeh P, Salimibeni M, Atashi M, Barbulescu M, Plataniotis K N and Mohammadi A 2019 Gaussian mixture-based indoor localization via Bluetooth low energy sensors, *IEEE SENSORS*, Montreal, QC, Canada, pp 1-4, doi: 10.1109/SENSORS43011.2019.8956950

[7] Mussina A, and Aubakirov C 2018 RSSI based Bluetooth low energy indoor positioning, *IEEE 12th International Conference on Application of Information and Communication Technologies (AICT)*, Almaty, Kazakhstan, pp 1-4, doi: 10.1109/ICAICT.2018.8747020

[8] Conti M 2017 Real time localization using Bluetooth low energy, *Bioinformatics and Biomedical Engineering. IWBBIO 2017. Lecture Notes in Computer Science* **10209** ed I Rojas and F Ortuno (Springer, Cham), https://doi.org/10.1007/978-3-319-56154-7_52

[9] Raspberry Pi 4 Tech Specs, Available at: https://www.raspberrypi.org/products/raspberry-pi-4-model-b/specifications/, Last visit on 28.07.2020

[10] SimpleLink Bluetooth low energy/Multi-standard SensorTag. Available at: http://www.ti.com/tool/ CC2650STK?keyMatch=CC2650STK&tisearch=Search-EN-everything&usecase=GPN. Last visit on 28.07.2020

**Acknowledgments**

The research, the results of which are presented in the present paper, have been carried out under TU-Varna Scientific Project "Research of the possibilities of integrating machine learning and Blockchain technologies for the Internet of Things", which is financed by the state budget.

# Selection of countermeasures against propagation of harmful information via Internet

**L A Vitkova, A P Pronichev, E V Doynikova, I B Saenko**

SPIIRAS, liniya 14-ya, 39, St.Petersburg, Russia

E-mail: vitkova@comsec.spb.ru

**Abstract**. Today, Internet allows accessing and propagation various information. Some of this information can be undesired for particular user or even harmful. While there are quite a large number of systems to respond to unwanted and harmful information and some well-established set of countermeasures, such as filtering, blocking access, and notification, existing systems are often based on one type of countermeasures. From our point of view, there is a need for the technique of automated selection of optimal countermeasures to counteract undesired or harmful information on the Internet depending on the type of such information and characteristics and needs of the protected system. In this paper we propose the set of interconnected models, including threat model, countermeasure model and information object model, and countermeasure selection technique for protection against harmful information in Internet using these models. The proposed technique uses single-criteria optimization on the basis of introduced countermeasure selection index and allows selecting applicable and optimal countermeasures for the particular system from specific threats. The application of the technique is demonstrated on the experiments.

## 1. Introduction

Today, Internet allows accessing and propagation various information. Some of this information can be undesired for particular user or even harmful. Harmful information can include annoying suggestions for gambling or lottery, a proposal, request and/or order to commit suicide, and/or description of suicide as a way to solve problems, data about production, transfer, sale or storage of child pornography, description of the production, development and use of drugs and psychotropic substances, and others. Some information can be forbidden for propagation by legislative acts, other can be specified as undesired by particular user. Today, parental control systems filter traffic based on security policies. Social media platforms provide protection from unwanted information for a child if the user has specified their age. This protection is also provided by some operating systems, such as iOS, by blocking access to pages with such content. Again, the user must specify the correct age. When implementing parental control security policies in a social network or operating system, the systems themselves only analyze messages and pages that the user interacts with. The tasks of countering the propagation of unwanted and harmful information, its distributor, and evaluating user behavior on such pages are not set or solved by social network platforms.

In fact, all existing tools implement the following actions: 1) checking information, 2) restricting access to a resource with unwanted information [1, 2]. While today there are quite a large number of systems to respond to unwanted information and some well-established set of countermeasures, such as filtering, blocking access, and notification, existing systems are often limited with above mentioned approach based on one type of countermeasures, i.e. blocking access. From our point of view, there is a need to classify countermeasures, and develop the models and technique for automated selection of optimal countermeasures to counteract unwanted or harmful information on the Internet depending on the type of such information and characteristics and needs of the protected system. In this paper we propose the set of interconnected models, including threat model, countermeasure model and information object (IO) model, and countermeasure selection technique for protection against harmful information in Internet. The proposed technique uses single-criteria optimization on the basis of introduced countermeasure selection index and allows selecting applicable and optimal countermeasures for the particular system from specific threats.

The paper is organized as follows. The second section provides the results of the related works and solutions research. The proposed models and technique are given in Section 3. The implementation and experiments are described in Section 4. The paper ends with conclusions and discussion about future research.

## 2. Related works

Parental control systems, anti-virus programs and methods, models and algorithms developed in this field are the first to protect against unwanted information.

Due to the huge volume of media content available from various sources (for example, TV, Internet, online games, etc.), parental controls are becoming increasingly important for censorship and access control of media content. Parental control systems traditionally allow users to restrict access to specific types of media content based on media characteristics, such as ratings, time, or resource type. In [3], mechanisms for filtering access to media content are proposed. The authors [4] in their work suggest a way to intercept information about the file at the moment when the file is saved to the device and can be run in single-user mode (offline). In [5], a method for controlling traffic and access to information in a network environment in a local network is proposed. In [6], a method for making decisions about parental control approval for user content is proposed, but with the parent's approval of the child's rights to upload a media file from an external website or resource.

Most existing systems for countering unwanted information on the Internet use methods based on the classification of web pages, resources, and the use of countermeasures such as lists (url blacklist). However, traditional methods are improved by using machine learning algorithms. In [7] the authors state that the blacklist is now completely ineffective in searching for both malicious URL variants and newly created URL's. In addition, creating and updating lists requires human input. Blacklisting is a time-consuming process. Machine learning solutions implicitly rely on the function development stage to extract characteristics. Such as: linguistic, lexical, contextual, or semantic attributes, URL string statistics, n-grams, word packets, link structures, content composition, DNS information, network traffic, etc. As a result, machine learning solutions must recognize new malicious URL's, and these systems are not yet accurate enough. The authors tested few deep learning methods such as recurrent neural networks, long-short-term memory, convolution neural networks and identity-recurrent neural networks. Algorithms and methods based on deep learning can extract objects automatically by accepting raw input texts. In the [8] shows a method for detecting malicious web pages by analyzing the domain and scanning it.

In fact, the black and white list policy is not really applicable to protect against unwanted information, for example in social networks, because the URL's of an IO contains not the attributes of an IO, but the attributes of the path of the object. The domain of the social network platform, the authors in it, and the messages are connected on the page of news portal. Adding the path of such a portal to the blacklist leads to excessive blocking.

The need to use algorithms for semantic text analysis leads owners of systems for countering unwanted information to the problem of analyzing extremely large amounts of information. Search and analytical systems and their clients (browsers) take the lead in developing such approaches. In the [9], a machine learning method is demonstrated that allows us to study a variety of messages in the social network Twitter, user moods, and conduct analytics. Such works are also popular in the field of marketing, economics and finance to assess consumer sentiment. In the [10] a method is proposed for evaluating the tone of messages in the social network Twitter, based only on smiles. In the [11], the Heracles platform for developing text analysis algorithms is proposed.

In addition, there are also many solutions that protect users of the Internet and social networks from unwanted information based on the analysis of the graph of relationships. Patent [12] is devoted methods for protecting a social network user from prohibited objects (text, photo, media file) that were previously saved in the database. This technical result is achieved by building a social graph, a cluster of accounts in a social network, and a database of prohibited objects that are associated with social graphs and accounts. The invention relates to parental control systems that restrict user access to social network resources in accordance with the rules set by the parent. In [13], a method for quickly identifying similar communities and social network participants is proposed. During the analysis process, the system selection criteria for similarity of profiles and generates a list of IDs accounts that meet these criteria. Finally, in the created social graph, edges characterize the direct or indirect connection between social network accounts.

A review of existing studies shows that most of the existing solutions are based on the methodology within which the IO is checked. If unwanted information is detected, access to the resource will be limited. Also, a review of relevant studies shows that today there is some well-established set of countermeasures, such as filtering, blocking access, and notification. However, existing systems for countering unwanted information use only one type of measure from the established set of countermeasures. Therefore, there is a need to classify countermeasures, and develop a model and methodology for automatically selecting optimal countermeasures to counteract unwanted information on the Internet.

## 3. Countermeasure selection technique

The developed technique is based on the analytical modelling of counteraction process and decision support methods. In process of analytical modelling we specified the object affected by the harmful information, the threats against this object, and countermeasures against these threats. The developed technique uses single-objective optimization for countermeasure selection index. The first subsection considers the countermeasures and their types, and their applicability against particular threats, as well as the models of countermeasure, threat and object. The second subsection describes the proposed technique.

*Countermeasures and models for countermeasure selection*

We understand the countermeasure as an action or tool for protection of object from information security threat, where information security threat is represented by harmful information. We limit object with IO in Internet. We outlined the following properties for information objects specification: object size; type of information object (harmful or not); and harmful object class. Based on the outlined properties we specified the object model as follows: $IO = <size, type, class>$, where *size* – is a class of information object, it can take values $\{s, m, l\}$, where $s$ – small (sio), $m$ – medium (mio), $l$ – large (lio); *type* – is a type of information object, it can take values $\{h, n\}$, where $h$ – harmful objects, $n$ – not harmful objects; *class* – is a class of harmful information object, it can take values {*cour, reg, lot, suic, por, dru, from6*, *from12*, *from16*, *from18*, *none*}, where *cour* – IO containing data earlier forbidden by court's decision; *reg* – IO containing data earlier blocked by authorized bodies; *lot* – IO containing suggestion for gambling or lottery; *suic* – IO containing proposal, request and/or order to commit suicide, and/or describing suicide as a way to solve problems, as well as endorsement of suicide; *por* – IO containing data about production, transfer, sale or storage of child pornography, the acquisition of child pornography; *dru* – objects containing description of the production, development and use of drugs and psychotropic substances and/or description of the cultivation locations of plants containing drugs and/or places of their sale and prices; from6 – IO containing information for children from 6 years old; from12 – IO containing information for children from 12 years old; from16 – IO containing information for children from 16 years old; from18 – IO containing information forbidden for children; *none* – not applicable (if *type* is $n$). The class of affected threats connects threat model and countermeasure model, while class of the information object connects countermeasure model and object model. Connections between the threats, countermeasures and object size are shown in table 1 on the example of drug information threat.

We analyzed possible countermeasures and outlined the following properties for their specification: the class of affected threats; implementation type; implementation agent; class of the information object; and complexity of implementation. Based on the outlined properties we specified the countermeasure model *rm* as follows: *rm=<rm_name, rm_thr, rm_obj, rm_impl, rm_agent, rm_cost>*, where *rm_name* – countermeasure type (message filtration for sio in browser; block mio/lio URL by internet provider; mio/lio URL filtration by antivirus/parental controls/browser/operation system; black SEO; user notification for sio/mio/lio by browser/antivirus/parental controls/operation system; administrator notification for sio/mio/lio; internet provider notification for sio/mio/lio); *rm_thr* – the class of affected threats (*cour, reg, lot, suic, por, dru, from6, from12, from16, from18, none*); *rm_obj* – class of the information object (small, medium, or large); *rm_impl* – implementation type (manual or automated); *rm_agent* – implementation agent (browser, internet provider, parental controls, antivirus, mobile application, operation system); *rm_cost* – complexity of countermeasure implementation (it depends on *rm_impl* and *rm_agent* and currently specified by the experts). The class of affected threats connects threat model and countermeasure model, the class of the information object connects countermeasure model and object's model, and agent and implementation type connect countermeasure model with protected information system. Connections between the threats, countermeasures and object size is provided in table 1 on the example of drug information threats.

**Table 1.** Mapping between the information objects size, threats and countermeasures.

| Threat class | Threat description | IO size | Countermeasure |
|---|---|---|---|
| **Drugs (Th4)** | Information object containing description of the manufacture, development and use of drugs and psychotropic substances and/or description of the cultivation locations of plants containing drugs and/or places of their sale and prices | SIO | Message filtration for SIO in browser |
| | | | User notification for SIO by browser |
| | | | User notification for SIO by antivirus |
| | | | User notification for SIO by parental controls |
| | | | User notification for SIO by operation system |
| | | | Administrator notification for SIO |
| | | MIO | Block MIO URL by internet provider |
| | | | MIO URL filtration by antivirus |
| | | | MIO URL filtration by parental controls |
| | | LIO | Block LIO URL by internet provider |
| | | | LIO URL filtration by browser |
| | | | Block LIO URL by parental controls |
| | | | LIO URL filtration by antivirus |

The countermeasures should stop propagation of harmful information. The technique for countermeasure selection is provided in the next subsection.

*Technique for countermeasure selection*

The developed technique gets as input the harmful information objects, the threats relevant for this object and the list of countermeasures specified using the models described in the subsection below. The technique includes two main stages: (1) selection of the applicable countermeasures considering the protected system, information object's and threats' characteristics; (2) selection of optimal countermeasures on the basis of integral countermeasure selection index. The technique incorporates the following steps:

1. Input information object *io* that have *type=h*, *class* is given as input from harmful information detection system.
2. Get *rm_agent* and other characteristics that depend on the protected system.
3. Determine the set of possible countermeasures *cms* for *io* considering *class, size* and *agent,* and characteristics of the protected system.
4. For each countermeasure *c* from *cms*:
   - Determine countermeasure selection index *rm_ind* as integral sum of weighted countermeasure parameters' values including implementation agent and implementation type (see countermeasure model above).

5. Select countermeasures *scms* from *cms* with max *rm_ind.*
6. Output *scms.*

As the result the technique outputs the set of optimal countermeasures against detected harmful information object for the particular system under protection.

## 4. Implementation and Experiments

The developed technique was implemented using Java 11. For the experiments the dataset containing 10000000 objects of different size (sio - 33,33%, mio - 33,34%, lio - 33,32%) containing different threats (*reg* – 14,9%, lot – 14,89%, suic – 59,51%, por - 14,89%, dru – 29,76%, from18 – 14,87%, other 44,67 – see section 3). The input data set used in the experiment is a file containing the strings represented by a json objects with the following structure: (a) "SIZE" – the size of the IO; (b) "id" – the unique identifier of the IO; (c) "threats": ["t1", "t2", "t3"...] – an array of threats contained in the object. In the output set, an array of countermeasures is added to the existing structure, consisting of the following objects: (a) "name" – the name of the countermeasure; (b) "complexity" – complexity of countermeasure implementation; (c) "priority" – priority of the threat; (d) "type" – type of implementation (automatic, manual, etc.); (e) "agents": ["a1"," a2"...] – array of agents to implement the countermeasure. For example, in case of input: "SIZE"="MEDIUM"; "threats"= "suicide"; and the system with characteristics: "agents"="ANTIVIRUS"; the output set will be: "countermeasures": "name"="FilterAntivirus","complexity"=3,"priority"=1,"type"="AUTO","agents"="ANTIVIRUS".

The analysis of the countermeasure selection component results allowed obtaining the following statistics: (1) The set of possible countermeasures included: (a) 3 countermeasures for 7,28% of information objects; (b) 4 countermeasures for 92,72% of information objects. (2) The countermeasures added to the set of possible countermeasures distributes as follows: (a) message filtration in browser - 31,72%; (b) filtration by antivirus - 44,37%; (c) filtration by parental controls - 0,87%; (d) administrator notification - 100,00%; (e) internet provider notification - 100,00%; (f) block by internet provider - 45,35%; (g) other - 14,03%. (3) The countermeasures added to the set of optimal countermeasures distributes as follows: (a) message filtration in browser - 31,72%; (b) filtration by antivirus - 44,37%; (c) filtration by parental controls - 0,87%; (e) administrator notification - 23,04%. In figure 1 the distribution of the technique operation results, i.e. distribution of the optimal countermeasures against simulated information objects containing harmful information (threats), is represented.

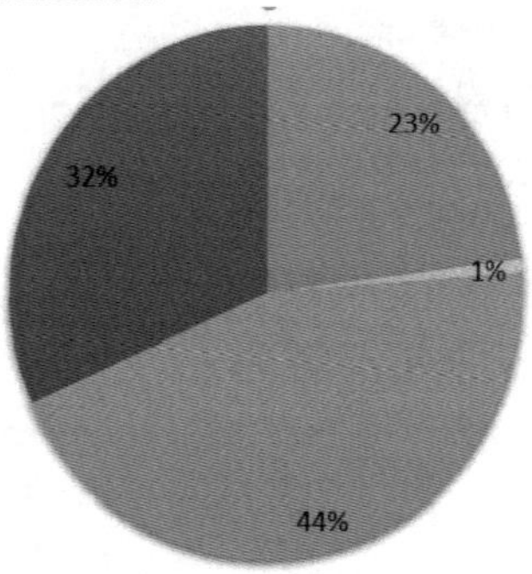


**Figure 1.** Distribution of optimal countermeasures for the system modelled in the experiment. Where 44.37% - the countermeasures implemented by antivirus filtering tools; 31.72% - filtering in the browser; 23.04% - notification to the resource administrator where the information object is located; 0.86% - filtering by the parental control system.

As part of the experiment, we used a set of threats, information signs, and objects specific to the legal field of Russia. However, any operator, whether a parent or an authorized body, can set its own rules and countermeasures. The proposed technique is flexible (can be implemented on the user's/internet provider's/authorized body's side) and automatically determines what countermeasures are the most or least available and optimal in the given conditions.

## 5. Conclusion

The paper considered the challenge of counteraction against propagation of undesired for particular user or even harmful information via Internet. The models of counteraction process, including the threat model, countermeasure model and information object model were developed. The technique for automated selection of optimal countermeasures to counteract against such information using the developed models is proposed. It considers the information type and characteristics and needs of the protected system. It uses single-criteria optimization on the basis of introduced countermeasure selection index and allows selecting applicable and optimal countermeasures for the particular system from specific threats. The technique is flexible and can be implemented on the user's, internet provider's or authorized body's side. The operation of the technique was tested on different types of objects, systems and countermeasures. For the experiment a dataset (~ 10 million) was used, containing information about objects with various threats. The experimental data showed the choice of optimal countermeasures distributed as follows: 44.37% - antivirus filtering tools, 31.72% - filtering in the browser, 23.04% - notification to the resource administrator; 0.86% - filtering by the parental control system. The experiment demonstrated the possibility of their implementation and applicability of the developed technique for the selection of countermeasures. The technique can be applied in public control systems (schools, educational institutions), in the corporate segment (assistance in solving the restriction of access to information resources in the company's network), as well as for personal use (parental control systems, filtering).

**References**

[1] Desnitsky V A, Kotenko I V and Parashchuk I B 2019, May methods of assessing the effectiveness of network content processing systems for detecting malicious information taking into account the elimination of uncertainty in the semantic content of information objects *2019 SCM.* (St. Petersburg, Russia), pp 41-44

[2] Kotenko I V and Saenko I B 2012 Developing the system of intelligent services to protect information in cyber warfare (*SPIIRAS Proceedings* vol 3(22)), pp 84-100.

[3] Shiang L 2015 Systems and methods for media detection and filtering using a parental control logging application. U.S. Patent 9,204,193, issued December 1, 2015

[4] Schepis A and Newstadt K 2018 Methods and systems for applying parental-control policies to media files. U.S. Patent No. 9,330,274. Washington, DC: USPTO, 2016

[5] Slemmer J B, Olsson S, Kreiner B, Jarboe A and Kleinfelter K 2018 Parental control in a networked environment. U.S. Patent Application 16,115,624, filed December 27, 2018

[6] Schepis A and Newstadt K 2016 Systems and methods for applying parental-control approval decisions to user-generated content. U.S. Patent 9,495,547, issued November 15, 2016

[7] Vinayakumar R, Soman K P and Poornachandran P 2018 Evaluating deep learning approaches to characterize and classify malicious URL's., *Journal of Intelligent & Fuzzy Systems*, **3** pp 34 1333-1343

[8] Agbefu R E, Hori Y and Sakurai K 2013 Domain information based blacklisting method for the detection of malicious webpages. *IJCSDF* , **2**(2), pp 36-48

[9] Mehta R P, Sanghvi M A, Shah D K and Singh A 2020 Sentiment analysis of tweets using supervised learning algorithms. *In ICTSCI proceedings*, Springer, Singapore, pp 323-338

[10] Sarma P K and Sethares W 2018 Simple algorithms for sentiment analysis on sentiment rich, data poor domains, *In Proceedings of the COLING*, pp 3424-3435

[11] Schouten K, Frasincar F, Dekker R and Riezebos M 2019 Heracles: a framework for developing and evaluating text mining algorithms, *Expert Systems with Applications,* **127**, pp 68-84

[12] Larkina A D and Tushkanov V N 2018 A way to limit user access to suspicious objects of a social network. Patent RU No. 2651252. Patent holders: Kaspersky Lab

[13] Fabrikant A and Sarma A D 2014 Pre-processing of social network structures for fast discovery of cohesive groups. U.S. Patent No. 8,909,646. Washington, DC: USPTO

**Acknowledgments**

The work is performed by the grant of RSF #18-11-00302 in SPIIRAS.

IOP Conf. Series: Materials Science and Engineering 1032 (2020) 012018 doi:10.1088/1757-899X/1032/1/012018

# On numerical method for modeling oil filtration problems in piecewise-inhomogeneous porous medium

**A Nematov[1], E Sh Nazirova[2] and R T Sadikov[3]**

[1] Tashkent University of Information Technologies named after Muhammad al-Khwarizmi, Associate Professor of Multimedia Technologies, Uzbekistan
[2] Tashkent University of Information Technologies named after Muhammad al-Khwarizmi, Head of the Department of Multimedia Technologies, DSc, Uzbekistan
[3] Assistant of the Department of Multimedia Technologies, Tashkent University of Information Technologies named after Muhammad al-Khwarizmi, Uzbekistan

E-mail: magistr_uz@bk.ru

**Abstract.** Oil and gas fields in natural conditions are rarely of homogeneous nature. Given the chaotic nature of changes in rock mass permeability within a single layer, its significant parts (areas) can be considered uniform in permeability. If significant in sizes homogeneous zones (regions) are distinguished within the formation, the parameters of such a macro-inhomogeneous formation significantly affect the characteristics of the filtration flows. The area of zonal inhomogeneity consists of the parts of different permeability. Within the same zone, the permeability is on average the same, but differs from the permeability of adjacent zones.

## 1. Introduction

The issues of filtration and well flow rate have been studied most widely for plane-parallel flows in homogeneous and piecewise-homogeneous formations of constant thickness in the presence of various boundaries of the recharge area, fluid discharge and change in the homogeneity of the formation (layer). The calculation of such flows was performed on the assumption that Forchheimer's postulate of vertical pressure constancy is fulfilled. For piecewise-homogeneous reservoirs, the fluid head is represented by a harmonic function, which made it possible to use the theory of functions of a complex variable, reducing the problem of studying flows in a reservoir to finding an analytical function in terms of flow - the complex flow potential. Within the framework of this theory, all works can be separated into three main groups:

1) when the boundary of homogeneous zones, discharge lines and feeding boundaries are straight;
2) when the boundaries are represented by circles;
3) when the boundaries differ from circles and straight lines:

There are known works in which boundary-value problems of conjugation are solved by the method of integral equations: in the theory of filtration, these are studies.

Filtration in heterogeneous fractured-porous media is considered in the context of the Barenblatt-Zheltov model. It is believed that the movement of fluid in fractures is described by the nonlinear two-term Forchheimer's law in the first form in the form allowed with respect to the filtration rate, and the movement of fluid in porous rock blocks is described by one of the most general nonlinear laws - the DHC law. The main filtration equation is obtained in the considered case and the problem under study is formulated. To solve it, it is proposed to use either the method of reduction to an integro-differential equation, or the Galerkin method with finite elements. Various options for its application are considered. Finally have been performed numerical calculations.

The paper presents a numerical simulation of two-phase filtration problems in fractured-porous media using a double porosity sample with a highly inhomogeneous permeability coefficient. A system of equations is given for the case of two-phase filtration without taking into account capillary and gravitational forces, which is a coupled system of equations for pressure and saturation in a porous medium with a system of cracks. Various options for setting the functions of the flow between the porous medium and cracks are considered. Numerical implementation for approximating velocity and pressure is based on the finite element method. To discretize the saturation equation by means of the method of introducing artificial diffusion, the classical Galerkin method with upstream approximation is used. The results of computational methods for a model problem using different flow functions are presented [1-3].

In the paper we considered the mathematical model and computational experiments to resolve the problems of bending of three-layer plates of complex configuration. The mathematical model and computational algorithm proposed in the paper are developed according to the Hamilton-Ostrogradky's variation principle, the Bubnov-Galerkin methods and the Rvachev R-functions. The numerical results are presented in tabular and graphical form. Numerical studies of the results obtained are considered using the R-function method in combination with the Bubnov-Galerkin variational method to resolve the problem of calculating the vibration of three-layer plates of complex configuration [4].

For the Analysis of Functioning of Technological Process of Filtering of a Suspension the adequate mathematical model is developed, numerical algorithm and the computing experiments on the computer are carried out. The results of computing experiments are illustrated as the diagrams [5].

The theory of mathematical models of fluid filtration in an anisotropic inhomogeneous porous medium based on the theory of generalized analytical functions and generalized potential is presented. Solved in the final form and numerically based on the method of discrete singularities three-dimensional and two-dimensional boundary-value problems of filtration of a homogeneous liquid and problems of the evolution of the interface of liquids of various physical properties (viscosity, density), which are of interest for the practice of developing oil-bearing (water-bearing) soil layers of complex geological structure and monitoring of groundwater pollution in such layers [6].

In this paper the Cauchy problem for nonlinear systems is considered. The conditions of existence of the solutions on time for the problem Cauchy are given. Furthermore the properties of the finite velocity of a propagation and localization of the disturbance, an asymptotic of self-similar solutions will be defined. The results of numerical solutions will be carried out and on the basis of calculations some necessary statements will be given [7].

This article [8] presents some of the results of many years of scientific research on the creation of a new method of non-invasive measurement of blood glucose (determination without obtaining blood) by studying the biophysical parameters of biologically active points. The authors proposed and substantiated the concept of creating models for an automated system for non-invasive measurement of blood glucose levels by electrical resistance of the skin at informative biologically active points. A functional diagram is proposed, technical means of implementation are outlined, and a methodology for creating an automated system for non-invasive blood glucose measurement is described in detail.

In the design and construction of tunnels, dams and other hydraulic engineering structures modeling of the groundwater flow is necessary. Deep bed filtration of suspension in a porous medium is considered. Forward flow of suspension periodically changes to back flow of clean water and vice versa. Partial lifting of retained particles at the moments of flow direction changes is modeled. Numerical analysis of suspended and retained particles concentrations is carried out. It is shown that at periodic change of flow detection the change of concentrations of suspended and retained particles rapidly reaches a periodic mode [9].

The paper develops a mathematical model of the process of purification of water from oil in a porous medium. Based on this model, the ability to clean a filtration device with a granulated coupling medium that recovers the filtration medium in a fine-grained sedimentation mode without disassembling or replacing the filtration element has been studied. In the course of theoretical research, a physical model of the process of water purification of oil-containing water in a granular fusion environment has been developed. Based on these models, the factors determining the effectiveness of cleaning are determined. After the implementation of the experimental plan for the first time, a mathematical model of the water treatment process containing the oil in the layer in the volume of the granulated medium was developed on the basis of the regression equation. This model allows the calculation of design and experimental parameters with granulated filter elements [10].

In applied mathematics, when solving a problem on a computer, a technological chain is implemented: the object of research - a mathematical model - an algorithm (numerical methods) - a program on a computer - a computational experiment - analysis (or comparison with experimental and other data). The object of mathematical technology is the computational part of this chain: a mathematical model is a numerical algorithm-program-calculation on a computer. The use of this technology makes it possible to analyze and predict non-stationary oil recovery processes in reservoir conditions and to control them.

## 2. Mathematical model

Mathematical model of oil filtration problem in a piecewise inhomogeneous porous medium is reduced to the integration of a partial differential equation of parabolic type

$$\beta\mu\frac{\partial P}{\partial t}=\frac{\partial}{\partial x}\left(k(x,y)\frac{\partial P}{\partial x}\right)+\frac{\partial}{\partial y}\left(k(x,y)\frac{\partial P}{\partial y}\right)-Q \qquad (1)$$

The boundary conditions for the integration of equation (1) are:

$$P(x,y)=P_H(x,y), \qquad t=0, \qquad (2)$$

$$-\frac{k(x,y)}{\mu}\frac{\partial P}{\partial n}=\alpha(P_A-P), \qquad (x,y)\in\Gamma, \qquad (3)$$

$$\oint_{s_i}\frac{k(x,y)}{\mu}\frac{\partial P}{\partial n}ds=-q_i(t), \quad (x,y)\in s_i, \quad i=\overline{1,n_q}, \quad Q=\sum_i^n q_i\delta. \qquad (4)$$

Here: $P$ is the pressure in the reservoir; $P_H$ - the initial reservoir pressure; $\mu$ - oil dynamic viscosity; $k$-permeability of the reservoir; $\beta$ - coefficient of elastic capacity of the reservoir; $q_i$ - the flow rate of the *i-th* well; $\delta=\delta(x-\xi_i)(y-\xi_j)$ - Dirac delta function.

To numerically solve problem (1) - (4), the following dimensionless variables are introduced:

$$P^*=P/P_0;\quad x^*=x/L;\quad y^*=y/L;\quad k^*=k/k_0;\quad \tau=\frac{k_0t}{\beta\mu L^2};\quad Q^*=\frac{Q\mu}{\pi k_0P_0h_0}.$$

Here $P_0$, $k_0$, - are some characteristic values of pressure and permeability of the formation, respectively.

Later, for simplicity, the sign "*" in the equations is omitted. Then, taking this into account, problem (1) - (4) in dimensionless variables is rewritten as follows:

$$\frac{\partial P}{\partial \tau}=\frac{\partial}{\partial x}\left(k(x,y)\frac{\partial P}{\partial x}\right)+\frac{\partial}{\partial y}\left(k(x,y)\frac{\partial P}{\partial y}\right)-\mathrm{Q}, \qquad (5)$$

$$P(x,y)=P_H(x,y), \quad t=0,\ (x,y)\in G, \qquad (6)$$

$$-\frac{k(x,y)}{\mu}\frac{\partial P}{\partial n}=\alpha(P_A-P), \qquad (x,y)\in\Gamma, \qquad (7)$$

$$\oint_{s_i} \frac{k(x,y)}{\mu}\frac{\partial P}{\partial n} ds = -q_i(t), \quad (x,y)\in s_i, \quad i=\overline{1,n_q}. \tag{8}$$

## 3. Numerical simulation

For the numerical solution of problem (5) - (8) by the differential sweep method, the filtration region $G$ with the outer boundary $\Gamma$ is covered by the grid region

$$\Omega_{xy\tau_k} = \left\{ \left(x_i = i\Delta h,\ y_j = j\Delta h,\ \tau_r = r\Delta\tau\right);\ i=\overline{1,N_j};\ j=\overline{1,M_i},\ k=\overline{0,N_\tau},\ \Delta\tau = \frac{1}{N_\tau} \right\}.$$

To obtain a differential-difference problem, the algorithmic concept of an implicit scheme of variable directions (longitudinal-transverse scheme) is used. The transition from the *r-th* time layer to the layer $r$+1 takes place in two stages with a step 0.5Δτ. The result is a successive solution of two systems of differential-difference equations.

$$\begin{cases} \dfrac{d}{dx}\left( k_j(x)\dfrac{dP_j^{(r+0.5)}(x)}{dx} \right) - \dfrac{1}{0.5\Delta\tau} P_j^{(r+0.5)}(x) = -\dfrac{P_j^{(r)}(x)}{0.5\Delta\tau} - \Lambda_y\left[k_j(x)P_j^{(r)}(x)\right], \\ \dfrac{d}{dy}\left( k_i(y)\dfrac{dP_i^{(r+1)}(y)}{dy} \right) - \dfrac{1}{0.5\Delta\tau} P_i^{(r+1)}(y) = -\dfrac{P_i^{(r+0.5)}(y)}{0.5\Delta\tau} - \Lambda_x\left[k_i(y)P_i^{(r+0.5)}(y)\right]. \end{cases} \tag{9}$$

Here

$$\Lambda_y\left[k_j(x)P_{i,j}^{(r)}\right] = \frac{k_{i-0.5,j}P_{i-1,j}^{(r)} - \left(k_{i-0.5,j} + k_{i+0.5,j}\right)P_{i,j}^{(r)} + k_{i+0.5,j}P_{i+1,j}^{(r)}}{h_{xy}^2},$$

$$\Lambda_x\left[k_i(y)P_{i,j}^{(r+0.5)}\right] = \frac{k_{i,j-0.5}P_{i,j-1}^{(r+0.5)} - \left(k_{i,j-0.5} + k_{i,j+0.5}\right)P_{i,j}^{(r+0.5)} + k_{i,j+0.5}P_{i,j+1}^{(r+0.5)}}{h_{xy}^2}.$$

The resulting system of differential-difference equations (9) is solved by the differential sweep method along each of the straight lines $x_i$ with initial conditions known at $\tau=\tau_r$, and then along each of the lines $y_j$, where the found values corresponding to the $r$+0.5-th layer are taken as the initial condition.

According to the differential sweep method, the solution of the first equation of the system of differential-difference equations (9) on the $r$+0.5-th time layer is derived by formulas:

$$P_j(x) = \frac{\gamma_j(x)u_j(x) - \alpha_j(x)w_j(x)}{\alpha_j(x)v_j(x) - \beta_j(x)u_j(x)};\ \frac{dP_j}{dx} = \frac{1}{k_j(x)}\cdot\frac{\gamma_j(x)v_j(x) - \beta_j(x)w_j(x)}{\alpha_j(x)v_j(x) - \beta_j(x)u_j(x)}. \tag{10}$$

Here the coefficients of the left-hand $\alpha_j(x)$, $\beta_j(x)$, $\gamma_j(x)$ and right-hand $u_j(x)$, $v_j(x)$, $w_j(x)$ sweep are found as solutions to the following Cauchy problems

$$\begin{cases} k_j(x)\dfrac{du_j(x)}{dx} = v_j(x), & u_j(0)=1, \\ \dfrac{dv_j(x)}{dx} = R_j u_j(x), & v_j(0)=0, \\ \dfrac{dw_j(x)}{dx} = F_j u_j(x), & w_j(0)=0; \end{cases} \quad \begin{cases} k_j(x)\dfrac{d\alpha_j(x)}{dx} = \beta_j(x), & \alpha_j(1)=1, \\ \dfrac{d\beta_j(x)}{dx} = R_j\alpha_j(x), & \beta_j(1)=0, \\ \dfrac{d\gamma_j(x)}{dx} = F_j\alpha_j(x), & \gamma_j(1)=0; \end{cases} \tag{11}$$

Here

$$F_j = -\frac{P_j^{(r)}(x)}{0.5\Delta\tau} - \Lambda\left[k_{i,j}P_{i,j}^{(r)}\right] - \delta_{i,j}q_{i,j}, \quad \mathrm{R}_j = \frac{1}{0.5\Delta\tau}.$$

The solution of the second differential-difference equation (9) on the $r$+1-th time layer with boundary conditions (6) - (8) is determined similarly.

Numerical integration of the Cauchy problem is carried out by the Runge-Kutta method using the normalization procedure of the sweep coefficients and the coefficients of this method.

In each step, when calculating vector $\vec{U}_{i+1}$, the normalized vector ($\vec{U} = (u, v, w)$ or $\vec{U} = (\alpha, \beta, \gamma)$) is substituted into the right-hand part of the system of equations instead of $\vec{U}_i$. The normalization procedure can be omitted if the chosen method steadily solves the Cauchy problem.

## 4. Computational experiments and their visualization

Based on the developed mathematical model, the Matlab program developed a software product called "Computer modeling of the filtration process of liquids in a non-homogeneous porous medium." The program includes blocks that include the filtration time of liquids, initial pressure, viscosity of the liquid, length of the layer in x, y, coordinates, layer thickness, well flow rate, permeability coefficients of the fragmented medium. Based on the parameters, that we entered in the program, when you press the button to calculate the indicators, the result will display images in 3D graphics and cut-outs, and write the numerical result in the file result.txt.

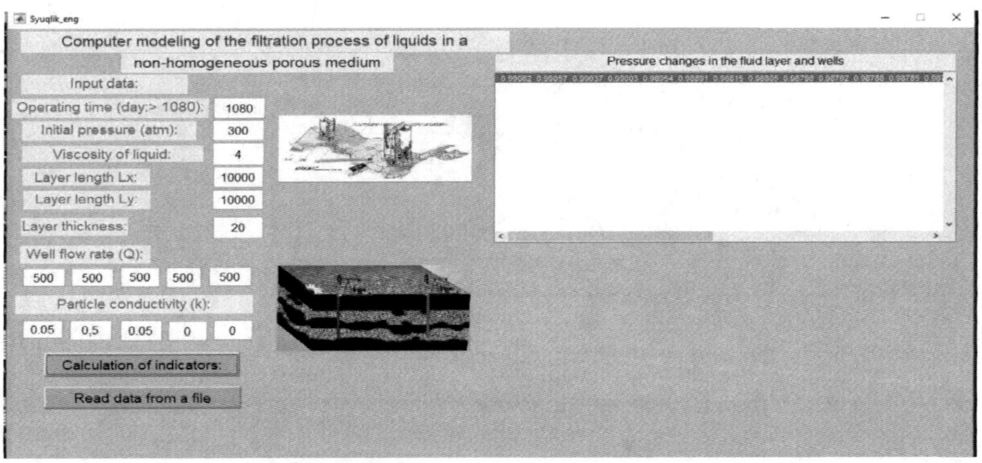


**Figure 1.** Matlab software product interface window.

In order to conduct computational experiments, the following initial data were considered: reservoir length $L$=10000 m; reservoir thickness $h$=20 m; initial reservoir pressure $P_н$=300 atm; permeability of reservoir k=0.5 and 0.05 Darcy; reservoir elastic capacity β=2.3 $10^{-5}$ $cm^2$/kg; oil viscosity μ=4 cps (centipoise).

Figures 2-5 show the results of calculation in the form of a graphical object.

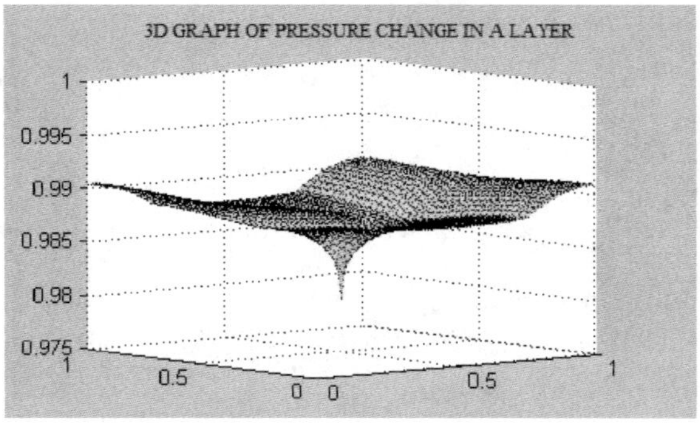


**Figure 2.** 3D graph of pressure distribution (lower profile)

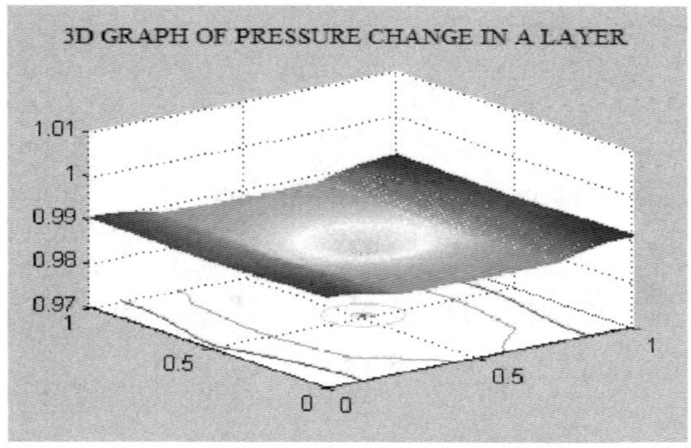


**Figure 3.** 3D graph of pressure distribution (upper profile)

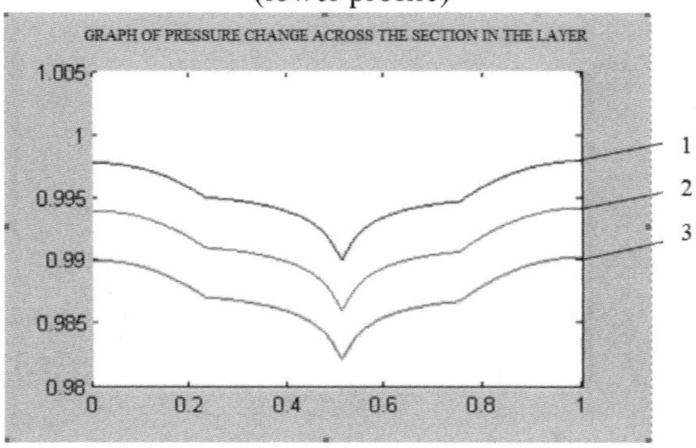


**Figure 4.** Graph of pressure distribution profiles in section $x$: 1-360.2-720.3-1080 days

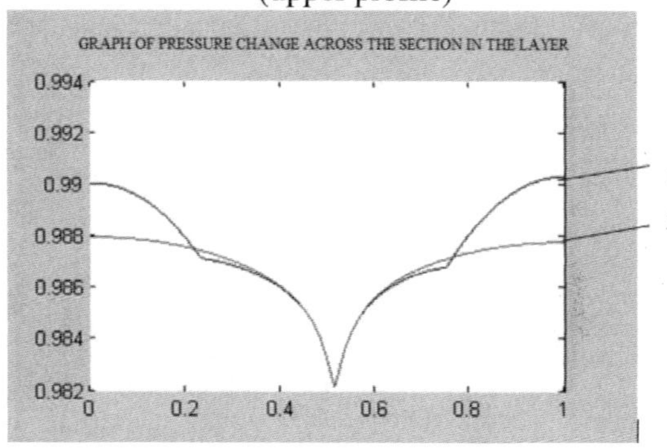


**Figure 5.** Graph of pressure distribution profiles in sections:1-$x$, 2-$y$, 1080 days

To ensure computing process control, an information array $I = \{\inf_{ij}\}$ $(i = \overline{1,N};\ \ j = \overline{1,M})$ is created.

In this case, the permeability values for each zone have the form:

$$G_1\text{: } k_1=0.05; \quad G_2\text{: } k_2=0.5; \quad G_3\text{: } k_3=0.05.$$

**Figure 6.** Filtration area in piecewise inhomogeneous porous medium

| | | | | | | | | | | | | | | | | | | | | |
|---|---|---|---|---|---|---|---|---|---|---|---|---|---|---|---|---|---|---|---|---|
| 1 | 1 | 1 | 1 | 1 | 2 | 2 | 2 | 2 | 2 | 2 | 2 | 2 | 2 | 2 | 2 | 3 | 3 | 3 | 3 | 3 |
| 1 | 1 | 1 | 1 | 1 | 2 | 2 | 2 | 2 | 2 | 2 | 2 | 2 | 2 | 2 | 2 | 3 | 3 | 3 | 3 | 3 |
| 1 | 1 | 1 | 1 | 1 | 2 | 2 | 2 | 2 | 2 | 2 | 2 | 2 | 2 | 2 | 2 | 3 | 3 | 3 | 3 | 3 |
| 1 | 1 | 1 | 1 | 1 | 2 | 2 | 2 | 2 | 2 | 2 | 2 | 2 | 2 | 2 | 2 | 3 | 3 | 3 | 3 | 3 |
| 1 | 1 | 1 | 1 | 1 | 2 | 2 | 2 | 2 | 2 | 2 | 2 | 2 | 2 | 2 | 2 | 3 | 3 | 3 | 3 | 3 |
| 1 | 1 | 1 | 1 | 1 | 2 | 2 | 2 | 2 | 2 | 2 | 2 | 2 | 2 | 2 | 2 | 3 | 3 | 3 | 3 | 3 |
| 1 | 1 | 1 | 1 | 1 | 2 | 2 | 2 | 2 | 2 | 2 | 2 | 2 | 2 | 2 | 2 | 3 | 3 | 3 | 3 | 3 |
| 1 | 1 | 1 | 1 | 1 | 2 | 2 | 2 | 2 | 2 | 2 | 2 | 2 | 2 | 2 | 2 | 3 | 3 | 3 | 3 | 3 |
| 1 | 1 | 1 | 1 | 1 | 2 | 2 | 2 | 2 | 2 | 2 | 2 | 2 | 2 | 2 | 2 | 3 | 3 | 3 | 3 | 3 |
| 1 | 1 | 1 | 1 | 1 | 2 | 2 | 2 | 2 | 2 | 2 | 2 | 2 | 2 | 2 | 2 | 3 | 3 | 3 | 3 | 3 |
| 1 | 1 | 1 | 1 | 1 | 2 | 2 | 2 | 2 | 2 | 2 | 2 | 2 | 2 | 2 | 2 | 3 | 3 | 3 | 3 | 3 |
| 1 | 1 | 1 | 1 | 1 | 2 | 2 | 2 | 2 | 2 | 2 | 2 | 2 | 2 | 2 | 2 | 3 | 3 | 3 | 3 | 3 |
| 1 | 1 | 1 | 1 | 1 | 2 | 2 | 2 | 2 | 2 | 2 | 2 | 2 | 2 | 2 | 2 | 3 | 3 | 3 | 3 | 3 |
| 1 | 1 | 1 | 1 | 1 | 2 | 2 | 2 | 2 | 2 | 2 | 2 | 2 | 2 | 2 | 2 | 3 | 3 | 3 | 3 | 3 |
| 1 | 1 | 1 | 1 | 1 | 2 | 2 | 2 | 2 | 2 | 2 | 2 | 2 | 2 | 2 | 2 | 3 | 3 | 3 | 3 | 3 |
| 1 | 1 | 1 | 1 | 1 | 2 | 2 | 2 | 2 | 2 | 2 | 2 | 2 | 2 | 2 | 2 | 3 | 3 | 3 | 3 | 3 |
| 1 | 1 | 1 | 1 | 1 | 2 | 2 | 2 | 2 | 2 | 2 | 2 | 2 | 2 | 2 | 2 | 3 | 3 | 3 | 3 | 3 |
| 1 | 1 | 1 | 1 | 1 | 2 | 2 | 2 | 2 | 2 | 2 | 2 | 2 | 2 | 2 | 2 | 3 | 3 | 3 | 3 | 3 |
| 1 | 1 | 1 | 1 | 1 | 2 | 2 | 2 | 2 | 2 | 2 | 2 | 2 | 2 | 2 | 2 | 3 | 3 | 3 | 3 | 3 |
| 1 | 1 | 1 | 1 | 1 | 2 | 2 | 2 | 2 | 2 | 2 | 2 | 2 | 2 | 2 | 2 | 3 | 3 | 3 | 3 | 3 |
| 1 | 1 | 1 | 1 | 1 | 2 | 2 | 2 | 2 | 2 | 2 | 2 | 2 | 2 | 2 | 2 | 3 | 3 | 3 | 3 | 3 |

**Figure 7.** Information array for permeability of porous medium

## 5. Conclusion

The developed mathematical software tool provides the calculation of basic indices of oil field development and the visualization of results of a computational experiment in a graphical form. The software tool provides numerical calculations in graphical visual form and can be used in design, analysis and forecasting of new oil field development.

**References**

[1] Ravshanov N, Nazirova E 2018 Numerical simulation of filtration processes of strongly polluted oil in a porous medium, *PONTE International Journal of Sciences and Research* 74(11), pp 107-116, doi: 10.21506/j.ponte.2018.11.10

[2] Fadeev SI 1978 On the numerical solution of linear boundary value problems for ordinary differential equations by the differential sweeping method. *Methods of spline functions, Computing systems* **75**, Novosibirsk

[3] Sadullaeva S A 2016 Numerical investigation of solutions to a reaction-diffusion system with variable density, *J. Sib. Fed. Univ. Math. Phys.*, **9**(1), pp 90–101, https://doi.org/10.17516/1997-1397-2016-9-1-90-101

[4] Nuraliev F, Anarova Sh, Amanov O, Jumaev S and Abdirozikov O 2020 Mathematical model and computational experiments for the calculation of three-layer plates of complex configuration, *J. Phys.: Conf. Ser.* **1546** 012094 doi:10.1088/1742-6596/1546/1/012094

[5] Ravshanov N and Shermatova G 2012 Computational experiment for the analysis of functioning of Technological process of filtering of a suspension European researcher. 2012. **2** (17) pp 114-119

[6] Piven VF 2015 *Mathematical models of fluid filtration* Oryol: Publishing house Oryol State University, Cartush, p. 408

[7] Sadullaeva Sh A and Khojimurodova MB 2018 Properties of solutions of the cauchy problem for degenerate nonlinear cross systems with convective transfer and absorption, *Algebra, Complex Analysis, and Pluripotential Theory. USUZCAMP 2017. Springer Proceedings in Mathematics & Statistics*, **264,** pp 183-190, https://doi.org/10.1007/978-3-030-01144-4_15

[8] Zayinidinov KN and Turapov UU 2016 Mathematical model of a non-invasive glucometer sensor, *Journal: Higher school*, **21**-1 pp 44-53

[9] Galaguz Y and Safin G 2016 Modeling of Particle Filtration in a Porous Medium with Changing Flow Direction, *Procedia Engineering* **153**, pp 157-161 https://doi.org/10.1016/j.proeng.2016.08.096

[10] Valeriy II, Solodova ES and Khlebnikova VV 2018 Mathematical model of oil-containing water purification process in the volume of granulated media, *Proceedings of the Institution of Mechanical Engineers, Part M: Journal of Engineering for the Maritime Environment* **233**(3), pp 879-885, https://doi.org/10.1177/1475090218794597

# Comparative analysis of prototypes for two-touch finger interfaces of smartwatch

**Y N Dimitrov[1], V P Aleksieva[1], H G Valchanov[1]**

[1] Computer Sciences Department, Technical University of Varna, Varna 9010, Bulgaria

E-mail: valeksieva@tu-varna.bg

**Abstract.** Two factors are important when the Smartwatch interfaces are developing - the small size of the display and the limited area for input control. The proposed by the authors prototype take them into account and aim to solve the "fat finger" problem and the difficult observation that affects the usability of the touch screen. It propose two fingers touch interface for smartwatch. The paper presents a comparison study of it for with two other prototypes. The results of complex evaluation of prototypes are given.

## 1. Introduction

Nowadays the smartwatches become more and more popular. According to the Canalys [1] the worldwide smartwatch sales reached 14.3 million units in Q1 2020, with an increase of 12% year on year, despite COVID-19. The main vendors and their market share are presented in figure 1.

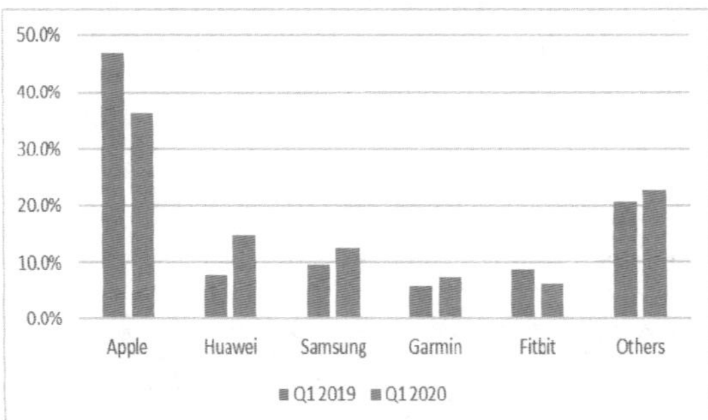


**Figure 1.** Market share of smartwatch by vendors.

The smartwatches have different features in addition of function watch - receiving calls or messages, making payments, counting the number of steps taken or monitoring the quality of sleep and heart rate. Most smartwatches are connected to a smartphones via Bluetooth or Wi-Fi. In all of these cases, there is a need to switch (and choose the proper function) from watch to another function. Smartwatches use mainly two methods for these switching functions – push buttons (like in classic watches) or touch screen (like in smartphones).

At the beginning of smartwatch era, the navigation was only with buttons (full push, half push, simultaneously push), but nowadays the most of the vendors use the touch AMOLED displays in smartwatches. For example, Apple Watch Series 5 has a touch display that is always on, but the display dims when the wrist is down, yet key features, like watch hands, remain visible at all times [2].

The watch face can be personalized by selecting apps or shortcuts to features (up to nine). Huawei watch GT2 46&42mm has 3D curved bezel-less watch face with touch AMOLED display [3]. In addition, some vendors use combination of technologies. For example, Samsung Galaxy Watch Active2 44&42mm has a touch AMOLED display, but adds a touch frame for quick control [4]. In smaller number of cases, the vendors as TicWatch, made the navigation with capacitive interface in bezel ring [5]. Touch-based interfaces are more efficient and intuitive, but when performed on a small surface such as a smartwatch display, there is a problem with the "fat finger" [6] - the user can't properly select the small target interface controls. Due to the fact that the user covers the screen with his finger, the visual control is lost to confirm the taken action.

In this paper, the authors present a prototype for a two-finger touch interface activation method for smartwatch and a comparison study of it and two other prototypes.

## 2. Related works

Many authors research and evaluate the touch-based interaction with smartwatch. Some researchers evaluate the touch or pressure technologies on the device components different from the display like devices' sides [7], band/strap [8], back side [9, 10]. The other ones have explored the area and surfaces around the device like touches on user's skin close to the wearable device [11] or using touch sensitive fabrics close to the device [12]. There are also researches about replacing standard touch interfaces with those based on taps [13] or joystick sensor control [14].

The authors in [15] research the interaction at the edge side of round smartwatches (with diameter of 60mm), where the sensors' position place is on the edge of round, but the two-finger multi-touch positions haven't been researched and evaluated.

## 3. Creation of two-finger touch interface prototype of the smartwatch

The process passed in four stages:

1. Choice of the dimensions of the 3D model

The authors designed and printed 3D model of smartwatch to observe the touch areas on the bezel. This mechanical prototype is as close as possible to real smartwatch according to its size and form. The diameters of the real smartwatches vary between 34.5mm and 58mm [2, 3, 4, 5], but 73% of them are between 42mm and 46mm. The height of the real smartwatches depends on kind of functions, but vary between 10,9mm and 16mm. Almost 80% of them are between 14mm and 15mm. Based on this statistics, the chosen dimensions of the 3D model are - diameter: 44mm, height: 15mm, bezel angle: 450, bezel width: 4mm; color: white; material: PLA (figure 2). A push buttons or a crown are not added. The surface is divided on 12 equal sectors - on the same places where are marked hours on the analogue watch. The model has lugs for standard 18mm watch strap (to be closer to the standard watch). The diameter of each sensor is 3mm and the space between them is equal.

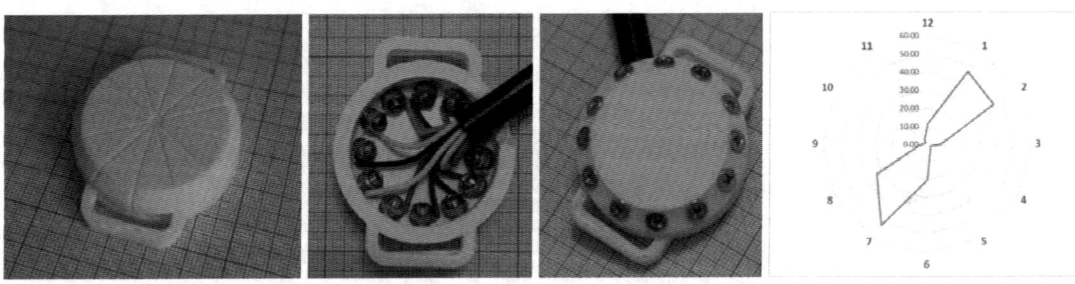


**Figure 2.** 3D Model for evaluation of two-finger touch interface activation.

2. Evaluation of the two touch sectors on the prototype

The authors register which two separate sectors on the device bezel are activated. The detailed results from this evaluation are presented from authors in [16], but in figure 2 is presented the radar chart of the number of touches per sensor (statistics is about users with right leading hand).

3. Comparison of activation of a function sequence for this prototype and prototype with buttons

The authors made this comparison [17] and the main conclusion is – the touch prototype outperformed the prototype with buttons in speed of operations, especially when the set of the interface commands is longer.

4. Evaluation of the prototype in comparison with the similar prototypes.

## 4. Comparison study of the authors' prototype with other prototypes with same interfaces

To be possible to make a comparative analysis of the authors' prototype with others, the same criteria for evaluation as in [14] and [15] are used. In [15], physical touch with the device edge was integral to the vast majority of inputs – only 17% are with two fingers. Participants also favored their dominant hands and use of their thumb and index fingers. The authors gave the results for eight ordinal directions, but in this comparison study only results for matching direction are used. In [14] the authors use multiple fingers to move the whole prototype, which is with a typical size of smartwatch. They showed that their prototype is close to the commercial smartwatches of the same size while input events are generated responsively (55-61ms) and accurately.

The experimental data for authors' prototype are presented in comparison with [14] and [15] in table 1.

**Table 1.** Comparison of prototypes for two fingers touch interfaces for smartwatch.

| Criteria | Prototype of [14] | Prototype of [15] | Authors' prototype |
|---|---|---|---|
| |  | 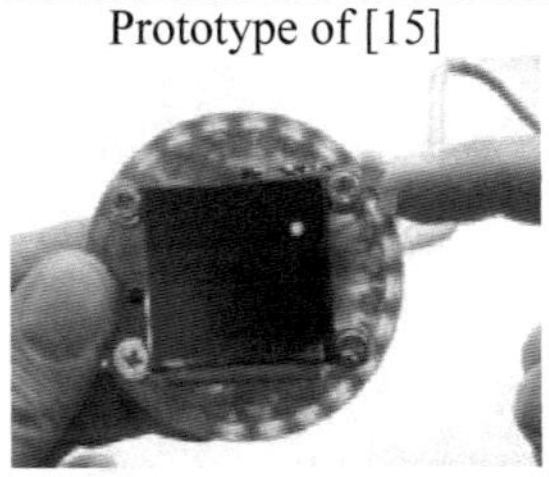 |  |
| Satisfaction of participants | 0.5 | 0.76 | 0.55 |
| Number of sensors | 12 (3 by side) | 24 (every 7.5 degrees) | 12 (every 15 degrees) |
| Touch time [ms] short sequence | 39 | 5.95 | 10.5 |
| Touch time [ms] long sequence | 61 | 50.37 | 14 |
| Number slip-offs | 0.37 | 0.02 | 0.33 |
| Number misses | 0/100 | 0.175 | 0.27 |
| Simple touch average error ratio | 0 | 0.1 | 0.6 |
| Complex touch average error ratio | 0 | 0.3 | 2.4 |
| Size [mm] | 49/46/16.5 (rectangle) | 60/18.85 (circle) | 44/15 (circle) |

Due to the obvious diversity of indicators for evaluating the two fingers touch interfaces, it is impossible to define a specific function for the quality of the proposed by authors' prototype. In such cases, average complex indicators are used. Based on the data, presented in the table 1, the average arithmetic (1) and average geometric (2) estimation is made.

$$R_a = \frac{\sum_{i=1}^{n} b_i d_i}{\sum_{i=1}^{n} b_i} \tag{1}$$

$$R_g = \left(\prod_{i=1}^{n} d_i^{b_i}\right)^{\frac{1}{\sum_{i=1}^{n} b_i}} \quad (2)$$

In (1) and (2) the normalized quantitative assessment of the **i**$^{th}$ indicator (measure of a criterion) is

$$0 \le d_i \le 1 \quad (3)$$

**n** – number of indicators, and $\mathbf{b_i}$ is the **i**$^{th}$ weighting coefficient, as

$$\sum_{i=1}^{n} b_i = 1 \quad (4)$$

These coefficients ($\mathbf{b_i}$) are selected on the basis of expert assessments or statistical methods. When the values cannot be obtained in this way (insufficient number of analyses or expert assessments in the research area), the coefficients are chosen to be equal.

Due to the heterogeneous nature of the selected criteria, the absolute values given in table 1 must be converted into relative values (dimensionless quantities). For example, for the prototype interfaces studied, if the average number of errors is higher, this means a worse implementation, while the shorter times for selecting the required function shows a better implementation. To obtain comparable results, the conversion into relative values is done so that the higher values for each of the criteria are considered better. If the complex estimations (arithmetic and geometric) are higher, this means more effective prototype. The complex evaluation according to a selected list of criteria gives a possibility to eliminate the shortcomings of the prototypes according to a single criterion.

The results of the comparative analysis are presented in table 2.

**Table 2.** The complex evaluation of prototypes for two fingers touch interfaces for smartwatch.

| Evaluation | Prototype of [14] | Prototype of [15] | Authors' prototype |
|---|---|---|---|
| Ra | 0.463 | 0.342 | 0.762 |
| Rg | 0.465 | 0.214 | 0.546 |

The results obtained for the complex evaluation present fact, that the prototype for two fingers touch interfaces for smartwatch proposed by the authors is better than these proposed in [14] and [15]. The main advantage of the authors' prototype is its standard size and less touch time in long sequence of touches.

## 5. Conclusions

This paper presents a prototype of smartwatch with two fingers touch interface and a comparative analysis of them and two others [14, 15]. A system of criterion for comparing two fingers touch interfaces for smartwatch has been proposed. Based on them, a complex evaluation of these prototypes was made. When examining the criteria separately, it can be seen that in the case of "Simple touch average error ratio" and "Complex touch average error ratio" the proposed prototype does not provide good results. Despite this fact, the complex assessment of prototypes presents the highest value for the authors' prototype and it is better than others.

The presented in the paper two fingers touch interface is suitable for smartwatches, designed for sports and fitness, as it offers the shortest time to activate a long sequence function.

The goal of the further research could be to develop a software, which will recognize the user body pose. Depending on of the body pose (straight, sitting, lying down), the smartwatch should provide the user with context menus with certain functions.

## References

[1] Canalys: Worldwide smartwatch shipments grew 12% in Q1 2020 despite coronavirus *https://www.canalys.com/newsroom/canalys-worldwide-smartwatch-shipments-q1-2020 Last visited on 15.07.2020*

[2] Apple watch series 5 Available at: *https://www.apple.com/apple-watch-series-5/design/* Last visited on 15.07.2020
[3] Huawei watch GT2 Available at: *https://consumer.huawei.com/en/wearables/watch-gt/* Last visited on 15.07.2020
[4] Samsung Galaxy Watch Active2 44 mm, Stainless Steel Available at: *https://www.samsung.com/bg/wearables/galaxy-watch-active2-r820n/SM-R820NSKABGL/* Last visited on 15.07.2020
[5] TicWatch C2 +: *https://www.mobvoi.com/eu/pages/ticwatchc2plus?gclid=EAIaIQobChMI8pHnu4PP6gIVAp3VCh2R-Q3cEAAYASAAEgLYgfD_BwE* Last visited on 15.07.2020
[6] Siek K, Rogers Y and Connelly K 2005 Fat finger worries: how older and younger users physically interact with PDAs, *Human-Computer Interaction - INTERACT 2005. Lecture Notes in Computer Science*, **3585**, ed M Costabile and F Paternò, (Springer, Berlin, Heidelberg) https://doi.org/10.1007/11555261_24
[7] Darbar R, Sen P and Samanta D 2016 PressTact: side pressure-based input for smartwatch interaction, *Proc. 2016 CHI Conf. Ext. Abstr. Hum. Factors Comput. Syst.* pp 2431–2438, https://doi.org/10.1145/2851581.2892436
[8] Ahn Y, Hwang S, Yoon H, Gim J and Ryu J 2015 BandSense: Pressure-sensitive Multi-touch Interaction on a Wristband *Proc. 33rd Annu. ACM Conf. Ext. Abstr. Hum. Factors Comput. Syst.* pp 251–254, https://doi.org/10.1145/2702613.2725441
[9] Lim S, Shin J, Kim S and Park J 2015 Expansion of smartwatch touch interface from touchscreen to around device interface using infrared line image sensors, *Sensors* **15**(7) https://doi.org/10.3390/s150716642 pp 16642–16653
[10] Xiao R, Laput G and Harrison C 2014 Expanding the input expressivity of smartwatches with mechanical pan, twist, tilt and click *Proc. 32nd Annu. ACM Conf. Hum. Factors Comput. Syst.* pp 193–196, https://doi.org/10.1145/2556288.2557017
[11] Zhang C, Bedri A, Reyes G, Bercik B, Inan OT, Starner TE and Abowd GD 2016 TapSkin: Recognizing on-skin input for smartwatches, *Proc. 2016 ACM Interact. Surfaces Spaces* pp 13–22, https://doi.org/10.1145/2992154.2992187
[12] Schneegass S and Voit A 2016 GestureSleeve: using touch sensitive fabrics for gestural input on the forearm for controlling smartwatches, *Proc. 2016 ACM Int. Symp. Wearable Comput.* pp 108–115, https://doi.org/10.1145/2971763.2971797
[13] Yoon H and Park S 2020 A Non-Touchscreen Tactile Wearable Interface as an Alternative to Touchscreen-Based, Wearable Devices *Sensors 2020* **20**(5) 1275 https://doi.org/10.3390/s20051275
[14] Yeo H, Lee J, Bianchi A and Quigley A 2016 Sidetap & slingshot gestures on unmodified smartwatches, *Proc. 29th Annu. Symp. User Interface Softw. Technol.* pp 189–190, https://doi.org/10.1145/2984751.2984763
[15] Oakley I and Lee D 2014 Interaction on the edge: offset sensing for small devices, *Proc. SIGCHI Conf. Hum. Factors Comput. Sys.* pp 169–178, https://doi.org/10.1145/2556288.2557138
[16] Dimitrov Y 2018 Two-finger touch interface activation method for wearable devices *Proc. ICEST 2018* ISSN: 2603-3259 pp 146-149
[17] Dimitrov Y, Aleksieva V and Valchanov H 2019 Wearable device interfaces comparison, *16th Conference on Electrical Machines, Drives and Power Systems (ELMA),* Varna, Bulgaria, 2019, pp 1-4, doi: 10.1109/ELMA.2019.8771545

**Acknowledgments**
The research, the results of which are presented in the present paper, have been carried out under TU-Varna Scientific Project "Research of the possibilities of integrating machine learning and Blockchain technologies for the Internet of Things", which is financed by the state budget.

# Objects distance measurement in augmented reality for providing better user experience

**N Neshov**[1], **A Manolova**[1]

[1] Radiocommunications and Videotechnologies Department, Technical University of Sofia, Bulgaria

E-mail: nneshov@tu-sofia.bg, amanolova@tu-sofia.bg

**Abstract.** In this paper we present an algorithm for providing better user experience in AR applications based on HMD (Head Mounted Display), pair of cameras and Deep Learning semantic segmentation. The user can view the scene, captured by the cameras using HMD along with augmented information regarding distances to all detected objects. The cameras are attached to the HMD and must be located as close as possible each other. Moreover one of the cameras is shifted at a known distance closer to the scene with respect to the other camera. Based on the traditional pinhole camera model and estimating the size of the projection of a given object in pixel coordinate systems in both of the cameras, we are able to calculate the distance from the cameras to the object. For location and calculation of the size of projected objects we used a semantic segmentation based on deep learning algorithm.

## 1. Introduction

In the recent years AR/VR (Augmented Reality/Virtual Reality) technologies are being used in wide range of applications and have a big impact over the user experience. Many HMDs (Head Mounted Display) provide immersive environment by embedding virtual content in a real world scenario. One of the main challenges in AR/VR is intuitive interactions with virtual content. Providing interaction in AR is related to correct interpretation of depth and distance from HMD to real objects. There is a large number of AR applications, where measuring the distances to the real objects is essential. The list includes, but is not limited to: prototyping large structures [1], helping facility managers by providing building information at place [2], improving the functional vision of people with near-complete vision loss [3], AR assisted intraoperative distance estimation [4], virtual furnishing [5], high precision manual chirurgical tasks [6] and etc. Yet many researches confirm that the distance underestimation is not completely understood. The "focal rivalry" or when the human eyes can't focus on both virtual and real objects simultaneously, leaves room for error when accomplishing high precision manual tasks. Inaccuracy in distance measurements leads to unacceptable errors, degrading the interaction performance. For example, differences in distance perception for real and virtual objects may affect pointing accuracy.

The goal of this paper is to propose a platform and an algorithm for AR application that provides the user with information regarding distances to some of the objects observed in the scene. The user wears a HDM (Oculus Rift in our experiments) with two cameras are attached to it, shifted at a known distance with respect to the scene. The application delivers video from the one of the cameras in real time to the HMD, computes distances to the objects and embeds them as a virtual label over the real scene. In this way, a HMD designed for VR can be used for AR, with certain limitations.

## 2. Algorithm description

On figure 1 is shown the main architecture of proposed system.

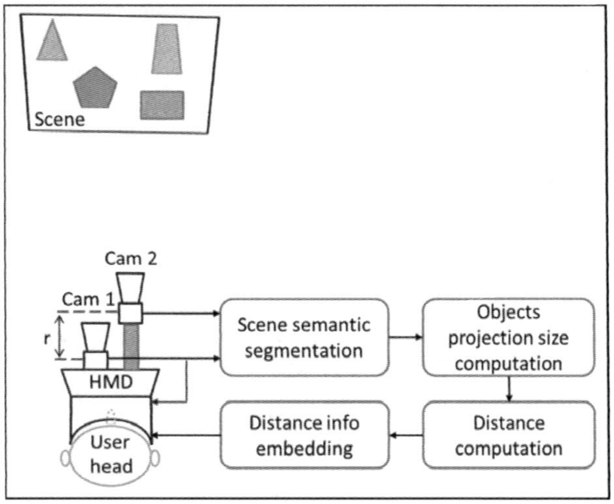


**Figure 1.** Overview of proposed system.

The two cameras attached to the HMD are shifted at a known distance *r* with respect to the scene simultaneously providing video frames. The first camera delivers video directly to the HMD, hence the user can view the real scene. Both video streams are passed to the first block (*Scene semantic segmentation*), where we apply DL algorithm based on the DeepLabV3 model with a ResNet-101 backbone [7] in order to distinguish different objects in the scene and find their respective edges. In the next step (*Objects projection size computation*), a bounding box for each object is constructed and the height of the bounding box is considered as the size of this object. The sizes for all objects act as input in the next step (*Distance computation*) of the algorithm. In order to calculate the distance in real units, we were inspired by the method described by S. Sayeed in [8], which utilizes the traditional pinhole camera model. The idea behind this method applied for the goal of this research is illustrated in figure 2. Let's assume that camera 1 is our reference camera (i.e. we want to calculate the distance *d* between camera 1 and object on the scene).

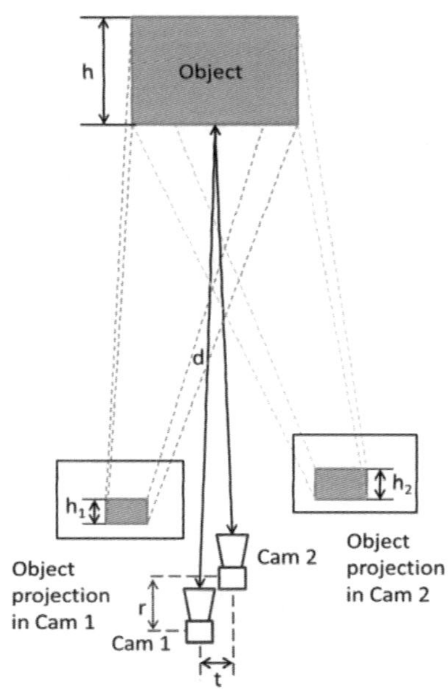


**Figure 2.** Cameras location and object projections.

As already mentioned, we assume that the two cameras are shifted between each other (in perpendicular direction with respect to the scene plane) at a distance $r$. Moreover we also assume that the distance $t$ between the two cameras (in parallel direction with respect to the scene plane) is much smaller than the distance from camera 2 (camera 1 respectively) and the object. Given an object with height $h$, the projection of this object to camera 1 (i.e. $h_1$) will be smaller than the projection of the same object to camera 2 (i.e. $h_2$). If the aforementioned assumptions are accomplished, we can build our mathematical formulation for calculating $d$ with the help of the model depicted in figure 3. We can write the following equations:

$$\frac{h}{d} = \frac{h_1}{f}, \tag{1}$$

and

$$\frac{h}{d-r} = \frac{h_2}{f}, \tag{2}$$

where $f$ is the focal length of the first and second camera (must be the same). If we divide equation 1 by the equation 2, and after conversions, we deduce the following for the desired distance $d$:

$$d = \frac{r}{1-\frac{h_1}{h_2}}, \tag{3}$$

where $r$ and $d$ are expressed in real units (i.e. meters, centimetres, etc.) and $h_1$ and $h_2$ - in pixels.

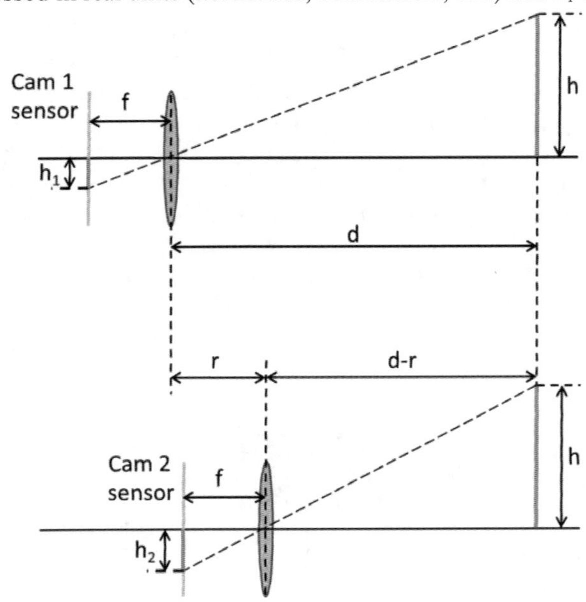


**Figure 3.** Geometrical representation of two shifted cameras for object distance measurement based on the pinhole camera model.

Based on the general diagram in figure 1 and once the distances to each object are calculated, the respective label is embedded as augmented information in the video frame (*Distance info embedding*).

## 3. Experiments

In this section we present the conducted experiments for the proposed algorithm. We investigate the absolute and relative errors in measurement using an object (bottle), which was placed at different distances $d$ with respect to the cameras. Moreover we assess the errors against different values of $r$. A total of 30 experiments were conducted. Given $r$, the precision of the measurements depends on the size of projected object, which in turn depends on the size of the object in the real scene, the distance from the object and cameras, and the resolution. Based on equation 3, the results show that for a given

$r$, and low resolution images and/or small projected objects (i.e. small values of $h_1$ and $h_2$), the value of $d$ becomes very sensitive when $h_1$ and $h_2$ change by even one pixel (since it depends on the ratio $h_1/h_2$). In this paper, all presented experiments are performed on images with fixed resolution of 1280 x 1024 pixels.

Figure 4 depicts two example images (captured by camera 1 and camera 2) of the object placed at certain distances, along with the mask images detected by the DL Semantic Segmentation algorithm [8] and the corresponding bounding box, which height (as mentioned earlier) determines the size of the object.

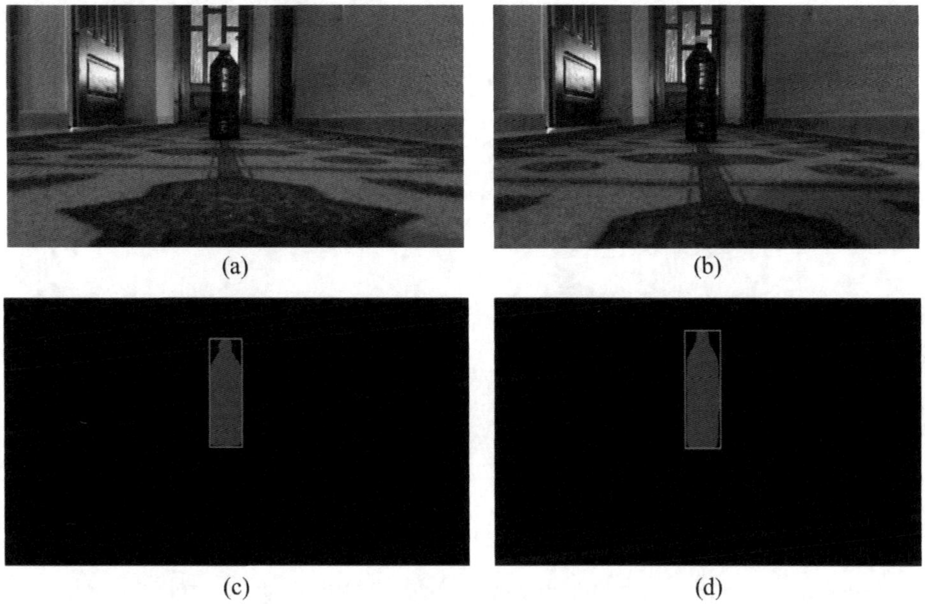


**Figure 4.** Camera 1 (a) and Camera 2 (b) images of object (their corresponding masks and bounding boxes – (c) and (d) respectively), placed at a distance $d = 94$ cm and $r = 10$ cm.

In table 1 are shown the absolute errors and in table 2 - the relative errors as a function of distance to the object $d$, and distance between the cameras – $r$ (see figure 2).

**Table 1.** Absolute errors at different distance $d$ from the object to Camera 1, and different distance $r$ between the cameras.

| Absolute error, cm. | | Distance to object - *d*, cm. | | | | | |
|---|---|---|---|---|---|---|---|
| | | 94 | 113 | 133 | 152 | 172 | 191 |
| Distance between cameras - *r*, cm. | 5 | 3.73 | 5.00 | 5.64 | 6.44 | 5.57 | 16.83 |
| | 10 | 2.81 | 2.94 | 5.40 | 6.82 | 11.85 | 22.00 |
| | 15 | 1.53 | 1.71 | 4.92 | 3.56 | 2.91 | 14.31 |
| | 20 | 1.14 | 1.38 | 2.06 | 1.16 | 3.86 | 8.60 |
| | 25 | 1.00 | 1.52 | 3.00 | 12.00 | 2.13 | 32.45 |

The results in table 1, table 2 and figure 5 show that the smallest error is observed when the object is near the cameras. As the distance increases, the average error increases. This behaviour is expected, because as the object moves away, its projections become smaller. Since the resolution in measured size is only one pixel, as the object gets smaller the step between two possible calculated distances is getting higher that leads to increase of the error. Therefore, the error also depends on the correct detection of the bounding box of the object.

**Table 2.** Relative errors at different distance *d* from the object to camera 1, and different distance *r* between the cameras.

| Relative error, % | | Distance to object - *d*, cm. | | | | | |
|---|---|---|---|---|---|---|---|
| | | 94 | 113 | 133 | 152 | 172 | 191 |
| Distance between cameras - *r*, cm. | 5 | 3.97 | 4.42 | 4.24 | 4.24 | 3.24 | 8.81 |
| | 10 | 2.99 | 2.60 | 4.06 | 4.49 | 6.89 | 11.52 |
| | 15 | 1.62 | 1.51 | 3.70 | 2.34 | 1.69 | 7.49 |
| | 20 | 1.22 | 1.22 | 1.55 | 0.76 | 2.25 | 4.50 |
| | 25 | 1.06 | 1.35 | 2.26 | 7.89 | 1.24 | 16.99 |

Figure 5 shows the same information as a graphical representation.

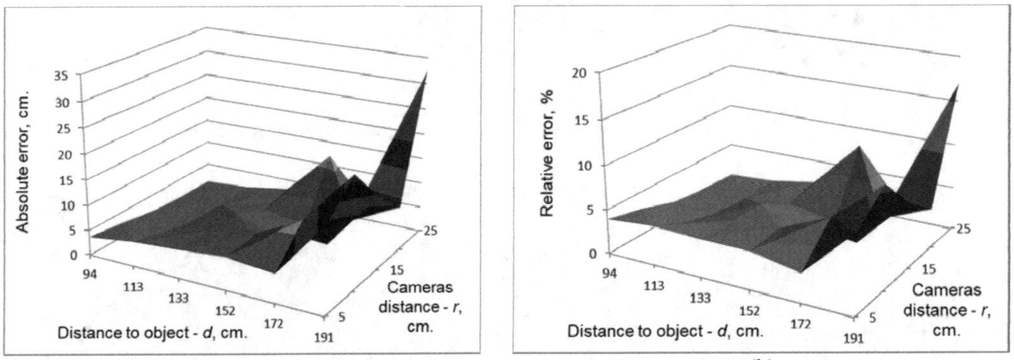


(a) (b)

**Figure 5.** Graphical representation of absolute (a) and relative errors (b) at different distance.

From the results in table 2, a conclusion can be made that the smallest relative error is below 1% and the largest one is below 17%. However, the average relative error is around 4%. Among all experiments performed, the error exceeds 5% only for 6 of all 30 cases. Figure 6 depicts an example image of 3 objects, shown to the user via the Oculus Rift. The algorithm embeds the relevant information: type of each object detected by DL Semantic segmentation, bounding boxes, and calculated distance. The real distance to each object is: Car at 100 cm, Bottle at 150 cm, and Monitor at 200 cm. The maximum error in this example is around 3 cm for the object at 200 cm. (< 2%). On figure 7 (a) and figure 7 (b) are shown the corresponding segmented masks for both of the images captured by camera 1 and camera 2 respectively.

**Figure 6.** Example image, shown to user via Oculus Rift, along with the labels embedded.

(a) (b)

**Figure 7.** Segmentation masks for images captured by Camera 1 (a) and Camera 2 (b) of 3 objects shown in figure 6.

## 4. Conclusions and future work

In this paper we propose an algorithm for distance to objects measurement using two cameras attached to an Oculus Rift HMD and placed at a known distance with respect to the scene. The algorithm presents to the user the following information: bounding box and label, type of recognised object as well as the distance to it. From the experimental results it can be concluded that the relative error does not exceed 5%. As a future direction we plan to investigate how to improve the DL algorithm for semantic segmentation for high resolution images. The intention is to tune up the algorithm to produce optimal accuracy for large distant objects while keeping the processing in real time.

## References

[1] Piekarski W and Thomas B 2003 Interactive augmented reality techniques for construction at a distance of 3d geometry", *in Proc. EVGE*, ACM , pp 19–28
[2] Irizarry J, Gheisari M, Williams G and Walker B 2013 Infospot: A mobile augmented reality method for accessing building information through a situation awareness approach, *Automation in Construction*, **33**:pp 11–23, doi: 10.1016/j.autcon.2012.09.002
[3] Kinateder M, Gualtieri J, Dunn M, Jarosz W, Yang and Cooper E 2019 Using an augmented reality device as a distance-based vision aid-promise and limitations, *Optom. Vis. Sci.*, **95**, No 9, pp 727–737
[4] Heinrich F, Schmidt G, Jungmann F and Hansen C 2019 Augmented Reality Visualisation Concepts to Support Intraoperative Distance Estimation, *in Proc. ACM Symp. on Virtual Reality Software and Technology (VRST)*, pp 1-2
[5] Swaminathan R, Schleicher R, Burkard S, Argurto R and Koleczko S 2013 Happy Measure: Augmented Reality for Mobile Virtual Furnishing, *International Journal of Mobile Human Computer Interaction*, **5**, No. 1, pp 16-44
[6] Condino S, Carbone M, Piazza R, Ferrari M and Ferrari V 2019 Perceptual limits of optical see-through visors for augmented reality guidance of manual tasks, *IEEE Transactions on Biomedical Engineering*, **67**(2), pp 411-419.
[7] Chen L, Papandreou G, Schroff F and Adam H Rethinking atrous convolution for semantic image segmentation, *CoRR*, vol. abs/1706.05587
[8] Sayeed S 2014.Distance to objects using single vision camera, *uploaded by Shahriar Sayeed*

**Acknowledgments**
The theoretical work was supported by the National Programme "Young Scientists and Postdoctoral Students", PMC No577, 17.08.2018, of MES Bulgaria and the experimental work was supported by the contract КП-06-Н37/8 from 06.12.2019: „Inference algorithms for semantic knowledge extraction based on deep architectures for context-aware holographic communication" of the Bulgarian Research Fund of the Ministry of Education and Science, Bulgaria.

IOP Conf. Series: Materials Science and Engineering 1032 (2020) 012021 doi:10.1088/1757-899X/1032/1/012021

# Technique for storing and automated processing of weather station data in cloud platforms

**S Sokolov[1], S Vlaev[2], M Chalashkanov[3]**

[1] University of Telecommunications and Post, Department of Information Technology, Sofia, Bulgaria, strahil.sokolov@gmail.com
[2] University of Telecommunications and Post, Department of Information Technology, Sofia, Bulgaria, svlaev@mail.bg
[3] University of Telecommunications and Post, Department of Information Technology, Sofia, Bulgaria, mike83@abv.bg

E-mail: strahil.sokolov@gmail.com

**Abstract.** In this paper we present an approach for design of cloud-based systems for storing and processing of sensor data from weather stations. In recent years there are abundant IoT devices that are capable of fairly precise data acquisition of local atmospheric parameters and reliable transmission over wireless network. The used dataset is acquired from a modern weather station model from a superior European manufacturer. 17 parameters are used for training neural network based LSTM model. In our research we have also compared several models and approaches for modeling this type of data. We propose a design of container based system for automated processing of the received data, visualization and cloud storage technique. We are comparing two types of open-source container technologies for implementing data processing. We have evaluated the performance and provide the set of criteria for choosing a container platform. Storing the data in the cloud is an important part and we provide our consideration for implementing a particular open source platform that could reduce costs fulfil the requirements for weather data storage. The designed system is expected to be cloud-provider agnostic. The proposed features make the system especially suitable for purposes such as local weather forecast; disaster prediction: hurricane, severe storms, floods, tornadoes, etc.; severe droughts with fire dangers; prediction of landslides. Depending on the availability of additional sensors the future development of the system may be extended with parameters of the soil. Training the model on a freely available data set is in our future plans for system development.

## 1. Introduction

There is a growing demand for IoT and Cloud based applications worldwide and it is considered that one area is bound to exist with the other. Therefore there is a need for modern approaches and platform which are able to provide a combination of automated data acquisition, secure cloud storage, algorithms for data processing and improved availability. Our aim is to enhance the current practices in the field of IoT Cloud.

There exist various modern data acquisition systems and wireless devices for weather data, which are used in meteorology and smart agriculture. Several designs are available, as the researchers in [1] have shown that automatic weather stations seem to be more efficient and user-friendly. An automatic weather station is a device that is capable of measuring and recording weather parameters using sensors and does not rely on human intervention. The measured data can be stored in the embeded data logger or can be transmitted to a centralized data storage facility though a supported type of communication channel. In case that the measurements are stored in a local data logger, the records must be manually copied to a computer at a later time for further processing. This is not an acceptable scenario, especially when the weather station is located at a remote location and usually operates in unattended mode. Therefore, the availability of communication system is an essential building block in an automated weather station. The modern technologies [2], [3] allow that such stations be built be built at a different location and maintained on the remote location with minimum effort to transmit data at regular intervals. The work showed that multiple weather parameters can be captured in real-time, at pre-programmed intervals [1]. This type of measurements are important to detect extreme events such as floods, hurricanes, tornadoes, as well as to issue timely warning messages to the public [4].

The authors in [5] show that the Mobile Cloud Computing is an emerging technology that involves a type of infrastructure in which data storage and processing are done outside the smartphone. IoT (Internet of Things) has been around as a concept technology and attracts ever-growing interest in the telecommunications domain, especially when IoT relates to wireless telecommunications. As the main purpose of IoT the authors describe the interaction and cooperation between physical entities and abstract objects that are being sent over the wireless protocols to act and serve as a combined entity [6], [7]. This enables the rapid growth and demand of resources in both Cloud Computing and Internet of Things. The authors have provided a survey most frequently used IoT and Cloud Computing technologies with a focus on the security issues of both technologies and claim that Cloud Computing technology improves the IoT operations. In addition, there are security challenges of the integration of IoT and Cloud Computing which need to be taken into consideration and avoided if possible.

Cloud computing is the abstract underlying platform [5] which provides computing resources, storage, services, and middleware that enable us to server applications to the public over the Internet. The IoT cloud computing is defined as combination of cloud computing technology with mobile devices and sensors. It is a multidisciplinary area and can also be referred to as mobile cloud computing. There exist two main perspective to the mobile cloud: a) based on infrastructure, and b) ad-hoc based. The concept that is based on infrastructure relies on the fact that the hardware infrastructure remains static and also provides services to the mobile users and potentially to other services. In our current research, this type of more static cloud is covered.

Several challenges have been listed in [5], [8] regarding cloud computing: One of them is Heterogeneity and is related to the wide variability of devices, operating systems, platforms, and services available and possibly used for new or improved applications.

Another challenge which is pointed often by authors in the field is performance. There are IoT Cloud applications that introduce specific performance and QoS requirements at several levels (with aspect to communication, computation, and storage) [9].

Further challenge is often reliability: adopting the IoT Cloud for mission-critical applications, reliability concerns typically arise e.g., in the context of smart mobility, users are often on the move and the moving device communication and connection is often intermittent or unreliable.

Volumes of data or the so called Big Data is another major concern within the IoT Cloud computing field. With the ever growing number of IoT devices, the demands for storage, access, and processing of the huge amount of data they produce also rises. [10]

Last but not least, the authors have mentioned monitoring of the IoT Cloud as a major challenge. It represents essential activity in Cloud environments and serves as basis for billing, capacity planning, for managing resources, SLAs, performance and security audit, troubleshooting.

Processing of the weather-station data in the cloud requires efficient algorithms and approaches which are able to extract relevant features and make precise classification of the inputs in a computationally efficient way. With the recent focus of deep learning methods, especially long short-term memory (LSTM) neural networks, there have been a number of interesting applications especially in the field of network traffic prediction [11]. The true value of deep learning methods [12] in traffic event forecasting is similar to the feed which the sensor data is providing from weather station. There is a need, according to the authors that these methods need to be further exploited in terms of the depth of the model architecture, the spatial scale of the prediction area, and the predictive power of spatial-temporal data.

The authors have proposed a deep stacked bidirectional and unidirectional LSTM (SBULSTM) neural network architecture is proposed, which considers both forward and backward dependencies in time series data, to predict network-wide traffic speed [13]. A bidirectional LSTM (BDLSM) layer is utilized to capture spatial features and bidirectional temporal dependencies from historical data. The authors propose a model which is able to handle missing values in input data by using a masking mechanism. Their model is scalable model and is applied on complex networks.

The rest of the paper is organized as follows: In chapter 2 we describe the used IoT Cloud architecture and its components; Chapter 3 provides details on the used approach for weather station data processing; Chapter 4 describes our approach for enhancing the performance of the system by using the selected components in comparison to traditional techniques; chapter 5 provides details on the used devices for weather data acquisition, experimental results and discussion.

## 2. Description of the proposed IoT Cloud architecture

The proposed architecture consists of input layer (the weather stations), the data is then transmitted over the Internet to our private cloud and the distribution layer (load balancer) which propagates the requests to the backend layer which is composed of Docker containers. The applications which run in Docker container have a small footprint and are relatively resilient. The bottom layer is the persistence layer and it is based on CEPH storage. It provides unmatched object storage, high availability and low cost of operation compared to many of the products which we have had experience with so far. More details will be provided below.

Below are some of the features which we are aiming to achieve via our design by incorporating elements of Cloud Computing technology for IoT and overcome some of the challenges that have been presented in the previous chapter:

- **Storage over Internet**: One of the features of the proposed architecture is that we achieve Storage provisioning through Software Defined Storage (SDS) mechanism [14]. The core technology which enables the persistent storage for the weather-station data and for the data processing algorithms storage is CEPH [14], [15], [16]. CEPH is an open-source software storage platform that implements object storage on a single distributed computer cluster, and provides simultaneously interfaces for object-, block- and file-level storage. Its main goal is completely distributed operation without a single point of failure, scalable to exabyte level, and freely available. CEPH replicates data and makes it fault-tolerant, [16] using off-the shelf hardware and requires no specific hardware support. The resulting storage subsystem is by design self-healing and self-managing with additional minimization of administration time and other costs.
- **Service over Internet**: The usage of Docker containers to provide backend services for data processing, presentation services for the processing results, our goal was to achieve high availability and continuous integration and delivery of the services. Docker containers being lightweight and faster than virtual machines are able to provide this type of flexibility [15], [17]. The applications that perform the weather-station data processing are Python based; they run in Docker containers and use the persistence storage layer to read the input values and store the predictions.
- **Applications over internet**: Our load balancing layer is based on HAProxy [15]. HAProxy is free software and open source. It is the tool which provides high availability load balancer service in our architecture for TCP and HTTP-based applications. It is able to distribute requests across multiple backend servers using high performance algorithms and is written in C. This makes this software famous for being fast and efficient.
- **Computationally capable**: The designed subsystem for weather station data processing is able to handle massive amounts of data fed by the weather stations. It is based on scalable architecture [15]. For automation of the environment while handling greater workload, we have used Red Hat® Ansible®. It is a famous IT tool for automating repetitive tasks that

transforms inefficient tasks of software release cycles into predictable, scalable, and simplified processes [18]. It is used in our design to provide automation for configuration management, application deployment, cloud provisioning and service orchestration.

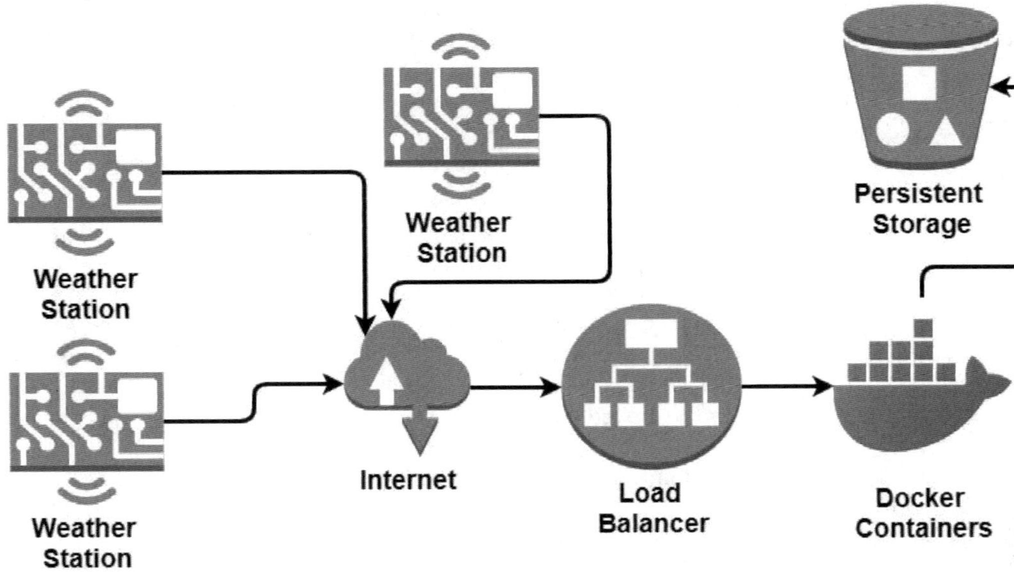


**Figure 1.** IoT Cloud architecture for weather station data processing.

## 3. Weather station data processing in the cloud

The proposed architecture on figure 1 relies on the LSTM scalable model [11] to process the input data and provide forecasting for the coming intervals of time. Each value or feature (such as temperature, solar radiation, wind speed) discrete time series from the used sensors can be represented as a vector with $n$ historical steps:

$$X^f = [X_{T-n}, X_{T-n-1}, X_{T-n-2} ... X_{T-2}, X_{T-1}], \quad (1)$$

where $f$ is the $f$-th feature, $T$ is the value of the feature at time $T$. The prediction of the values at the time $T$ will be represented by the following matrix:

$$X_T^P = [X^1 X^2 X^3 .... X_P]^T, \quad (2)$$

where $P$ is the number of measurements which are provides by the weather station.
Thus, the current snapshot of the data at time $T$ will be designated as (3):

$$X_T^P = \begin{bmatrix} X^1 \\ X^2 \\ \vdots \\ X^P \end{bmatrix} = \begin{bmatrix} X_{T-n}^1 X_{T-n+1}^1 \cdots X_{T-2}^1 X_{T-1}^1 \\ X_{T-n}^2 X_{T-n+1}^2 \cdots X_{T-2}^2 X_{T-1}^2 \\ \vdots \\ X_{T-n}^P X_{T-n+1}^P \cdots X_{T-2}^P X_{T-1}^P \end{bmatrix} \quad (3)$$

RNN is a class of powerful deep neural network using its internal memory with loops to deal with sequence data. The architecture of RNNs, which also is the basic structure of LSTMs, is illustrated below. For a hidden layer in RNN, it receives an input vector $X_T^P$, and generates the output vector $Y_T$. RNNs exhibit the superior capability of adapting themselves to predict nonlinear time series problems [11], Some of the RNNs are bound to reach the vanishing during the Backpropagation (BP) coefficient learning, and thus, become incapable of learning from long time lags, or keeping track of long-term dependencies. It is necessary to consider the above problems when using RNNs and use any of the advanced techniques to deal with them, such as the Long short term memory LSTM architecture and Gated Recurrent Unit (GRU) architecture. According to [11] LSTMs have been shown to work better on sequence-based tasks with long-term dependencies. GRU represents a simplified LSTM architecture being fairly new – is used in the context of machine translation. Out of the many existing variants of LSTM, we have chosen to use the proposed algorithm [19] for processing of the weather station data, i.e. standard LSTM architecture.

## 4. Performance considerations while using cloud infrastructures

The standard approach to building a software delivery process involves supporting a large number of virtual machines, with their own environments and versions. They must be constantly available, which means that they constantly occupy resources, even when they do not perform any tasks, but simply idle. With the proposed dynamic infrastructure, everything is exactly the opposite. Containers are used instead of regular virtual machines. The advantage is that many containers can run on a single virtual machine without much interference. The containers needed to perform a particular processing task – that can be scheduled – are created when they are required, rather than working constantly, borrowing resources. All versions and variables in the environment can be used dynamically, and after the task is completed, the created resources are destroyed. New ones are created as needed.

## 5. Experimental results and discussion

In our experiments we used data from a weather station ATMOS 41 [3]. It has 12 weather sensors in a single device for measuring atmospheric conditions. It fulfills one of the prerequisites for being able to provide continuous deployment in severe conditions and climates, such as deserts or mountains. Two of the 17 features used in building our experimental model are precipitation and solar radiation. The device manufacturer claims high accuracy on these measurements, i.e. 0.017 mm resolution means it is said to accurately measure even the smallest of rainfalls and even heavy dew events that other rain gauges might omit. In addition, the device's anemometer is also claimed to be very accurate even at low wind speeds due to the lack of moving parts. The attached accelerometer also contributes to the station's level and accuracy.

To measure the performance gain while using of the proposed IoT Cloud system components in comparison to the traditional infrastructure, we have prepared an experiment of multiple deployments of the architecture on our private cloud infrastructure. The speedup which we were able to achieve of container-based deployment vs. virtual machine-based is more than 2 times measured at up to 10 runs of the automated infrastructure deployment.

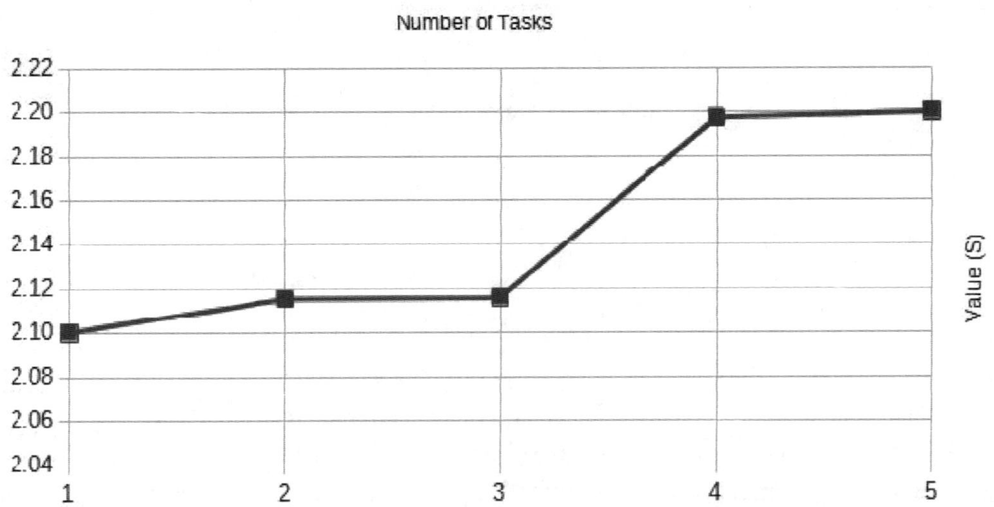


**Figure 2.** Speedup coefficient with 10 runs of the deployed architecture.

The calculated Speedup coefficient $S=T_{hypervisor}/T_{containers}$ is measured as the ratio between the time needed to deploy a number of instances (*Tasks*) with the runtime components required to operate this environment on a hypervisor and a container platform. The advantages of the container platform can be inferred from figure 2.

## 6. Conclusion

A design was proposed for a container based IoT Cloud capable system that delivers automated processing of the received data, visualization and cloud storage. We are comparing two types of open-source container technologies for implementing data processing. Performance evaluation has been calculated and we provide the set of criteria for choosing a container platform. Storing the data in the cloud is an important part and we provide our consideration for implementing a particular open source platform that could reduce costs and fulfil the requirements for weather data storage. The designed system is expected to be cloud-provider agnostic. The proposed features make the system especially suitable for purposes such as local weather forecast; disaster prediction: hurricane, severe storms, floods, tornadoes, etc.; severe droughts with fire dangers; prediction of landslides. Depending on the availability of additional sensors the future development of the system may be extended with parameters of the soil. Training the model on a freely available data set is planned for the next steps of the system development.

## References

[1] Weerasinghe, R, Aroos M, Pannila A, Jayananda M and Sonnadara D 2011 Construction of an automated weather station for ground level weather measurements, *IESL Annual Transactions*, pp 450-455

[2] Tenzin S, Siyang S, Pobkrut T and Kerdcharoen T 2017 Low cost weather station for climate-smart agriculture, *9th International Conference on Knowledge and Smart Technology (KST)*, Chonburi, pp 172-177, doi: 10.1109/KST.2017.7886085

[3] ATMOS 41 Weather Station, Available at: https://www.metergroup.com/environment/products/

[4] Peek, S and Peek D 2004 Cellular weather station and computer system using the public cellular data telephone system and internet for controlling irrigation and method of use, U.S. Patent US6343255B1, issued January 6, 2004

[5] Stergiou C, Psannis K, Kim B and Gupta B 2016 Secure integration of IoT and cloud computing, *Future Generation Computer Systems*, **78**, Part 3, 2018, pp 964-975, ISSN 0167-739X, doi: https://doi.org/10.1016/j.future.2016.11.031

[6] Schuler T, Dunse T, Østby T and Hagen, J 2014 Meteorological conditions on an Arctic ice cap—8 years of automatic weather station data from Austfonna, *Svalbard. Int. J. Climatol.*, **34**, pp 2047-2058, doi: https://doi.org/10.1002/joc.3821

[7] Royapoor M and Roskilly T 2015 Building model calibration using energy and environmental data, *Energy and Buildings* **94**, pp 109-120, doi: https://doi.org/10.1016/j.enbuild.2015.02.050

[8] Vines, R. L. Krutz R. Dean, and Krutz R. L.. *Cloud security: A comprehensive guide to secure cloud computing*. Wiley Publishing, Inc, 2010.

[9] Singh A and Chatterjee K 2017 Cloud security issues and challenges: A survey, *Journal of Network and Computer Applications*, **79**, pp 88-115, doi: https://doi.org/10.1016/j.jnca.2016.11.027.

[10] Ivanova E, Iliev T and Stoyanov I 2020 Modelling and simulation of big data networks, *7th International Conference on Energy Efficiency and Agricultural Engineering (EE&AE 2020)*, Ruse, Bulgaria, 2020, pp. 1-4

[11] Sundermeyer M, Schlüter R, Ney H 2012 LSTM Neural networks for language modeling. interspeech, doi: https://doi.org/10.1016/0165-6074(89)90269-X.

[12] Stoyanov I, Iliev T, Mihaylov G, Evstatiev B and Sokolov S 2018 Analysis of the cybersecurity threats in smart grid University of telecommunications and post, Sofia, Bulgaria, *IEEE 24th International Symposium for Design and Technology in Electronic Packaging (SIITME)*, Iasi, 2018, pp 90-93, doi: 10.1109/SIITME.2018.8599261.

[13] Cui Z, Ke R, Pu Z and Wang Y 2018 Deep bidirectional and unidirectional LSTM recurrent neural network for network-wide traffic speed prediction, *arXiv preprint arXiv:1801.02143.*

[14] Vlaev S, Sokolov S and Vukadinoff M 2018 Scaling and optimizing cloud storage, *International Journal "Information Theories and Applications*, **25**, no 3, pp 280-289

[15] Sokolov S, Idiriz O, Vukadinoff M and Vlaev S 2020 Scaling and automation in cloud deployments of enterprise applications, *Journal of Engineering Science and Technology Review Special Issue on Telecommunications, Informatics, Energy and Management*, pp 103 - 106, ISSN: 1791-2377, 2020

[16] Weil S 2007 Ceph: reliable, scalable, and high-performance distributed storage, *PhD diss.*, University of California, Santa Cruz

[17] Merkel D 2014 Docker: lightweight Linux containers for consistent development and deployment, *Linux journal*, no. 239: 2

[18] Sokolov S, Iliev T and Stoyanov I 2019 Analysis of cybersecurity threats in cloud applications using deep learning techniques, *42nd International Convention on Information and Communication Technology, Electronics and Microelectronics (MIPRO)*, Opatija, Croatia, 2019, pp 441-446, doi: 10.23919/MIPRO.2019.8756755.

[19] https://github.com/zhiyongc/Stacked_Bidirectional_Unidirectional_LSTM [last accessed September 2020]

**Acknowledgments**

This work has been sponsored by Grant NID-1/17.04.2020 - "Implementation and adoption of cloud services for the teaching process of the University of Telecommunications and Post".

# Software-defined wireless urban network

**O V Streltsov[1], Y G Birka[1], V Y Gnatenko[1], V S Sytnikov[1] and P V Stupen[1]**

[1] Computer Systems Department, Odessa National Polytechnic University,
1 Shevchenko av., Odessa, 65044, Ukraine

E-mail: streltsov.o.v@opu.ua

**Abstract.** Software-defined wireless urban network (SD-WUN) is a new network subclass for implementation in the "Smart City" project. SD-WUN can be offered as a single technical solution, built based on the most relevant technologies in the world of computer networks and the Internet. The theoretical model of existing technological solutions interaction is one of the best ways to eliminate the problems of the "Smart City" project mentioned in the work and offers one of the possible ways of its development in the future. The analysis is based on assessments of the prospects for the development of existing methods for implementing computer network nodes and the Internet of things (IoT), and their comparison with each other. One of the most pressing problems of the "Smart City" project implementation is the best choice from the point of view of network technologies. An urgent problem in the field of network technologies is the problem of the transition of network addressing from the IPv4 address space to IPv6. It is obvious that the Smart City project is associated with the popular IoT technology (Internet of Things) due to the progression of the number of devices. Another key point is the need to implement a project based on a high-speed wireless network, excluding wired switching. It is obvious that the network should possess such properties as security, distribution of computing resources, high data transfer speed and ease of use and support of the network.

## 1. Introduction

The modern society development, is fully consistent with the concept of Industry 4.0 in terms of technical equipment. Requirements for Internet technologies and their components, computer systems and software components cause the improvement of theoretical and practical developments in accordance with this concept. The modern production processes development is moving towards full automation with the real time management of all processes, taking into account external conditions changing. Internet technologies and their components must ensure reliable communication between the production process components and personnel. And this makes it necessary to improve the development of components that could carry out a complex restructuring of their characteristics, self-tuning and self-learning in accordance with this concept [1].

The Internet of Things technology (IoT) use the Internet for information exchange not only between humans, but between all kinds of "things": machines, devices, sensors, etc. On the one hand, sensor-based things can exchange and process data without human intervention, and on the other hand, a person can actively participate in this process, for example, when it comes to "smart home" or "smart production".

A variety of IoT is the Industrial Internet of Things, (IIoT). It opens the direct path to the fully automated production facilities creation. The key equipment components are equipped with various sensors, actuators and controllers; the collected data is processed and sent to the appropriate enterprise services, than allows the staff to make informed and balanced decisions quickly. But the maximum task is to achieve such enterprise automation level that make machines work without the participation of people possible in all areas. The personnel role is reduced to monitoring the machines operation and emergency situations responding [1-3].

Such problems are found in various human activity areas, including robotics, various mobile platforms and critical application systems [4-7].

The human environment and urban infrastructure development face similar challenges. Today, one of the most pressing tasks of the Smart City project is the best network technologies choice. In this regard, an urgent problem in the network technologies field is the problem of switching network addressing from the IPv4 address space to IPv6. It is obvious that the Smart City project is associated with the popular IoT technology, due to a sharp increase in the number of devices. In addition, an important point is the need to ensure the project implementation based on a high-speed wireless network without wired switching. It can be argued that the network for the project should have such properties as:
- safety and security,
- computing resources distribution,
- high data transfer rate,
- ease of use and network support.

## 2. Problem analysis

### *2.1. Technology development options*

The “Smart City” project implementation primarily involves the creation of a long-range wireless network in accordance with the concept "Industry 4.0", based on the use of sensors, a large number of network addressed devices, with a set of network protocols and related technologies. An analysis of technology development options, their relevance and pace of development, allowed to distinguish the following:
- IPv6 addressing;
- the prospect of wireless networks through 5G;
- the use of sensor-based LoRaWAN networks (Long Range Wide-Area Network).

### *2.2. LoRa, as an implementation of IoT*

One of the most relevant technological solutions for IoT is the LPWAN (Low-power Wide-Area Network) wireless technology, which is a wireless network with low energy costs (figure 1).

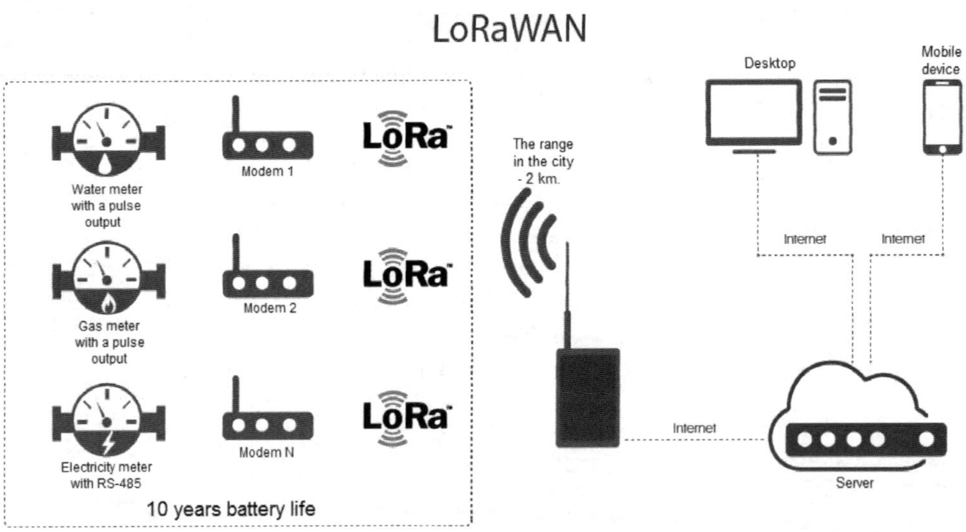


**Figure 1.** Principle LoRaWAN functioning.

Protocol LoRaWAN, which implements the network topology "star" is a specification LPWAN [8]. Radius network is about 1-2 km in such a case. A stable signal is expandable up to 4 km without significant obstruction signal, but with a possible loss of packets [9]. The low energy consumption of the sensors used in the network is another advantage of LoRaWAN.

### *2.3. IPv4 and IPv6 Addressing Dilemma*
Modern statistics (figure 2) indicate that 20% of the global volume of Internet access devices are connected to the network using IPv6 [10, 11].

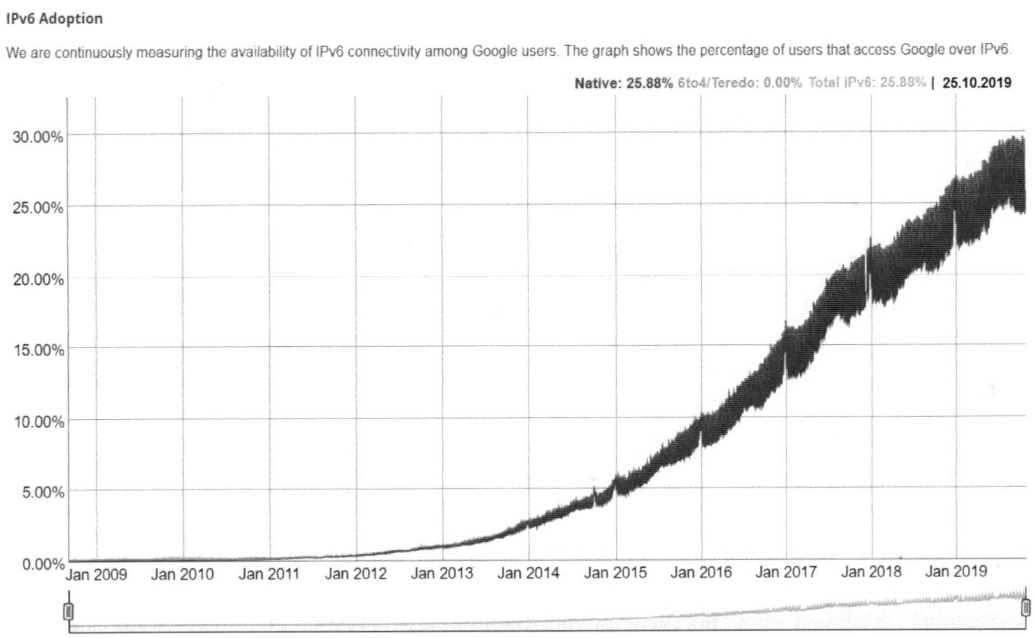


**Figure 2.** IPv6 usage statistics in the world.

In comparison with IPv4 addressing, IPv6 deprives the network of certain bottlenecks that need to be addressed: NAT (Network Address Translation), the need to use VLAN (Virtual Local Area Network), the complexity of managing network security rules. IPv6 addressing has many advantages over IPv4 addressing, however, the challenges of global transition hinder its rapid implementation. The main problem is that IPv4 requires maintaining backward compatibility in building networks, so the transition requires large financial investments.

Small network providers do not have the appropriate resources to conduct relevant experiments and cannot actively contribute to the global transition.

The generally accepted fact today is that the global transition is an irrevocable event. Currently, the United States, India, the countries of Central Europe and Asia are leading the world in implementing IPv6. Regardless of the pace and complexity of adapting a new type of addressing, a global transition in the near future will occur under the conditions of a constant increase in the number of IoT devices and equipment with Internet access. If the transition of existing networks is difficult to establish the reasons and consequences of using IPv4 destination network equipment, the city specified for the project "Smart City" should be built by using IPv6.

### *2.4. Technologies of wireless long range*
With current trends in the world of network technology, high-speed wireless networks are a priority. This is due to the large amount of media traffic and QoS (Quality of Service) requirements. 5G is the single and perhaps the most rational solution for today. An approximate solution is WiMAX

(Worldwide Interoperability for Microwave Access), but the technology loses competition and is selective in cases of use, being in the 3rd generation category.

WiMAX wins the competition due to the range and frequencies at which it operates. But in the modern world, high speeds and support for the quality of service are required (table 1). However, IoT assumes low energy consumption, and the principles of functioning of 5G will favorably affect the implementation of the “Smart City” project [12].

As to 5G, active discussions are ongoing. The subject of discussion is electromagnetic safety. 5G operates at high frequencies, which may not be safe for human health until proven. In conditions of uncertain security, a vivid precedent was the case in April 2019 in the Swiss canton of Geneva, where the regional government imposed a moratorium on the use of 5G.

The vulnerability of 5G is that the declared figures of the transmission speed are guaranteed in the absence of any interference. Experiments have shown that bad weather conditions strongly affect communication quality. Additionally, it should be noted that in conditions of a city saturation with devices and high-rise buildings, the signal may be lost.

**Table 1.** WiMAX and 5G comparison.

| Parameter | WiMAX | 5G |
|---|---|---|
| *Throughput* | 25-75 Mbit/s | 1-35 Mbit/s |
| Radius | 25-80 Km | Unlimited (depends on station density) |
| Frequency band | 1.5-11 GHz | 30-300 GHz |
| QoS | Non priority | Provided |
| Energy consumption | Middle | Low |

### *2.5. Technologies of software-defined networks*

Modern computer networks have many disadvantages. It's a constant need to optimize the information flow, complex security organization, a large number of protocols to be integrated in the set up. All it needs improvement and organizing simplification.

Software-defined networks (SDN) - is the principle of building networks, in which the entire network management mechanism is placed on a separate software base, and all traffic management is based on special protocols that operate on the concept of "data channel" and can perform various actions with it (allow, prohibit, redirect, rewrite fields in packages, etc.) [13]. In fact, the controller determines the network management rules based on real-time events, as well as the specialized applications operation (for example, emulation of STP or routing protocols, depending on the channels state in real time). The end result is transferred to the OpenFlow protocol switches in a flow-tables containing information about where, how and what traffic to pass. On the one hand, this approach gives great flexibility in network management, on the other - significantly simplifies network administration. This promotes virtualization in network management.

An important question - is building a network security system based on OpenFlow. The controller actually centralizes the intelligent functions of Firewall, IPsec and other common network security systems. The problem of security and its organization is the most energy-intensive and requires large resources.

## 3. Research results.

### *3.1. Objective*

Search and formation of a unified technical solution and strategy for the implementation of network communication tools in the Smart City project, based on analysis and comparison of the complex of available technologies within the concept of "Industry 4.0".

### *3.2. The choice of technological solutions*

Taking into account the specifics of the project and a result of a comparative analysis of the most promising solutions, a new subtype of networks is proposed as a common set of technological solutions for creating an urban wireless network of a wide range based on mobile communications (SD-WUN).

The proposed technical solution provides for the implementation of public Internet through 5G. It is planned to reserve communication channels with frequencies for the functioning of sensors based on LoRaWAN and the TCP protocol. Quality of Service (QoS) is provided according to the needs for media traffic based on the UDP protocol.

The implementation based on IPv6 involves the subsequent exclusion of IPv4 and NAT, and the structuring of network space without using VLANs, with a simplification of security policy, reduction to the use of AAA protocol and multi-level Firewall, with the transfer of security under the control of the network operating system.

The technical solution involves the construction of a SDN. This will give the network the flexibility to allocate resources through an open architecture in which various components can be combined. The technology involves dynamic cloud services, representing a computer network of physical objects. Such implementation is based on a network operating system.

A wide range of frequencies using 5G allows for a large range of different techniques and combinations of device types. With increasing amounts of the new wave device type, popularity acquires RFID technology. The network assumes the use of passive, semi-active, active RFID tags with a range of up to 300 meters. This makes it possible to localize objects in space, access control and contributes to the implementation of IoT functions.

### *3.3. Technical Solution*

The proposed technical solution takes into account the fact that IPv6 addressing allows you to structure addresses by geographic location. Since the length of the network address is 128 bits, which gives an almost inexhaustible limit of address space and the possibility of networking [14]. The solution is to use IPv6 geography not only to identify the city, but also to determine the area or neighborhood. This improves the structure of the urban network, and makes it easier to solve problems that are not solved by routing protocols. The density of devices makes it possible. Smart City is to use to 2000000 devices per 1 square kilometer, which corresponds to the use of a large number of stations that will relay the wireless signal.

Each 5G signal propagation zone has its own radius, within which addresses of a certain range are used, which is reserved for geographical affiliation, an example of which can be a residential city block. This makes it possible to structure in the management of the SD-WUN network (figure 3). 5G provides a low energy requirement, and introduces the possibility of implementing many technical innovations, such as augmented reality and others that require high data rates of large amounts of data in a short period of time.

The urban network should focus on the SDN. This creates great advantages, making the network flexible and distributed. The network does not require large resources to manage it. This choice caused great richness of personal use devices, and IoT organizing means.

SDN will give it flexibility in the resources allocation through an open architecture, which can combine different components. The technology implements a computer network of physical objects, which provides dynamic cloud services. The implementation is based on a network operating system.
A significant advantage of this networking - is managing the program module and artificial intelligence. This ensures that there is no data flows distribution problem, as the system monitors channels in real time, and shifts the responsibility to choose the shortest path. This creates real-time routing dynamics.

The Internet of Things in this project version can be embodied in the full range of industries and reproduction prospects. This is a smart transport, smart buildings, communication with industry, web crawling things with artificial intelligence.

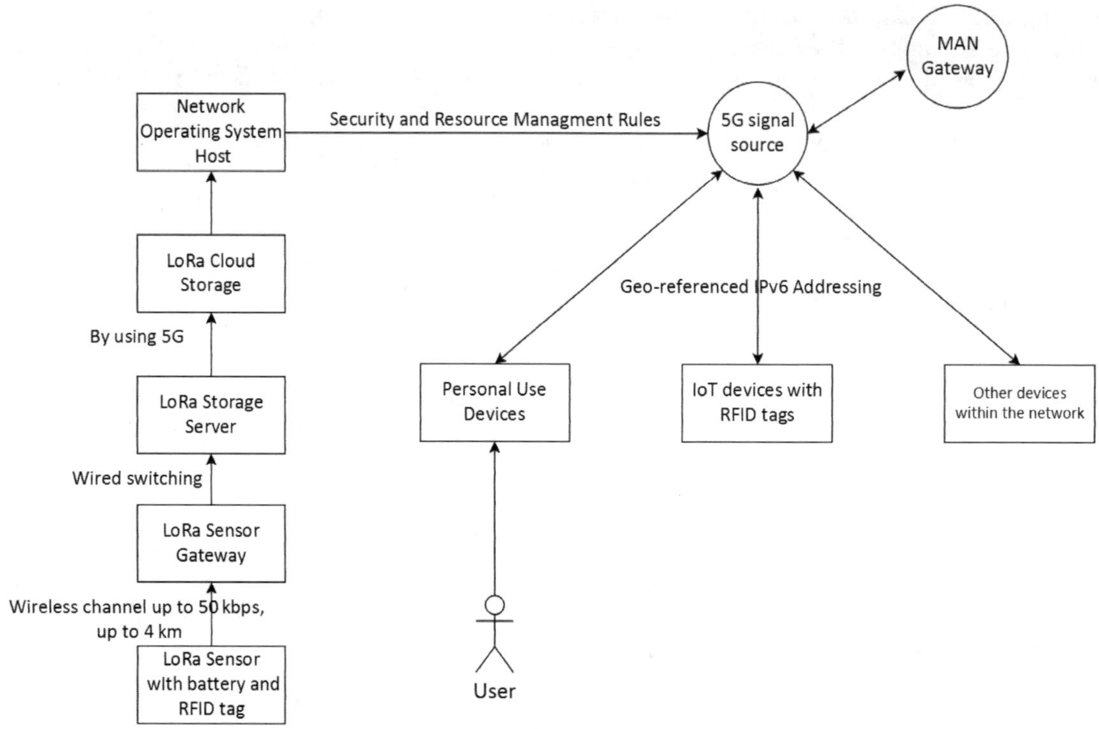


**Figure 3. The principle of SD-WUN operation.**

Radio frequency identification allows geographic identification of smart city objects. This technology can be used very variably and widely, especially given the project needs, such a solution will be in great demand.

The geographical distribution of network address space creates an opportunity to plan global network re-equipment processes. This simplifies the planning of certain network events, simplifies the network perception.

The quality of service is ensured, which makes it possible to work with a large amount of high-quality media traffic.

Such project implementation brings the city into a single autonomous mechanism that is distributed and self-serving. The city observes its events, optimizes processes, creates comfort and makes planning.

The proposed model increases the network technologies efficiency due to greater disclosure of project functions in more convenient ways.

During the simulation, two scenarios of operation of individual nodes of the system were considered. The first – is the operation of a smart home in a smart city, and household management. The second – is the operation of radio frequency identification, and its use for project purposes. Both models show positive results and confirmed the effectiveness of these technologies.

## 4. Conclusions

Based on the analysis of information sources for a variety of solutions that can be used in the field of network technologies and IoT, more promising ones with potential for further development are selected. The proposed network implementation option for the Smart City project is a city-specific software-defined wireless network (SDWN), as a subtype of 5G-based SDN networks, with leading LoRa technology, as the main IoT implementation technology, using RFID for identification of communication devices.

The main advantage of the proposed network model is the ability to solve all the relevant problems in this area with a single, universal technical solution.

The main problem in implementing the model is the cost of the project and a number of factors affecting the model. Significant influence factors are the narrow specialization and compatibility of the technologies used in the model.

Further research requires the creation and verification of a large project model with the reproduction of the full range of proposed technologies. Namely, the reproduction of the test model in a real test layout within the boundaries of a wireless local area network with software control, and the implementation of radio frequency identification using the Internet of Things. At present, only the theoretical substantiation of the SD-WUN network subclass has been performed.

It is also necessary to explore the possibility of implementing a technology stack, on a global scale, based on the idea of communication between cities with SD-WUN.

**References**

[1] Grangel-González et al. 2017 The industry 4.0 standards landscape from a semantic integration perspective, *22nd IEEE International Conference on Emerging Technologies and Factory Automation (ETFA)*, Limassol, pp 1-8, doi: 10.1109/ETFA.2017.8247584

[2] Santos B, Charrua-Santos F and Lima T 2018 Industry 4.0: An Overwiew, https://www.researchgate.net/publication/326352993_Industry_40_an_overview

[3] Industry *4.0*, available: https://www.cognex.com/ru-ru/what-is/industry-4-0-machine-vision/development (in Russian)

[4] Semenov S, Voloshyn D and Ahmed A 2019 Mathematical model of the implementation process of flight task of unmanned aerial vehicle in the conditions of external impact, *International Journal of Advanced Trends in Computer Science and Engineering*, **8(1),** pp 7-13

[5] Zhuravska I, Musyienko M and Tohoiev O 2019 Development the heat leak detection method for hidden thermal objects by means the information- measuring computer system. *CEUR Workshop Proceedings*, **2353**, pp 350-364

[6] Musiyenko M, Denysov O, Zhuravska I and Burlachenko I 2016 Development of double median filter for optical navigation problems, *Proceedings of the 2016 IEEE 1st International Conference on Data Stream Mining and Processing*, DSMP 2016, **7583535**, pp 177-181

[7] Ukhina H, Sytnikov V, Streltsov O, Stupen P and Yakovlev D 2019 Transfer Function Coefficients Influence on the Processing Path Bandpass Frequency-Dependent Components' Amplitude-Frequency Characteristics Properties at the NPP TP ACS, *Conference Proceedings of 2019 10th International Conference on Dependable Systems*, Services and Technologies, DESSERT 2019, Leeds; United Kingdom; pp 193-196 doi: 10.1109/DESSERT.2019.8770050

[8] LoRaWAN™ *What is it?* A technical overview of LoRa® and LoRaWAN™ - *Technical Marketing Workgroup* 1.0. LoRa® Alliance. 2015

[9] Ray B 2018 *What Is LoRaWAN?* [Technical Breakdown] [Online] – available at: https://www.link-labs.com/blog/what-is-lorawan

[10] Common *Problems When Deploying IPv6*, 2012 [Online] – available at: https://docs.oracle.com/cd/E23824_01/html/821-1453/ipv6-troubleshoot-2.html

[11] Google *IPv6 Statistic* [Online] – available at: https://www.google.com/intl/en/ipv6/statistics.html#tab=per-country-ipv6-adoption

[12] Mihaylov G, Iliev T, Bikov T, Ivanova E, Stoyanov I and Keseev V 2018 Test cases and challenges for mobile network evolution from LTE to 5G, *41st International Convention on Information and Communication Technology Electronics and Microelectronics (MIPRO)*, pp 0449-0452, doi: 10.23919/MIPRO.2018.8400085

[13] Loew S 2017 *Advantages of software defined networking (SDN),* [Online] – available at: https://www.fidelus.com/software-defined-networking-advantages/

[14] Loshin P 1999 *IPv6: theory, protocol, and practice.* Library of Congress Cataloging-in-Publication Data

# Analysis of the energy recovery possibilities of energy from lavender straws after a steam distillation process

**D Chakyrova[1], N Doseva[1]**

[1] Department of Thermal engineering, Technical University of Varna, 1 Studentska Street, Varna 9010, BG

E-mail: chakyrova_d@tu-varna.com

**Abstract.** In recent years, the area sown with lavender has increased in Bulgaria, which is then processed to essential oils by steam distillation. Plant residues after the steam distillation process are a major environmental problem and, at the same time, a valuable source of biofuels. Thus, the processing of lavender waste solves the problem of environmental pollution and generates energy that can be used for various purposes. This paper analyzes the possibilities for obtaining energy from lavender biomass. The processes of direct combustion and pyrolysis with their advantages and disadvantages are analysed and the yields of the obtained products are predicted by analytical calculations. For the purposes of this study, an experiment was conducted to determine the pyrolysis products of lavender straws and their quantitative ratios, which were compared with analytical calculations. The results show that the energy obtained can ensure the implementation of the steam distillation process and the residual energy can be used for other purposes.

## 1. Introduction

In recent decades, both the processing and the use of biomass as a fuel in various industries, in households and in transport have increased. The reason for this is Directive 2009/28 / EU on renewable energy (recast in December 2018), which establishes a comprehensive policy for the production and promotion of energy from renewable sources and requires the EU to meet at least 20% of its total energy needs. with energy from renewable energy sources until 2020.

Lavender oil is one of the most common essential oils in the world. Bulgaria has joined France as a world leader in its cultivation and production of lavender oil for several years. By 2019 the areas sown with lavender in Bulgaria are 18,000 ha according to the data presented in [1].

The extraction of lavender essential oils is carried out by a distillation process. In Bulgaria the method of steam distillation is used, which is extremely energy intensive. In the agricultural areas where the distilleries are built, near the areas sown with lavender, the sources of energy for the steam generator are electricity from the national power grid or oil. This contributes to the high cost of the final product and additional environmental pollution.

In the aspect of the above, the processing of the waste straws of lavender after a process of steam distillation to extract the essential oils in it leads to the utilization of energy in them, cheaper the final product and protection of the environment from pollution.

Data for processing biomass of different types and comparing the methods for obtaining energy from it were found in the literature [2, 3, 4, 5]. No data for the treatment of waste lavender stalks by pyrolysis and comparison with other methods for obtaining energy from them were found.

The main goal of the present study is to analyze lavender straws as a source of energy. This requires an assessment of their properties, which define them as fuel.

In order to determine the energy potential of biomass, it is necessary to establish its energy value. As an indicator of the energy contained in biomass is often used the HHV. The atomic ratio [6] to estimate the HHV of fuels is used.

When the ratio of cellulose, hemicellulose and lignin in the biomass is known, its behavior during pyrolysis can be predicted [7], [8] and [9].

Another important point is to determine the yields of pyrolysis products. This can be done analytically or experimentally [10, 11, 12, 13].

The thermochemical (pyrolysis, gasification, esterification and direct combustion) way of conversion of biomass into liquid, gaseous and solid fuel is analyzed, comparing only pyrolysis and combustion.

This study analyzes the properties of biomass and the products of its processing into secondary fuels, which characterize them as such. For this purpose, literature data were used: for proximate and ultimate analysis of waste lavender stalks as biomass [14], [15]; for analytical models of the share distribution of biomass pyrolysis products [8], [10], [11] and [12]; for the composition of the obtained biogas ; for analytical dependences for the determination of HHV and LHV on lavender straws and biomass pyrolysis products [11], [13]; for analytical energy balance models [6], [11], [13]. The process of processing the lavender waste straws by pyrolysis was experimentally studied, and a gas analysis of the obtained biogas was performed in order to compare the predicted results of the analytical calculations.

## 2. Materials, methods and data used

### *2.1. Analytical research.*

Next equations presents some of the models proposed in the literature for analytical determination of the proportions of the products of biomass pyrolysis.

Models for determining the shares of individual products [8, 10, 11, 12]:

$$Y_{char} = Y_{ce} f_{char,ce} + Y_{hc} f_{char,hc} + Y_{lg} f_{char,lg} \quad (1)$$

$$Y_{vol} = 1 - Y_{char} \quad (2)$$

$$1 = g_{vp} + g_{gg} + g_{char} \quad (3)$$

$$g_{vapor} = \frac{W}{100};\ g_{char} = (1 - g^{0}_{gas}).\left(1 - \frac{W}{100} - \frac{A}{100}\right) + \frac{A}{100} \quad (4)$$

$$g_{gas} = g^{0}_{gas} \cdot \left(1 - \frac{W}{100} - \frac{A}{100}\right);\quad g^{0}_{gas} = \left(0.85 - e^{1.2 - \left(\frac{t_{pyr}}{143}\right)}\right) \quad (5)$$

$$Y_{H_2,F} \approx Y_{H,F}\ ,\ Y_{CO,F} \approx Y_{O,F} * \frac{28}{16}\ ,\ Y_{CH_4,F} = -2{,}18.10^{-4} + 0{,}146 Y_{CO,F} \quad (6)$$

$$\frac{Y_{H_2,F}}{Y_{CO_2,F}} = 3.10^{-4} + \frac{0{,}0429}{1 + \left(\frac{T}{632}\right)^{-7{,}23}} \quad (7)$$

Next equations presents some models of determine of the HHV of biomass and its pyrolysis products.

In the case of wet biomass having a working humidity $W_{wb}$, the heat obtained from the combustion of 1 kg of fuel is due solely to the dry biomass fraction, i.e. (1- $W_{wb}$). Then, according to [13]:

$$Q_{HHV} = (1 - W_{wb}).Q_{HHV},\ [MJ/kg] \quad (8)$$

Modeling of the HHV of biomass and pyrolysis gas are shown in next equations [13, 11]:

$$Q_{HHV} = 0{,}352.C + 1{,}162.H - 0{,}111.O + 0{,}063.N + 0{,}105.S\ [MJ/kg] \quad (9)$$

$$Q_{HHV} = 0{,}349.C + 1{,}178.H - 0{,}103.O - 0{,}015.N + 0{,}101.S - 0{,}021.A\ [MJ/kg] \quad (10)$$

$$Q_{HHV}=0{,}341.C+1{,}323.H-0{,}120.O-0{,}12.N+0{,}68.S-0{,}015.A\ [MJ/kg] \quad (11)$$

$$LHV=0{,}127.CO+0{,}108.H_2+0{,}358.CH_4\ [MJ/m^3], \quad (12)$$

where (9), (10) and (11) are for biomass and (12) is for gas.

On the other hand, the expression for LHV of biomass, taking into account its working humidity, acquires the form [18]:

$$Q_{LHV}=(1-W_{wb}).Q_{HHV}-2{,}447.[W_{wb}+9.(1-W_{wb}).H]\ [MJ/kg], \quad (13)$$

where: H-mass fraction of hydrogen in the dry biomass.

Next equations presents the energy balance models [6, 11, 13].

$$\begin{aligned} &A.cp_aT_0+F.cp_fT_0+W.H_0+F.HHV+Q_{ext}= \\ &=(c_{CO}.g_{CO}+c_{CO_2}.g_{CO_2}+c_{CH_4}.g_{CH_4}+c_{H_2}.g_{H_2}+c_{O_2}.g_{O_2}+c_{N_2}.g_{N_2}).T_g+ \\ &+(1-x_g).W.H_g+P_c.q_c+Q_{gasif}+Q_{loss}+Q_{product} \end{aligned} \quad (14)$$

$$Q_{phh}^{gas}=c_{gas}g_{gas}\left(t_{pyr}-t_0\right)+c_{steam}g_{steam}\left(t_{pyr}-t_0\right)+rg_{steam} \quad (15)$$

$$Q_{phh}^{char}=c_{char}g_{char}\left(t_{pyr}-t_0\right) \quad (16)$$

$$Q_{pyr}=\Delta H_s+\Delta H_r \quad (17)$$

$$\Delta H_s=\dot{m}_{tyre}\left(\int_{T_0}^{T_{reac}}c_{p,tyre}dT\right) \quad (18)$$

$$\Delta H_r=\dot{m}_{tyre}.(553-3142.\mu_{char}), \quad (19)$$

where: F - the amount of fuel; A - the amount of air; W - the amount of steam; P - the quantity of char produced; $Q_{gasif}$ - the net heat of the reaction; $Q_{loss}$ - total heat loss; $q_c$ - heat of combustion of char; $(1-x_g).W$ - the part of the steam remaining in the gas which leaving the reactor; H - enthalpy; c - specific heat.

### *2.2. Experimental studies of the biomass pyrolysis process.*

The laboratory system is shown in figure 1.

The main purpose of the experimental part is to obtain process data that are missing in the literature. To achieve this goal, an experiment was conducted in which the waste lavender stalks were subjected to pyrolysis after a steam distillation process.

A discrete amount of feedstock is placed in the pyrolysis reactor and heated. The vapor obtained from the decomposition of the raw material leaves the system by passing through a cooler and a hydro-seal, which does not allow atmospheric air to enter the system.

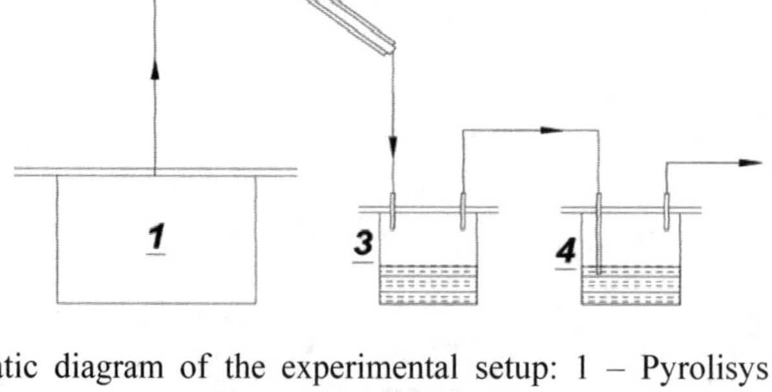


**Figure 1.** Schematic diagram of the experimental setup: 1 – Pyrolisys reactor; 2 – Heat exchanger; 3 – Liquid tank, 4 reservoir - hydraulic shutter.

The experiment was performed at a reactor temperature of 350 - 400°C. The amount of gas, pyrolysis oil and solid residue obtained after completion of the experiment is taken into account. The resulting pyrolysis gas is analyzed to determine its elemental, qualitative and quantitative composition.

## 3. Results and discussions

*3.1. Conversion processes.*
As part of this study, a technical and environmental assessment was developed for two of the thermochemical methods of treatment of lavender waste: direct combustion and pyrolysis.

*3.2. Structure and chemical composition of biomass from lavender stalks.*
Lavender grown in Bulgaria (Lavandula angustifolia mill) is a traditional type of lavender.

The content of cellulose, hemicellulose and lignin differs significantly when comparing their content in the flowers and stems of lavender after the distillation process - table 1.

**Table 1.** Content of lignin, cellulose and hemicellulose in spent lavender stalks according to [14].

| | | **Mass fraction of lignin (dry basis), %** | **Mass fraction of cellulose (dry basis), %** | **Mass fraction of hemicellulose (dry basis), %** |
|---|---|---|---|---|
| **Winter** | Leafs | 21.7±0.5 | 12.7±2.5 | 13.1±1.4 |
| | Stalks | 25.4±3.2 | 13.9±3.9 | 33.6±5.2 |
| **Summer** | Leafs | 17.1±2.7 | 3.6±1.7 | 19.9±0.8 |
| | Stalks | 23.1±3.6 | 13.0±0.9 | 42.7±1.7 |
| | Flowers | 23.6±2.6 | 12.6±2.1 | 22.4±2.1 |

Table 2 presents the proxymate analysis of lavender, according to the authors of [16].

**Table 2.** Proxymate analysis of lavender according to [16].

| **Biomass/component, wt %** | VM | FC | M | A | Total |
|---|---|---|---|---|---|
| **lavender** | 66,7 | 15,3 | 10,2 | 7,8 | 100 |

**Table 3.** Working composition of lavender stalks after the distillation process [14].

| **Composition** | N | C | Cl | H | O | S | Ash | Humidity, wb% | HHV, MJ/kg |
|---|---|---|---|---|---|---|---|---|---|
| **Mass share (db%)** | 1,3 | 48,1 | 0,2 | 5,8 | 37,8 | 0,1 | 6,7 | 61 | 19.570 |

For comparison, there are studies [17] in which the HHV and ash content after the process of burning straws of lavender, thyme, sage, fennel and thistle were evaluated experimentally. The HHV was determined as initially, the biomass was dried to 22% humidity, and then, using a laboratory dryer was dried to 4.5% humidity. The cited publication presents results according to which the HHV of lavender stalks is 20.66 MJ/kg.

A higher content of liglin compared to that of cellulose implies a higher yield of solid residue, a higher content of volatile substances implies a higher amount of biooil and synthetic gas. The contradiction of the expected results requires an experiment in the laboratory in order to find the relationship between the process parameters, the quantities of pyrolysis products obtained and their distribution by species.

An experiment was performed with lavender stalks, after steam distillation at a temperature in the pyrolysis reactor 350°C - 400°C. The obtained results are presented in table 4 and table 5, being compared with those of the analytical calculations.

Table 4 presents the results for product yields from the experiment and compares with the predicted values calculated from the described models in the literature – (1), (2), (3), (4), (5), (6), (7),. No publications have been found in the literature on experimental studies of the process of pyrolysis of lavender stalks. For this reason, in table 4 compares the results of the experimental studies in this article with those of other authors, selecting biomass with a similar composition (cellulose, hemicellulose and lignin).

Table 5 presents the result of the gas analysis of the obtained pyrolysis gas.

**Table 4.** Yields of pyrolysis products.

| Autor | Biomass | Tempe-rature, °C | Structural composition (wt. %) | | | Pyrolysis products | | | | | |
|---|---|---|---|---|---|---|---|---|---|---|---|
| | | | Cell. | Hem. | Lignin | Char | | Oil | | Gas | |
| | | | | | | Pred. | Exp. | Pred. | Exp. | Pred. | Exp. |
| This peper | Lavender straw | 350 | 38,2 | 21,5 | 40,3 | 22,8 | 35 | - | 55 | 77,2 | 10 |
| Demirbas [7] | Tea waste | 450 | 33,2 | 23,3 | 43,5 | 27,9 | 38 | - | - | 72,1 | - |
| Mohammed [2] | Napier gras | 450 | 38,75 | 19,76 | 26,99 | 18,8 | 46 | - | 28 | 81,2 | 26 |

**Table 5.** Composition of the obtained pyrolysis gas and the calculated LHV.

| Component | $H_2$ | CO | $CO_2$ | $H_2S$ | $CH_4$ | LHV, MJ/kg |
|---|---|---|---|---|---|---|
| vol % | 1,9 | 12,3 | 21 | 790 ppm | 64,8 | 26,418 |

The results of the calculated values of the HHV of lavender stalks and pyrolysis products are shown in the table 3 and table 5, using the models in (9), (10), (11) and (12).

The main problem in compiling the energy balance of the pyrolysis process is the determination of the heat required for the realization of the gasification process. Different authors offer different approaches.

According to the author of [6], the calculation of the heat required to perform endothermic reactions in the pyrolysis process is 1,683 kWh/kg raw, which is comparable to the values obtained from the calculation by formulas in table 3 - 2,003 kWh/kg raw.

The analysis of the results in tables 1 to 5 shows that the heat obtained from the burning of 1 kg of waste lavender stalks is 5.436 kWh. The heat obtained from the combustion of pyrolysis products per 1 kg of lavender waste stalks is 9,996 kWh. From the point of view of ecology and environmental pollution, pyrolysis is a gentle technology for obtaining energy from the waste stalks of lavender, which has become increasingly important in recent years.

Table 6 compares the characteristics of the energy balance of the two biomass conversion processes from lavender stalks.

**Table 6.** Comparison of conversion processes.

| Process | Income from incineration, kWh/kg raw | Costs, kWh/kg raw | Profit, kWh/kg raw | Ecological effect |
|---|---|---|---|---|
| Direct combustion | 5,436 | - | 5,436 | $NO_x$, $SO_2$, dioxins |
| Pyrolysis | 9,996 | Pyrolysis 2,003<br>Drying 1,153 | 6,84 | No harmful substances are released |

## 4. Conclusions

The possibilities for conversion of biomass from lavender waste straws from steam distillation process into energy have been studied. Analytical models and experiments were used. The results are as follows:

1. Yields of pyrolysis products of lavender straws have been calculated using analytical models. The results are compared with those of the conducted experiment and those of other authors conducted with biomass with a similar composition (cellulose, hemicellulose and lignin). All results are similar.

2. With the help of analytical models, the calorific value of the waste straws of lavender and the obtained pyrolysis gas is calculated. The results are similar to those of other authors.

3. The energy analysis of the process of combustion and pyrolysis of waste lavender straws shows the advantages of conversion by pyrolysis: 1) the heat from the burning of pyrolysis products covers the costs of the process (3.156 kWh/kg raw) and remains (6.84 kWh/kg raw), which can be used for other purposes. From the direct burning of the lavender stalks is obtained (5,436 kWh/kg raw); 2) the

burning of pyrolysis products does not pollute the environment, while the burning of lavender straws releases $NO_x$, $SO_2$ and dioxins.

The energy analysis concluded that the energy contained in the lavender waste stalks after the steam distillation process could be used, preferably by pyrolysis, which is an environmentally friendly technology. Extracting energy through pyrolysis from waste lavender stalks leads to economic and environmental benefits.

## References

[1] InteliAgro http://inteliagro.bg
[2] Mohammed I Y et al. 2015 *BioResources*, **10** (4), pp 6457-6478
[3] Yevich R and Logan J A 2003 *Global Biogeochemical Cycles,* **17**(4), 1095-doi: 10.1029/2002GB001952
[4] Wanga L et al. 2012 *Energy Procedia* **20**, pp 20-29
[5] Patel B and Gami B 2012 *Iranica Journal of Energy & Environment*, **3** (2), pp 123-128
[6] Basu P 2010 *Biomass Gasification and Pyrolysis: Practical Design and Theory*, Elsevier, p 365
[7] Demirbas A 2004 Effects of Temperature and Particle Size on Bio-Char Yield From Pyrolysis of Agricultural Residues, *Journal of Analytical and Applied Pyrolysis* **72**(2), pp 243–248
[8] Wang S Guo X Wang K Luo Z 2011 *Journal of Analytical and Applied Pyrolysis*, **91** (1), pp 183–189
[9] Fegbemi L Khezami L and Capart R 2001 *Applied Energy*, **69** (4), pp 293–306
[10] Sharma A K Ravi M R and Kohli S http://web.iitd.ac.in/~ravimr/Publications/IndianJournals/sesi-pyrolysis.pdf
[11] Suhotckij A 2012 *Secondary energy resources. Alternative and renewable energy sources,* Minsk 978-985-530-137-1 (in Russian)
[12] Neves D Thunman H Matos A Tarelho A and Gómez-Barea A 2011 Progress in Energy and Combustion Science, **37** (5), pp 611 – 630, https://doi.org/10.1016/j.pecs.2011.01.001
[13] Atsonios K et al. 2015 *International Journal of Thermodynamics*, **18** (4), pp 267-275
[14] UNEP Technologies for Converting Waste Agricultural Biomass to Energy [Report]. - Osaka: United Nations Environmental Programme, 2013
[15] Lesage-Meessen L, Bou M et al. 2015 *Applied Microbiology and Biotechnology* **99**, pp 3375–3385
[16] Vassilev S et al. 2010 *Fuel* **89**, pp 913–933
[17] Zabinski A, Sadowska U and Wcislo G J 2015 *Agricultural Engineering*, **4** (156), pp 139-145
[18] Acar A Ayanoglu A 2012 *Energy Education Science and Technology Part A*, pp 749-758.

## Acknowledgments

The authors are grateful to Eng. Slavcho Georgiev, who within the project BG16RFOP002-1.005-0017-C01 "Development of new methods and technologies as a stand-alone mechatronic product, or embedded as part of a mechatronic unit in order to improve the efficiency, quality and environmental friendliness of production" under OP"Innovation and Competitiveness" 2014 - 2020 provided invaluable assistance in the development and financing of this study.

# Bin weather data of Varna and their influence on the seasonal efficiency of heat pumps with energy source outdoor air

**D Chakyrova**[1], **N Doseva**[1], **N Kalcheva**[2]

[1] Thermal engineering Department, Technical University of Varna, 1 Studentska str., Varna, 9010, Bulgaria
[2] Software and Internet Technologies Department, Technical University of Varna, 1 Studentska str., Varna, 9010, Bulgaria

E-mail: d_chakyrova@tu-varna.bg

**Abstract.** The seasonal transformation coefficient (SCOP) of a heat pump with a heat source outside air as a function of the temperature of the latter is studied. The bin method was used to determine the climatic data for the city of Varna. Outdoor air temperature data were used and statistical analysis of the data was performed. The average frequency and cumulative distribution of temperature bins values in the period from 2005 to 2019 in the city of Varna is calculated. The results are presented using graphs and tables. In the data stability assessment report, the standard deviation and coefficient of variation were used. The SCOP of a heat pump heating installation operating in Varna has been calculated. All results were compared with those of other studies of the authors under different conditions and with reference data.

## 1. Introduction

In recent years, heat pumps with a outdoor air as heat source increasingly used. Determining the design load of heating systems with air-to-water heat pump depends largely on the parameters of the outside air in the place where the system operates. The analysis of the heat load of the existing buildings is the basis for making rational decisions to increase the energy efficiency of the system. For this reason, before configuring the system, it is necessary to estimate the amount of energy needed to heat and cool the building in typical weather conditions. These quantities can be much more difficult to estimate than the design loads or the required system capacity, which are usually calculated under a set of design conditions. The many factors such as solar energy, internal heat sources and partial load of the system that affect energy needs vary during the forecasting period and make the calculations that take into account all options quite complex.From reference literature [1], there are two widely used methods for assessment the energy needs of buildings: the degree-days method and the method based on the hourly distribution of the outside air temperature (the so-called bin temperature method). A characteristic feature of both methods is that they are based on models that consider the steady-state heat flow throughout the building envelope.

A significant disadvantage of the degree-days method is that it assumes the efficiency of the heating source is constant and does not depend on the dynamically hourly changing of the external air parameters.

The methodology provided in the legislation of the Republic of Bulgaria for the design of building air conditioning systems covers to a greater extent conventional heat sources and has some missing links when it comes to heat pumps with outdoor air as heat source. Thus, the thermoeconomic efficiency of the designed system remains in question.

The Bulgarian methodology for calculating the heating load of buildings provides for the use of external temperature, which occurs with a limited duration during the heating period. Most of the time, the heat pump operates in part load mode. For this reason, part load analysis is important for the correct choice of HVAC system.

The stationary approach laid down in the Bulgarian normative documents for calculating the thermal load of buildings is improved with the help of the so-called bin method set in EN 14825 – 2019 [2]. A strength of the bin temperature method is that it is able to estimate the part load of the heat source in HVAC systems with heat pumps [3, 4, 5, 6, 7]. The standard [2] allocate Europe in three climate zone: warmer, average, and colder and suggest the bin temperature data. Bulgaria falls into two of them: average and colder.The use of standardized three climatic zones for Europe and outdoor air temperature bin introduces some inaccuracies in the calculations for the different geographical areas of the same zones.

In [8] the authors published a study which shows that there is a discrepancy in determining the seasonal efficiency ratio (SCOP) for different cities in Bulgaria when using climate bins from EN 14825 and those from the specific climate data. Data on the outside air temperature for one year - 2011 were used, which is disadvantig of this study.

Nevertheless, this article argues that such identification of the climatic features of individual regions of Europe can be misleading, especially in the case of analysis and optimization of air-to-water heat pump systems. It is necessary to develop a mathematical approach to obtain realistic climatic data for the distribution of temperature in the bin as in [9, 10, 3]. This is the main goal of the present study. Also, the SCOP of an air-to-water heat pump with the temperature bins for Varna defined in this article has been calculated and [8], with literature data for these bins from [11] and for Average climate zone from [2]. The rezults are compared.

The analysis is performed for a heat pump with outdoor air as a heat source, integrated with an electric heater as a backup system, used for building located in Varna. The Building Energy Signature (BES) and Declared Heat Capacity (DHC) data as in Annex H, [2] was used for calculations. Some of the data are: Design Temperature for heating, $T_{designh}$ -10°C; Bivalent Temperature, $T_{biv}$ - 6°C; Capacity of the unit at $T_{biv}$ (A-6fW33) - 9,7kW; Design Load Heating, $P_{designh}$ - 11,46kW; Declared capacity of the unit at $T_{designh}$ 7,8kW; Climate – Average; Operation Limit Temperature (TOL) -10°C; Capacity at TOL - 7,8kW; Supplementary heater – Electric.

## 2. Climate data

In order to refine the calculation of SCOP in this article a mathematical processing of the measured temperatures by airport meteorological station ("latitude 43°14' N, longitude 27°50' E, altitude 70 m) for the period from 2005 to 2019 in the city of Varna was made.

In this paper, the dry-bulb temperature bin data for the city of Varna, Bulgaria are determined by an assessment methodology based on monthly outdoor temperatures. The data are calculated from -24.51°C to 42.49°C with a step of 1 degree, in the period from 2005 to 2019 in 8 daily 3-h shifts, and are presented in tabular form.

For the needs of the study are calculated:

- the absolute frequency of temperatures in the interval between -24 and 42 $^0$C;

$$\sum_i Nbin_i = n, \tag{1}$$

where: $Nbin_i$ - number of hours of different values of the outdoor dry-bulb temperature; n - total numbers of hours;

- the relative frequency;

$$v_i = \frac{Nbin_i}{n} \tag{2}$$

$$\sum_i v_1 = 1 \tag{3}$$

- the relative cumulative frequencies, as a ratio of the accumulated cumulative frequencies $c_i$ and the total number of measured values n:

$$\gamma_i \frac{c_i}{n} \tag{4}$$

$$c_i = \sum_{j=1}^{i} Nbin_j \tag{5}$$

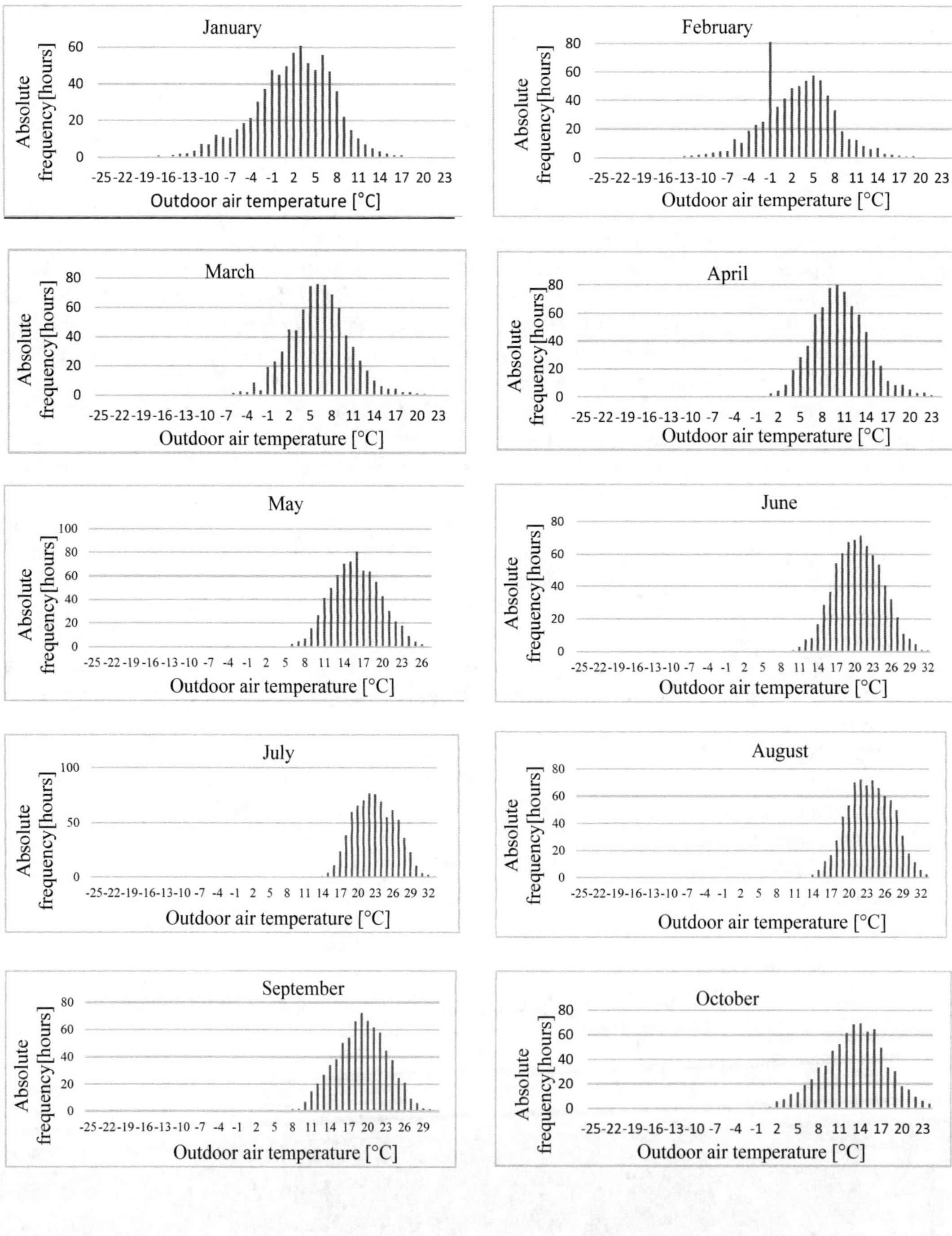

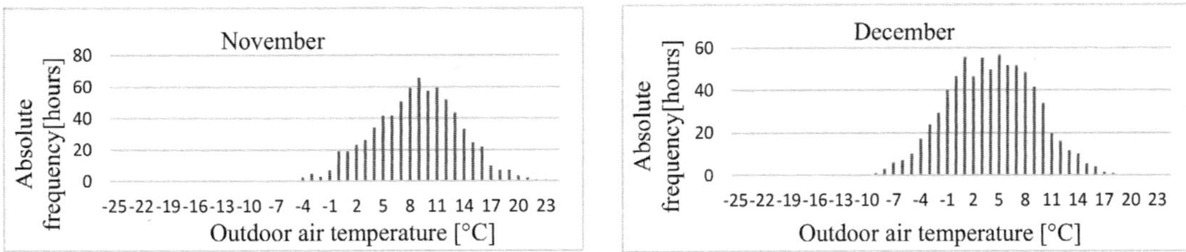


**Figure 1.** Average absolute frequency distribution of Nbin values for Varna city from 2005 to 2019 year.

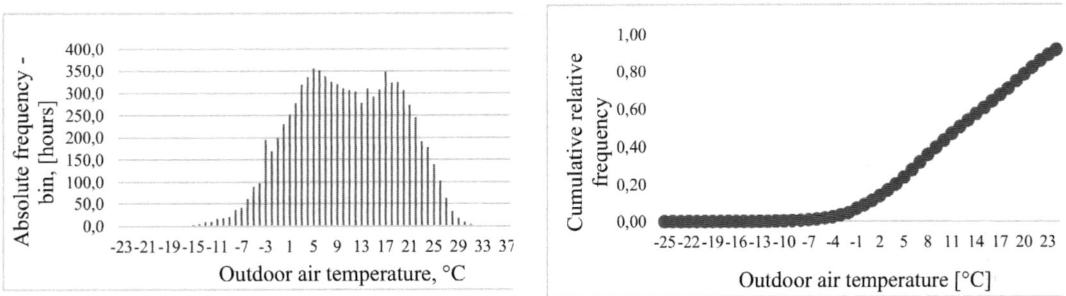


**Figure 2.** a) Annual total bin data for Varna; b) Cumulative relative frequency.

## 3. SCOP of Heat pump air-to-water in heating mode

For the calculations, the data from [2] for the Building Energy Signature (BES) and Declared Heat Capacity (DHC) data as in Annex H, "Calculation Example for $SCOP_{on}$ and $SCOP_{net}$ - Application to a fixed capacity air-to-water heat pump used for floor heating", are used. A similar approach is used in the calculations in [8].

The mean seasonal COP of the heat pump, $SCOP_{net}$ and of the whole system composed of electric air-to-water heat pump and electric heaters, $SCOP_{on}$, are evaluated according to the following equations:

$$SCOP_{net} = \frac{Q_{HP}}{E_{HP,us}} \tag{6}$$

$$SCOP_{on} = \frac{Q_b}{E_{HP,us} + E_{bu}} \tag{7}$$

Where $Q_{HP}$ and $Q_b$ is the heat produced, $E_{HP,us}$ and $E_{bu}$ is the electricity consumed by the heat pump and the additional electric heater, respectively.

## 4. Results and discussion

The results of the calculations of the temperature bins for a climatic year in Varna, as well as their cumulative distribution are presented in figure 1-2.

With the help of meteorological data for the temperature measured 8 times every 3 hours for each 24 hours. The hours in which the temperature is observed within 1°C around the integer values are summarized and arranged in bins.

Yearly total bin data for Varna are shown in figure 2. Temperatures between -17°C and 35°C are observed, and the frequency of very low and very high values is extremely low. The bins have the highest relative frequency of 4% between 7°C and 15°C during the heating and between 20°C and 23 °C during the cooling.

Figure 3 presents the cumulative distribution of yearly bin data for Varna, which is constructed using the cumulative relative frequency (figure 2).Using the presented curve and taking into account the limits (i.e. 15 °C and 24 °C) of the heating and cooling period, their duration is determined. From figure 3 is determined for Varna - 4,582 hours for heating and 1,566 hours for cooling.

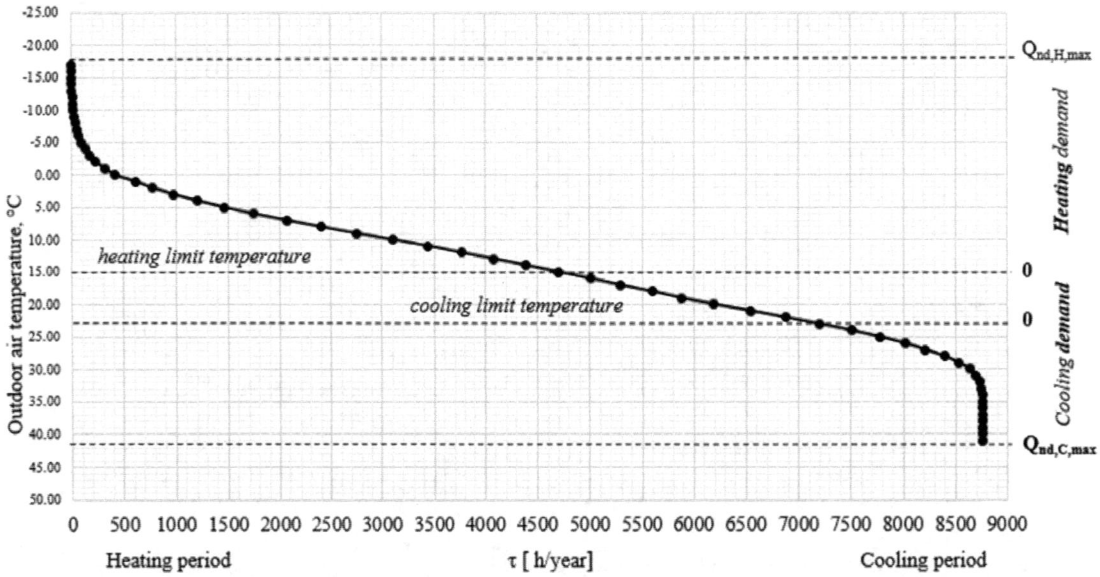


**Figure 3.** Cumulative distribution of yearly total bin data for Varna city from 2005 to 2019 year. **Note**: $Q_{nd,C,max}$ and $Q_{nd,H,max}$ are the maximum cooling, heating demand for building, respectively.

Figure 4 compares the results of the distribution of temperature bins for Varna with those of [8] and [11] only for the heating period. The figure shows that there is good coverage between the bins defined in this article for the climate year from 2005-2019 and those from [11], which are determined more than 50 years ago. It can be concluded that in terms of the duration of the heating period and the distribution of temperatures by hours there are no significant changes in the climate. The difference in the results of [8] is due to the fact that the bins were taken only for one year, which was obviously warmer.

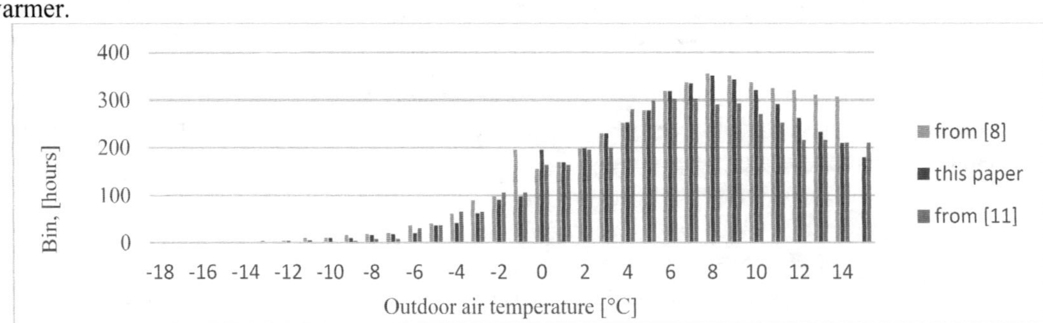


**Figure 4.** Bin temperature distribution in heating period and in temperature interval -18°C and 15°C.

Table 1 presents the results of the calculated SCOPs and compares them with those calculated in [8] and using the data from [2, 11]. The methodology from [2] was used for the calculations.

**Table 1.** Seasonal COP.

| | Source | Annual heating demand, kWh | Annual energy input including electric back up heater, kWh | $SCOP_{on}$ | $SCOP_{net}$ |
|---|---|---|---|---|---|
| 1 | Chakyrova et. al.[8] | 20 910 | 5 705 | 3.66 | 3.67 |
| 2 | [11] | 17 752 | 4 893 | 3.63 | 3.65 |
| 3 | This paper | 16 979 | 4 717 | 3.57 | 3.62 |
| 4 | EN 14825-2019 [2] | 23 558 | 6 461 | 3.60 | 3.65 |

## 5. Conclusion

The aim of this study was to calculate ambient dry-bulb temperature bin data for Varna by using a reliable estimation methodology, based on monthly-average outdoor temperatures and solar clearness index. Due to space limitations, only yearly total $N_{bin}$ values (h/year) for Varna are presented in tabular form and in figure. The temperature bins range between -18 °C and 40 °C with the most frequent are the ones between 7°C -15 °C and 20°C -23 °C. The required energy for heating and cooling of buildings is determined using the cumulative relative frequency (figure 1b), the cumulative distribution of the annual data for the total bin (figure 2) and the duration of heating and cooling periods for Varna.

## References

[1] ASHRAE Handbook. Fundamentals. SI edition, Ashrae, 2017
[2] European standard EN 14825: 2019
[3] Naldi C et al. 2015 Climate influence on seasonal performances of air-to-water heat pumps for heating, *Energy Procedia*, **81**, pp 100 – 107
[4] Stignor C H and Walfridson T 2019 Nordsyn study on air-to-air heat pumps in humid Nordic climate, *Nordic Council of Ministers*, http://dx.doi.org/10.6027/ TN 2019:504
[5] Meng H et al. 2018 Research status and application of airsource heat pump technology, Thermal Science and Engineering https//pdfs.semanticscolar.org/aca5.
[6] Lacourt A 2014 Requirements for seasonal Efficiency for air-conditioning units, *The REHVA European HVAC Journal*, **5**, pp 32-35
[7] Nouvel R, Cotrado M and Pietruschka D 2015 European Mapping of Seasonal Performances of Air-source and Geothermal Heat Pumps for Residential Applications, *Conference: CISBAT 2015*, At: Lausanne, Swiss
[8] Chakyrova D, Rusev D and Doseva N, 2020 Estimation of seasonal efficiency of air-to-water heat pump used in heating mode for different climatic zones in Bulgaria, *Int. J. Eng. Sci. Technol. Rev. Special Issue on Telecommunications, Informatics, Energy and Management*, , http://www.jestr.org/index.php?option=com_content&view=article&id=69&Itemid=116
[9] Papakostas K, Tsilingiridis G and Kyriakis N, 2008, Bin weather data for 38 Greek cities, *Appl. Energy*, **85**, pp 1015–1025
[10] Papakostas K 1999 Bin weather data of Athens Greece, *Renewable Energy* **17**, pp 265-275
[11] Climate guide for the Republic of Bulgaria, 1983, **3**, ed M Kyuchukova

# Experimental study and analysis of hydraulic characteristics of diaphragm (chamber) gas flow meter

**P G Tsankov[1], NA Georgeva[1], S G Delcheva[1]**
[1]Trakia University, Faculty "Technics and Technology" – Yambol, 38 "Graf Ignatiev" str., 4600 Yambol, Bulgaria

E-mail: ptsankov@abv.bg

**Abstract:** In the conditions of life, the most common household gas flow meters are diaphragm (chamber) flow meters. From a technical point of view, hydraulic and durability characteristics are important A basic requirement that must be met by measuring instruments used for industrial or domestic applications is to be as accurate as possible. The diaphragm flow meter type BK G2.5T was experimentally studied in the work. The hydraulic (resistance) characteristic $\Delta p = f(Q)$ of this type of flow meters is obtained and it is derived the analytical dependence of this characteristic - square parabola and means pressure drop at a nominal flow rate – $\Delta p_m$ = 0,4 mbar. The established pulsation character of the operation of the diaphragm flow meter is analyzed by the introduced one the characteristic of the pressure pulsations $\delta_p = p_{max} - p_{min} = f(Q)$, as for the specific flow meter the pulsations are $\delta_p$ = 0,1…0,2 mbar. The original results from obtained researches will be useful for constructors for optimization of the design of the gas systems.

## 1. Introduction

Natural gas is increasing its importance for meeting the energy needs in the various fields of industries and households. This is necessitated both by its wide distribution and by its ecological characteristics [1, 2, 3]. To measure the consumption of natural gas, various constructions of gas flow meters have been created. The extensive classification of gas flow meters based on the principle of operation includes diaphragm (membrane), rotary (volumetric), turbine, differential, Coriolis, vortex, ultrasonic and others [4, 5, 6, 7].

Practically the most common flow meters are diaphragm (membrane) due to their wide exploitation in the home [8, 9]. When installing diaphragm flow meters in domestic gas supply installations, it is necessary to take into account the hydraulic characteristics of these devices and the specifics of their work, specified in more detail in point 2. Designers of gas systems are faced with the lack of sufficient information about these characteristics - often set a generalized pressure drop for a group of flow meters, this drop is not associated with the specific flow rate [3, 6, 7].

The aim of the present work is to clarify some specific features of the hydraulic resistance characteristics and the principles of work of diaphragm flow meters.

## 2. Principle of operation and main characteristics of diaphragm flow meters

Diaphragm flow meters are used to measure gas consumption in a wide range - from 1.6… 160 $m^3/h$, but in the practice are used mainly flow meters with small flow rate - 1.6 ... 10 $m^3/h$. Structurally diaphragm flow meters consist of several assemblies [7] (figure 1). There is a huge variety of designs of diaphragm flow meters [5, 10, 11], many protected patents for various units - valves, actuators, and registering mechanisms, etc.

The diaphragm flow meter is a volumetric flow meter with a periodic filling and emptying of variable volume chambers, in which the movement for this filling and emptying is mechanically transmitted to a counter, and the flowing gas drives it. The main elements of the measuring device (figure 1) are:

- two flexible membranes or diaphragms 3, dividing together with the housing 1 and the cover 2 the whole device into **four chambers (measuring compartments)**;
- valves 5 for controlling and directing the gas flow when filling and emptying the chambers;
- a suitable connection and measuring mechanism 4 for maintaining the diaphragms, **synchronizing the valves** and recording the number of cycles.

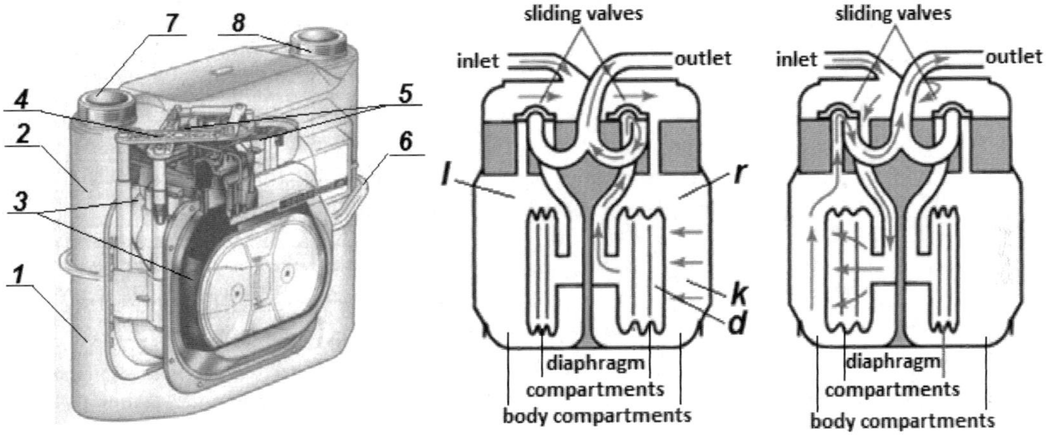


(a) (b)

**Figure 1.** Diaphragm flow meter with sliding valves and built-in temperature corrector (a) device: 1 - housing; 2 - cover; 3 - membranes; 4 - measuring mechanism; 5 - gas distribution sliding valves; 6 - tightening hoop; 7 - gas inlet; 8 - gas outlet (b) principle of operation - left (l) and right (r) compartments, in which there are two chambers: diaphragm (d) and body (k).

The movement of the membrane is regulated in such a way that the total displacement in the successive cycles, realizing the filling and emptying of the chambers, is the same.

The operation of the diaphragm flow meter is connected at any time with filling the one chamber, emptying the other chamber from the same compartment, and pausing for the other two chambers located in the adjacent compartment. For example, for the process shown in figure 1b, left position: the housing chamber (k) is filling and the diaphragm chamber (d) of the right compartment (r) is emptying. At the same time, the chambers (d-k) in the left compartment (l) are at rest [7]. The gas distribution valves move in synchrony and occupy one of the four possible positions.

The improvement of the constructions of the diaphragm flowmeters is in the direction [12], [13], [14]: achieving a minimum resistance to the gas flowing and at the same time driving the device; achieving maximum density and wear resistance of the valves; increasing the accuracy of measurement and installation of temperature correctors in the diaphragm flow meter.

The main characteristics of diaphragm flow meters relate to their accuracy - the dependence of the error of the measured amount of gas on the flow rate, pressure or temperature of the gas [13], [15], [2], [7] – figure 2a.

From a hydraulic point of view, however, the dependence of the pressure loss during different flow through the diaphragm flow meter is interesting - the characteristic $\Delta p = f(Q)$ – figure 2b [7]. When designing domestic gas systems, the allowable pressure drop in the whole system (without flow meter) is sometimes $[\Delta p] < 2$ mbar [16], [3], [6]. Although considered to be a small fraction of the gas pressure used to operate the flow meter, $\Delta p = 2$ mbar for flow meters up to G10 is set in samples of prospectus characteristics for pressure losses in the flow meters. This shows: the flow meter is a large hydraulic resistance; it is not commented that the hydraulic losses in it depend on the flow rate (the characteristics $\Delta p = f(Q)$ are often absent); the flow meter can have a significant effect on lowering the pressure in front of the specific fuel consumer.

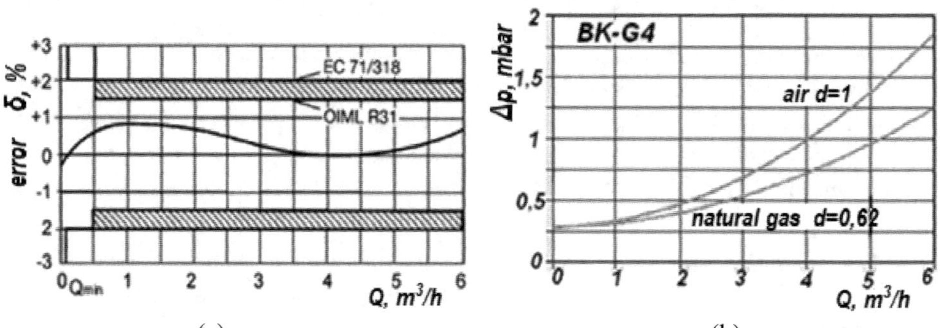


(a) (b)

**Figure 2.** Characteristics of diaphragm flow meter: dependence of the error (accuracy) with an admissible limit of the measured amount of gas from the flow rate - for flow meter BK - G2,5 - $\delta = f(Q)$; b) pressure drop in flow meters type BK-G4 - $\Delta p = f(Q; d)$.

The hydraulic characteristic is also interesting due to the fact that the diaphragm flow meter is a volumetric type of flow meter. As described above, its working principle is based on discretion or portion work. When switching the sliding valves, which achieves the sequential filling and emptying of the chambers, moments of stopping and starting of the respective flows should be expected. The interruption at the time of switching can be expected to result in pressure and flow fluctuations. This problem is not commented in the commercial - production characteristics of different companies and in the study of diaphragm type flow meters.

## 3. Installation and methodology for investigation of the hydraulic characteristics of a diaphragm flow meter

In connection with the implementation of the planned study of the hydraulic characteristics of a diaphragm gas flow meter $\Delta p = f(Q)$ a laboratory installation was built – figure 3.

The supply of the installation is with a reciprocating compressor *1*. It should be specified that the working gas is air, which requires recalculation of the obtained hydraulic characteristic of the flow meter for another working gas (for example, natural gas $d = \rho_{ng} / \rho_a = 0.62$).

The tested diaphragm flow meter *10* is type BK - G2,5 T, for which the nominal flow rate is 2,5 $m^3/h$. The flow control during the test is performed by means of a valve *14* in the nominal range of the flow meter - from $Q_{MIN}$ to $Q_{MAX}$ (i.e. from 0.3 to 4 $m^3/h$). Preliminary control of the current volume flow is carried out with a flow meter - rotameter *13*.

With the pressure reducing valve *7* the required working pressure in the system for the tested diaphragm flow meter *10* is adjusted, controlled by the two-pipe liquid manometer *9*. Tests were performed at two values of the inlet pressure – $p_{i1}$ = 27,8 mbar and $p_{i2}$ = 38,9 mbar.

During the study, the two important hydrodynamic parameters were measured and calculated: the volumetric air flow rate Q (measured by the volumetric method: by the volume W measured with the tested diaphragm flow meter *10* and the time τ for its flow, measured with a stopwatch 17) and the pressure drop Δp (losses of pressure) in the diaphragm flow meter measured with a differential pressure gauge *11*.

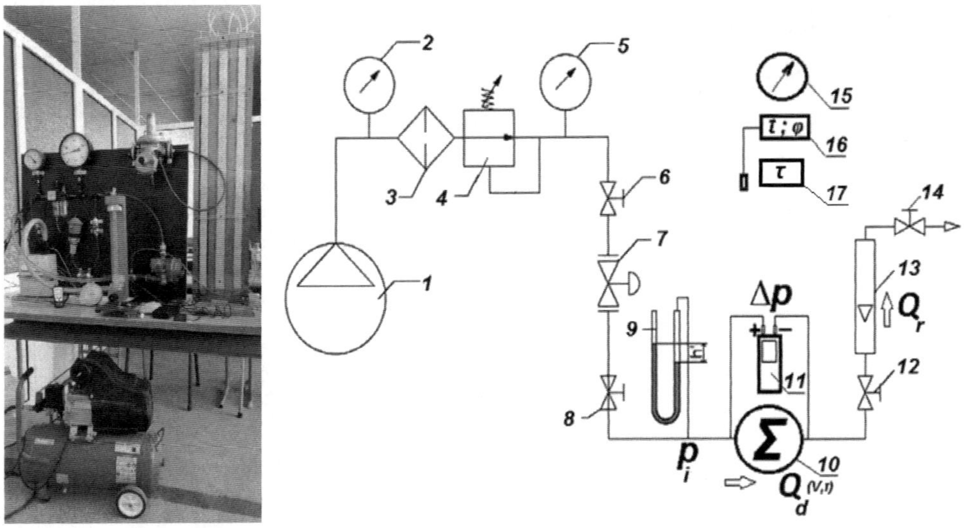


**Figure 3.** Laboratory installation - construction and general scheme of the stand for testing of gas elements and diaphragm flow meter: 1 - compressor; 2 - manometer; 3 - filter; 4 - pressure reducing valve; 5 - manometer; 6 - shut-off valve; 7 - pressure reducing valve (low pressure); 8 - shut-off valve; 9 - two-pipe liquid manometer (pi); 10 - diaphragm flow meter (Qd); 11 - differential manometer; 12 - regulating valve; 13 - flow meter – rotameter (Qr); 14 - regulating valve; 15 - barometer – aneroid (pa); 16 - thermometer – hygrometer (t; ϕ); 17 – stopwatch (τ).

The air flow required for the recalculation - atmospheric pressure ($p_{atm}$) and ambient air temperature ($t_a$) are measured with a barometer - aneroid *15* and a thermometer - hygrometer *16*.

During the study, at a constant flow of air flowing through the flow meter *10*, the sequential filling and switching of the volumes (chambers) caused pressure pulsations, the pressure changes from $p_{min}$ through $p_m$ to $p_{max}$, and vice versa. The measurement of these pulsations required a creative assessment in the reporting. The obtained results prove good correspondence and tendency of the characteristics, regardless of the subjective nature of part of the assessment. The frequency of the pulsations is directly related to the flow volume.

The obtained results gave grounds for introduction of additional hydraulic characteristic of the diaphragm flow meter- besides the classical dependence - the **resistance characteristic** $\Delta p_m = f\,(Q)$, a dynamic characteristic estimating the **pressure pulsations** depending on the flow which we introduced. This characteristic is defined as:

$$\delta_p = \Delta p_{max} - \Delta p_{min} = f\ (\ Q\ ), \tag{1}$$

where $\Delta p_{max}$ and $\Delta p_{min}$ - maximum and minimum pressure drop in the volume flow meter at a constant flow rate Q.

Another characteristic is defined - **relative deviation of the pressure pulsations**, which determines what part of the pressure pulsations make up the measured average pressure drop:

$$\Delta = \left(\frac{\delta_p/2}{\Delta p_m}\right).100 \tag{2}$$

## 4. Results of the research and analysis of the obtained hydraulic and pulsation characteristics

The results of the study of diaphragm flow meter type BK-G2.5 T at two different inlet pressures are shown in figure 4. The graphs of both studies show three characteristics - these are the dependences of the maximum, average and minimum pressure drops in the flow meter during different flow *Q* - $\Delta p_{max}$, $\Delta p_m$, $\Delta p_{min} = f(Q)$.

The actual characteristic Δp (*Q*) is the mean curve and actually expresses the mean pressure drop $\Delta p_m$ in the flow meter at the respective flow rate Q. The other two graphical dependencies show the upper and lower ranges of the change in pressure drop as a function of flow. Regardless of the complexity of the studies, the characteristics $\Delta p_{max}$ and $\Delta p_{min} = f(Q)$ are symmetrically located with respect to the mean resistance characteristic and have a clearly defined quadratic dependence.

The main characteristics have a pronounced parabolic (quadratic) character - proved by those describing the correlation curves and the high value of the correlation coefficient $R^2 > 0.99$ in both studies (at different inlet pressures).

Describing the correlation curves are very close, which is confirmed by the combined graphs Δp ($Q$) for the two inlet pressures given in figure 5. This shows that the inlet (operating) pressure in the gas system does not affect from the performance of the flow meter and allows to write a generalized equation for the hydraulic characteristics of the studied diaphragm flow meter BK-G2,5T:

$$\Delta p = 2{,}3.Q^2 + 6{,}1\ .\ Q + \ 12{,}5 \quad (3)$$

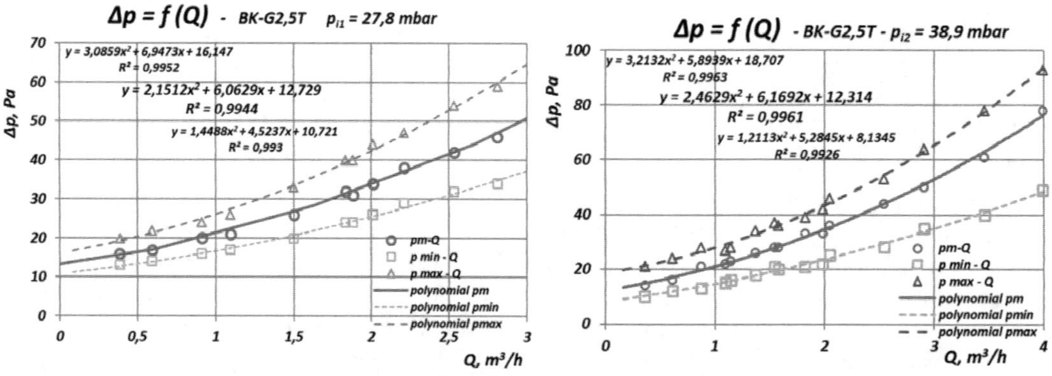


**Figure 4.** Characteristics Δp = $f$ ($Q$) of diaphragm flow meter BK-G2,5T at pi1 = 27,8 mbar and pi2 = 38,9 mbar).

As a generalized practical assessment required in the design of gas systems, it can be said that the pressure loss in this type of flowmeter is in the range of Δp = 15÷ 80 Pa = 0.15÷ 0.8 hPa = 0.15÷0.8 mbar, as the minimum static resistance of the flow meter (at $Q$ = 0) is about $\Delta p_0 = 12$ Pa.

Another important conclusion of the study concerns the pressure pulsations $\delta_p$ determined depending on (1). The experimental results of this study are presented in figure 6 as a dependence of $\delta_p = f(Q)$ - i.e. give the change of the pressure pulsations with the increase of the flowing volume flow of the gas.

The dependence analysis shows the indisputable linear increase of the pressure pulsations with the increase of the flow rate. In addition, with an increase in the working pressure, there is a slight increase in these pulsations - as evidenced by the correlations describing the process (increasing coefficient of linear dependence).

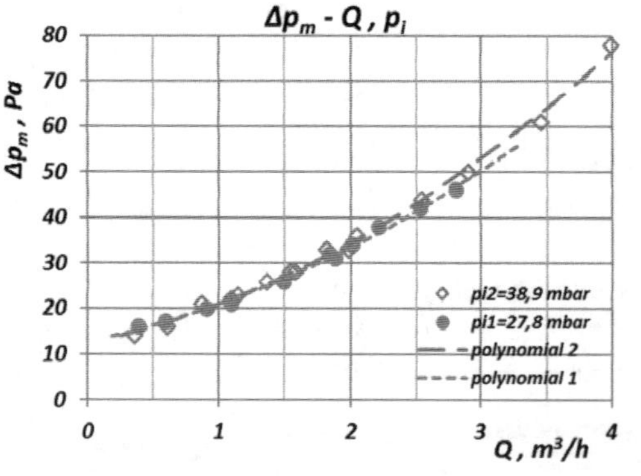


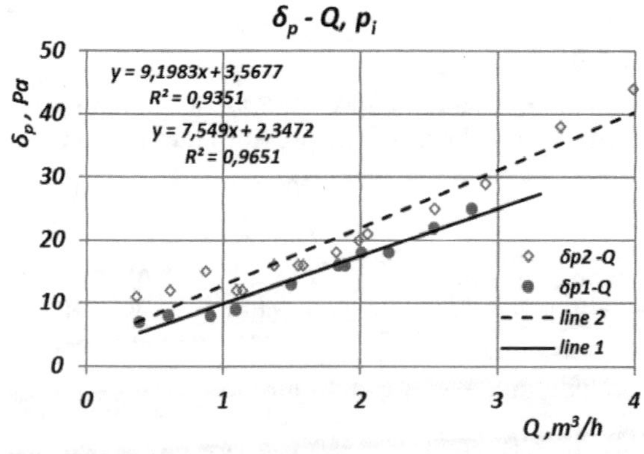


**Figure 5.** Characteristics of the diaphragm flow meter BK-G2,5T Δp = $f$ ($Q$) at inlet

**Figure 6.** Characteristics of pressure pulsations $\delta_p$ = $f$ ($Q$) at different inlet pressures –

pressures $p_{i2}$ = 38.9 mbar and $p_{i1}$ = 27.8 mbar. $p_{i1}$ = 27.8 mbar and $p_{i2}$ = 38.9 mbar.

In absolute terms, the pressure pulsations at a nominal flow rate of $Q$ = 2.5 $m^3/h$ reach $\delta_p$ = 22÷25 Pa, at an average pressure drop of $\Delta p_m$ = 42÷44 Pa. The estimate for the share of the pulsations in the total pressure drop, given by the relative deviation of the pressure pulsations Δ (determined by (2)), shows that in the specific case it varies in the range Δ = 25÷30%.

Pressure pulsations are a significant part of the total losses and should be reported. For example, the pressure drop in the diaphragm flow meter in this case (at nominal flow rate) should be defined as:

$$\Delta p = \Delta p_m \pm (\delta_p / 2) = 43 \pm (24/2) = 43 \pm 12 \ Pa \quad (4)$$

The question of the influence of pressure pulsations $\delta_p$ on the overall operation of the gas system is interesting. Despite the relatively high relative share of pulsations in relation to the total pressure drop (Δ = 25÷30%), they constitute in absolute terms $\delta_p$ = 0.1÷ 0.2 mbar, which relative to the operating pressure in gas systems $p_i$ = 20÷50 mbar is relatively small. Given the volume of gas located from the flow meter to the consumer, which volume is a kind of accumulator and damper of these ripples, a reasonable assumption can be made - the impact of pressure ripples is insignificant. This conclusion - the influence of pressure pulsations on the operation of the entire gas system and possible recommendations for the location of the diaphragm flow meter should be checked experimentally.

Nevertheless, the proven conclusion about the pressure drop Δp ($Q$) and the pressure pulsations $\delta_p = f(Q)$ should be taken into account when evaluating and analyzing the operation of the gas systems.

## 5. Conclusions

Experimental results of the hydraulic characteristic Δp ($Q$) of a common diaphragm flow meter design have been obtained from our own studies. A correlation dependence (3) describing this characteristic with a high value of the derived correlation coefficient, which has a quadratic character. At nominal flow, the pressure losses in this type of flow meter are $\Delta p_m$ = 0,4 mbar.

The pulsation character of the pressure change in the diaphragm flow meter has been established. This pulsation is evaluated by the introduced characteristic "pressure pulsations" - $\delta_p$. A linear proportionality of the pressure pulsations $\delta_p$ depending on the flow rate $Q$ has been established. In relative and absolute form, the pressure pulsations are respectively - Δ = 25÷30 % and δp = 0,1÷ 0,2 mbar. The results of the research can be used in the design, analysis and evaluation of the operation of gas systems.

## References

[1] Treloar R 2010 Gas Installation Technology, 2nd Edition, *Wiley Blackwell*, p 496
[2] Nikolov G 2007 Distribution and use of natural gas, *Economics*-Sofia, p 510 (in Bulgarian)
[3] Staskevich N, Severinets G, Vidgorchik D 1990 Handbook of gas supply and gas use, *Leningrad, Nedra*, p 762 p. (in Russian), ISBN 978-9-549-20231-1
[4] Dane Enrich 2001A guide to metering technologies, *ASHRAE Journal*, **43**, Iss. 10, p 33
[5] Karyakin E et al 2013 Industrial gas equipment: a guide, *Saratov 6th edition, revised and supplemented, Gazovik*, (in Russian), ISBN 978-5-9758-1454-8
[6] Tsankov P 2018 Collection of solved tasks and reference book on gas supply, *Sliven, Obnova*, p 246 (in Bulgarian), ISBN9 78-619-7445-09-1
[7] www.kromschroeder.ru; www.fiorentini.com; http://gastechnika.eu; www.elster-instromet.com; www.emerson.com; www.madas.it; www.gasfitterbasics.com
[8] Crainic M 2010 A short history of diaphragm gas meters, Part I Description of the prior art, Installations for Buildings and Ambiental Comfort, *Conference XIX*, Timisoara - Romania 15-16 April 2010, pp172-188
[9] Crainic M 2010 A shorth history of diaphragm gas meters, Part II Evolution of modern diaphragm gasmeters, *Installations for Buildings and Ambiental Comfort Conference XIX*-edition, Timisoara – Romania, 15-16 April 2010, pp 189-215

[10] Wang H, Hang Y, Li S, et al 2012 Temperature and pressure compensation of diaphragm gas meter, *Gas Heat*, **32** (1), pp. B13–B15, B31
[11] Gavra L, Crainic M, Pupsa P, Popa B 2012 Residential Smart Gas Meters, *10th International Symposium on Electronics and Telecommunications(ISETC)*, 15-16 Nov.2012, pp 37-40
[12] Li X, Deng X, Zhang S, Zhan J, Wu L 2018 Metrological performance of diaphragm gas meters with different use age, *The Journal of Engineering*, 7th International Symposium on Test Automation and Instrumentation (ISTAI 2018), pp 1-4
[13] Ficco G 2014 Metrological performance of diaphragm gas meters in distribution networks, *Flow Meas. Inst.*, **37**, pp 65–72
[14] Crainic M, Popa S and Pupşa P 2007 Temperature compensated diaphragm gas meters, Installations for Buildings and Ambiental Comfort, *Conference XIV*, Timisoara, Romania
[15] Cen K, Wang H, Yuan, X et al 2009 Inspection and analysis of measurement indication error for diaphragm gas meter in use, *Gas Heat*, **29** (11), pp B11–B14
[16] GAZTEK-Design of building gas installations. *Sofia*, 2008, p 30 (in Bulgarian)

# Analysis of energy consumption in the industrial sector in the European Union

**Nely Georgeva[1], Petko Tsankov[2]**

[1]Trakia University, Faculty "Technics and Technology" – Yambol, 38 "Graf Ignatiev" str., 4600 Yambol, Bulgaria, e-mail: nely.georgieva@trakia-uni.bg
[2]Trakia University, Faculty "Technics and Technology" – Yambol, 38 "Graf Ignatiev" str., 4600 Yambol, Bulgaria, e-mail: ptsankov@abv.bg

E-mail: nely.georgieva@trakia-uni.bg

**Abstract.** The European Union (EU) pursues a coherent policy to reduce the carbon footprint of human impact on the climate, with one of the main strategies being to reduce the use of energy resources. On the other hand, the EU is the world's largest importer of energy and making it highly energy dependent which is reduced the competitiveness of Member States' economies. Reducing the energy resources used leads to political, economic, social and environmental benefits. All this explains the EU's motivation to make energy efficiency a key goal in development policy for the Union. At the level of the European Union a number of normative documents have been adopted, related to increasing the energy efficiency of the two largest areas of energy consumption - buildings and manufacturing systems. In the share of energy consumption in the EU, manufacturing production has a significant share, which represents the potential for increasing energy efficiency. This article analyses the dynamics of energy intensity of the European economy and consumption of energy in the manufacturing sector of the EU member states, compares the situation in Bulgaria. The efficiency of the undertaken measures for stimulating the increase of the energy efficiency in the sector is analysed. Conclusions from the study are formulated

## 1. Introduction

The Energy Efficiency Directive 2012/27 / EU sets the EU target for energy efficiency by 2020, which required a reduction of primary or final energy consumption by 20%, compared to the projected levels for 2020 according to the baseline scenario of the European Commission since 2007. In December 2018, the European Parliament and the Council of the European Union adopted the revised Energy Efficiency Directive (EED) [1]. The revised EED sets that the energy efficiency target for the EU should be a minimum of 32,5 % till 2030. In this document, the energy efficiency is identified as a strong driver for achieving climate goals. Reducing energy consumption is set as a key building block in the Commission's proposal for an "European strategic long-term vision for a prosperous, modern, competitive and climate-neutral economy", presented in November 2018 [2].

As a result of the EU's consistent policy between 1990 and 2016, energy consumption was reduced by almost 2%, greenhouse gas emissions by 22%, while GDP increased by 54% [3]. However, the achieved results are not sufficient to meet the objectives set in the adopted strategic documents. Unstable momentum to improve energy efficiency is a matter of serious concern. The analysis of the available data shows that the improvements in the energy intensity of the global economy are slowing

- the improvement of 1.2% in 2018 is about half of the average observed since 2010. This supposes to make greater efforts to an intensification of the existing measures.

The analysis of primary and final energy consumption shows that for 2018 primary energy consumption for the EU was 4.9% above the target of efficiency energy for 2020 and 22.0% above the 2030 target and final energy consumption for the EU was 3.2% above the efficiency target for 2020 and 17.0% above from the 2030 target [4]. One of the sectors with a large share in energy consumption is an industry.

The aim of this study is to analyse the dynamics of energy consumption and energy intensity of the European economy in the manufacturing sector of the EU member states, Bulgaria as one of the major energy consumers and comparison between both.

## 2. Exposure

Various data points can be used as indicators for comparison and assessment of energy consumption: gross internal energy consumption, primary and final energy consumption, energy intensity, etc. Depending on the needs of the research different indicators can be used individually or in combination to clarify the existing dependencies, to highlight the regularities, and to draw conclusions about the trends.

In the present study, the main indicators used final energy consumption and energy intensity. According to the Eurostat glossary, final energy consumption is the total energy consumed by end users and it excludes energy used by the energy sector, including deliveries, and transformation. It also excludes fuel transformed in the electrical power stations of manufacturing auto-producers and coke fuel transformed into blast-furnace gas where this is not part of overall manufacturing consumption but of the transformation sector. Energy intensity is one of the indicators to measure the energy needs of an economy and is the ratio between gross domestic energy consumption (GIEC) and gross domestic product (GDP), calculated per calendar year. The GIEC is calculated as the sum of the gross domestic consumption of the five energy sources: solid fuels, oil, gas, nuclear and renewables. It is often used as an approximation of energy efficiency.

According to Eurostat, final energy consumption has been increasing slowly since 1994, reaching its highest level - 991 Mtoe in 2006 [5]. The reductions in 2011-2012 (figure 1) can be partially attributed to the reduced economic production due to the economic crisis, expressed by 0.4% contraction of real GDP in 2012, and continuing until 2014 reduction - of meteorological conditions. After 2014, the trend changed as the final energy consumption in the EU again marked a positive growth, reaching 939 Mtoe in 2018, which is only 0.1% lower compared to the 2017 consumption.

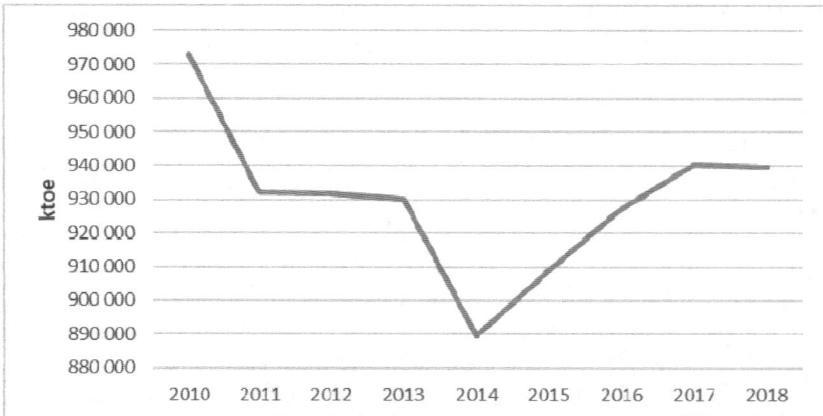


**Figure 1.** Final energy consumption in EU for 2010-2018, source Eurostat

According to Eurostat (published on 06.06.2020), the largest final energy consumers in the EU (figure 2) are Germany, France, England, Italy (figure 2 shows the countries with annual consumption near or above 1000 ktoe) [8]. The information is summaries for the period 2009-2018 and clearly identifies the largest energy consumers in Europe.

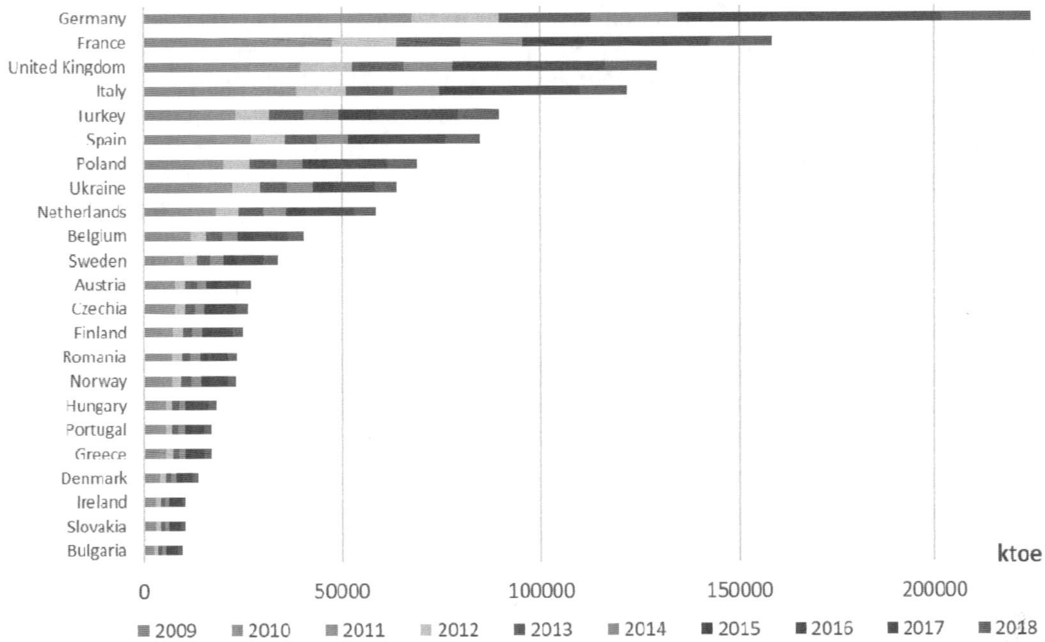


**Figure 2.** Final energy consumption by country in EU 2009-2018, source Eurostat.

A breakdown of energy consumption by economic sectors in the EU shows an increase in the final energy consumption, which is growing at a slower pace than GDP. This is an indicator of the relative decoupling of energy consumption from economic growth, which leads to a positive trend towards increasing the efficiency of European economies. For the period 2014 and 2017, the largest increase in energy consumption was observed in buildings (+ 8.3%), followed by transport (+ 5.8%) and industry (+ 4%). For 2018 there is a decrease in energy consumption for almost all sectors except for the Transport sector. Figure 3 shows that industry is the third largest sector in terms of energy consumption, which justifies the importance of monitoring and increasing energy efficiency in this industry. Figure 3 shows that industry is the third largest sector in terms of energy consumption after transport and households, which justifies the importance of monitoring and increasing energy efficiency in this industry.

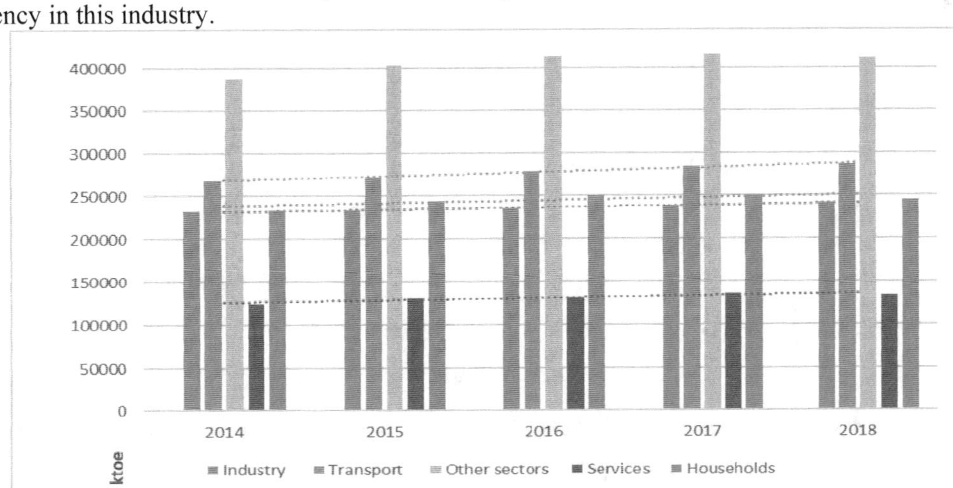


**Figure 3.** Final energy consumption by sectors for the period 2014-2018 in EU, source Eurostat.

The state of the final energy consumption by sectors for Bulgaria is presented in figure 4, according to the data indicated in the Integrated Plan in the field of energy and climate of the Republic of Bulgaria 2021-2030, Annex 3, published on 21.02.2020 [7]. Thanks to the policy pursued in recent years for increasing the energy efficiency of buildings, the energy consumption in this sector is lower than in the industry sector, opposite to the situation in the EU. Despite the crisis that our economy has gone through, the energy consumption in the sector is growing except in 2014, when there was a sharp decline, after which it began to grow again [8]. One of the explanations for this fact is the high energy consumption of production sector, which can be clearly established by the next indicator - energy intensity. According to the latest data published by the National Statistical Institute in 2018 [9], there is a slight decline only in the transport sector, while in other sectors the upward trend.

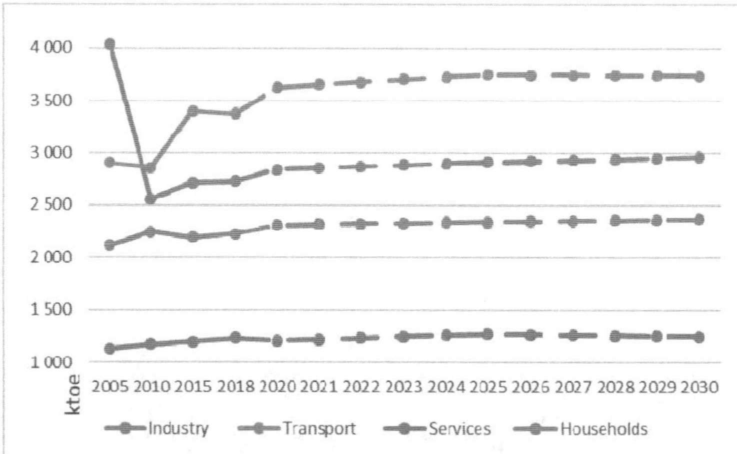


**Figure 4.** Final consumption by sectors for the period 2005-2015 and a forecast for the period up to 2030 for Bulgaria.

The energy intensity shows how much energy is needed to produce a unit of GDP and thus is a good indicator for assessing energy consumption as well as for analyzing trends. In the period 1990-2018 in the EU there is a steady trend of decreasing energy intensity - 37% in total (1.7% per year). There are various reasons for the improvement in energy intensity: the general shift from industry towards a service-based economy in Europe, performing less energy-intensive activities and production methods within the industry, the closure of inefficient units, or the use of more energy-efficient appliances. According to Eurostat data (figure 5), all Member States have reduced the energy intensity in the period, including Bulgaria, although our economy ranks first in this indicator.

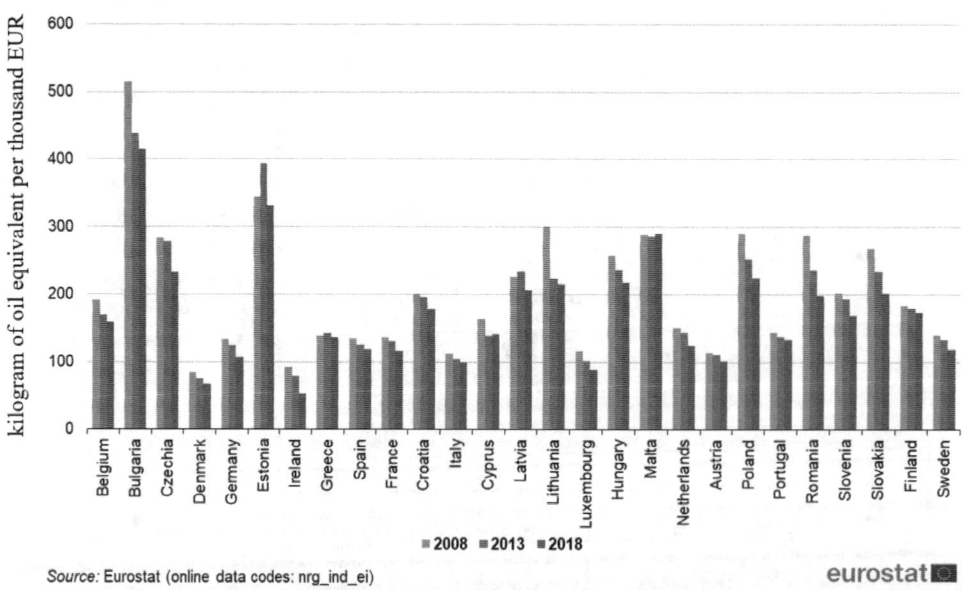


**Figure 5.** Energy intensity of the economies of the EU Member States for selected years.

The analysis of the energy intensity by sectors in the European Union shows fluctuations in the trend of the indicator but the general tendency is for its decrease (figure 6). A more gradual decline is in the Industry sector, followed by the Transport sector.

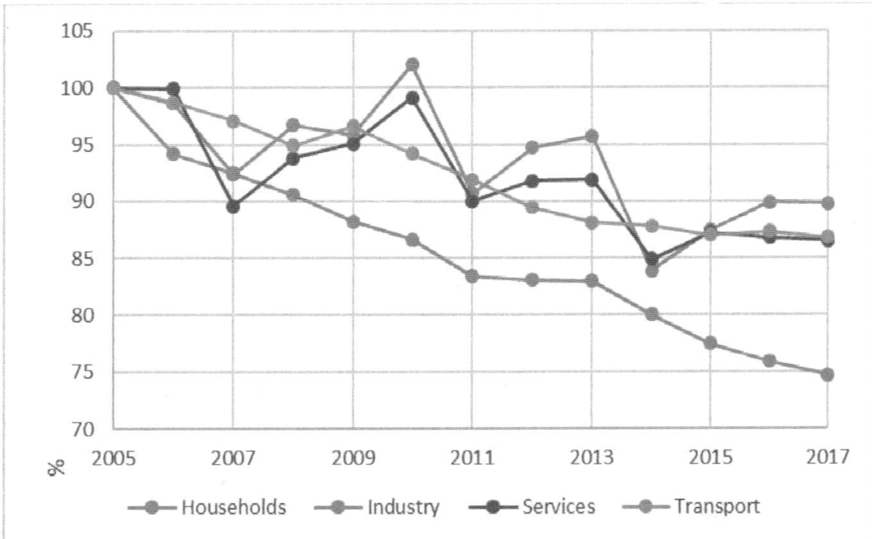


**Figure 6.** Final energy intensity by sector for the EU in base year 2005, *Source: European Environment Agency (EEA).*

According to data from Odyssee And Mure Databases [10], the two sectors of the Bulgarian economy with the highest energy intensity are the Industry and Transport sectors. Figure 6 presents a comparison of the index levels in Bulgaria and in the EU for the period 2010-2017 as the industry is the second largest sector according to the energy intensity. This determines its low economic efficiency. Regardless of a slight decline in the energy intensity of the manufacturing sector over the review period, it remains well above EU rates.

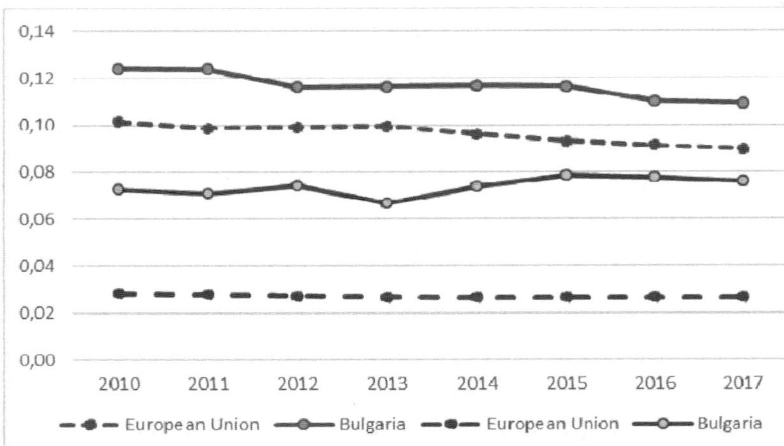


Energy intensity of industry (at purchasing power parities) Energy intensity of transport

**Figure 7.** Comparison of energy intensity in the Transport and Industry sectors in Bulgaria and the EU. *Source: Odyssee And Mure Databases.*

## 3. Conclusion

In the present study, the authors analysed the statistics provided by various sources in the field of energy consumption in Europe and the energy intensity of the manufacturing sector as a potential opportunity to raising the energy efficiency of the European and national economies. Some differences in the published information in the sources were found, as well as differences in the indicators on which consumption is reported. An attempt to solve these problems was realized through the Odyssee-Mure project which is co-ordinated by ADEME with the technical support of Enerdata and Fraunhofer.

The analysis clearly shows that a constant decrease in energy consumption since 1990 is a result of a consistent long-standing EU policy to reduce energy consumption and increase energy efficiency as a part of legislative measures to improve the climate and reduce the carbon footprint on the nature. There is a difference between the current state and the historically proven dependence of increasing energy consumption and increasing economic growth. This is observed in all EU member states.

The analysis indicates that, unfortunately, the achievement of the targets for 2020 and 2030 will most likely not be met.

In terms of final energy consumption, transport and households were found to be the most energy-intensive sectors in the EU, with Industry remaining in third place. In Bulgaria, thanks to the policy for increasing energy efficiency in buildings, pursued in recent years, the Households sector is in third place.

The analysis of energy intensity shows that in the EU the main energy-intensive sectors are households and services, in Bulgaria they are transport and industry.

In summary, the Industry sector as one of the major consumers of energy has considerable potential for energy efficiency in this sector. This also outlines the direction in which efforts should be directed to improve energy efficiency in the next period 2021-2030.

As proved above, the manufacturing branch ranks as one of the major sectors with high energy consumption and high energy intensity, which make urgently legislative measures have to be taken to ensure the continuous monitoring and estimation of energy consumption in the sector, ensuring proper measures, which need to be taken towards improving the state in order to reach the 2030 targets.

## References

[1] Directive (EU) 2018/2002 of the European Parliament and of the Council of 11 December 2018 amending Directive 2012/27/EU on energy efficiency (Text with EEA relevance.) PE/54/2018/REV/1

[2] Clean Planet for all A European strategic long-term vision for a prosperous, modern, competitive and climate neutral economy, Communication from The Commission to The European Parliament, The European Council, The Council, The European Economic And Social Committee, *The Committee of the Regions and The European Investment Bank*, Brussels, 28.11.2018, COM (2018) 773 final

[3] Trends and projections in Europe 2019, Tracking progress towards Europe's climate and energy targets, *EEA Report* **No 15**/20191994, ISSN 1977-8449

[4] Energy consumption in 2018, Newsrelease, 26/2020 - 4 February 2020, available at https://ec.europa.eu/eurostat

[5] Energy data 2020 edition, ISBN 978-92-76-20629-3 ISSN 2600-237X, doi: 10.2785/68334

[6] Simplified energy balances [nrg_bal_s], 06.2020, https://appsso.eurostat.ec.europa.eu

[7] Integrated plan in the field of energy and climate of the Republic of Bulgaria 2021 – 2030 - project, *Ministry of Energy, Ministry of Environment and Water*, Republic of Bulgaria, available at https://ec.europa.eu/energy/sites/ener/files/documents

[8] Ivanov K 2020 Analysis of the energy consumption of the industrial sector in the Republic of Bulgaria, *ARTTE* **8**, No. 2, ISSN 1314-8788 (print), ISSN 1314-8796 (online), doi: 10.15547/artte.2020.02.007, pp 133-140

[9] Bulgaria 2020, brochure, *National Statistical Institute*, Published on: https://www.nsi.bg/

[10] Odyssee and Mure, https://www.odyssee-mure.eu/project.html

# The gaps of renewable energy legislation in Bulgaria

**T D Stankova**[1], **D S Toneva**[2]

[1,2] Department of Ecology and Environmental protection, Technical University of Varna 9000, Bulgaria

E-mail: stankova.todorka@gmail.com

**Abstract.** Nowadays when the climate changes are of a worldwide vital importance, the European Union is on the way to trach the 2020 targets for smart and sustainable development. The set in 2007 targets are on the way to be even exceeded till the end of 2020 when EU is expecting to report at least a 20% greenhouse emission reduction from their levels in 1990, 20 % gross final energy consumption by renewable energy sources and 20% energy efficiency improvement. That satisfactory result isn't enough to reflect the climate changes with the speed of their consequences. Namely, because of this the EU has put in place a legislation frames to achieve new climate and energy targets up to 2030. While the EU is working on fulfilling a decarbonisation strategy and insuring a legislative framework via Directive 2009/28/EC of the European Parliament and of the Council of 23 April 2009 on the promotion of the use of energy from renewable sources for all EU member states, Bulgarian parliament transposed the Directive into a Law for energy from renewable sources which from 2011 after its' first acceptance till 2019 was changes 18 times, 4 of which in only 1 year (2013). This paper underlines some of the legislative gaps between the Directive 2009/28/EC and the Bulgarian legislation, the contraries between the EU and Bulgarian kind of fulfilment of a main target to stimulate the investments in renewable energy and on the other side the contraries in the internal Bulgarian regulation papers which negatively reflect the investment intends in the country.

## 1. Introduction

The energy mix of all EU member states is changed significantly over the last 10 years when the RES energy is growing thanks to the implemented proactive measures for RES promotion. The set in 2007 targets are on the way to be even exceeded till the end of 2020 when EU is expecting to report at least a 20% greenhouse gas emission reduction from their levels in 1990, 20 % gross final energy consumption produced by renewable energy sources and 20% energy efficiency improvement [1]. That satisfactory from statistical point of view result isn't enough to reflect the climate changes with the speed of their consequences. Despite the fact that the nature is "getting a break" from humans because of COVID 19 disease economy reflections [2], 2020 is announced by WWF as a critical year for our future and for the climate [3]. The first quarter of 2020 had been recorded as the hottest year in the history till now. One of the facts that proof it is the unprecedented heatwave that hit Siberia with temperatures reaching 30°C where the normal one is 0°C. Or this is a temperatures up to 10% above average [4].These are just some of the evidences which remind that the climate change fight must go on. The European RES policy is still developing and new aims to reach by 2030 are at least 40% cuts in greenhouse gas emissions (from 1990 levels), 32% share for renewable energy and 32.5% improvement in energy efficiency [5].

While the European parliament is working on fulfilling a decarbonisation strategy and insuring a legislative framework in the renewable energy filed as one of the major sector which contribute to the climate change, the Bulgarian parliament is conducting an opposite direction policy. This paper aims to highlight the main legislative gabs regarding RES sector development based on comparison and analysis of the EU and the Bulgarian's renewable promotion policy and legislation.

## 2. Discussion

The main political stream of the EU regarding energy production and consumption is strongly influenced by the need of a rapid change of the total attitude of production and consumption especially when structuring the national and the EU energy mix. The EU common policy regarding renewables in the last 20 years integrates the environmental protection goals as an attempt to oppose and adapt to climate change. As an emphasis of these efforts the Directive 2009/28/EC promoting renewables is brought in the EU regulation [6]. EU counsel, taking into account the geographic location of the countries and estimated renewables' potential, the structure of the national energy mix and the preparedness of the grid of the EU member state etc., set in some derogation regarding starting point and defined 2020 RES target levels for some of the EU countries. Bulgaria is among these countries with 2020 target of 16 % - share in gross final energy consumption of energy produced by renewable source. Figure 1 shows the reached percentage of gross final energy consumption by renewable energy by the EU (28) up to 2020 where only 9 countries among which Bulgaria is, fulfill their 2020 target. Most of the rest 19 EU member states are even not close to their target [7].

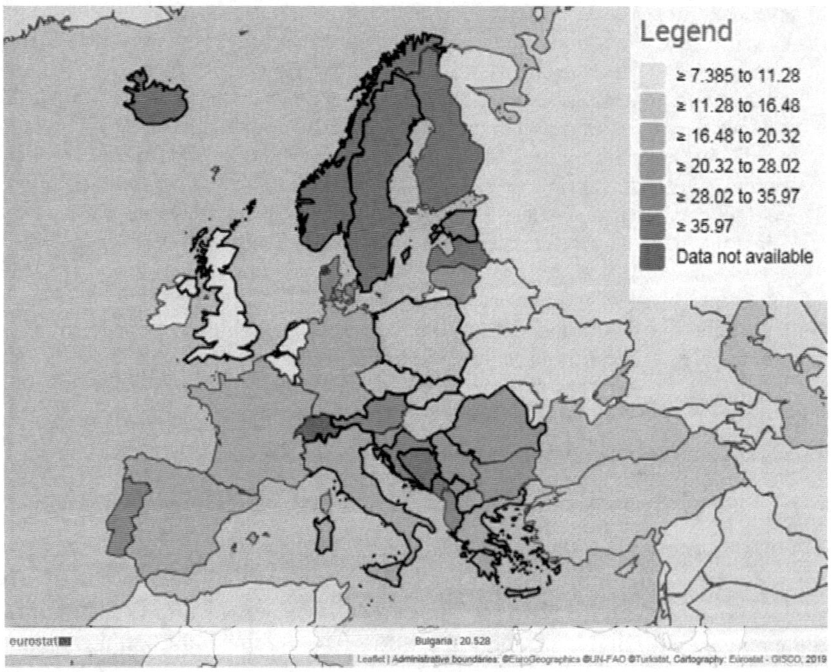


**Figure 1.** Renewable energy sources' share in gross final energy Consumption, 2020 ( EU, 28).

In Bulgaria, the Directive had been transposed into a Law for energy from renewable sources (LERS) with 6 months delay [8]. This instantly means time shortage for reaching the 2015 and 2020 RES aims.

The Energy from renewable sources act aims are set in Article 2: The primary objectives of this Act are as follows:

1. promotion of production and consumption of energy produced from RES;
2. promotion of production and use of biofuels and energy from RES in transport;

3.creating conditions for integrating gas from RES in the natural gas transmission and distribution networks;

4. creation of conditions for inclusion of heating and cooling from RES in heating transmission networks;

5. providing information regarding the possible support schemes…;

6. creating conditions for achieving sustainable and competitive energy policy and economic growth through innovation, and implementation of new products and technologies;

7. creating conditions for achieving sustainable development at regional and local levels;

8. creating conditions for increasing the competitiveness of small- and medium-size enterprises by production and consumption of electricity, heating and cooling from renewable sources;

9. security of energy deliveries, supplies and technical safety;

10. environmental protection and restricting climate change;

11. improving the living standards of the population through economically efficient use of energy from renewable sources [8].

The Act regulates the Governors in the renewable energy field. Since 2011 after the LERS first acceptance from the Bulgarian Parliament until 2019, the law had been changed 18 times, 4 of which just in only 1 year - 2013. Last changes from 21 May 2019 are in force from the same date. The numerals changes in this law create preconditions for not attempted mistakes and errors in act implementation beside the insecurity and the fluctuations in the RES policy implemented in Bulgaria.
In accordance to the Directive 2009/28/EC and the LERS, the Minister of Energy has to report to the EU the overall national target progress and achievement in every 2 years till the final one in 2020. As mentioned before the 2020 target for Bulgaria is 16% (based on 9,4% share in 2005) RES in gross final energy consumption when 20% is the target for the most EU member states. This target is broken down to: 20,8% share of electricity from RES in gross final energy consumption, 23,8% share of heating and cooling from RES in gross final consumption of energy form heating and cooling and 10,8% share of energy consumption in the transport sector [9].

Till June 2020 four national progress reports regarding the Directive 2009/28/EC are published. In the Forth National progress report regarding the period 2015-2016 (issued in December 2017), 18,8 % RES in gross final energy consumption is announced as reached in 2016. The reported achievements till 2016 are presented on Table 1 [10]. Thus the percentage for heating and cooling exceeds the specific 2020 target. Meanwhile the achievement regarding the energy share percentage for transport is much lower than the target. Namely because the state had accepted that the 2020 target is reached in 2016, the conducted policy isn't stimulating the investment in the renewable energy field as it should be according to the Directive 2009/28/EC.

**Table 1.** The sectoral and overall shares of renewable energy sources, Bulgaria 2011-2016.

| | **2011** | **2012** | **2013** | **2014** | **2015** | **2016** |
|---|---|---|---|---|---|---|
| RES-H&C [1] | 24,9% | 27,5% | 29,2% | 28,3% | 28,6% | 30% |
| RES-E [2] | 12,9% | 16,1% | 18,9% | 18,9% | 19,1% | 19,2% |
| RES-T [3] | 0,8% | 0,6% | 6,0% | 5,8% | 6,5% | 7,3% |
| **Overall RES share** | **14,3%** | **16,1%** | **19,0%** | **18,0%** | **18,2%** | **18,8%** |

[1] Share of renewable energy in heating and cooling: gross final consumption of energy from renewable sources for heating and cooling
[2] Share of renewable energy, used for electricity: gross final consumption of electricity from renewable sources for electricity
[3] Share of renewable energy in transport: final energy from renewable sources consumed in transport

The following facts in the Fourth progress report are taken into account:

In the part for counting the Biomass gross final energy consumption all the 59% of the population of the country who use wood mainly for household purposes are included [10]. The use of a wood stove is not a part of the renewable energy sources if it do not meet the sustainable criteria of paragraph 1 of Article 5 of the Directive. From table 1 can be seen that using RES for heating and cooling (where the wood for household use is counted as a biomass) for 2016 is 30% and forms more of the percentage for electricity and transport.

Another concern is regards the used data in the report. The state use data from the National Statistic Authority (NSA) and Ministry of Agriculture instead of using exact data form the Energy System Operator (ESO) who is operating the grid, the export and import, the balance, the production and consumption of each producer in the energy mix. The data used form the NSA is different measurement unit (m3, instead of ktoe or MW,MWh) and in most part of the report the different types of renewable systems data is in different unit, which makes the data not comparable, respectively it put another doubt about the way of counting the percentage of gross final energy consumption.

The contraries between the LERS and the Directive are in result of the announced reached target. According to Article 18(1) the benefits provided by LERS as: "Production of electricity from renewable resources etc…, shall be encouraged by: "item 6 - purchasing of the produced electricity…; item 7- setting a preferential price for renewable energy up to 10MW…; item 8- setting a preferential prices for renewable energy from biomass…, are canceled with the same article but with paragraph (2) which says that the benefits under art.18(1) item 6,7 and 8 shall not apply to energy facilities for production of electricity from renewable sources, the connection of which is applied for after the date of the report of the Minister of Energy, in which it is reported that the overall national target under, is achieved. This means that all renewable energy projects' applications after 2012, when Bulgaria declare 16,1% achievement, which is just 1 year after the law for renewable energy was adopted, had to be canceled.

Meanwhile for that application, each investor had to do two advance payment:

- A payment of 2500Euro per MW under art. 23 (8) which payment stay in favour of the grid operator;
- A payment on the amount from 12 500 up to 25 000 Euro per MW whether the project is more than 5MW by Article 29(1), when the investor apply for grid connection.

Talking about energy project with 200MW capacity, this means that investor have to pay 3 mln. Euro in advance for a project which future is not clear. Nevertheless this tax is payable even if the producer do not use preferential prices or feed in tariffs. From the other side other electricity producers using gas or convention sources are not obliged to pay it at all.

Another contrary is that under Article 23(2) Application can be list only after the Energy and water regulatory commission (EWRC) accepts available grid connection capacity. Under Article 22(5) every year till 30-th of June the Energy and water regulatory commission must publish on their web-site the available substation capacity. Available grid connection was not accepted and published on the web-page of the EWRC after 2012 till nowadays [11].

According to the LERS the advance payment should be used for the purpose of the grid connection. Thus another legislative gab is at hand. Once the investors is responsible for the project grid connection to the border of his property under art.27(1). Under point (2) after the border of the property the expenses for the grid connection is on the account of the grid operator [6].

Looking backward in the RES field for many years, the Bulgarian state practice double finance of some of the projects, mostly photovoltaic. The State once provided national grands and finance up to 80% the investments in photovoltaic projects and secondly they provide preferential prices to the same investors who developed projects with money of the resident taxpayer. In addition to that the preferential prices for energy from photovoltaic projects (PV) were artificially kept very high for a period of 4 years at levels between 400 and 300 Euro per MWh. At that time the law for renewable energy provided long term guaranteed contracts for a period of 25 years for PV project. Most of the installed PV projects are in exploitation before the reduction of that price in 2014. In comparison to the PV projects' the price per MW for wind energy projects varies between 98-48 Euro per MW with 20 years long-term guaranteed contract [11]. If for 2016, 699MW is the installed wind energy capacity, the PV project total capacity for the same period is 1028 MW [12]. As it is shown on Figure 2 in percentages, the installed electricity capacity share in Bulgaria up to 2019 is 8% for the PV projects, wheatear the wind projects take 5% [13]. This is one of the reasons why the end consumer is now paying a very high-electricity price in comparison with the standard of the country. According to the stakeholders, the RES energy is the blamed for the high price of the electricity in Bulgaria. Even this fact contradicts to the main idea of promotion of energy production and consumption.

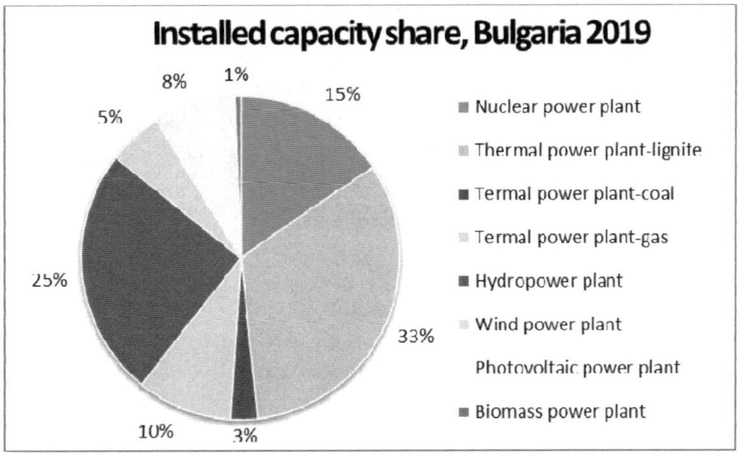


**Figure 2.** Installed capacity share, Bulgaria 2019

Another reason for the higher price of electricity is the higher and also double taxation of RES energy producers under the Energy Act in force since 9 December 2003. According to the act all producers of electricity pay both transmission and access tax to the grid, but RES energy for electricity pay higher taxes that brought another inequality in the energy sector [14]. The Energy and water regulatory commission (EWRC) of Bulgaria annually accepts the taxes. EWRC's statement for accepting the different taxation among energy producers is the nonpermanent production of the renewable sources and the need of balance. For that reason on the top of each taxes payable by the renewable energy producer, there is a separate balance tax, which aim to compensate the nonpermanent production. In fact, instead of promotion of RES the Bulgarian practice shows privilege position of producers of nonrenewable energy of electricity.

Even, if we can accept that the announce percentage is reached 18,8% and exceed the target of 16% the Bulgarian state can use the advantage of the Directive 2009/28/EC which allow statistical transfer of renewable energy to other EU member states[6]. The method for statistical transfer isn't working in Bulgaria in practice till nowadays. This means that the population of the Bulgarian state consumes all renewable energy and just the cheaper energy form the nuclear power plant and energy from thermal power plant is exported. This makes another concern about controlling the percentage reported in the progress report. If you want to announce a bigger percentage in gross final energy consumption, you could do not sell RES energy to other EU member states and keep the whole RES energy for internal use. But, when the state announce that the target is achieved already in 2016, then why the statistical methods are not working? By making the statistical transfer, the state could reach both social and economic effects by selling the exceeded percentage of RES energy abroad. By using the geographical resources economically efficiently the state will be able to insure the energy security and independency as well as to become a key player in the renewable energy market, as by 2020 there are 10 EU Member States that cannot reach their target[12].

## 3. Conclusion

In the field of RES sector, there still are contradicting requirements and energy legislatives act's born gabs. The RES sector in Bulgaria still needs to overcome barriers as double taxation and complicated permitting procedure. Thus, the main aims of LERS- creating conditions for achieving sustainable and competitive energy policy and economic growth through innovation, and implementation of new products and technologies, creating conditions for achieving sustainable development at regional and local levels, creating conditions for increasing the competitiveness of small- and medium-size enterprises by production and consumption of electricity, heating and cooling from renewable sources, are not satisfied yet. The lack of accepted available grid connection capacity (since 2012) together with the non-announces preferential price (concerning the middle and large-scale projects), RES are substantial weakness in the practice of implementation of RES policy in Bulgaria.

The bottom line of promotion of RES is partially compromised by the persisting legislative gabs in RES sector and the existing inequalities between energy producers in Bulgaria. On our opinion, based on presented facts and findings of this research the unstable and incoherent practice of regulation of RES promotion implementation still is at hand in our country. The results of that kind of policy implementation is visible in the progress reports where the installed capacity of 699MW of wind power announced in the 3-th progress report (for the period of 2013) do not increase in the following 4-th and 5-th progress reports. This once more reaffirmed the results of the written above policy providing.

Highlighting the gabs and barriers faced by RES sector could be helpful in the context of overcoming the insufficiency of administrative capacity in the field of RES in Bulgaria.

In order the State to follows the declared in the integrated plan policy [15] the identified gabs in the renewable energy framework should be overcome with active participation of RES producers organization, so the further development of energy sector to be insured. All measures should be taken at an early stage and a consistent policy should be sustainably implemented to attracting new investors, through an adopted RES promotion policy that complies with the main objectives of the Directive 2009/28/EC and the newly announced EU 2030 RES targets.

**References**

[1] Nikolaev A and Konidari P, 2017, D*evelopment and assessment of renewable energy policy scenarios by 2030 for Bulgaria, Renewable energy, 111 (2017) 792-802,* p 792-801

[2] Molintas D 2020 Analysis of Coronavirus and carbon emissions, MPRA Paper No. 98858, p 1-7

[3] World Wildlife Fund 2020, Climate overview https://www.worldwildlife.org/stories/2020-a-critical-year-for-our-future-and-for-the-climate

[4] Ciavarella A, Cotterill D, Scott P, Kew S, Philip S, Oldenborgh G, Skalevag A, Lorenz P, Robin M, Otto F, Hauser M, Seneviratne S, Lehner F, Shirshov P, Zolina O 2020, Prolonged Siberian heat of 2020, World Weather Attribution, Published online: https://www.worldweatherattribution.org/siberian-heatwave-of-2020-almostimpossible-without-climate-change

[5] European Counsel 2020, *2030 climate and energy framework,* Published online: https://ec.europa.eu/clima/policies/strategies/2030_en#tab-0-0

[6] Directive 2009/28/EC of the European Parliament and of the Council of 23 April 2009 on the promotion of the use of energy from renewable sources https://eur-lex.europa.eu/legal-content/EN/TXT/PDF/?uri=CELEX:32009L0028&from=EN

[7] Law for energy from renewable energy sources, https://www.lex.bg/laws/ldoc/2135728864 Ciavarella A, Cotterill D, Scott P, Kew S, Philip S, Oldenborgh G, Skalevag A, Lorenz P, Robin M, Otto F, Hauser M, Seneviratne S, Lehner F, Shirshov P, Zolina O 2020, *Prolonged Siberian heat of 2020, World Weather Attribution,* Published online: https://www.worldweatherattribution.org/siberian-heatwave-of-2020-almostimpossible-without-climate-change

[8] Eurostat statistic data for RES sector in the EU28 member states 2020, Published online: https://ec.europa.eu/eurostat/databrowser/view/t2020_rd330/default/map?lang=en

[9] National renewable energy action plan 2011, Ministry of Economy, Energy and Tourism, https://www.google.bg/url?sa=t&rct=j&q=&esrc=s&source=web&cd=&cad=rja&uact=8&ved=2ahUKEwiuk5vm48LrAhULy4UKHc2TB60QFjAAegQICBAB&url=https%3A%2F%2Fwww.me.government.bg%2Flibrary%2Findex%2Fdownload%2Flang%2Fen%2FfileId%2F575&usg=AOvVaw3kXXsEYzFfpZc1jB-sWvtf

[10] Fourth National progress report in the promotion and use of energy from renewable sources, https://ec.europa.eu/energy/files/documents/memberstateprogressreportstranslatedintoenglish20172018zip_en, p 45, p 67

[11] Toneva D, Stankova T 2018 Development of wind energy projects in Bulgaria – challenges and opportunities, *Proceeding in 1-st international conference Applied computer technologies, Macedonia*, p 60-63

[12] Report from the commission to the European Parliament, The Council, The European Economic and Social Committee and the Committee of the Regions, Bruseel, 9.4.2019 COM (2019) 225final, p 6, available at: https://ec.europa.eu/commission/sites/beta-political/files/report-progress-renewable-energy-april2019_en.pdf
[13] Energy system operator, Statistical annual data sheet, 2019, Sofia, p 3
[14] Energy Act 2020, *Ministry of Energy and Water,* https://lex.bg/laws/ldoc/2135475623
[15] Integrated plan in the field of energy and climate of the Republic of Bulgaria 2021-2030, p 21, https://www.me.government.bg/files/useruploads/files/national_energy_and_climate_plan_bulgaria_clear_22.02.20.pdf

**Acknowledgments**
Part of the results of the present research were obtained under the National program "Young scientists and postdoctoral students", financed by the Ministry of Education and Science of Bulgaria.

# Achievement of maximum energy efficiency through vector control of induction motor electric drive

**S Rachev[1], G Ivanova[2]**

[1] Department of Electrical Engineering, Technical University of Gabrovo, 4 Hadgi Dimitar str., Gabrovo 5300, Bulgaria
[2] Department of Electrical Engineering, Technical University of Varna, 1 Studentska str., Varna 9010, Bulgaria

E-mail: sratchev@mail.com

**Abstract.** Electric motors are the predominant driving devices globally, consuming electricity. For this reason, research aimed at minimizing their energy consumption is very relevant. The paper concerns an applied energy-saving method - vector control of an induction motor. With regard to this, a necessary condition for achieving maximum efficiency is to choose the optimal value of the magnetizing current. If the magnetizing current is not controlled properly, the result may be higher electricity consumption from a frequency-controlled motor than that of a direct-on-line power supply. In this case, a specific electric drive with certain parameters is analyzed. The research is based on a mathematical model, representing a system of differential equations, which is transformed and solved by appropriate software. The aim is to look for a minimum of electrical power losses. The physical side of the topic is treated exclusively without analyzing the economic benefits of the introduction of energy saving measures.

## 1. Introduction

Electrical machines come in a vast range of sizes and types. Their torque available on the shaft varies from mNm for micromachines to MNm for large machines [1].

The electric motors work under their own inherent operating characteristics. However, in many applications some desirable characteristics are obtained by use of control equipment. For most electric drives, the static and dynamic characteristics of electric motors can be matched to the load characteristics and so the transition from electrical to mechanical energy is more productive and inexpensive [2].

In turn, three-phase induction motors (IM) cover machines with rated powers from less than 100 W to 20 MW [3].

Indeed quite not large improvements in their efficiency would contribute to measurable energy savings. The criterion for optimizing the electric machine can be to improve its performance. Users have other criteria, such as small initial price.

The synchronous speed of IM depends on the frequency of the supply voltage and the number of poles of the machine [4].

Generally speaking, electric drives regulate the speed of the motor by changing the frequency of the current that supplies the motor. Nowadays, the two most often applicable methods for speed regulation are frequency and flux vector control. The flux vector drive adjusts both the absolute value and the way of the magnetic flux in IM [5].

At the same time, it is clear that the more complex a control is, the higher the cost [4].

The pumps are particularly suitable for possible reduction of electricity consumption - they are not always required to operate at maximum capacity. This means that there is a possibility of energy savings if the speed of the motor can be changed according to the load demand. In other words, there are possible energy savings if the motor speed can be varied under the operating load [6].

It is needed to be explored the electro-mechanical and electro-magnetic processes occurring in the dynamic behaviour of IM in order to get a clearer idea of their work. The specific goal of the study is to obtain the losses of electric power in the IM for the pump unit in cases of direct-on-line (DOL) operation, frequency and indirect vector control and to compare them in terms of savings obtained.

## 2. Some considerations

The electromagnetic loads of AC electric machines (value of magnetic induction, current density, etc.), energy losses and heating is determined not by the active but by the apparent power, as the value of the magnetic flux in the motor depends on the total voltage and not only on its active component.

IM are suited for DOL operation. DOL drives are not controllable, in the sense that they can be turned on and off only. Such drives stress both network and machinery and may be energy inefficient. An IM direct torque control (DTC) drives are a good solution for many applications [7].

The rotating magnetic field of the stator generated during the supply voltage crosses the rotor windings, and an electro-motive-force is induced in them by mutual induction. To set up the mutual flux, IM draws magnetizing current. The absolute value of this current is very high due to the great reluctance of the air gap of the motor [8].

Complete adjustability of the electric drive can be ensured if electromagnetic torque control is provided. This makes separate coordinate control of the vectors included in the selected equation of the electromagnetic torque imperative. The vector control drive is able to react promptly to variations during operation and to control torque, which distinguishes it from a scalar control controller for various mechanisms. A special feature is that an active motor model for vector control is applied for the inverter switching circuit.

The motor current can be divided into two portions: flux current (as the exciting current of a DC motor) and in-phase current (as the armature current of a DC motor). By changing the voltage supplied to the motor is controlled the flux current. The in-phase current depends in direct proportion to the load. The vector control retains the V/Hz core and adds additional blocks around the core to refinement the electric drive behaviour [5].

With vector control, direct and indirect methods are possible. Vector sensor-less control generally means control without any speed and flux sensor.

It is necessary to emphasize that the changes of the mechanical quantities and the reaching of the established values proceed more slowly than those of the electrical ones [9].

The dynamic mechanical characteristics giving the relationship between the shaft torque and the angular velocity is usually quite different from the static mechanical characteristics.

Typical examples of variable torque mechanisms are pumps. The pumps can be operated quite well with a simple open-loop, i.e. sensor-less inverter-fed IM drive.

The power losses in all windings of an IM in established non-transient mode with mains supply are [10]:

$$P_{loss} = i_{sd}^2 R_s + i_{sq}^2 \left(R_s + R_R\right), \tag{1}$$

where $i_{sd}, i_{sq}$ – stator current components (magnetizing and quadrature one, respectively); $R_s$ – stator phase resistance; $R_R$ – rotor resistance in modified inverse equivalent circuit.

The following dependencies must be added

$$R_R = b^2 R_r \tag{2}$$

$$b = L_m / L_r , \tag{3}$$

where $b$ – factor of proportionality; $R_r$ – rotor resistance in basic $T$-shape equivalent circuit; $L_m$ – mutual inductance; $L_r$ – rotor inductance.

The main feature of vector control is that the magnetizing current determines the magnitude of the flux coupling of the rotor field and is kept invariant. Each resistance moment corresponds to a value of the magnetizing current when the power losses are minimal. In this regard, the minimum power losses will be determined as follows [10]:

$$P_{loss}^{*} = T_m \frac{R_S + \gamma^2 R_R + \gamma^2 R_S}{\gamma L_M p}, \tag{4}$$

where $\gamma^2 = R_S / (R_R + R_S)$.

The studies are based on the phase basic $T$-shape equivalent circuit with its corresponding parameters, previously implemented in [11], which is subsequently converted as inverse equivalent circuit for frequency/vector control (where motor leakage inductances absent).

## 3. Results and findings

Applying software simulations, it is useful to observe all quantities and to study the influence of various changing parameters. Simplifying assumptions are used: there is no asymmetry between the phase windings; the interspace between the movable and immovable part of the motor is fixed; the rotor parameters are adjusted to the primary circuit. The nonlinearity of the equations is due to the multiplication of independent variables or their functions. As is known, the differential equations of an AC machine contain variable coefficients. The direct coordinate transformation equations are applied in order to eliminate variable coefficients. The coordinate system used is orthogonal.

The electromechanical system of an electrically driven pump is described by a mathematical model based on a system of five differential equations - four for model stator/rotor currents and fifth equation is the motion equation. The particulars of the electric motor considered are the same used in [11] and can be seen in table 1. The dynamic simulation studies of the IM pump drive have been carried out by means of purposeful software. Three possible cases have been developed: DOL operation, control while maintaining V/Hz$^2$ = const. and when applying indirect vector control.

**Table 1.** Induction motor data and electric equivalent circuit parameters

| Description | Data |
|---|---|
| Type, Designation | ATM 100LBH4 |
| Rated power ($P_{rated}$) | 3.0 kW |
| Rated stator voltage ($V_{rated}$) | 400 V |
| Operating frequency ($f$) | 50 Hz |
| Line stator current ($I_l$) | 6.3 A |
| Efficiency | 86.5% |
| Pole pair number | 2 |
| Rotor speed ($N_r$) | 1450 rpm |
| Power factor | 0.79 |
| Rotor torque of inertia ($I_r$) | 0.04 kg·m$^2$ |
| Stator resistance $R_s$ | 1.68 Ω |
| Rotor resistance $R_r'$ | 1.686 Ω |
| Stator leakage reactance $X_s$ | 2.564 Ω |
| Rotor leakage reactance $X_r'$ | 2.559 Ω |
| Magnetizing reactance $X_m$ | 62.287 Ω |

Characteristics of total motor electrical power losses and magnetizing current at DOL starting are depicted in figure 1 and figure 2, respectively.

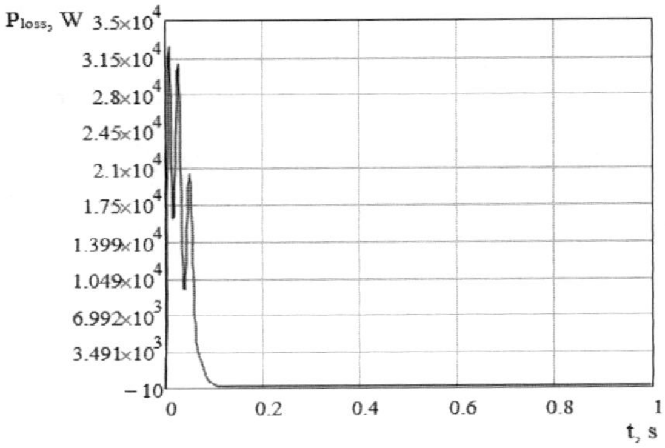


**Figure 1.** Electrical power losses during DOL starting.

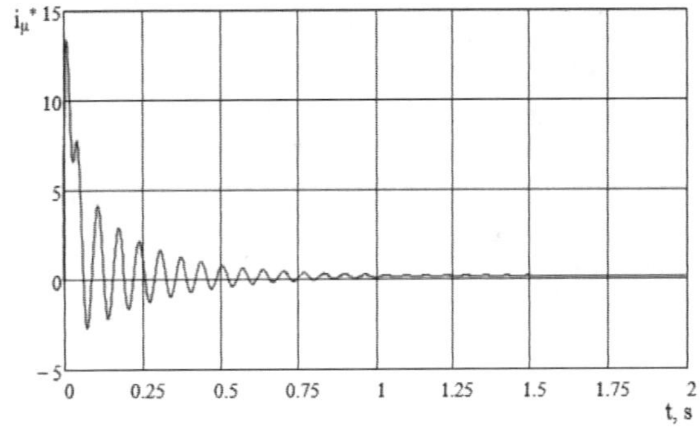


**Figure 2.** Magnetizing current during DOL starting.

Figure 3 and figure 4 present the electrical power losses while maintaining V/Hz$^2$ = const. and when applying indirect vector control respectively.

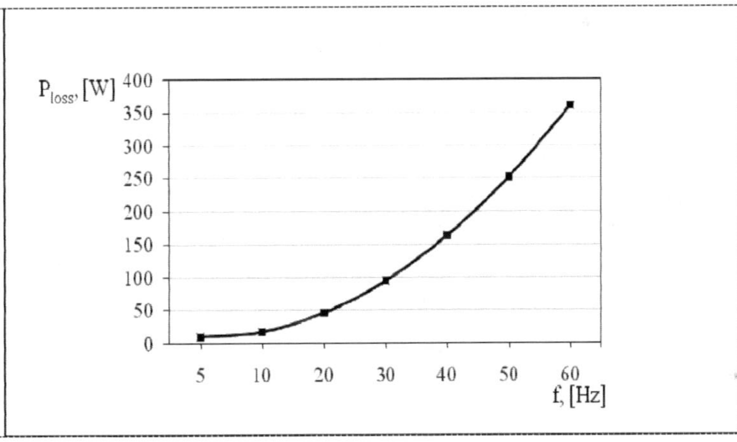


**Figure 3.** Electrical power losses while maintaining V/Hz$^2$ = const.

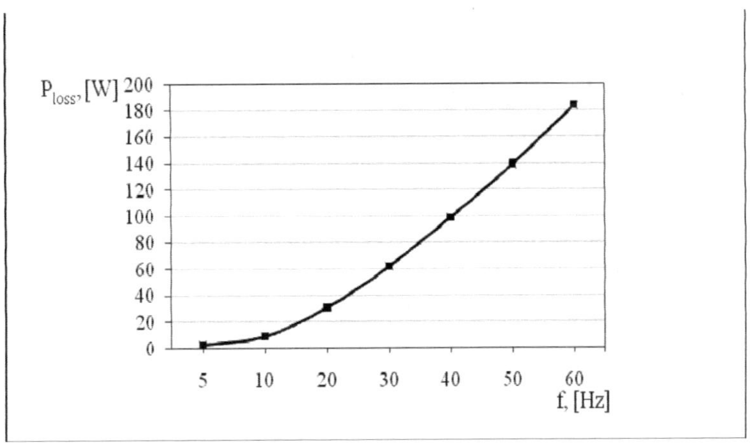


**Figure 4.** Electrical power losses when applying indirect vector control.

Some numerical values in model studies are interesting: electrical power losses in the established regime at DOL starting – 351.460 W; electrical power losses in the established mode of operation while maintaining $V/Hz^2$ = const. – 251.104 W; electrical power losses in the established mode of operation when applying indirect vector control – 138.374 W.

## 4. Conclusion

The possibility of energy saving of the pump drive with adjustable speed has been studied in the research. The results of the studies carried out are obtaining numerical values for the losses of electric power in a specific IM for a pump unit under different ways of control. The simulation results can be validated by performing measurements of operating such pumping systems. Unfortunately, in many cases such pumps operate in places with certain levels of security where access for outsiders is restricted or prohibited in its entirety.

Research and design of such systems as the electric drive based on vector controlled IM represents rather difficult task.

Speed regulation of pump systems IM powered by inverter leads to significant economic benefits due to reduced energy costs. In electromechanical energy conversion, the reduction of electrical power losses when applying frequency / vector control is noticeable.

## References

[1] Friswell M I, Penny J E T, Garvey S D, and Lees A W 2010 *Dynamics of Rotating Machines* (Cambridge University Press) p 9
[2] Kumar V, Behera R K, Joshi D, and Bansal R 2020 *Power Electronics, Drives, and Advanced Applications* (Taylor & Francis Group, LLC) p 455
[3] Wach P 2011 *Dynamics and Control of Electrical Drives* (Springer-Verlag Berlin Heidelberg) p 110
[4] Hughes A 2006 *Electric_Motors_Drives Fundamentals, Types and Applications Third edition* (Oxford Great Britain) pp 221, 279, 299, 374
[5] Petruzella, F D 2016 *Electric motors and control systems* (McGraw-Hill Education) pp 266, 268
[6] Miller R, Miller M R 2013 *Industrial Electricity and Motor Controls* (McGraw-Hill Education) p 280
[7] Pyrhönen J, Hrabovcová V, Semken R S 2016 *Electrical machine drives control: An introduction* (John Wiley&Sons Ltd,) p 429
[8] Sahdev S K 2018 *Electrical machines* (Cambridge University Press) p 743
[9] Hindmarsh J, Renfrew A 1997 *Electrical machines and drives systems* (Butterworth-Heinemann) p 13, 177
[10] Borisevich A V 2015 *Energy-saving vector control of asynchronous motors: status overview and new results* (Moscow: Infra-M) pp 21-22
[11] Rachev S, Dimitrov L, Ivanov I, Karakoulidis K 2017 Study the effects of no nominal conditions on the performance of high efficiency induction motor, *International Conference on ENERGY and ENVIRONMENT (CIEM)*, Bucharest, pp 192-196, doi: 10.1109/CIEM.2017.8120772.

## Acknowledgments

This paper has been implemented with the support of the Project BG05M2OP001-1.002-0023 Competence Center ‘Intelligent Mechatronic, Eco- and Energy-Saving Systems and Technologies’ financed by the Operational Program ‘Science and Education for Intelligent Growth’ of the European Regional Development Fund.

# Influence of buoyancy forces in multi-storey buildings on the efficiency of a regenerative air handling unit with heat recovery

**A Penev[1] and L Tsokov[1]**

[1] Department of Thermal and Refrigeration Engineering, Technical University of Sofia, 8 St. Kliment Ohridski, 1756 Sofia, Bulgaria

E-mail: gartz@tu-sofia.bg and lubo@tu-sofia.bg

**Abstract.** Various solutions are currently being offered to modernize the ventilation in old residential buildings. In small apartments, it is very appropriate to use decentralized ventilation units - for example with regenerative heat exchangers. Theoretical and experimental studies of such devices show a significant impact of the operational mode of regenerators (determined by the environment) on their efficiency. The aim of this article is to demonstrate the impact of the buoyancy forces in multi-storey buildings on the efficiency of a regenerative air handling unit for decentralized residential ventilation with heat recovery. Numerical solutions are presented for close to the real operating modes, airflow rates and air temperatures of the supply and the exit air, and the temperature ratio.

## 1. Introduction

Fixed regenerative heat exchangers are used in decentralized ventilation systems for heat recovery of the exhaust air, which is associated with the energy economy of the microclimate systems and the provision of thermal comfort and quality of the indoor air. Recently, local ventilation systems for individual rooms, 1-2 storey buildings, and small houses have started integrating ventilation units into the outside wall of the room, using regenerative heat exchange. Compared to industrial regenerative heat exchangers, which operate at almost constant parameters, regenerators in the field of HVAC operate at variable environmental parameters - temperature, humidity, and pressure difference ($\Delta P_b$) in buildings, leading to different flow rates of hot and cold airflow through the regenerator. This $\Delta P_b$ between the indoor pressure and outdoor pressure in the multi-storey building is the pressure of stack, created naturally by air density differences, which drives a ventilation air rate in the building. Unfortunately, this stack effect in multi-story buildings influences the efficiency of the regenerators on the different levels of the dwelling and it is difficult to realize effective work of all regenerator's units. It was noticed in [1] that during the winter there are practically periods in which no heat recovery occurs, and this is confirmed by airflow temperature measurements, [2]. For this reason, it was necessary to study the different airflow into the matrix of regenerator at different floors of the dwelling induced by mechanical fans and natural ventilation in different operating conditions using Computational Fluid Dynamics (CFD) simulations.

The article presents a CFD model of air flows with the boussinesq approximation (i.e. created $\Delta P_b$) in a multi-story building coupled with mechanical ventilation with the regenerator's fans. The computational results were used to investigate the ventilation airflow rates, the thermal and pressure field in the dwelling, and the efficiency of the regenerators. The results of our simulations show the mutual influence between the thermal and aerodynamic mode of the building ventilation and the thermal and aerodynamic mode of the regenerators.

## 2. Model statement

*2.1. Geometry and mesh*

The three-dimensional geometries of the matrix and the building are shown in Figure 1. The building has got 5 identical rooms on different levels connected to one common chimney vent with height H=14 m and a free-open square channel 0.2 m x 0.2 m. The main computational domain involves the volume of the rooms and the chimney connected by outlets with a diameter of 0.15 m. The matrixes are presented as single domains connected to the main computational one with inlet diameters 0.093m.

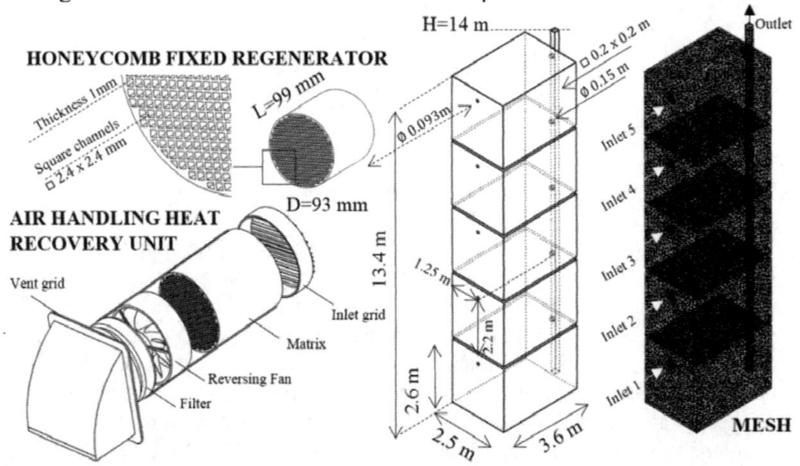


**Figure 1.** Model statement of the regenerator and the building.

*2.2. Computational model and assumptions*

The Finite-volume method and software Fluent, ANSYS, [3, 4] was used to solve the differential equations model. The velocity, pressure, and temperature fields of the fluid in the volumes are computed by the system of equations for mass, momentum, and energy conservation. The model describes a three-dimensional airflow induced by mechanical ventilation and buoyancy forces and a radiative underfloor source heating. The boundary conditions are set constant, and the process is steady-state, the heat fluxes of all wall surfaces of the building are (adiabatic), except the radiation heat flux from the floors to the apartments, and also the air change rate by infiltration is zero. The values of atmospheric parameters are constant and the wind around the building is ignored. The physical properties of airflow are fixed except the density whose deviation creates the forces of $\Delta P_b$. The thermodynamic relation between air density and temperature is the ideal gas equation of state. The additional model for turbulent airflow is SST-k-ω coupled with a "P1 model" which includes the computational radiation equation, [5].

The air handling unit (AHU) is modeled as an individual domain with a constant pressure loss coefficient $f_{loss} = 20$, equation (1):

$$f_{loss} = \Delta P/(\rho v^2/2), \tag{1}$$

where $\Delta P$ is the airflow pressure drop, Pa; ρ – the airflow density, kg/m$^3$; v – the velocity at the minimum flow area, m/s.

The AHU domain is connected to the main airflow computational domain as an interface between it and the cells simulating the fan. The fan is considered to be infinitely thin, but the pressure rise across it is presented as a function of the airflow velocity through it. The correlation between the fan pressure and airflow velocity is given as a polynomial function, equation (2). The swirl tangential and radial velocity of the stream is assumed zero.

$$\Delta P_{FAN} = 2.5126v^2 - 20.845v + 30.849 \quad (2)$$

The constant static pressure of zero Pascals was specified at all wall inlets and the outlet of the chimney. The numerical computations were performed at an external temperature of 275 K and an atmospheric pressure of 101 325 Pa. The sources of heat are placed on the floors of the rooms with a constant surface temperature of 300 K and the flow of heat towards the room is modeled by convection and radiation.

The graph in figure 2 shows a fan curve and three resistance curves of the ventilation system defining three duty-points of the fan - A, B and C. Duty-point B presents the modes of the fan in ideal temperature conditions when the temperature in the dwelling is equal to the external temperature and buoyancy forces are neglected. Duty-points A and C are present when the inside temperature is higher than the outside one, which leads to buoyancy forces and displacement of the resistance curve of the system. Point C is the duty-point of the fan when the airflow direction is towards the room - air supply direction. Point A is the duty-point of the fan when the airflow direction is to the outside - air exhaust direction.

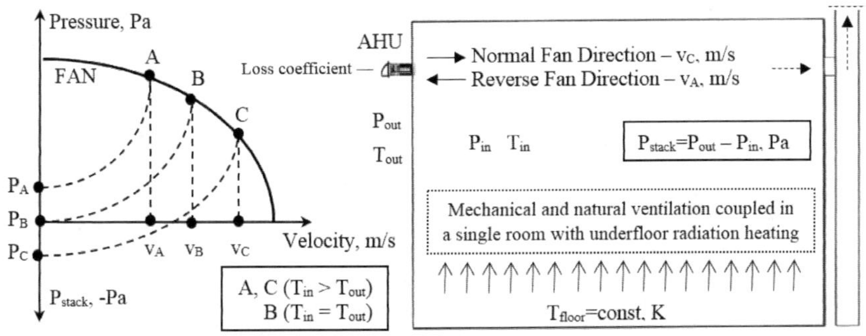


**Figure 2.** Problem statement of computational model

### *2.3. Thermal properties of air and honeycomb ceramic matrix*

The average dry air thermo-physical properties at pressure p=101 325, Pa for the temperature range 275≤T, K≤300, and the properties of the matrix are given in Table 1.

**Table 1.** Thermal-physical properties of dry air and honeycomb ceramic matrix.

| Fluid / Unit | $\rho$, kg/m$^3$ | $c_p$, J/kgK | $\lambda$, W/mK | $\mu$, Pa.s |
|---|---|---|---|---|
| Dry Air | $\rho$= p/RT | $c_p$=1004,2 | $\lambda$=0,02607 | $\mu$=1,704e-5 |
| Cordierite | $\rho$=1500 | $c_R$=900 | $\lambda$=1,20 | – |

## 3. Numerical results

### *3.1. Results of airflow distributions*

The presented calculations describe separately temperature and aerodynamic state of the system for three operational modes of the AHU's fans - supply air direction, exhaust air direction, and not working fan. All states were calculated when the inside temperature is higher than the outside temperature. For each of the flow directions in these temperature conditions, pressure and velocity fields in the building are given. For reasons of comparison, the same distributions were calculated when the fans of the AHU were not working, and only buoyancy forces drove the airflow in the building. For the last case, the distribution of the inside temperature is presented too, to show the effect of natural ventilation in case of free opened inlets (AHUs without ventilation close valves).

The pressure distributions for the three AHU operating modes and the distribution of temperature in the building in the case of a non-working fan can be seen in figure 3. The figure shows the pressure fields in the building for each operational mode of AHU.

The results for the velocity distribution inside the building and the AHU are shown in figure 4. The airflow rates for each room under the respective conditions can also be seen in this figure.

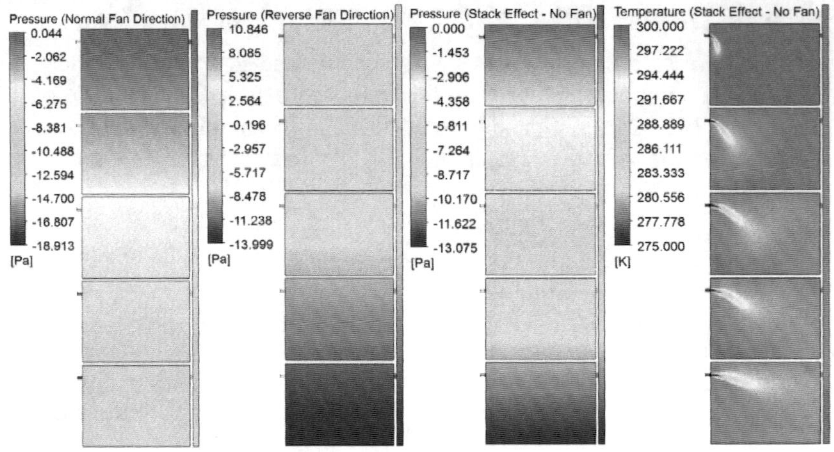


**Figure 3.** Pressure distributions at different modes of AHU.

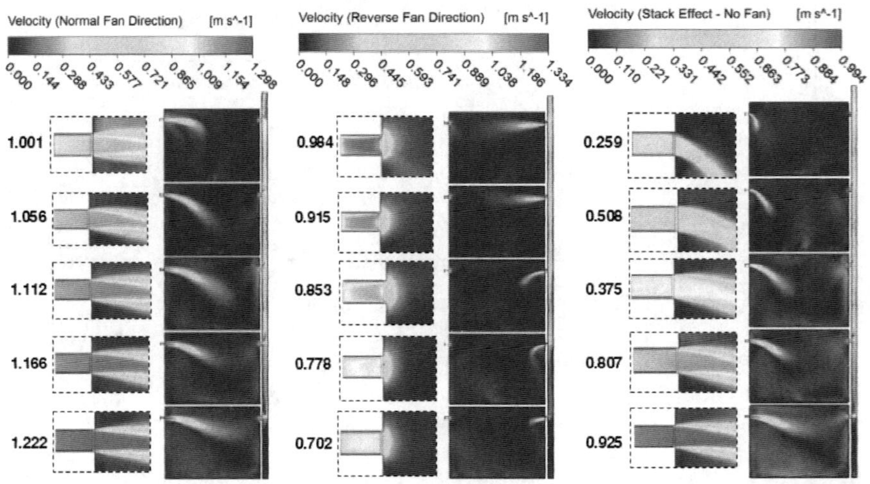


**Figure 4.** Velocity distributions at different modes of AHU and average inlet/outlet air velocity.

### *3.2. Results of the flow temperature in AHU*

The change of the velocity in the regenerator matrix respectively the magnitude of the airflow rate affects the air change rate in the apartment and, also, it strongly affects the temperature ratio of the regenerator. The variations of the inlet and outlet temperatures of the regenerators on three different levels are presented in figure 5.

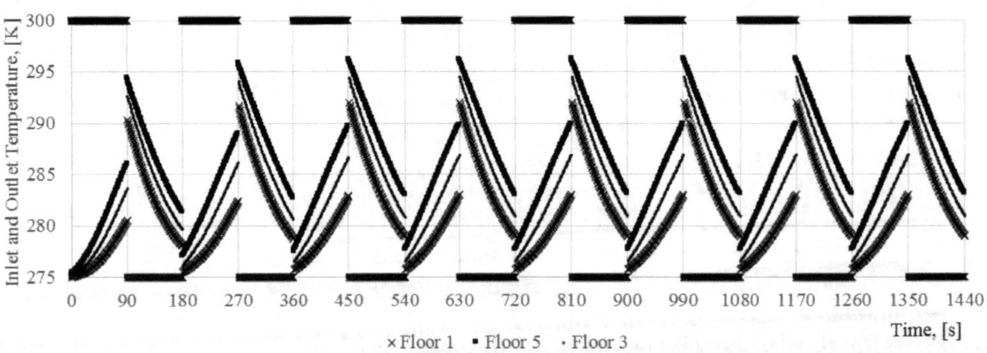


**Figure 5.** Outlet temperatures during the operating of the regenerator at different floors.

The curves in figure 5 were calculated using the model presented in, [6]. The graphs present the inlet and outlet air temperatures for a mode of the regenerators with half-periods of 90 seconds.

The regenerative air handling unit reaches the quasi-steady-state regime after 8 switches.

The values of the temperature ratios for both directions of airflow in AHU are presented in table 2.

**Table 2.** Temperature ratio during the hot and cold half-period, %.

| Floor | $\Theta_{h,1}$ | $\Theta_{c,1}$ | $\Theta_{h,2}$ | $\Theta_{c,2}$ | $\Theta_{h,3}$ | $\Theta_{c,3}$ | $\Theta_{h,4}$ | $\Theta_{c,4}$ | $\Theta_{h,5}$ | $\Theta_{c,5}$ | $\Theta_{h,6}$ | $\Theta_{c,6}$ | $\Theta_{h,7}$ | $\Theta_{c,7}$ |
|---|---|---|---|---|---|---|---|---|---|---|---|---|---|---|
| 1 | 91,98 | 31,61 | 86,69 | 36,15 | 85,26 | 37,14 | 84,92 | 37,36 | 84,85 | 37,42 | 84,83 | 37,43 | 84,82 | 37,43 |
| 2 | 86,00 | 38,58 | 81,28 | 41,78 | 80,14 | 42,49 | 79,88 | 42,65 | 79,82 | 42,69 | 79,81 | 42,70 | 79,80 | 42,70 |
| 3 | 86,00 | 40,32 | 77,50 | 46,11 | 75,30 | 47,36 | 74,80 | 47,65 | 74,68 | 47,71 | 74,66 | 47,72 | 74,65 | 47,73 |
| 4 | 86,00 | 42,24 | 74,22 | 50,15 | 71,05 | 51,86 | 70,32 | 52,24 | 70,16 | 52,33 | 70,12 | 52,35 | 70,11 | 52,36 |
| 5 | 80,47 | 48,41 | 68,92 | 55,26 | 65,99 | 56,74 | 65,33 | 57,07 | 65,18 | 57,14 | 65,14 | 57,16 | 65,13 | 57,16 |

The temperature ratios were calculated by equations (3) and (4):

$$\Theta_h = (T_{h,i} - T_{h,o}^*) / (T_{h,i} - T_{c,i}) \quad (3)$$

$$\Theta_c = (T_{c,o}^* - T_{c,i}) / (T_{h,i} - T_{c,i}), \quad (4)$$

where the indexes "i " and "o" are inlet and outlet. The indexes "h" and "c" are for hot and cold airflow; $T_{h,i}$, $T_{c,i}$ – the inlet temperatures of the hot and cold airflow, K; $T_{h,o}^*$, $T_{c,o}^*$ – the outlet average temperatures during the hot and cold half-periods, K.

## 4. Conclusion

The proposed CFD models can be used for numerical experiments to optimize the operation of the regenerator and to find suitable components integrated into the AHU according to the geometry and interior of the building. Unbalanced operating modes of regenerators, influenced by natural ventilation, lead to poor thermal efficiency, different air exchange rates on different floors of the building and cannot be ignored in multi-storey buildings. This leads to the need to select a new control system and components for this type of decentralized ventilation modules.

## Reference

[1] Koiv A, Mikola A, and Simson R 2015 *Ventilation solutions in renovated apartment buildings in cold climate conditions* 7(03), (Tallin University of Technology, Estonia) p 129

[2] Koiv A, and Mikola A 2013 *Ventilating of old apartment buildings (*Recent Advances in Energy and Enviroment System, Estonia) p 6

[3] Anderson J D, and Wendt J 1995 *Computational fluid dynamics* **206**, (New York: McGraw-Hill, USA) p 332

[4] Ansys I 2011 *ANSYS FLUENT theory guide* **15317**, (USA) p 514

[5] Hussain S, and Oosthuizen P H 2012 *Numerical investigations of buoyancy-driven natural ventilation in a simple atrium building and its effect on the thermal comfort conditions*. (Applied Thermal Engineering, Canada) **40**, pp 358-372

[6] Penev A, Tsokov L, and Penkova N 2020 *A conjugate heat transfer in a fixed-bed ceramic regenerator for air conditioning applications*. (Journal of Chemical Technology & Metallurgy, Sofia), **55**(5), pp 1132-1139

## Acknowledgments

This article is subject of the study of the author's PhD dissertation. The authors would like to thanks the Research and Development Sector at the Technical University of Sofia for the financial support.

# Hydraulic power take off system for wave energy utilization

**R Velichkova[1], M Pushkarov[1], R A Angelova[2], I Simova[1], D Markov[1], I Denev[3] and P Stankov[2]**

[1] Department of Hydroaerodynamics and hydraulic machines, Technical University of Sofia, 8 Kl.Ohridski Blvd, Sofia 1000, Bulgaria
[2] Centre for Research and Design in Human Comfort, Energy and Environment (CERDECEN), Technical University of Sofia, 8 Kl. Ohridski Blvd, Sofia, Bulgaria
[3] Department of Mechanical Engineering and Heat Engineering, Faculty of engineering and pedagogy, Technical University of Sofia, 59 Burgasko shose Blvd, Sliven, Bulgaria

E-mail: rvelichkova@tu-sofia.bg

**Abstract.** Wave energy from seas and oceans is among the most promising and abundant energy sources on the Earth. There are different designs of conversion systems for the utilization of the wave energy, resulting from the different ways of the wave energy absorption, and depending on the location characteristics. The paper proposes a new hydraulic power take-off system (PTOS) that can be used in various types of wave energy conversion systems.

## 1. Introduction

The energy from the tides is inexhaustible and environmentally friendly, but it requires a serious investment to build a tidal power plant. Its technology requires connections to the power grid. Also, tides are not a constant phenomenon; they occur over some time. All these are among the reasons why tidal energy is still not fully utilized. However, according to scientific reports, tidal energy has great potential to thrive in the world of alternative renewables.

Different concepts for wave energy conversion (WEC) exist; over 1000 WEC techniques have been patented in Japan, North America, and Europe. Despite the considerable variation in design, WEC systems are generally categorized by location and type [1-3].

The energy power take-off subsystem (PTOS) of each wind-wave energy converter (WWEC) is defined as the mechanism by which the energy absorbed by the primary converter is transformed into electricity. The primary wave energy converter (PWEC) can be, for example, a chamber with an oscillating water column or a point absorber – buoy, floating platform. PTOS is of great importance for any WWEC, as it directly affects not only the efficiency of the conversion of the absorbed energy into electricity, but also preconditions the size, mass, and structural dynamics of the wind energy converter.

Through its direct impact on all components of WWEC and on its operation (figure 1), PTOS also influences the total average costs for electricity generation [4]. The characteristics of PTOS affect the efficiency of converting the energy of the waves, and consequently – the annual amount of energy produced. PTOS has an immediate effect on the capital costs of WWEC, as its relative quota is about 20-30% [5]. The reliability of the PSOE determines the annual availability of WWEC, i.e., the energy produced and the operating and maintenance costs [4].

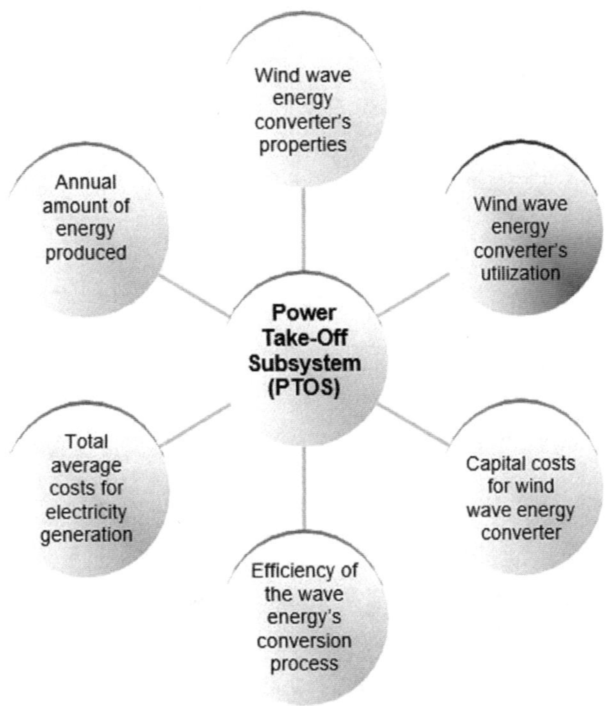


**Figure 1**. Influence of the Power Take-Off Subsystem on the energy production chain

There are many PTOS types, and the choice of PTOS for a particular WWEC is of crucial importance. The systematic comparison of the different PTOS types is difficult, however, for at least two reasons:

- the limited amount of available data;
- a particular PTOS can be used in a maximum of two types of WWEC.

Five main types of PTOS are applied to convert sea waves' energy into electricity under real conditions: air turbines, water turbines, hydraulic converters, direct mechanical converters, and direct electrical converters.

The paper aims to present a new version of a hydraulic station for converting the sea wave energy. The principle of hydraulic converters is presented. The main ways to convert the energy of the sea and ocean waves into electricity are discussed. The proposed hydraulic scheme is described in detail.

## 2. Hydraulic converters

When PWEC is based on the direct motion of a solid body by waves, as in the case of point absorber and retarder, the conventional rotary generators are not directly applicable, and an intermediate element is needed: most often a hydraulic converter. The hydraulic converter is used effectively in low-frequency periodic impacts with high forces. In this situation, the direction of energy movement is the opposite of that of the hydraulic drive systems. The energy of the wave-driven solid body increases the pressure of the working fluid in the hydraulic system, which supplies a hydraulic motor, to whose rotor the electric generator is connected.

A diagram of a PTOS with a hydraulic converter is presented in figure 2. In this case, a point absorber (buoy, float), firmly connected to the piston of a hydraulic cylinder, is used as a PWEC. Under the action of the waves, the point absorber and the piston in the hydraulic cylinder respectively, move reciprocating, and the working fluid moves in a closed circular circuit from the hydraulic cylinder to the hydraulic motor and vice versa. Hydraulic accumulators for high and low pressure are included to reduce the pressure pulsations of the working fluid at the inlet and outlet of the hydraulic motor. Rotary radial-piston hydraulic motors are most often used for PTOS with a hydraulic converter, as their behaviour in low-speed, high-load applications have been thoroughly studied.

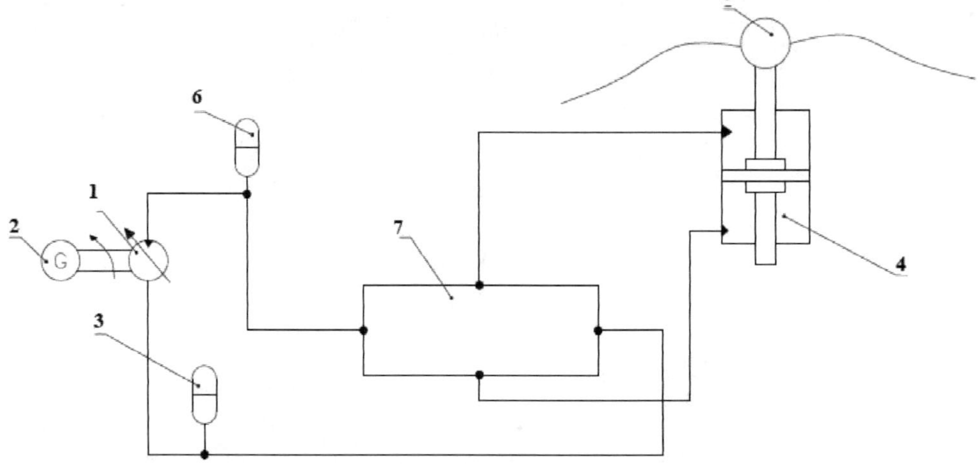


**Figure 2**. Exemplary scheme of PTOS with a hydraulic converter: 1 – Hydraulic motor, 2 – Generator, 3 – Low-pressure accumulator 4 – Actuator, 5 –Buoy, 6 – High-pressure accumulator, 7 – Check valves.

The use of PTOS with a hydraulic converter is associated with solving some specific problems, which have to be considered when designing a PTOS with a hydraulic converter.

The hydraulic system elements must be carefully selected in terms of their performance and efficiency both individually and as a system, as well as in terms of the environmental impact of the system. Due to the change of the sea waves energy in a wide range, gas accumulators must be included in the hydraulic system to absorb energy at peak moments and smooth out the pressure change at the inlet of the hydraulic motor. Rotary radial-piston hydraulic motors with digital control [6] have been developed to increase their efficiency under partial load and to facilitate their control [7]. The hydraulic converters have many moving parts, and the piston seals wear out during operation, which can substantially increase the maintenance costs. Attention must also be paid to the PTOS protection in extreme conditions, when the stroke of the piston may exceed its design value and destroy the system. One possible solution is to provide a piston stroke limiter [8] or to use a radial hydraulic piston [9].

### 3. Direct mechanical converters

The schematic diagram of PTOS with the direct mechanical conversion of the wave energy into electricity is presented in figure 3. The mechanical energy of a wave-driven oscillating solid body (buoy) is converted into electrical energy using an additional mechanical fitting that drives a rotary generator. This mechanical fitting can be a gearbox, reels, and ropes. The rotary mechanical fittings usually include a flywheel to accumulate or release energy to smooth out short-term changes in the wave energy.

One of the advantages of this PTOS type is that the wave energy conversion into electrical energy involves three steps: i.e., the energy of the waves is converted into mechanical energy of the body, which is transformed into mechanical energy of the shaft, which in turn is transmuted into electrical energy. On the other hand, the mechanical converter performs countless operating cycles, and its reliability needs to be checked and proven.

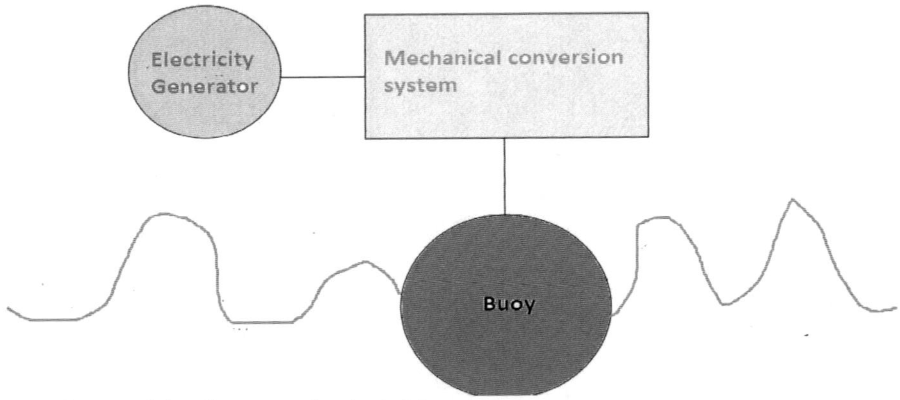


**Figure 3.** Scheme of the direct mechanical drive system.

## 4. Direct electrical converters

Advances in both the technology for permanent magnets production and power electronics have made the PTOS that directly convert the wave energy into electricity a desirable solution. A scheme of the PTOS is presented in figure 4. The energy of the reciprocating motion of a wave-driven solid body (buoy, float) is converted directly into electricity using a linear electric generator [10, 11]. A rod is attached to the bottom of the buoy. The head, which consists of an even number of permanent magnets, mounted one below the other, is in the lower part of the rod. Every two adjacent permanent magnets have the opposite polarity. The waves create a reciprocating motion of the rod, together with the head, which moves inside a stator. In the stator, the electric current is induced.

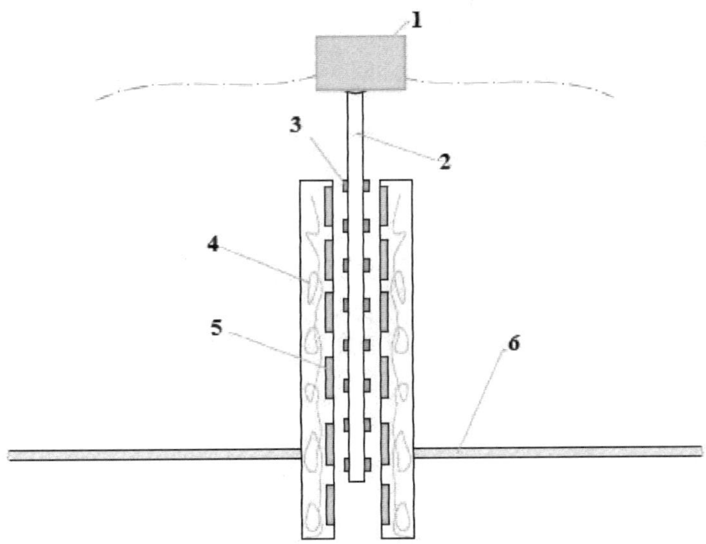


**Figure 4**. Diagram of a direct linear drive system for wave energy conversion: 1 – Buoy; 2 – Translator, 3 – Magnets, 4 – Stator, 5 – Coil, and 6 – Platform.

As a result, the wave energy is converted directly into alternating electric current, whose frequency and voltage are continually changing. Therefore, before connecting to the electrical network, it is needed to transform the alternating current into direct current firstly; then it is converted into alternating current with a given frequency and voltage, which can be done passively or actively [5]. It is required to constructively create and guarantee the existence of a minimum air gap between the rod head and the stator coil in all possible conditions.

**5. New hydraulic system for converting the energy of the ocean and sea waves**

New PTOSs are continually being created and studied, as the wave climate in the various water basins has characteristics that change in a wide range. As a result, there is no technology, which could be universally applied.

Figure 5 shows the design of a new hydraulic station to convert the energy of the ocean and sea waves into electricity.

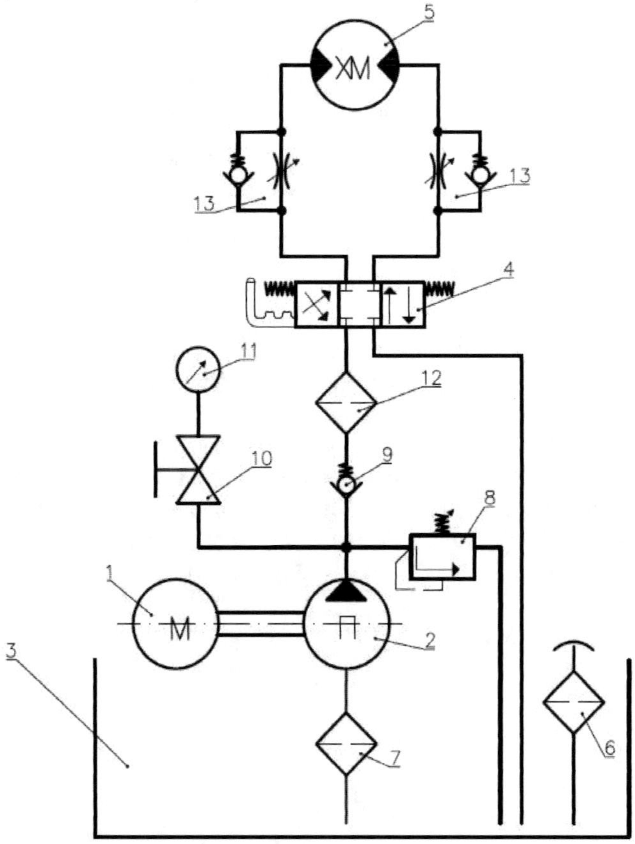


**Figure 5**. The designed new system for wave energy conversion: 1 – electric motor, 2 – gear pump, 3 – hydraulic oil tank, 4 – 4/3 hydraulic road distributor with manual control, 5 – hydro motor, 6 – filter breather, 7 – suction (submersible) filter, 8 – safety overflow valve, 9 – backflow preventer valve, 10 – shut-off valve, 11 – pressure gauge, 12 – pressure filter, 13 – regulating throttle with backflow preventer valve for pipe installation.

The motor (1) is fitted on a P profile, which is welded to the tank (3) cover and is coupled to the gear pump (2) through a finger coupling with a rubber working body. On one side of the gear pump, the suction pipe is mounted through a flange connection, and the suction filter (7) is attached to it. On the other side of the gear pump, the inflator line is mounted again using a flange connection. The backflow preventer valve (9), the safety overflow valve (8), the shut-off valve (10), and the pressure gauge (11) are joined to the inflator line. The pressure filter and 4/3 hydraulic road distributor with manual control (4) are also assembled alongside the inflator pipeline. After the hydraulic distributor with manual control, two regulating throttles with a backflow preventer valve for pipe installation (13) are mounted. The hydraulic system is connected with the hydro-motor (5) through flexible hoses. The system is driven by the rotation of the electric motor, which provokes rotation of the gear pump through the finger coupling. The gear pump sucks the working fluid from the tank and injects it into the line under pressure. The safety overflow valve, installed immediately after the gear pump, is set to the maximum permissible operating pressure of the gear pump. The pressure gauge and the shut-off valve monitor the pressure in the inflator line. A pressure filter for purification of the working fluid and the backflow preventer valve are also installed on the pressure pipeline. After the hydraulic elements, the working fluid reaches the 4/3 hydraulic road distributor with manual control and the hydraulic motor, which must be driven. Two regulating throttles with a backflow preventer valve are mounted in the two directions of rotation to adjust the speed of the hydraulic motor.

## 6. Conclusion

The paper presents the main ways to convert the energy of the sea and ocean waves into electricity. A new hydraulic system for this type of conversion is proposed. The new system can be applied in all devices that consist of turbines used to capture the waves' energy in real conditions.

## References

[1] Clément A *et al* 2002 *Renew. Sust. Energy Rev.* **6,** pp 405–431
[2] Velichkova R *et al* 2018 Utilzation of wave energy by hybrid system. *In Proceedings of the 2018 EFEA*, doi: 10.1109/EFEA.2018.8617056
[3] Markov D *et al* 2017 On the utilization of the wind waves energy along the Bulgarian Black Sea coast. *Proceedings of the 56th Science Conference of Ruse University*, Book **1.2**, pp 73–79
[4] Ross D 1995 Power from the waves , *Oxford University Press*, Oxford, UK.
[5] Salter S H 1974 *Nature* **249**, pp 720–724
[6] Pelc R and Fujita R M 2002 *Mar. Policy* **26,** pp 471–479
[7] Trostmann E 2019 Water hydraulics control technology, *Routledge, Abingdon-on-Thames*, UK
[8] Polinder H and Scuotto M 2005 Wave energy converters and their impact on power systems. *Proc. of 2005 Int. Conference on Future power systems*, pp 1–9
[9] Leijon M *et al* 2006 *Renew. Energy* **31**, pp 1309–1319
[10] Gao Y *et al* 2016 Energy **95**, pp 99-109
[11] Faiz J and Nematsaberi A 2017 IET Renew. Power Gen. **11**, pp 1163-1176

**Acknowledgments**
This study is part of the project „Utilization of wave energy by hybrid system", KP06-H37/30, funded by the Bulgarian Science Fund of the Ministry of Education and Science.

# Experimental investigation of the refrigeration cycle of water-to-water heat pump

**Z D Kolev[1], P G Mushakov[1], and S Y Kadirova[2]**

[1] Department of Heat, Hydraulics and Environmental Engineering, University of Ruse, 8 Studentska Street, Ruse, Bulgaria
[2] Department of Electronics, University of Ruse, 8 Studentska Street, Ruse, Bulgaria

E-mail: skadirova@uni-ruse.bg

**Abstract.** The refrigeration cycle of a reversible water-to-water heat pump has been investigated at different operation parameters of a laboratory installation. The investigation has been accomplished in cooling working conditions. The real operation cycles of the pump have been plotted in "logp-h" diagram by determination of definite thermodynamic parameters of the refrigerant. For this purpose, COOL PACK software has been used. The investigated refrigeration cycles are characterized by different switch off temperature of the heat pump. Based on the obtained results, the following parameters characterizing the heat pump's efficiency have been determined: specific heat of vaporization, specific compression work, specific heat of condensation, and coefficient of performance.

## 1. Introduction

Nowadays, due to the energy and environmental crisis, the use of renewable energy requires reduction of the environmental pollution. The increasing environmental pollution and its influence on people's lives has led to design of systems that are environment-friendly as well as paying attention to energy consumption. One of these systems, which is environment-friendly and low energy consumption, is the heat pump system [1].

Currently, renewable energy resources are becoming alternative options due to their special advantages; especially biomass and geothermal energy require attention in the worldwide for their abundant and widely distributed features [2,3].

In the context of sustainable development, the use of underground water source heat pump air conditioning system is a major feature of nowadays air conditioning applications. The energy saving effect of the ground water heat pump air conditioning installations is obvious, and this type of systems have the advantages of small environmental pollution, cover the requirements of cooling and heating, require low investment and operation cost, simple operation of the system and high water temperature. As well they are with high operating stability, and that is a new developing area in the environmental protection field [4].

Through the heat pump technology, natural low-grade energy stored in the soil, water, air or waste heat energy from industries and daily lives, is supplied for building of cooling/heating and hot water serving [5].

As presented in study [6], heat pumps are also applied in systems of greenhouses cooling. Greenhouse cooling systems are with three different types of heat pumps: air source-air supply, water source-air supply and water source-water supply types, are assessed by hourly calculations related to a simple heat balance model. Nowadays, heat pumps are used worldwide in various applications, both commercial and residential buildings, or in industrial applications. On the other hand, on the market can be found units from small size (3-5 kW) up to large size (>200 kW) [7,8,9].

One of the important parameters for cooling systems is the type of refrigerant. Principally, refrigerants used in cooling systems damage the atmosphere and cause global warming. For these purpose, alternative refrigerants which are not harmful to the environment and atmosphere must be used [10].

Different protocols and regulations lead to using refrigerants with both zero Ozone Depletion Potential and low Global Warming Potential. These new limitations cause the progressive phase-out of HFC and to their replacement by the 4th generation of refrigerants based on HFO mixtures. Alternative refrigerants with low Global Warming Potential are in development process for residential heat pumps, air-conditioners and heat pump water heaters, since R410A, R407C and R134a have Global Warming Potential of 2088, 1650 and 1430, respectively [11].

In study [12] a novel cycle is regulated by the average density, is investigated based on cycle simulation. The used refrigerants are: pseudo azeotropic mixture R410A and zeotropic mixture R407C.

The ever-increasing energy consumption of buildings has caused a great interest in improving the seasonal characteristics of heat pumps [13,14].

The aim of the research is to investigate experimentally the refrigeration cycle of a water-to-water heat pump installation.

## 2. Methodology

### *2.1. Laboratory installation*

The investigation has been realized by a laboratory heat pump installation of water-to-water type, presented in figure 1.

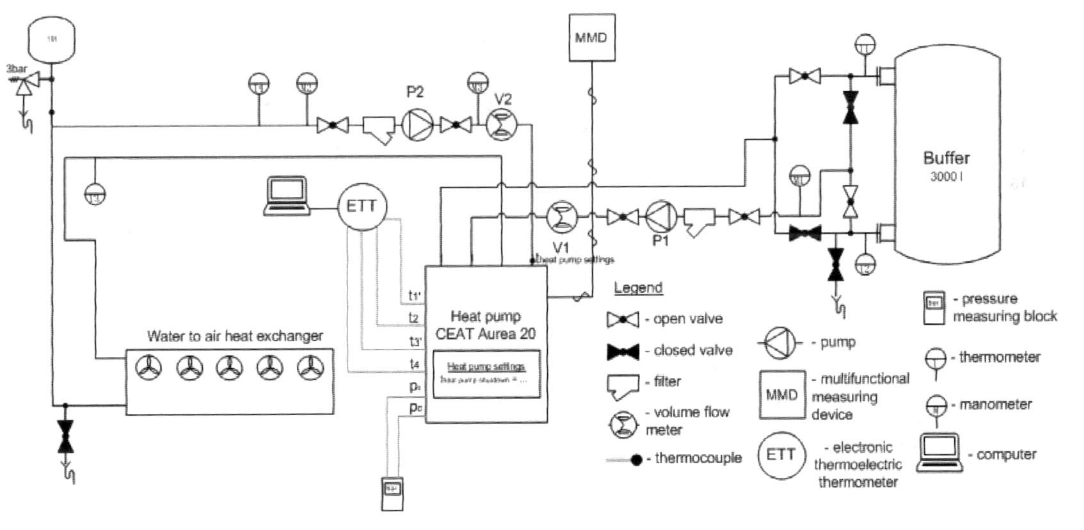


**Figure 1.** Principal scheme of the laboratory installation.

In the cooling regime of the system, the working of the heat pump is regulated with set of shutting down temperature of the heat pump. This is the temperature of the water coming from the heat consumer and entering the evaporator of the heat pump.

In order to ensure a constant temperature of the fluid coming from the buffer into the heat pump, the conditions for stratification in the buffer are met.

The heat pump is reversible, which allows to be studied in summer and winter modes. Water buffer is used as the heat source. The heat consumer is a two-stage water-to-air heat exchanger.

### *2.2. Building of the refrigeration heat pump cycle*

The refrigeration cycle is built on the basis of measured temperatures and pressures of the refrigerant in the laboratory system in figure 1. Figure 2 presents the theoretical refrigeration cycle of heat pump in "logp-h" diagram.

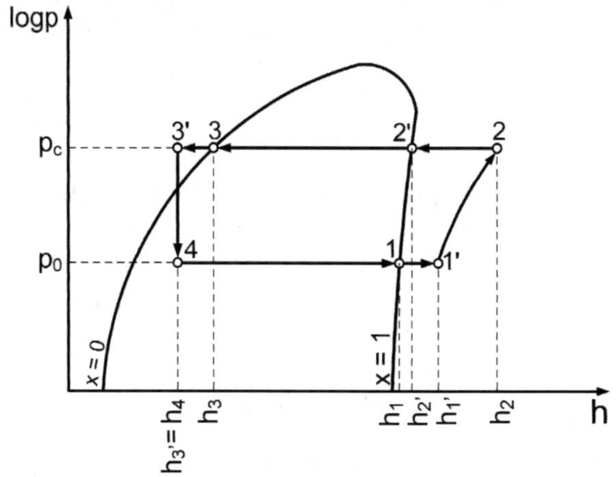


**Figure 2.** Theoretical heat pump cycle.

The vaporization pressure is $p_0$ and the pressure of condensation is $p_c$.

The processes building the refrigeration cycle are:

1-1' – process of preheating of the vapours at constant pressure;
1'-2 – adiabatic process of vapours compression;
2-2' – process of vapours cooling;
2'-3 – process of condensation at constant pressure and at constant temperature;
3-3' – precooling process of the liquid at constant pressure;
3'-4 – throttling process at constant enthalpy;
4-1 – process of vaporisation at constant pressure and at constant temperature.

### *2.3. Determination of the heat pump parameters*

The parameters determining the efficiency of the heat pump can be calculated by equation 1÷4. The reported enthalpy values at the respective intersection points in the cycle are used (figure 2).

The specific heat of the vaporization process is estimated by the equation:

$$q_0 = h_1 - h_4 \tag{1}$$

The equation for the specific compression work is:

$$l_a = h_2 - h_{1'} \tag{2}$$

The specific heat of condensation process is calculated by:

$$q_c = h_2 - h_3 \tag{3}$$

The equation for the coefficient of performance is:

$$COP = \frac{q_0}{l_a} \tag{4}$$

## 3. Results

Table 1 presents the values of the parameter for setting up the heat pump ($t_{heat\ pump\ swutdown}$), as well as the measured parameters ($t_{1'}$ - $p_c$), on the basis of which the refrigeration cycles are built. The investigated regimes are characterized by a different setting of the shutdown temperature of the heat pump.

**Table 1.** Setting up the heat pump and values of measured parameters.

| | $t_{heat\ pump\ swutdown}$, $^oC$ | $t_{1'}$, $^oC$ | $t_2$, $^oC$ | $t_{3'}$, $^oC$ | $t_4$, $^oC$ | $p_0$, *bar* | $p_c$, *bar* |
|---|---|---|---|---|---|---|---|
| **Regime A** | 12.5 | 10.8 | 71.0 | 34.4 | -0.6 | 4.76 | 15.21 |
| **Regime B** | 14.0 | 11.7 | 69.9 | 34.7 | 0.3 | 4.92 | 15.30 |
| **Regime C** | 15.5 | 12.4 | 68.4 | 34.8 | 1.4 | 5.29 | 15.40 |
| **Regime D** | 17.0 | 13.2 | 66.9 | 34.8 | 2.6 | 5.44 | 15.40 |

The setting of the heat pump is according to the temperature of the water returning from the heat exchanger, at which temperature the heat pump is getting switched off. The presented values of the measured temperatures and pressures refer to the moment of switching off the heat pump.

Figure 3 presents the variation of the measured temperatures over time during the operation of the heat pump in Regime A. Temperatures values are measured at intervals of 1 s.

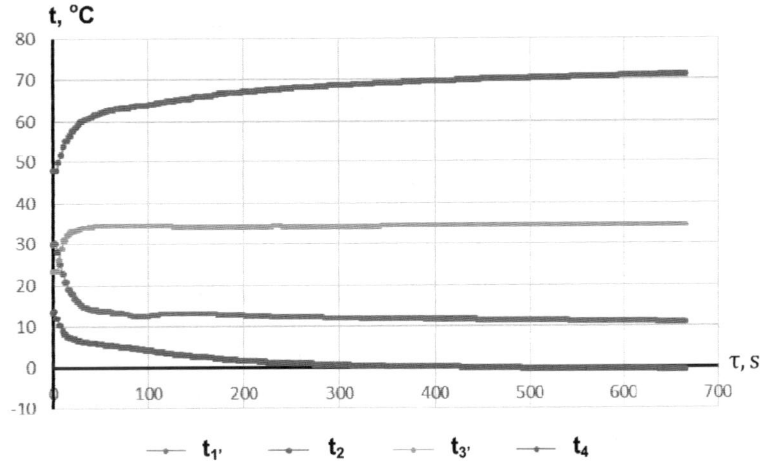


**Figure 3.** Variation of the temperatures in Regime A.

The presented graphical dependences for the alteration of the temperatures measured in the cycle cover the period of operation of the heat pump in the respective regimes. The measured temperatures in table 1 represent the endpoints of the graphical dependences.

Figure 4 presents the refrigeration cycles built in the "logp-h" diagram in the studied regimes. The cycles are built on the base of the temperatures and pressures measured at the time of shutting down the heat pump (table 1).

Since the refrigerant R407C is a zeotropic mixture, the isotherms in "logp-h" diagram have a slope, i.e. they are not parallel to the isobars. As shown in the diagram, the evaporation of the refrigerant at constant pressure, proceeds with a increase in the temperature while the process of condensation at constant pressure - with an decrease in the temperature.

The heat pump switch off temperature increases, and the curve of refrigeration cycle shifts upwards in the diagram. The relatively higher change in the evaporation pressure compared to the condensation pressure is because of the temperature range of the refrigerant evaporation is controlled by the respective setting of the switch off temperature of the heat pump.

Table 2 presents the enthalpy values of the specific enthalpy taken from the diagram at the respective intersection points in the cycle, and the calculated parameters presenting the efficiency of the heat pump.

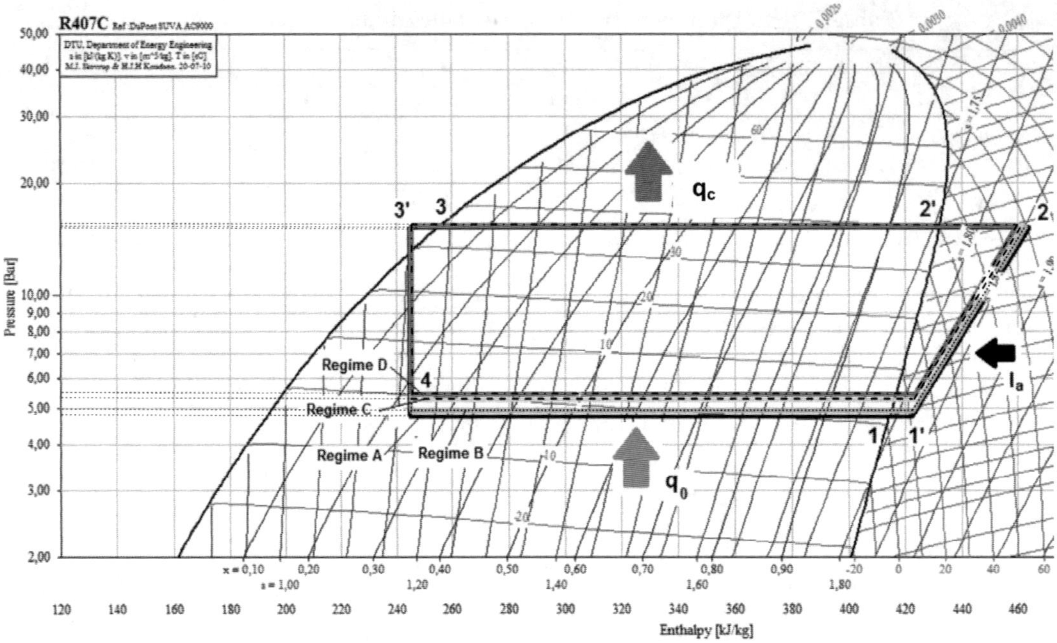


**Figure 4.** Refrigeration cycles of heat pump in "logp-h" diagram.

**Table 2.** Specific enthalpies taken from the software, and calculated heat pump efficiency parameters.

| | $h_1$, kJ/kg | $h_{1'}$, kJ/kg | $h_2$, kJ/kg | $h_3$, kJ/kg | $h_4$, kJ/kg | $q_0$, kJ/kg | $l_a$, kJ/kg | $q_c$, kJ/kg | COP - |
|---|---|---|---|---|---|---|---|---|---|
| **Regime A** | 414.54 | 422.15 | 464.35 | 244.73 | 254.74 | 159.80 | 42.20 | 219.62 | 3.79 |
| **Regime B** | 415.27 | 422.75 | 463.01 | 245.21 | 255.10 | 160.17 | 40.26 | 217.80 | 3.98 |
| **Regime C** | 416.23 | 422.69 | 461.37 | 245.42 | 255.59 | 160.64 | 38.68 | 215.95 | 4.15 |
| **Regime D** | 416.71 | 423.11 | 459.76 | 245.42 | 255.59 | 161.12 | 36.65 | 214.34 | 4.40 |

From the data observed from figure 4 and table 2 follows that with increase of the shutdown temperature, the increase of the coefficient of performance is due to the increase of the specific heat of vaporization, and the decrease of the specific compression work on the other hand.

The application of the results obtained in this study for real installations must be in accordance with the specific features of these installations, the temperature of the heat source and the temperature in the cooled buildings.

## 4. Conclusions

The refrigeration cycle of a reversible water-to-water heat pump is built at the end of four different regimes of operation of the laboratory installation, according to the presented methodology. The operation regimes have been realized at different settings of the heat pump.

The variation of the temperatures in time, used for building the refrigeration cycle in the respective periods of operation of the heat pump, has been studied.

According to the presented methodology the following parameters related to the efficiency of the heat pump, have been calculated: specific heat of vaporization, specific compression work, specific heat of condensation and coefficient of performance.

The vaporization and condensation processes in the refrigeration cycle have been realized with a change in temperature, since the refrigerant is a zeotropic mixture and the individual fractions in its composition have different temperature of change of the physical state, at constant pressure.

It has been found that when regulating the switch off temperature of the heat pump, the vaporization pressure changes significantly more compared to the condensation pressure.

It can be concluded that with the increase of the heat pump switch off temperature, the refrigeration cycle shifts upwards in thc "logp-h" diagram.

Last but not least, based on the results can be generalized that with increasing the shutdown temperature of the heat pump, the efficiency of its operation increases.

**References**

[1] Maddah S, Goodarzi M and Mohammad R S 2020 Comparative study of the performance of air and geothermal sources of heat pumps cycle operating with various refrigerants and vapor injection, *Alexandria Engineering Journal*, https://doi.org/10.1016/j.aej.2020.07.009

[2] Li H, Zhang X, Liu L, Wang S and Guoqiang Z 2017 Proposal and research on a combined heating and power system integrating biomass partial gasification with ground source heat pump, *Energy Conversion and Management,* **145** pp 158-168, https://doi.org/10.1016/j.enconman.2017.04.090

[3] Zlateva P and Dimitrov R 2019 Possibilities of using algae for renewable energy production in the Black sea region, *11th Electrical Engineering Faculty Conference (BulEF)*, 9030765, doi: 10.1109/BulEF48056.2019.9030765

[4] Kang Z, Zhou X, Zhao Y, Wang R and Wang X 2017 Study on optimization of underground water source heat pump, *Procedia Engineering*, **205** pp 1691-1697, https://doi.org/10.1016/j.proeng.2017.10.353.

[5] Ni L, Dong J, Yao Y, Shen C, Qv D and Zhang X 2015 *A review of heat pump systems for heating and cooling of buildings* (Renewable Energy: Elsevier) **84** pp 30-45

[6] Nyers M N and Nyers A J 2011 COP of heating-cooling system with heat pump, *IEEE 3rd International Symposium on Exploitation of Renewable Energy Sources (EXPRES)*, Subotica, pp 17-21, doi: 10.1109/EXPRES.2011.5741809.

[7] Okushima L, Mears D R, Sase S, Takakura T, Moriyama H, Furuno S and Ishii M 2014 Capacities for heat pump cooling systems for greenhouses in Japan *Journal of the Society of Agricultural Structures* **45**(2)

[8] Ferrara G, Ferrari L, Fiaschi D, Galoppi G, Karellas S, Secchi R and Tempesti D 2015 A small power recovery expander for heat pump COP improvement, *Procedia Engineering*, **81** pp 1151-1159, https://doi.org/10.1016/j.egypro.2015.12.140

[9] Hinov N, Petkov T, and Ibrishimov H 2019 Modeling of the thermal and electromagnetic processes in the induction water heating system, *IEEE XXVIII International Scientific Conference Electronics (ET),* Sozopol, Bulgaria, pp 1-6, doi: 10.1109/ET.2019.8878577.

[10] Altinkaynak M, Olgun E and Şahin A Ş 2019 Comparative evaluation of energy and exergy performances of R22 and its alternative R407C, R410A and R448A refrigerants in vapor compression refrigeration systems *El-Cezerî Journal of Science and Engineering* **6**(3) pp 659-667

[11] Pardo P and Mondot M 2018 Experimental evaluation of R410A, R407C and R134a alternative refrigerants in residential heat pumps *17th International Refrigeration and Air Conditioning Conference at Purdue 2498*

[12] Tsubaki K and Mitsutake Y 2015 Ground source heat pump system controlled by refrigerant average density, *Procedia Engineering,* **105** pp 153-159, https://doi.org/10.1016/j.proeng.2015.05.051

[13] Zlateva P and Yordanov K 2019 Experimental study of heat pump type air-water for heating system performance *E3S Web of Conferences* **112**, 01007, https://doi.org/10.1051/e3sconf/201911201007

[14] Zlateva P, Yordanov K, Tudorache A and Cirtina LM 2020 An analysis of energy resources in Bulgaria and Romania, *21th International Symposium on Electrical Apparatus and Technologies (SIELA)*, pp 1-4, doi: 10.1109/SIELA49118.2020.9167132

# Fly ash zeolites as a dual system for simultaneous $CO_2$ capture and its utilization in the production of synthetic gas

**D M Zgureva[1], S V Boycheva[2] and K B Filipov[2]**

[1] College of Energy and Electronics, Technical University of Sofia, 8 Kliment Ohridski Blvd, Sofia, Bulgaria
[2] Department of Thermal and Nuclear Power Engineering, Technical University of Sofia, 8 Kliment Ohridski Blvd, Sofia, Bulgaria

E-mail: dzgureva@tu-sofia.bg

**Abstract.** The deepening of global environmental problems caused by carbon emissions ($CO_2$) generated by various industries requires the development of technological solutions to limit their emitting into the atmosphere. Electrical power plants that use organic fuels as a primary energy source are serious emitter of $CO_2$. The current legislation prerequisites the implementation of technologies for carbon capture and utilization (CCU - Carbon Capture and Utilization) situated at the plant site. In our previous studies, we successfully applied the zeolites synthesized from fly ash obtained from combustion of lignite coal as solid adsorbents for selective capture of $CO_2$ in equilibrium and dynamic conditions. The aim of this study is upgrading of the fly ash zeolite utilization in a dual system for simultaneous carbon capture and $CO_2$ utilization for production of synthetic gas. Experiments were performed in a stainless steel laboratory reactor Büchiglasuster Miniclave 100 in equilibrium conditions at $CO_2$ pressure of 0.25 MPa with different duration. The solid samples saturated with $CO_2$ and reaction products were extracted from the reactor and subjected to gravimetric analyses and Furier-transform infrared spectroscopy (FTIR). The characteristic vibrations of $C_2H_{2n+2}$ molecules were observed onto FTIR spectra of the samples after their exposure for long time.

## 1. Introduction

Simultaneously with the industrial peak and the development of technologies, the parameters of the environment are also changing, as for the last 50-60 years the International Energy Agency (IEA) reports more than a twofold increase in the generated carbon dioxide ($CO_2$) [1]. The higher concentration of $CO_2$ into the atmosphere is considered as a key factor influencing the deepening of global warming, which leads to rising ground surface temperatures [2]. The transport sector and the energy production have the biggest share as $CO_2$ emitters due to the oxidation of organic fuels during the combustion processes. The contribution of hydrocarbon fuels (coal, oil, and gas) as primary energy sources in the world gross electricity production for 2018 is 63.9 %, and the coal has the biggest share of 38 %. Because of the necessity of diversification in the energy sector, the complete abandonment of fossil fuels is not economically and technically feasible. This requires intensive development and research of technological solutions to reduce $CO_2$ emissions into the atmosphere from Thermal Power Plants (TPP). Two main approaches are simultaneously explored: Carbon Capture and Storage (CCS) and Carbon Capture and Utilization (CCU). The CCS is based on the capture and concentration of pure $CO_2$, followed by its liquefaction and transportation for storage in deep geological formations, which is associated with high expenses for real scale application. The CCU approach is suitable for implementation on site of the TPP, and it proposes the utilization of pre-captured $CO_2$ in commercial products. For both technologies, the stage of $CO_2$ separation from the flue gas flows has been comprehensively investigated. It has been found that for the existing TPPs the process of physical adsorption is the most efficient from an economic and technical point of view [3-4]. In our previous studies, zeolites synthesized from fly ash obtained from combustion of lignite coal were successfully applied as solid adsorbents for selective capture of $CO_2$ in equilibrium and dynamic conditions [5-6]. Expanding the use of fly ash zeolite (FAZ) in the CCU system will be a cost-effective technological solution to overcome two environmental challenges for TPPs: huge amounts of fly ash waste disposal and greenhouse gas emissions. According to IEA statistical reports, only 0.15 Gt out of a total of 32.5 Gt $CO_2$ generated in 2017 are utilized in various sectors [1]. The direct use of $CO_2$ amounts to $20.10^6$ $tCO_2$, with the share of food packaging production being 50 %, industrial gases 35 %, and carbonated beverages 15 %. The indirect use of $CO_2$ amounts to $130.10^6$ $tCO_2$ for 2017 and is mainly due to its conversion into substances and raw materials, which is in most cases at the research stage. A promising CCU technology is the conversion of $CO_2$ into hydrocarbon fuels, and there are several research installations for the production of synthetic gas and oil that utilized $10.10^3$ $tCO_2$ in 2017.

This study is a part of comprehensive scientific study devoted to the FAZ application in CCU for production of synthetic gas. The broad surface, specific micro-mesoporous structure of FAZ, the presence of intermetallic inclusions, and the high content of iron in two oxidation extents ($Fe^{2+}$ and $Fe^{3+}$) are prerequisites for enhanced ability of these materials to capture and catalyze the reaction of $CO_2$ with the silanol groups from the zeolitic framework for the production of $C_nH_{2n+2}$. Preliminary results show the potential of FAZ to act as a triple system adsorbent-carrier-catalyst at pressure of 5.5 MPa, which can be applied for both $CO_2$ capture and its catalytic conversion [7]. The aim of this study is to evaluate the potential of FAZ in low pressure hydrocarbonization to optimize the process parameters required to scale the laboratory experiment.

## 2. Materials and methods

The FAZ exposed to $CO_2$ were obtained by double-stage alkaline treatment of fly ash (FA) from lignite coal and coal mixtures combusted in TPPs. Three different FAs were used as raw materials and they were collected from the electrostatic precipitators of TPP AES Galabovo ($FA_{AES}$), TPP Dimitrovgrad ($FA_{DM}$) and iron-enriched separate from the filter's bottom part ($FA_{Fe\text{-}S}$). The applied synthesis procedure was performed in a following consequence: a high temperature pre-fusion stage at 550 or 800 °C of mixtures of 10 g FA and 5g NaOH used as an alkaline activator; crushing of the cooled agglomerate and dillution in 100 ml distilled water; ultrasonic homogenization of the reaction slurry for 15 minutes; hydrothermal activation stage was carried out at 90 °C for 4 hours. The synthesized powder materials were extracted by filtration, washed and dried. The synthesis products were subjected to X-Ray diffraction analyses and Scanning electron microscopy for structure and morphology identification. It was observed that the resulting fly ash zeolites (FAZ) were characterized by a dominant faujasite structure (FAU), similar to that of commercial Na-X zeolite, and were composed of nanosized and submicron crystals. As a major factor determining the adsorption processes, the textural properties of the samples were also studied by $N_2$-physical adsorption technique. The standardized BET model was applied to the adsorption branch of isotherms in the region of monolayer adsorption ($p/p_0$=0.05-0.15) for the specific surface area calculation ($S_{BET}$, $m^2/g$). The volume of micropores ($V_{micro}$, $cm^3/g$) was calculated by t-plot modelling of adsorption isotherms, and the volume of mesopores ($V_{meso}$, $cm^3/g$) by applying of BJH equation to the desorption branch. The textural properties of the all samples are summarized in table 1.

Experiments for $CO_2$ adsorption and its catalytic conversion in hydrocarbons were performed in a stainless steel laboratory reactor Büchiglasuster Miniclave 100 in equilibrium conditions at pressure of 0.25 MPa for 18 hours (Case 1) and 114 hours (Case 2). The initial weight of all samples was 0.2 g and after their exposure, the gravimetric analyses were performed by analytical scales with an accuracy of ±0.0002 g. The samples were also studied by Fourier transform infrared spectroscopy (FTIR) on a Brucker Tensor 37 apparatus to identify the $CO_2$ physisorption and chemisorptions products.

**Table 1.** Samples subjected to $CO_2$ exposure.

| | Raw FA | Tf °C | $S_{BET}$ m²/g | $S_{external}$ m²/g | $V_{total}$ cm³/g | $V_{micro}$ cm³/g | $V_{meso}$ cm³/g | $d_{micro}$ Å |
|---|---|---|---|---|---|---|---|---|
| **Sample 1** | AES | 550 | 334 | 109 | 0.2196 | 0.0894 | 0.1302 | 13.78 |
| **Sample 2** | AES | 800 | 404 | 113 | 0.2819 | 0.1058 | 0.1761 | 13.80 |
| **Sample 3** | AES | 800 | 160 | 118 | 0.2087 | 0.0182 | 0.1905 | 11.78 |
| **Sample 4** | DM | 550 | 421 | 108 | 0.3027 | 0.1234 | 0.1793 | 14.06 |
| **Sample 5** | Fe-S | 800 | 134 | 118 | 0.1432 | n/a | 0.1432 | n/a |

## 3. Results and discussions

The results from gravimetric analyses are shown in table 2, where δm in g is the change between the initial and final sample weight. The percentage increase in sample weight after exposure compared to baseline was given for the two case studies (table 2). In case 1, where the exposure time is shorter, a significant increase in the weight of the exposed samples is observed, which is indicative for the accumulation of $CO_2$ into FAZ by physical adsorption or hemisorption. The intensive change of samples' mass for Case 1 could be explained with physical or chemical reaction between $CO_2$ and FAZ during the exposure. The captured amount of $CO_2$ assuming physical adsorption is presented in table 2. The results for Samples 1, 3 and 4 correspond well to our preliminary investigations for equilibrium $CO_2$ adsorption onto FAZ at a pressure range of 0.0001-0.1 MPa, described by the Langmuir model [5]. Sample 5 is characterized by the lowest value of the specific surface (table 1), but it shows the highest amount of captured $CO_2$ (table 2). Since sample 5 has the highest content of iron oxide phases (hematite $\alpha$-$Fe_2O_3$, magnetite $\gamma$-$Fe_3O_4$), it can be assumed that iron oxides contributed to additional hemisorption of $CO_2$. A number of studies show that iron oxides have active sites exposed at the surface, which can react with the gaseous molecules of $CO_2$, which leads to formation of siderite [8]. The $CO_2$ capture of Sample 2 is unexpectedly low compared to the large specific surface area of this zeolite. Also, this value is significantly lower than that measured in the adsorption tests at equilibrium conditions at lower pressure. Our previous studies show that the FAZ obtained by the melting step at 800 °C and subsequent ultrasonic homogenization are characterized by a fine dispersion of the iron oxide phases in the zeolite matrix and a significant amount of ionic iron [9]. In these zeolites from coal ash, high catalytic activity has been found in oxidative processes. This leads us to assume that in Samples 2 and 3, the measured low value from the gravimetric analysis does not show a low potential for $CO_2$ retention of these FAZs, but rather this result is due to catalytically stipulated $CO_2$ conversion process. During the catalysis temporary complexes are formed, then they decompose and the catalyst is recovered which leads to the maintaining of constant total mass of the system. For all samples from Case 2 at which long exposure in $CO_2$ ambient was performed, almost a slight change in mass is observed. To establish the transformation of $CO_2$ in hydrocarbons onto the FAZ surface the FT-IR spectra were analyzed and they are plotted on figure 1.

**Table 2.** Results from gravimetric analyses of samples exposed to $CO_2$ at 0.25 MPa.

| | Case 1 τ=18 hours | | | Case 2 τ=144 hours | |
|---|---|---|---|---|---|
| | δm g | Weight increase % | $CO_2$, mg/g | δm g | Weight increase % |
| **Sample 1** | 0.0639 | 31.95 | 319.5 | 0.0008 | 0.40 |
| **Sample 2** | 0.0247 | 12.35 | 123.5 | 0.0012 | 0.60 |
| **Sample 3** | 0.0206 | 10.30 | 103.0 | 0.0008 | 0.40 |
| **Sample 4** | 0.0735 | 36.75 | 367.5 | 0.0005 | 0.25 |
| **Sample 5** | 0.1028 | 51.40 | 514.0 | 0.0007 | 0.35 |

In the low wavelength region (400-1100 $cm^{-1}$) could be observed the typical vibration for zeolite structures in different structural cages [10]. Non intensive peaks in this region for Sample 5 confirm the lowest zeolitic yield established by XRD and surface analyses. The IR-spectra of the all samples reveal a wide adsorption band at 2800-3700 $cm^{-1}$ where typical vibration for O-H group (3200-4000 $cm^{-1}$) and C-H bonds (2800-3200 $cm^{-1}$) are situated. Adsorption bands corresponding to C-O from physically adsorbed $CO_2$ in region of 2100-2300 $cm^{-1}$ and of 1500-1800 $cm^{-1}$ are observed in the IR-spectra for the all studied FAZs. $CH_2$ scissor vibrations are appeared in the region 1300-1500 $cm^{-1}$. Two separate peaks although with low intensity are noticeable at Samples 2, 4 and 5, and are barely hinted at the other two patterns. IR spectra are indicative of the beginning of a catalytic process of conversion of $CO_2$ to hydrocarbons, but for a more definite result it is necessary to conduct further studies at higher pressures.

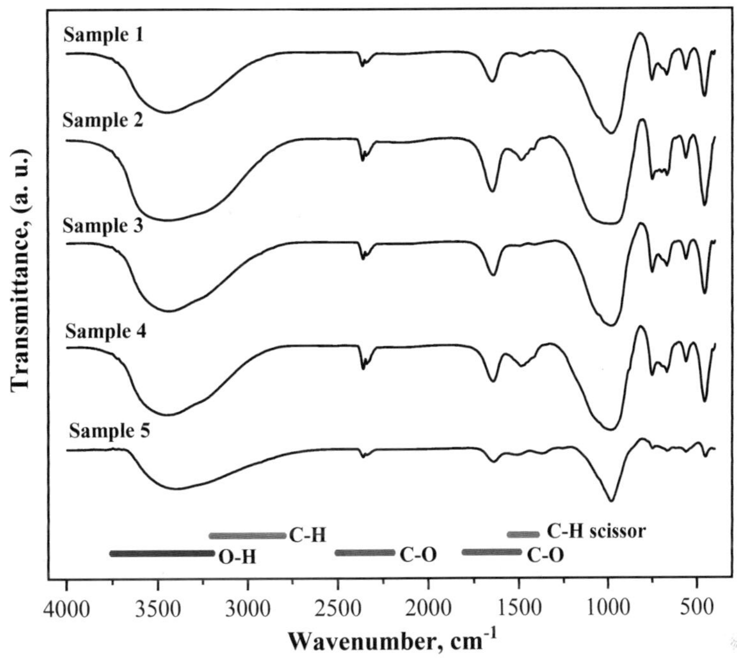


**Figure 1.** FT-IR spectrum of samples exposed to $CO_2$ for Case 2 and regions of bands vibration for O-H, C-H, and C-O.

Regarding to the reaction's mechanism it is supposed that the hydrogen radicals for the catalytic hydrogenation of CO2 are provided by acidic Brønsted centers in FAZ structure, while the catalytic performance of Fe2+/Fe3+ active centres is ascribed to the reduction of iron oxides forming the intermediate iron carbides. The mechanism of CO2 catalytic conversion to hydrocarbons onto FAZ is shown on figure 2.

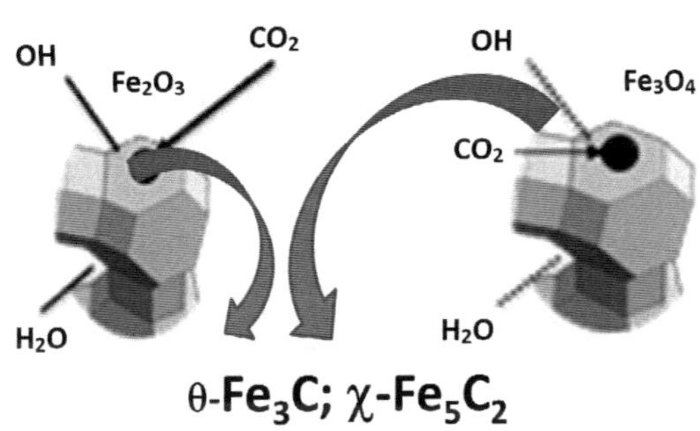


**Figure 2.** Mechanism of $CO_2$ catalytic conversion to hydrocarbons onto FAZ.

## 4. Conclusion

The experimental investigation of $CO_2$ conversion of hydrocarbons at low pressure were performed at equilibrium condition for 18 hours and 114 hours. At the short time of exposure, the physical adsorption is established for the all studied FAZ confirmed by gravimetric and FT-IR analyses. The beginning of catalytic reaction is observed for the sample with $S_{BET}$=404 $m^2/g$ and highest content of hematite due to the high temperature treatment during the synthesis procedure. The longer time of $CO_2$ exposure leads to the appearance of C-H band vibration on FT-IR spectra and a slight change in the weight of the system is also observed due to the catalytic reaction. It is established that the well-developed surface of materials intensifies the reaction of $CO_2$ conversion to hydrocarbons. For the clarification of mechanism and quantification of obtained results additional studies will be performed. The proposed mechanism of hydro-carbonization is similar to those at the Sabatier reaction for which is required high energy supply ($\Delta H$ = -164.9 kJ/mol) and free hydrogen. The combined system for carbon capture and unitization in the production process of synthetic gas is low energy approach for CCU installation situated at the power plant site. The further optimisation of FAZ application as triple system for adsorbent-carrier-catalyst will be directed to experimentally investigation of the effect of pressure and time on the $CO_2$ catalytic conversion processes.

## References

[1] IEA Statistics 2018 *$CO_2$ emissions from fuel combustion: Overview*
[2] Lu J, Cheng M, Ji Y, Hui Z 2009 *J. of Fuel Chemistry and Technology*, p 740
[3] Gibbins J and Chalmers H 2008 *Energy Policy* **12** 4317
[4] Yang H *et al* 2008 *J. of Environmental Sciences* **20** 14
[5] Zgureva D 2016 *Coal Combustion and Gasification Products* **8** 54
[6] Zgureva D and Boycheva S 2020 *Sustainable Chemistry and Pharmacy* **15** 100222
[7] Boycheva S and Zgureva D 2019 *Proc. Int. Conf. on Environmental Science and Technology (Rhodes)* 605
[8] Mendozaa E Y M, Santos A S, López E V, Drozd V, Durygin A, Chen J, Saxena S K 2019 *J. of Materials Research and Technology* **8**(3) 2944
[9] Popova M, Boycheva S, Lazarova H, Zgureva D, Lázár K, Szegedi A, 2019 *Catalysis Today* in press
[10] Yunnan M, Chunjie Y, Aref A, Xiumei Q, Chunyu Z, Dan L 2104 *Advanced Powder Technology* **25**(2) 495

## Acknowledgments

This research was supported by the Scientific Research Sector of Technical University of Sofia through a research project "Perspective young researches" № 201PR0020-10.

# Studies on the thermochemical energy storage in the coal ash zeolite/water system

**S V Boycheva[1], D M Zgureva[2], I K Marinov[1], S Miteva[1], I Yankov[1], A Ivanov[1], and A N Asenov[1]**

[1] Department of Thermal and Nuclear Engineering, Technical University of Sofia, 8 Kl. Ohridsky Blvd., Sofia 1000, Bulgaria
[2] College of Energy and Electronics, Technical University of Sofia, 8 Kl. Ohridsky Blvd., Sofia 1000, Bulgaria

E-mail: sboycheva@tu-sofia.bg; dzgureva@gmail.com

**Abstract.** Deterioration of the air quality and the greenhouse effect requires replacement of the conventional fuels with renewable energy sources (RES). The main disadvantage of RES is the intermittent production of energy during the day and the different seasons. This requires efficient energy storage solutions. Affordable technologies applicable to household-scale heating are thermochemical energy storage systems (TES). TES are usually applied for concentration and storage of solar energy. TES based on zeolites as heat storage media use the exothermic effect of water adsorption. The preferred storage medium is commercial zeolite 13X, which is characterized by high porosity and a large specific surface area. In these systems, the charging is carried out at 130-180 °C and the energy storage efficiency is over 90 %. In the present study, storage media of zeolite X synthesized by the alkaline conversion of coal fly ash was studied in a laboratory assembled TES as a possible substitute of zeolite 13X. The results revealed a micro-mesoporous structure for coal fly ash zeolites (CFAZ) in contrast to the microporous 13X. The charge of TES with CFAZ can be carried out at lower temperatures, while its discharge is enough exothermic. The utilization of coal ash as raw materials for synthesis of heat storage media contribute to many environmental benefits.

## 1. Introduction

Thermal energy storage systems (TES) allow for more efficient use of solar heating, which in the long run will limit the use of fossil fuels and will contribute to many environmental benefits. TES have been known in practice for a long time, but the development of efficient and cost-effective systems with high energy storage density is a challenge from a research and practical point of view. Based on their principle of storage, three types of TES are known: latent, sensible and thermochemical. The last technology attracts the most serious research interest as it provides the highest thermal energy storage density with minimal energy losses that allow high thermal power to be delivered by a compact volume [1]. The thermochemical energy storage is based on the reversible chemical reactions, hydratation/dehydratation and adsorption/desorption cycles charged by thermal supply [2]. Although characterized by high energy density, most reversible chemical reactions take place under catalytic control, which limits the wide practical applicability of these TES. In addition, thermochemical storage places a number of requirements on the kinetics and extent of the chemical reactions, absence of secondary interactions, easy separation of the reaction products, the safety and environmental compatibility of the chemicals, non-corrosive media, market availability, etc. [3]. As a result, storage media, whose thermochemical cycles are provided by atmospheric components: moisture or oxygen, are of great research and applied interest because they can operate as open systems [4]. One of the most emerging technologies to charge, store and discharge thermal energy is adsorption/desorption thermochemical cycles [5]. Adsorption technology for energy storage requires porous media with highly developed specific surface, as it based on the physical retention of the adsorbate molecules on the surface of the adsorbent. The adsorption step is exothermic and represents the discharging mode of the TES to release the energy, while the charging mode is performed as an endothermic desorption step to store the energy. The most studied materials as media for adsorption energy storage are zeolites, thanks to their chemical and thermal resistance, environmental compatibility, and their wide availability as natural minerals and synthetic products. These minerals are characterized by a highly developed specific surface allowing to adsorb a significant amount of water, which retains in their pores. In particular, the commercial zeolite 13X is considered to be one of the most efficient energy-saving media, which provides energy density above 200 $MW/m^3$ [6]. The charging of TES with zeolite 13 X is carried out at 130-180 °C, and the energy storage efficiency is over 90 % [7]. Other adsorbents are also broadly studied, such as activated alumina, silica gel, zeolites impregnated with hydroscopic salts, composites, etc. [5,6]. In addition to high storage density and discharge temperature, the most suitable energy storage material has to meet a number of other requirements, such as low regeneration temperature, low commercial cost and not to harm the environment.

The present research aims to study the potential of zeolites derived from coal ash as heat storage media in thermochemical energy storage systems. Because coal ash zeolites are relatively new materials developed to utilize coal ash to reduce landfills, the opportunities for their applications have not yet been fully explored. So far, coal fly ash zeolites (CFAZ) have been found to have a lower surface area than their pure synthetic counterparts, but sufficient to exhibit high adsorption and catalytic capacities [8]. Moreover, they are distinguished by a mixed micro-mesoporous crystal structure, which provides rapid kinetics for mass transfer processes, as well as they contain iron oxides transferred from the raw coal ash, the presence of which provides them stronger surface unsaturation and further favors the adsorption phenomena [8]. However, in the scientific literature there are no data on the energy storage capacity of these materials, which provokes our interest in this study, and in this point of view the findings here are an absolute novelty. Although we assume that the energy storage density is lower in CFAZ compared to pure synthetic zeolites, taking into account the environmental benefits of smart utilization of coal ash and the low cost of the material, an integral positive result can be expected. Since analogues of commercial zeolite 13X can be synthesized from coal ash under moderate conditions, the subject of our research is CFAZ of type X.

## 2. Experimental

### *2.1. Synthesis and characterization of thermochemical energy storage material*

CFAZ of X type was synthesized by long-term alkaline conversion of lignite coal fly ash at ambient conditions. This synthesis procedure named quasi natural crystallization (QNC) has been described in details in Ref. 9. Although it takes a long time (10-12 months), QNC has undeniable advantages as a method for producing zeolite X by alkaline conversion of coal ash, as it is performed with zero energy consumption, does not require specific equipment and can be performed on ash collection points. Since the zeolite X crystallizes as a metastable phase and requires moderate temperatures, the QNC technique provides a high yield of this zeolite phase. The type of the zeolite phase obtained by alkaline conversion of coal ash is proved by X-ray diffraction, the morphology of the sample was examined by scanning electron microscopy, and its surface characteristics, which are crucial for adsorption applications, were studied by $N_2$-physisorption. The results from material characterization reveal predominant zeolite phase in the resulting product of type X with microcrystalline morphology and a specific surface area of 261 $m^2/g$.

### *2.2. Laboratory set-up of thermal energy storage system*

The laboratory TES assembled for these studies was presented schematically in figure 1 and demonstrated in figure 2. The test storage material with weight of 3.45 g was loaded into a heat-insulated vertical cylindrical adsorption column with a fixed bed. The height of the fixed layer is 19 mm and the diameter is 22 mm. The height of the laboratory column is 60 mm. To prevent the passage of CFAZ particles into the effluent, a filter of glass fibers is mounted at the outlet of the adsorption column. In the charging mode, a hot dry air stream is passed through the adsorption column filled with CFAZ by means of a vacuum pump mounted at the end of the gas line, which desorbs the water accumulated in the pores of the zeolite. The air is heated and dried by passing through a heat-resistant vessel filled with a drying agent and heated to temperature 60-80 °C by a heater with thermoregulation. The charge air flow temperature was measured by a thermocouple NiCr-Ni (K). In a discharge mode, humid air with ambient temperature is passed through the adsorption column, created by bubbling air through a vessel filled with distilled water. The relative humidity (RH) of the discharge air flow is accurately measured with RH-converter "Comeco" with a working range 0-100 % RH, accuracy <± 2.5% RH, output signal 4-20 mA, a measuring probe length of 100 mm. In discharge mode, the temperatures of the air flows at inlet and outlet of the column were measured by a thermocouple NiCr-Ni (K). The flow rates of the charging and discharging air flows were precisely measured and regulated by Rotameter KDS-E Kobold with the following characteristics: air measurement range 3-30 Nl/h, accuracy ± 3 %, analog output 4-20 mA, working pressure up to 420 bar and temperature up to 130 °C. The data from measured temperatures, flow rates and relative humidity are transmitted to Process transmitter RMA42 Endress&Hauser. RMA 42 processes data in different ranges of current, voltage, resistance, as well as a wide range of thermoresistors and thermocouples for which it is pre-configured. The measurement signals are converted from analog to digital format, processed digitally in the device, then converted into analog signals and transmitted to different outputs. The data are collected and stored by Universal Data Manager Ecograph T RSG 30, Endress & Hauser, which records, visualizes and monitors all analog or digital input signals, from where they are transferred to PC.

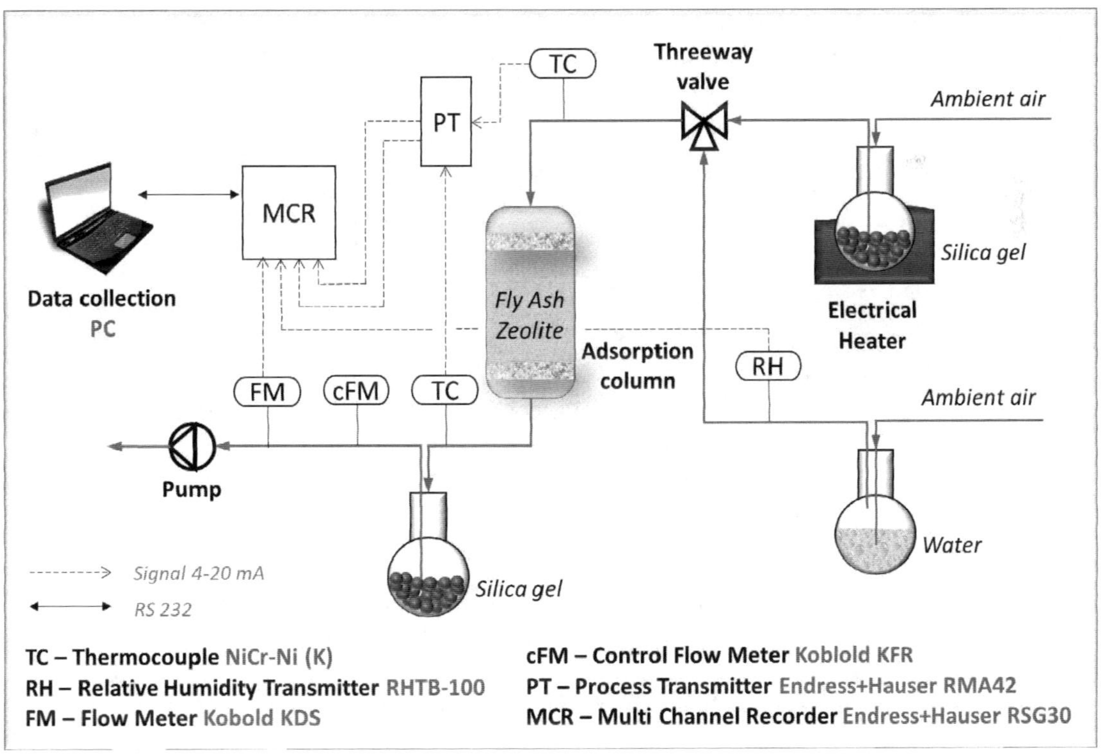


**Figure 1.** Schematic diagram of the laboratory thermochemical TES set-up.

**Figure 2.** Image of the laboratory thermochemical TES.

## 3. Results and Discussions

First, the laboratory experiment with the TES system was performed in a charge (desorption) mode to release water from the zeolite structure. Charging was performed at low temperatures in the range of 60-80 °C for 30 min. The dynamic temperature curves are plotted in Figure 3. Since the outlet air flow temperature is significantly lower than the inlet air flow, it can be argued that an endothermic process involving the release of water from CFAZ takes place in the adsorption column. Low temperature regeneration of the adsorbent would be a significant advantage for this thermochemical TES. The dynamics in the relative humidity of the ambient air and in the charging and discharging modes is presented in figure 4.

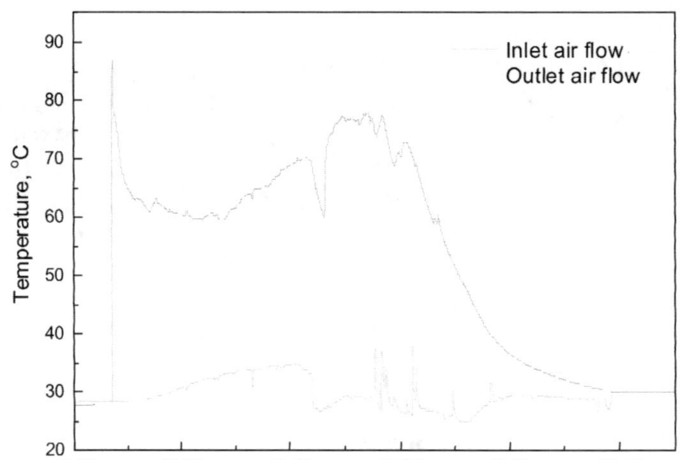


**Figure 3.** Thermochemical TES charging mode.

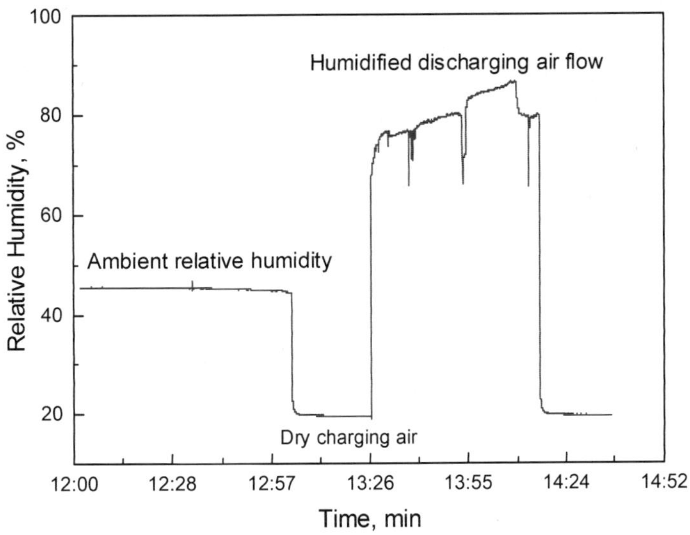


**Figure 4.** Relative humidity of the ambient air and in the charging and discharging modes.

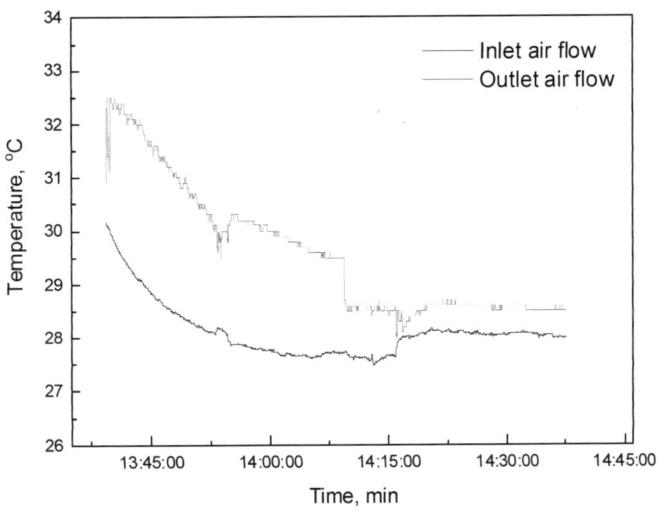


**Figure 5.** Thermochemical TES discharging mode.

However, for a higher efficiency of TES, a sufficiently high regeneration temperature is required, looking for a compromise not to use excess energy. Since the outlet temperature remains low, we can assume that complete regeneration has not been achieved under these conditions. The optimum regeneration temperature is to be determined by subsequent studies. For charging the real TES the enough energy could be supplied by solar collectors or by recovering of waste heat [10]. The adsorption column outlet temperature as a function of time in the discharging mode at RH 80-85 % is plotted in figure 5. The charging and discharging flow rates are kept 30±2 Nl/h. It has been established that flow rates have a significant impact on the energy density of the system. On the one hand, lower gas velocities favor mass transfer processes, and on the other increase heat dissipation.

The optimal air flow rate for charging and discharging TES is also to be investigated. The results of this measurement show that discharging the TES, the outlet air flow temperature ($T_{out}$) is higher than that of the inlet air ($T_{in}$) during the test time of 60 min, and after 30 min from the beginning, the temperature difference $\Delta T= T_{out} - T_{in}$ decreases significantly, but still by the end of the recording remains positive and no equilibration is reached. From the experimental data, the energy density of the studied TES was calculated according to the mathematical models for adsorption thermal storage, described by Kohler and Müller [11]. The calculation parameters and the obtained results are summarized in table 1.

**Table 1.** TES operating parameters and energy storage density

| Parameters | $V_{ads}$, $m^3$ | H, kg/kg | $d_{HA}$, kg/$m^3$ | $m_{HA}$, kg | $C_{p,HA}$, kJ/kgK | Q, (kJ) | ED, kWh/$m^3$ |
|---|---|---|---|---|---|---|---|
| Values | 7.22 $10^{-6}$ | 0.39 | 0.86 | 0.015 | 2.64 | 3.52 | 135 |

ED – energy storage density
Q – energy realized from TES discharge
$V_{ads}$ – volume of the adsorbent bed
$d_{HA}$ – humid air density
$C_{p,HA}$ – heat capacity of the humid air
$m_{HA}$ – mass of the humid air passes into the column
H – humidity per kilogram dry air

## 4. Conclusions

In this study a laboratory set up for thermochemical energy storage by adsorption/desorption cycles was assembled and applied to investigate the heat storage capacity of zeolite X prepared by alkaline conversion of coal ash. The obtained results for the energy storage density in the coal fly ash zeolite/water system are promising at the applied low TES charging temperature. The experimentally obtained energy storage density of 135 kWh/$m^3$ is in the range of 86-203 kWh/$m^3$ reported for zeolite 13X. The further development of coal ash zeolites as heat storage media will improve simultaneously the waste management of solid by-products and the TES economic efficiency.

## References

[1] Alva G, Lin Y and Fang G 2018 An overview of thermal energy storage systems, *Energy*, **144**, pp 341-378, ISSN 0360-5442, https://doi.org/10.1016/j.energy.2017.12.037

[2] Farulla GA, Cellura M, Guarino F and Ferraro M 2020 A Review of Thermochemical Energy Storage Systems for Power Grid Support, *Applied Sciences*, **10**, 3142, https://doi.org/10.3390/app10093142

[3] Sarbu I, Sebarchievici C 2018 A Comprehensive Review of Thermal Energy Storage, *Sustainability*, **10**, 191; doi:10.3390/su10010191

[4] Mukherjee A, Majumdar R, Saha SK, Kumar L and Subramaniam C 2019 Assessment of open thermochemical energy storage system performance forlow temperature heating applications *Applied Thermal Engineering*, **156**, pp 453-470

[5] Vasta S, Brancato V, La Rosa D, Palomba V, Restuccia G, Sapienza A and Frazzica A 2018 *Nanomaterials*, **8**, 522; doi:10.3390/nano8070522.

[6] van Alebeek R, Scapino L, Beving MAJM, Gaeini M, Rindta CCM and Zondag HA 2018, *Applied Thermal Engineering*, **139**, pp 325-333

[7] Hauer A 2007 In: Paksoy H.Ö. (ed.) Adsorption systems for TES—design and demonstration projects, *Thermal Energy Storage for Sustainable Energy Consumption, Springer*, pp 409-427

[8] Popova M, Boycheva S, Lazarova H, Zgureva D, Lázár K and Szegedi Á 2019 *Catalysis Today*, doi: 10.1016/j.cattod.2019.06.070

[9] Zgureva D, Boycheva S, Behunová D and Václavíková M. 2020 Journal *of Environmental Engineering (US)*, **146** (8), art. no. 04020081, doi: 10.1061/(ASCE)EE.1943-7870.0001752

[10] Dicaire D and Tezel FH 2013 *International Journal of Energy Research*, **37**, pp 1059-1068

[11] Kohler T and Müller K 2017, *Energy Science and Engineering*, **5**(1), pp 21–29

## Acknowledgments

This research was supported by the Scientific Research Sector of Technical University of Sofia through a research project "Student Innovation Hub" № 201IHUB0004-02.

# Methodology for assessing the residual resource during the operation of high-temperature units

**A Nikiforov[1], E Prikhodko[1], A Kinzhibekova[1] and A Karmanov[1]**

[1] Thermal power engineering Department, Toraighyrov University, Lomov st. 64, Pavlodar, Republic of Kazakhstan

E-mail: {aleke4599, john1380, akmaral70, aman270685}@mail.ru

**Abstract.** The article analyses the main methods for determining the residual resource of these aggregates. Considering the imperfection of the above methods for estimating the residual resource, a method was developed for determining the residual resource of the linings of high-temperature units. The main parameters, which influence the residual resource of the lining, were determined during the operation of the units. In order to take into consideration the influence of temperature stresses on the residual resource of the sintering furnace, we examined a schedule for heating the furnace after an overhaul and calculated the temperature stresses that occur during heating and compared them with the acceptable values. Considering the temperature of the working medium, which exceeded the calculated one – was made according to the records of the operation logs. In general, the obtained result reflects a picture of the actual operation of the sintering furnace between its repairs. For its implementation into real operation, it is necessary to combine a number of measuring processes into a single information complex: a continuous measurement of the lining temperatures during heating; a burner performance (from the position of the torch profile and fuel oil spills); the quality of the resulting technological cakes, etc.

## 1. Introduction

According to the definition, the residual resource of a high-temperature unit is the total operating time of the unit (in the number of hours, smelts, etc.) from the moment of the last control of its condition (or repair) until the moment of reaching the limit state, which is characterized by the equipment shutdown to prevent an emergency or its inexpedient operation.

Thus, the assessment of the residual resource is primarily necessary during the operation and in the supervision of the operation. In the supervision of the operation, the assessment of the residual resource acquires utmost importance, since the trouble-free operation of high-temperature units is directly proportional to the traumatism rate, that is, a threat to the life and health of the enterprise employees.

Moreover, the residual resource assessment is carried out to determine the real residual value of the equipment [1]. The analysis of the specifics of the operation of high-temperature units shows that using the standard values of the equipment depreciation from reference books, it is possible to obtain data on its condition and residual resource that do not reflect the real situation at all.

The important aspect of determining the residual life of the equipment is its operation with parameters exceeding those allowed by the manufacturer. At the same time, even a small period of time when working in off-design modes can significantly affect the duration of the unit's working campaign. The literature provides data on the study of factors affecting the wear of the lining and the development of operating recommendations to increase the working campaign of the units [2].

The distinctive feature of the lining operation is that for the most part of it, the main reason for taking it out for repair is the unsatisfactory condition of the lining. Lining destruction is mainly caused by thermal activity, and the frequency of the various stages of the wear mechanism depends on the thermal field in the lining [3]. In this case, the residual thickness of the lining reaches a critical value, exceeding of which can lead to the destruction of the enclosing structures of the equipment of the high-temperature unit. As a result, in this article we will consider the operation of those high-temperature units, the preventive maintenance schedule of which is determined by the state of the lining.

## 2. Formulation of the problem

The problem of determining the residual resource of high-temperature equipment is divided into two components. First, it is necessary to decide on a method for determining the residual resource that is most suitable for a given unit, or to develop our own method. Secondly, it is planned to determine the parameters of the factors affecting the wear of the lining, which must be considered when determining the residual resource by any method.

## 3. Theory

A wide range of industrial high-temperature units supposes a significant number of methods for determining the residual resource of these units, taking into account the specifics of the operation of a particular unit. In this connection, we carried out the following gradation of the above mentioned methods.

First, it is the determination of the residual resource by the method of expert assessments. The method involves the participation of highly qualified specialists (experts) in the field of operation of high-temperature units to assess the current state of the unit according to the developed scale. The scale does not have a digital expression and, as a rule, contains no more than five forms of the equipment state, such as HAZOP [4]. The experts make their judgments on the basis of either their personal inspection of the equipment or an inspection report. The method is also widely used to analyze the hazard of an accident in the design of industrial facilities.

The use of the expert assessment method has a number of advantages:

- there is no need to use complex mathematical tools;
- the installation of additional measuring equipment is not required on the analyzed high-temperature units;
- the simplicity of the method, which allows the expert to carry out work to determine the residual resource without long preliminary preparation.

Among the significant disadvantages, the following can be distinguished: the subjectivity of the expert in assessing the technical condition of the equipment is possible, which can arise both as a result of a mistake of the expert himself and as a result of the impossibility to assess the condition of the equipment without using additional diagnostic tools (or measurements during the operation). An additional disadvantage is the limited choice of forms of the equipment state, since the actual state of the equipment may not be fully described by the chosen form.

Secondly, this is the determination of the residual resource by the calculation method. This is a classic method according to which the residual resource is determined by the statistical data: the operating time of the equipment after the last repair, mean time to failure (MTTF) and others [5].

The method can give adequately accurate values when the unit is operating without significant changes in the operating parameters, a wide volume of constantly updated data for a given high-temperature unit when calculating MTTF, no significant changes in the material design, materials used, etc.

The lack of accurate data on MTTF, as well as the presence of non-stationary operating modes, which significantly affect the wear of the unit elements, reduces the reliability of calculating the residual resource.

Thirdly, these are instrumental methods for determining the residual resource [6]. In the literature, they have different names, but the essence of the methods is as follows. During the operation of the equipment, the necessary parameters are measured; their mathematical processing is made, taking into account the available statistics (for determining and constant correction of MTTF). Further, during the operation, the residual resource of the unit is determined.

These methods include the following [7]:

- assessment of the residual resource by the loading cycles;
- diagnostics of the residual resource of the equipment by changing the parameters of its technical condition;
- residual resource assessment by changing the technological parameters;
- diagnostics of the residual resource according to the development of corrosion damage;
- residual resource assessment based on the changes in the quality indicators and the equipment performance;
- residual resource assessment based on the equivalent operating temperature;
- residual resource assessment based on the given stresses and equivalent pressure;
- residual resource assessment by the load asymmetry coefficient.

It is worth noting the high reliability of the results obtained (when using measurements on the operating equipment). The disadvantages include the use of a significant number of sensors, as well as taking into account a limited number of parameters when assessing the residual resource by any of the methods.

Thus, based on the analysis of the existing methods for determining the residual resource, the following conclusions can be made. The most preferable is the instrumental method, which makes it possible to assess the residual resource with a sufficiently high accuracy directly during its operation. The application of this method supposes not only the presence of a significant amount of initial data measured on-line, but also the corresponding mathematical apparatus for their processing.

Taking into account the main disadvantage of the instrumental methods for assessing the residual resource - a limited number of parameters in its assessment, one more method was developed for determining the residual resource of the lining of high-temperature units, the essence of which is as follows.

Let us select the parameters that determine the residual resource of the lining:
the magnitude of the arising temperature stresses exceeding the ultimate strength of the used refractory materials and their duration;
the excess of the smelt temperature over the calculated value and the duration of contact of the smelt with the lining;
acidity (or basicity) of the slag exceeding the calculated value and the duration of the influence of the slag with the acidity (or basicity) of the slag exceeding the calculated one;
other factors.

The above parameters can be quantified to determine the residual resource of the lining, including directly during its operation.

In the developed method, the value of the wear rate of refractory materials of the lining of high-temperature units is proposed to be adjusted by coefficients that take into account the deviations of the operating parameters from the calculated ones according to the following formula:

$$K_B = \frac{B_{exc} \cdot \tau_1 + B_{calc} \cdot \tau_2}{B_{calc} \cdot \tau_{total}}, \tag{1}$$

where $B_{exc}$ – a parameter the value of which at a given time exceeds the calculated value;
$B_{calc}$ – a calculated value of the considered parameter;
$\tau_1$ – time (duration) of the effect of the parameter $B_{exc}$, h;
$\tau_2$ – time (duration) of the effect of the parameter $B_{calc}$, h;
$\tau_{total}$ – the total time (duration) of the effect of the parameter $B$ on the lining, h.

The product of the correction factors will give a general correction factor, which will show how much the mean time to failure (MTTF) decreases.

For a number of units, a more important indicator is the value of the wear rate of refractory lining materials per unit of time (or per one smelt), taking into account the deviation of the operating parameters from the calculated ones and is determined by the formula, $\upsilon_{exc}$, mm / smelt:

$$\vartheta_{exc} = \vartheta_{\text{calc}} \cdot \mathrm{K}_{\sigma} \cdot \mathrm{K}_{t} \cdot \mathrm{K}_{A}, \quad (2)$$

where $\upsilon_{calc}$ – the design rate of the wear of refractory materials of the lining of casting ladles, mm/smelt;

$K_\sigma$ – the accounting factor for the magnitude of the arising temperature stresses;

$K_t$ – the accounting factor for the excess of the smelt temperature over the calculated value and the duration of contact of the smelt with the temperature exceeding the calculated value;

$K_A$ – the accounting factor for the acidity (or basicity) of the slag exceeding the calculated value and the duration of the influence of the slag with the value of the acidity (or basicity) of the slag exceeding the calculated one.

In this case, the residual resource $n_{exc}$ (in the number of smelts), taking into account the correction factor for any moment of time, will be determined by the formula:

$$n_{\text{пр}} = \frac{\delta_{ini} - \delta_{min}}{\vartheta_{exc}}, \quad (3)$$

where $\delta_{ini}$ – initial lining thickness, mm; $\delta_{min}$ – rejection thickness (the minimum lining thickness at which the lining on the casting ladle changes), mm.

Consequently, the method for determining the residual resource is as follows. Based on the statistical data, the parameters influencing the value of the lining residual resource are determined. Further, according to various methods, the calculation (or measurement) of the digital values of the selected parameters is performed. A number of parameters require initial temperature data, which must be recorded during the operation of the equipment. For this purpose, during the construction of the lining of thermal units we propose to install temperature sensors in the lining at specified distances from its inner surface. The number of sensors and the distance from the inner surface should be selected according to the operational limitations connected with the possibility of emergency situations (leakage, metal leakage, etc.) [8].

To take into account the parameter "the magnitude of the arising temperature stresses exceeding the ultimate strength of the used refractory materials and their duration", the calculation of the temperature stresses is performed based on the data obtained (temperature distribution). These values of the temperature distribution (when determining the temperature of the inner surface of the lining) can be used to determine the parameter "excess of the smelt temperature over the calculated value and the duration of contact of the smelt with the temperature exceeding the calculated value with the lining." Then, after calculating the correction factors, the residual lining resource is determined.
The implementation scheme of the developed method is shown in figure 1.

The developed method for determining the residual life of high-temperature equipment allows continuous monitoring of both the thermal state of the lining and its residual life. In this case, one of the determining factors of the destruction - temperature stresses exceeding the ultimate strength of the material – are calculated according to the developed scheme based on numerical methods.

## 4. Experimental results

Let us determine the residual life of the sintering furnace lining based on the developed method.

Sintering furnaces are rotary drum kilns, where gases, brickwork and processed material participate in heat exchange, but due to the rotation of the furnace and the movement of the material, the heat exchange scheme is complex. The material receives heat from both sides. The open surface of the material receives heat from the gases by radiation and convection, and from the brickwork – by radiation. The closed surface of the charge receives heat from the brickwork of the furnace by radiation through a thin gas layer that separates the material from the brickwork. The heat from the surface layers into the deeper, inner layers is transferred during mixing of the material by pouring and moving it; therefore, the issue of heat distribution into the depth of the material layer for these furnaces is solved very simply, regardless of the heat transfer process.

A special feature of the furnace is its length – 100 meters, of which 90 meters are lined with fireclay bricks. This feature imposes a limitation on the heating rate of the furnace. Due to the fact that at high heating rates the elongation of the furnace body with the coefficient of the thermal expansion of steel is $10 \cdot 10^{-3}$ m / (m · °C) and the elongation of the brickwork along the working surface of a brick is $5 \cdot 10^{-3}$ m / (m · °C), a situation may arise when the steel of the furnace body will not "keep pace" with the expansion of the lining fireclay. Thus, the heating rate must be selected so that the temperature on the inner surface of the lining during the entire heating process does not exceed the temperature on the outer surface of the lining by more than two times.

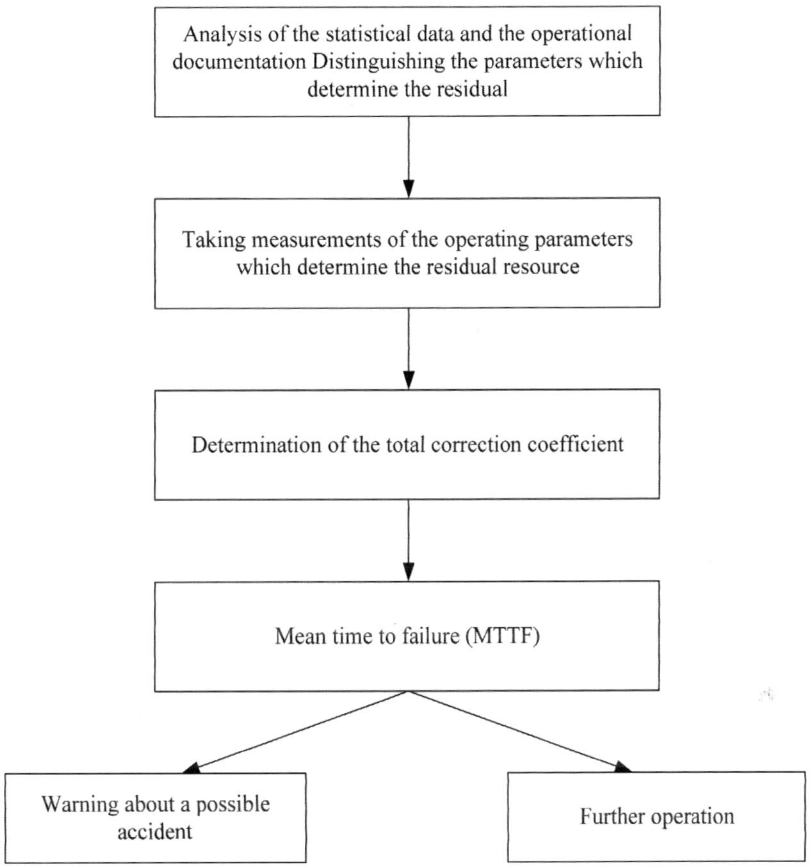


**Figure 1.** Diagram of the methodology for determining the residual service life of high-temperature equipment.

The initial data for the calculation will be as follows. The duration of the working campaign of the furnace before the overhaul is 48 months. After that, the furnace is shut down for the overhaul within 45 days. Periodically, during the operation of the furnace between major overhauls, it becomes necessary to carry out routine repairs (30 days), which consist in replacing part of the lining. The frequency of current repairs depends on the operating conditions and can be equal from once every three years to once a year.

Analyzing the parameters, the deviations of which during the operation of the sintering furnace can lead to an increase in the wear rate of refractory lining materials, the following can be distinguished:

- temperature stresses from high rates of temperature rise during heating (after the overhaul) and cooling (during current repairs);
- working medium temperatures exceeding the calculated ones. In this case, this factor can be justified, firstly, by the spillage of the fuel oil on the lining during the heating process; secondly, a high sintering product temperature. The effect of the high temperature of the

sintering product can be ignored due to the formation of a skull on the inner surface of the lining;

- a collapse of the skull and exposure of part of the lining to the flame of the torch. The influence of this factor is possible if the torch profile is incorrect (for example, the high temperature at the torch root can lead to the collapse of the front rings of the skull), which can be taken into account in calculations based on the sintering product quality. The analysis of the sintering product quality during the furnace operation after the overhaul shows that this factor can be neglected;
- chemical effect of aggressive working environment on the lining. We do not take into account this point, similar to the previous point.

Due to the large length of the furnace, the effect of the torch radiation and the significant effect of temperature stresses do not take place on the entire length of the furnace, but only on the first zone about 5-15 meters long. This is the zone: starting from 4-5 meters to 20-30 meters from the hot edge of the furnace. More than 95 % of all the furnace emergency shutdowns and current repairs are related to the lining of this zone. All the measurements and calculations given in the work refer to the lining of the zone from 5 to 20 meters from the hot edge of the furnace.

To take into account the influence of temperature stresses on the residual life, it is necessary to consider the heating schedule of the sintering furnace after major repairs, as well as to calculate the temperature stresses arising during heating and make their comparison with allowable values.
Let us consider the heating schedule for a high-temperature unit (a sintering furnace) (figure 2).

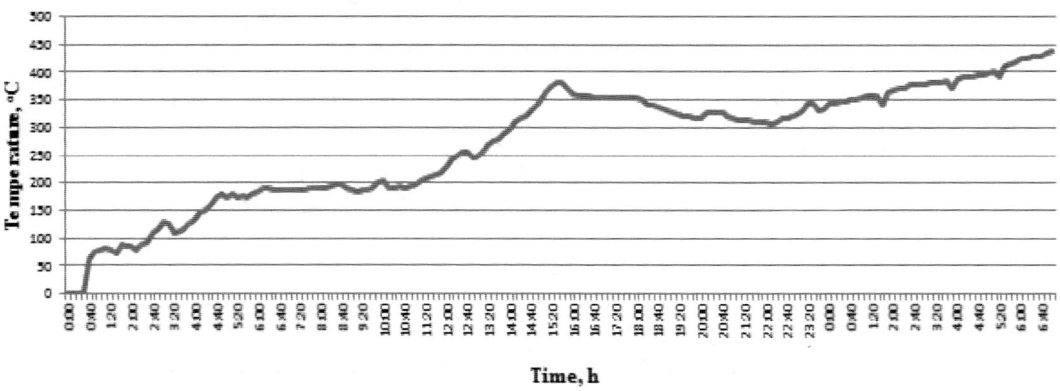


**Figure 2.** Heating schedule of the sintering furnace after the overhaul.

The heating graph shows that from 0 h 30 min to 1 h 10 min there is a jump in temperature, the average heating rate is 2 °C / min. After that, the heating rate is 0,44 °C / min up to 4 hours 40 minutes. Then a temperature exposure (shelf) is carried out for 6 hours 30 minutes. Then the heating temperature is raised at an average rate of 0,61 °C / min to the temperature of 381 °C, after which an exposure (shelf) is again made for relaxation of temperature stresses for 13 hours. And at the end of the heating process, the temperature is raised by 0,27 °C / min to the temperature of 438 °C.

Let us calculate the temperature fields in the lining of the sintering furnace based on the available heating schedule according to the method described in [9]. At the known temperatures along the lining cross section, we find the values of temperature stresses. When heated from the average temperature, in the high-temperature section a compression stress ($\sigma_{compr}$) arises, whereas in the low-temperature section a tensile stress ($\sigma_{tens}$) does.

Compression and tensile stress graphs are shown in figure 3. The compressive strength of the material used is 25 MPa; the tensile strength is 5 MPa. For clarity, compressive and tensile strengths are shown as horizontal lines.

The calculations show that the heating process is carried out at speeds exceeding the allowable level. Thus, the maximum value of the compression stress at the time 0 h 40 min is 1,4 times bigger than the allowable value. According to the tensile stresses, there are seven periods in which the arising tensile stresses are higher than the allowable level, while the maximum tensile stress at the time 0 h 40 min is 3,2 times bigger than the allowable value.

Next, we will take into account the second factor - the temperature of the working environment exceeding the calculated one, according to the records of the operating logs. On this furnace, the fuel oil spillage was recorded on the inner surface of the lining during the heating process due to the unstable operation of the nozzle. Let us estimate the duration of the impact of the burning fuel oil on the lining as 20 minutes. In the process of the furnace operation until now (within 18 months), the sintering temperature has not exceeded the calculated one.

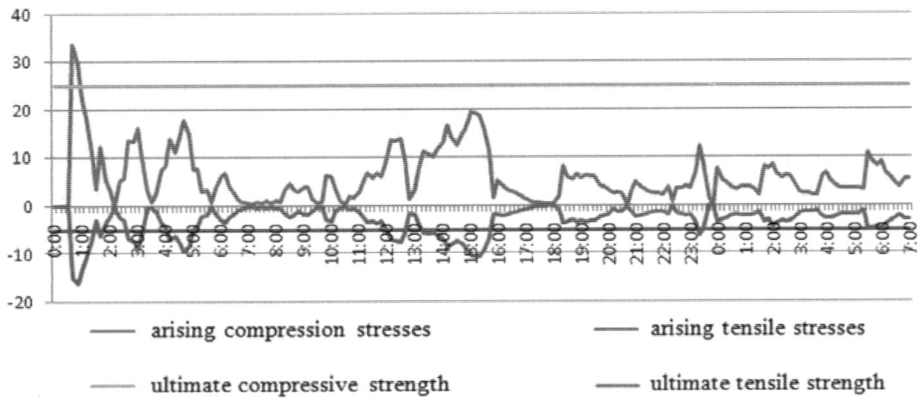


**Figure 3.** Arising temperature stresses.

The analysis of the sintering product quality during the furnace operation after a major overhaul shows that there were no significant deviations in the quality of the product for 18 months, in accordance with which we do not consider the effect of the flame of the torch and the aggressive working environment on the part of the lining.

We calculate the residual resource of the sintering furnace lining using two correction factors: $K_\sigma$ and $K_t$.

The coefficient of accounting for the magnitude of the arising temperature stresses $K_\sigma$, we find by the formula:

$$K_\sigma = \frac{\sigma_{exc} \cdot \tau_1 + \sigma_{calc} \cdot \tau_2}{\sigma_{calc} \cdot \tau_{total}},$$

where $\sigma_{exc}$ – the average value of stress, which for the considered period of time exceeds the allowable one according to the operating conditions of the lining, MPa;

$\sigma_{calc}$ – the tension value allowed under the conditions of the lining operation, MPa;

$\tau_1$ – the length of time during which the tension value exceeded the allowable one according to the operating conditions of the lining, h;

$\tau_2$ – the length of time during which the tension value did not exceed the allowable one under the operating conditions of the lining, h;

$\tau_{total}$ – duration of the action of temperature stresses on the lining, h.

For calculations, the arithmetic mean value of the stresses in the considered time interval was used with a time step of 10 minutes.

Let us calculate the values of the correction coefficients separately for the compression stresses $K_\sigma^{comp}$ and tensile stresses $K_\sigma^{tens}$, which are equal to 1,0029 and 1,13, respectively.

The accounting factor for the excess of the temperature of the medium over the calculated value of $K_t$ and the duration of the contact of the medium with the temperature exceeding the calculated value is found by the formula:

$$K_t = \frac{t_{exc} \cdot \tau_1 + t_{calc} \cdot \tau_2}{t_{calc} \cdot \tau_{total}},$$

where $t_{exc}$ – the average temperature value, which for the considered period of time exceeds the allowable one according to the operating conditions of the lining, °C;

$t_{calc}$ – temperature value which is allowable under the operating conditions of the lining, °C;

$\tau_1$ – the length of time during which the temperature value exceeded the one allowed by the operating conditions of the lining, h;

$\tau_2$ – duration of time during which the tension value did not exceed the allowable one under the operating conditions of the lining, h;

$\tau_{total}$ – duration of temperature action on the lining, h.

When calculating the accounting factor for the excess of the temperature of the medium over the calculated value of $K_t$ and the duration of the contact of the medium with the temperature exceeding the calculated value, we obtain its value equal to 1,016.

By multiplying the obtained correction factors, we obtain the value of the total correction factor of 1,15. This value shows a 15 % reduction in mean time to failure (MTTF).

## 5. Discussion of the results

In general, the obtained result reflects a picture of the actual operation of the sintering furnace between its repairs. Exceeding the ultimate strength of the lining material is a determining factor in the operation of rotary kilns. To a lesser extent, the lining resistance is affected by the temperature rise of the lining. This corresponds to the calculations performed: the value of the correction factor for stresses (compression and extension) is 1.133, while the value of the correction factor for temperature is 1.016

For the implementation of the method into real operation, it is necessary to combine a number of measuring processes into a single information complex: a continuous measurement of the lining temperatures during heating; a burner performance (from the position of the torch profile and fuel oil spills); the quality of the resulting sintering products, etc.

In this case, the main problem is to obtain quality data on the operation of the furnace. These data can be obtained by any available and technologically safe method and used to calculate the residual resource criteria. The method can be translated into a computer programming language and, when temperature sensors are connected to a computer, can be used at the workplaces of operators of high-temperature units.

## 6. Conclusions

A method for determining the residual life of high-temperature equipment has been developed, which makes it possible to calculate the duration of the unit's trouble-free operation during the operation of the equipment. The method is distinguished by simplicity and reliability in calculating temperature stresses and determining the criterion of the residual resource. The use of this method makes it possible to move from a deterministic forecast of the residual resource to its probabilistic assessment, which is the background information for making motivated decisions to improve the reliability and safety of the equipment operation.

Further use of the developed technique will consist in creating a computational complex for assessing the wear rate and residual resource of a number of high-temperature units: arc and rotary furnaces, casting ladles, etc

## References

[1] Lucko G 2003 *A statistical analysis and model of the residual value of different types of heavy construction equipment* PhD thesis (Civil Engineering) Virginia State University and Polytechnic

[2] Silva S N, Vernilli F, Justus S M, Marques O R, Mazine A, Baldo J B, Longo E, Varela J A 2005 *Ironmaking & Steelmaking* **32**:6, pp 459-467

[3] Parra R, Verdeja L F, Barbes M F, Goni C 2006 Furnace Lining Analysis and Design by Mathematical and Physicochemical Modeling *Conference: Sohn International Symposium / (Advanced Processing of Metals and Materials: Principles, Technologies and Industrial Practice)* At: San Diego, California, (EEUU), pp 561-576

[4] Frank C B 2015 *HAZOP: Guide to Best Practice* (Elsevier), p 168
[5] Sijacki Z V 2015 RCM in power plant practice illustrated on observation of material aging and defining of component life exhaustion *International Conference Power-GEN Middle East* / (Belgrade: University of Belgrade), pp 11-12
[6] Vengrinovich V L, Vintov D A and Dmitrovich D V 2014 *AIP Conference Proceedings,* 1581 1222
[7] Kazakov V S and Alekseev A S 2015 *Bulletin of the Bryansk State Technical University* **2** (46) p 32
[8] Nikiforov A S and Prikhod'ko E V 2005 *Refract And Ceram* **46**, pp 360-363
[9] Nikiforov A S, Prikhod'ko E V, Kinzhibekova A K, Kucher E O and Niconov G N 2015 *Method for determining thermal liner state of high-temperature unit* KZ 30372-A4 G01K-013/00

**Acknowledgments**
This research has been funded by the Science Committee of the Ministry of Education and Science of the Republic of Kazakhstan (Grant No. AP08855901).

# Modeling the influence of the characteristics of renewable organic materials on the energy performance of the boiler

**A Nikiforov[1], E Prikhodko[1], A Kinzhibekova[1] and Sh Nurkina[1]**

[1]Thermal power engineering Department, Toraighyrov University, Lomov st. 64, Pavlodar, Republic of Kazakhstan

E-mail: akmaral70@mail.ru; sholpan_shupeeva@mail.ru

**Abstract.** The object of the modeling is the process of burning organic waste, depending on the nature of the origin of the fuel, its chemical and fractional composition, density and lower calorific value. Processing the results of experimental study using analysis of variance will allow to estimate the variance caused by each factor individually and determine the level of influence of the factors on the output parameters of the process. It is necessary to obtain not only the form of this dependence, relying on the experimental data, but also to assess the influence of each factor on the lower calorific value. The analysis of the obtained dependence shows that the airflow speed and the fractional composition of this fuel have the greatest influence on the efficiency of the boiler plant. Checking the significance of the coefficients showed that all the coefficients of this model are significant. The regression dependencies were obtained that estimate the lower calorific value and the efficiency of the boiler which burns briquettes from organic matter; the analysis of the obtained dependencies was carried out, the significance of the coefficients included in the model was checked; the composition factor has the greatest influence on the lower calorific value.

## 1. Introduction

Renewable organic materials are beginning to be widely used as fuel in the global energy sector. Organic substances used as fuel can be found in abundance at a short distance from the planned heat generation plant. At the same time, one of the main issues is the technology and mechanism for obtaining thermal energy with maximum efficiency.

Thus, in the works [1-3], the combustion characteristics of fuel briquettes made from organic waste were investigated. The work [4] shows the influence of the composition of fuel briquettes on their thermophysical and mechanical properties. The influence of the composition on the properties of briquettes is assessed by carrying out measurements on the obtained briquettes

Information on the properties of the obtained briquettes is important not only for further improving their properties (increasing strength, calorific value, etc.), but also for choosing boilers where the briquettes will be burned most efficiently, as well as for designing new boilers for burning fuel briquettes of this composition [5].

The analysis of the operation of boilers on briquettes from organic waste shows that their combustion is similar to the fuel recommended by the manufacturer [6]. Moreover, experimental measurements have shown an overall increase in thermodynamic efficiency from 86,73 % to 94,08 % and increase in maximum combustion efficiency from 91 % to 96,2 %, while emissions of CO and $NO_x$ have been significantly reduced.

In addition, an important issue that is reflected in the literature is the study of the briquette combustion process. The rate of the chemical oxidation reaction mainly depends on the oxygen concentration, which, in turn, depends on the density of the briquette. Thus, the properties of the briquette affect the rate and degree of burnup, which is important for the combustion of a certain type of briquettes in various boilers [7].

Numerical modeling of the processes in the boiler is carried out with the help of software (for example, FLUENT). Using software, three-dimensional dynamic models of combustion in the boiler are obtained. Based on the modeling results, the influence of various factors on the efficiency of the boiler operation is determined [8, 9].

Processing the results of the experimental study of the systems using analysis of variance allows for a complex multifactorial process to evaluate the variances caused by each factor separately, to determine the level of influence of the studied technological factors on the output parameters of the process [10].

Analysis of variance makes it possible to answer the question whether the factor under study influences or does not influence the parameters of the product, but it does not allow determining either the degree or the nature of this influence. Correlation and regression analysis can be used to resolve these latter issues.

Experimental planning methods can be widely used in any experimental research, if it is possible to actively influence the experimental conditions, to change the input parameters at will. In this case, the task is to increase the efficiency of researches, i.e. to obtain, with the minimum possible number of experiments, the most information about the process under study in order to describe it with the greatest accuracy.

## 2. Formulation of the problem

The process of burning fuel briquettes from organic waste depends on many factors, namely: on the origin of the fuel, their chemical and fractional composition, density, lower calorific value, etc. All of these factors are closely interconnected, and in this regard, some difficulties arise in modeling the combustion of renewable organic materials.

## 3. Theory

The considered aspects of incineration processes of renewable waste fit well into the general strategy of systemic analysis as applied to sufficiently large systems. In this case, mathematical modeling is selected as a method for studying the object, and its main principle is decomposition, i.e. decomposition of a complex (large) system into simpler ones that interact with each other and within the framework of a general model. Despite all the complexity of the proposed way, it seems natural and correct, since it is part of a systematic approach to considering the behavior of large systems [10].

An objective indicator of fuel mass is its lower calorific value, which is directly related to such characteristics as density and strength.

The dependence on these factors can be considered as random and analyzed using probabilistic methods. For effective analysis, it is necessary to identify the relationships between the factors and present them in quantitative form - in the form of a mathematical model:

$$\sigma = \sigma(x_1, x_2, \ldots, x_n), \tag{1}$$

where $\sigma$ – this is the lower calorific value, kJ/kg;

$x_1, x_2, \ldots, x_n$ – factors influencing the result.

It is necessary to obtain not only the form of this dependence, based on experimental data, but also to assess the influence of each factor on the lower calorific value.

Based on the analysis of the assessment of the influence of various factors on the lower calorific value, we choose:

$x_1$ – the ash content, %;
$x_2$ – the density, kg/m$^3$;
$x_3$ – the ratio of the constituents in the briquetting mass, % / %;
$x_4$ – the relative humidity, %.

From physical considerations, we can assume that the relationship of the expression (1) has the following form:

$$\sigma(b,x) = b_0 + \sum_{i=1}^{n} b_i x_i + \sum_{i=1}^{n}\sum_{k=i+1}^{n} b_{ik}x_i x_k + \sum_{i=1}^{n}\sum_{k=i+1}^{n}\sum_{l=k+1}^{n} b_{ikl}x_i x_k x_l + \dots + \\ + b_{12\ldots n}x_1 x_2,,, x_n \quad (2)$$

where $b_0$– a free term equal to the output at x=0;

$b_i$ – a regression coefficient indicating the influence of the factor on the process;

$b_{ij}$ – regression coefficients that determine the degree of the influence on the interaction process of the factors.

Having determined the regression coefficients of this equation, we will get an idea of the influence of the studied factors on the lower calorific value of the fuel. The regression coefficients are calculated using the following formulas [10]:

$$b_i = \frac{1}{N}\sum_{j=1}^{N} \overline{\sigma}_j\, x_i^j, i = 0,..,n, b_{ij} = \frac{1}{N}\sum_{j=1}^{N} \overline{\sigma}_j\, x_i^j \ldots x_\mu^j, i, \mu = 1,2,..,n, \quad (3)$$

where $\overline{\sigma}_j$ – the value of the process mean output in j-variant;

$x_i^j$ – the factor value in j-variant.

At each experimental point we carry out two experiments. To calculate the estimates of the coefficients, we will use the arithmetic mean of these observations for each point $\tilde{y}^i = \frac{\overline{\sigma}^{i1} + \overline{\sigma}^{i2}}{2}$. Further, based on the experimental data, we calculate the estimates of the coefficients by the formulas (3) and obtain the form of the interpolation formula. The last step is to assess the significance of the regression coefficients and check the adequacy of the model.

To assess the significance of the regression coefficients, it is necessary to find their sample variance $S^2[b_i]$. The significance of the regression coefficients is determined by the inequality

$$b_i \succ S[b_i] \cdot t_{cr}(f), \quad (4)$$

where $t_{cr}(f)$ – the Student's coefficient for a given confidence p and the number of degrees of freedom $f = (\nu - 1) \cdot N$.

Thus, the mathematical model of this process can be described quite accurately using the interpolation formula (2), taking into account the assessment of the influence of each of the experimental factors on the target value.

Based on the order of the estimation of the coefficients when using this design, it is possible to draw up an experiment design and obtain an experiment design matrix.

Another equally important task is researches related to the efficiency of the combustion of fuels from renewable organic waste in boiler furnaces.

An objective indicator of the efficiency of using this fuel in the furnaces of boilers of low and medium power is its efficiency (i.e. coefficient of performance), which is directly related to such characteristics as the excess air coefficient, layer height and fractional size.

At the same time, a number of factors influence the final value of the boiler plant efficiency (its coefficient of performance). The dependence of the efficiency on these factors can be considered as random and analyzed using probabilistic methods. For effective analysis, it is necessary to identify the relationships between the factors and present them in quantitative form - in the form of a mathematical model

$$\eta = \eta(x_1, x_2, \ldots, x_n), \quad (5)$$

where η – this is the efficiency of the boiler plant, %;
$x_1, x_2, \ldots, x_n$ – the factors that influence the result.

Based on the analysis of the assessment of the influence of various factors on the efficiency, presented earlier, we choose the following as the experimental factors:
$x_1$ – the size of fractions of the fuel briquettes, mm;
$x_2$ – the excess air ratio in the combustion chamber;
$x_3$ – the airflow speed, m/s;

$x_4$ – the height of the fuel layer on the fire-grate, mm.

It is necessary to obtain not only the form of this dependence, based on the experimental data, but also to assess the influence of each factor on the value of the efficiency. From physical considerations, we can assume that relationship (5) has the form of an equation (2).

Having determined the regression coefficients of this equation, we will get an idea of the influence of the studied factors on the value of the efficiency. The calculation of the regression coefficients is carried out according to the formulas (3). The significance of the regression coefficients is determined by the inequality (4).

Let us find an interpolation formula that could be used to describe the dependence of the efficiency on the four factors listed above (without taking into account the pair interactions)

$$\tilde{\eta}(b,x) = b_0 + b_1x_1 + b_2x_2 + b_3x_3 + b_4x_4. \tag{6}$$

Using the above methodology, we will determine the regression coefficients of the equation to study the influence of the determining factors on the efficiency of the furnace.

## 4. Experimental results

Table 1 demonstrates the design of experiment to study the influence of various factors on the lower calorific value of renewable organic materials and its results. The values of the process outputs were obtained as a result of the experimental and calculated studies.

**Table 1.** The design of experiment №1.

| Planning | | | | | The process output |
|---|---|---|---|---|---|
| $x_0$ | $x_1$ | $x_2$ | $x_3$ | $x_4$ | Lower calorific value, kJ/kg |
| + | + | + | + | + | 18702.4 |
| + | - | + | + | - | 18924.3 |
| + | + | - | + | - | 18681.5 |
| + | - | - | + | + | 18694.1 |
| + | + | + | - | - | 18631.2 |
| + | - | + | - | - | 18652.2 |
| + | + | - | - | - | 18639.6 |
| + | - | - | - | + | 18631.3 |

The interpolation formula for the dependence of the lower calorific value on the process factors is as follows:

$$\hat{y} = 18708.40 - 30.90x_1 + 32.99x_2 + 56.06x_3 - 29.85x_4. \tag{7}$$

Table 2 shows the design of experiment to study the influence of various factors on the efficiency of the boiler when burning renewable organic materials and its results. The values of the process outputs were obtained as a result of the experimental and calculated studies.

**Table 2.** The design of experiment №2.

| Planning | | | | | The output |
|---|---|---|---|---|---|
| $x_0$ | $x_1$ | $x_2$ | $x_3$ | $x_4$ | η, % |
| + | + | + | + | + | 82.0 |
| + | - | + | + | - | 83.0 |
| + | + | - | + | - | 85.0 |
| + | - | - | + | + | 81.0 |
| + | + | + | - | - | 81.0 |
| + | - | + | - | + | 82.0 |
| + | + | - | - | + | 82.5 |
| + | - | - | - | - | 80.0 |

Thus, the interpolation formula for the dependence of the boiler efficiency on the process factors has the following form:

$$\tilde{\eta} = 82.06 + 0.56x_1 - 0.06x_2 + 0.69x_3 - 0.19x_4. \quad (8)$$

**5. Discussion of the results**

The analysis of the obtained dependence shows that the airflow speed and the fractional composition of the fuel have the greatest influence on the efficiency of the boiler plant.

The factor $x_3$ (composition, that is, the ratio of combustible components in the initial fuel mass) has the greatest influence on the lower calorific value (of all the values under consideration). Moisture, ash and density have almost the same effect on the lower calorific value but to a lesser extent.

Let us check the adequacy of the model and the significance of the quadratic effects in both experiments.

In the first case N = 8, ν = 2. Let us calculate the values of the output variable in points and determine the vector $\hat{Y}$. To check the adequacy of the resulting model, we calculate the estimate of the variances of observation errors, which we calculate using the sum of squares of errors according to

$$S_l = \sum_{i=1}^{8}\sum_{j=1}^{2}\left(\overline{\sigma}^{ij} - \overline{\sigma}^{i}\right)^2 = 13.18 \quad (9)$$

with the number of degrees of freedom $\varphi_2 = N(\nu - 1) = 8(2-1) = 8$ according to the formula

$$S^2 = S_l \cdot (\nu \cdot \varphi_2\varphi)^{-1} = 0.82. \quad (10)$$

$$S = 0.91. \quad (11)$$

Let us find the sum of squares characterizing the inadequacy of the model

$$S_D = \sum_{i=1}^{N} \nu \cdot \left(\overline{\sigma}^{i} - \hat{\sigma}^{i}\right)^2 = 1.65. \quad (12)$$

Hence follows the formula

$$F = (S_D/\varphi_1) \cdot (S_e/\varphi_2)^{-1} = (1.65/1) \cdot (13.18/8)^{-1} = 1.00 \quad (13)$$

In the table of 95 % Fisher–Snedecor distribution [11] we find the value $F_{cr} = 5.32$ for $1\text{-}P = \alpha = 0.05$. Here α is a given level of significance for testing the adequacy hypothesis. In our example $F = 1.001 \prec F_{cr} = 5.32$.

Thus, the obtained process model is adequate.

Now let us check the significance of the found regression coefficients. To do this, we determine the Student's coefficient at a given level of significance α = 0.05 and with eight degrees of freedom $t_{cr}(f) = 2.306$.

For the variances of the estimates of the coefficients, we obtain

$$S_i^2 = S^2 \cdot N^{-1} = 0.103; \; S_i = 0.321. \quad (14)$$

Here with

$$s_i \cdot t_{cr}(f) = 0.321 \cdot 2.306 = 0.740. \quad (15)$$

Checking the significance of the coefficients showed that all the coefficients of this model are significant. Therefore, the regression equation, taking into account this remark, will remain the same.

Let us check the adequacy of the model and the significance of the quadratic effects in the second experiment. In this case N = 8, ν = 2. Let us calculate the values of the output variable in points and determine the vector $\hat{Y}$. To check the adequacy of the resulting model, we calculate the estimate of the variances of observation errors, which we calculate using the sum of squares of errors according to

$$S_l = \sum_{i=1}^{8}\sum_{j=1}^{2}\left(\overline{\sigma}^{ij} - \overline{\sigma}^{i}\right)^2 = 56 \quad (16)$$

with the number of degrees of freedom $\varphi_2 = N(\nu - 1) = 8(2 - 1) = 8$ according to the formula

$$S^2 = S_1 \cdot (v \cdot \varphi_2\varphi)^{-1} = 3.5; \ S = 1.871. \quad (17)$$

Let us find the sum of squares characterizing the inadequacy of the model

$$S_D = \sum_{i=1}^{N} \nu \cdot \left(\overline{\sigma}^i - \hat{\sigma}^i\right)^2 = 15.6 \quad (18)$$

Hence follows the formula

$$F = (S_D/\varphi_1) \cdot (S_e/\varphi_2)^{-1} = (15.6/1) \cdot (56/8)^{-1} = 2.23 \quad (19)$$

In the table of 95 % Fisher–Snedecor distribution [11] we find the value $F_{cr} = 5.32$ for 1-P = α = 0.05. Here α is a given level of significance for testing the adequacy hypothesis. In our example $F = 2.23 < F_{cr} = 5.32$.

Thus, the obtained process model is adequate.

Now let us check the significance of the found regression coefficients. To do this, we determine the Student's coefficient at a given level of significance α = 0.05 and with eight degrees of freedom $t_{cr}(f) = 2.306$ [11].

For the variances of the estimates of the coefficients, we obtain

$$S_i^2 = S^2 \cdot N^{-1} = 0.4375; \quad S_i = 0.661. \quad (20)$$

Here with

$$s_i \cdot t_{cr}(f) = 0.61 \cdot 2.306 = 1.525 \quad (21)$$

Checking the significance of the coefficients showed that in the second experiment all the coefficients of this model are also significant. Therefore, the regression equation, taking into account this remark, will remain the same.

## 6. Conclusions

Thus, the analysis of the modeling results allows us to draw the following conclusions:

- the regression dependencies were obtained that assess the lower calorific value and the efficiency of the boiler which burns briquettes from organic matter;
- the analysis of the obtained dependencies was carried out, the significance of the coefficients included in the model was checked;
- factor $x_3$ (composition, i.e. the ratio of combustible components in the initial mass) has the greatest influence on the lower calorific value (from the considered values). Moisture, ash and density have almost the same effect on the lower calorific value but to a lesser extent;
- the airflow speed and the fractional composition of briquettes have the greatest influence on the efficiency of the boiler plant;
- the adequacy of this model was verified, which showed that the proposed models are adequate, because the errors characterizing the accuracy of the model do not exceed the errors of observations.

## References

[1] Ajimotokan H, Ehindero A, Ajao K, Adeleke A, Ikubanni P and Shuaib-Babata Y 2019 *Scientific African* **6** e00202 https://doi.org/10.1016/j.sciaf.2019.e00202.

[2] Wongwuttanasatian T and Sakkampang C 2016 Combustion characteristics and emission of briquette fuel from biomass mixed with glycerin *Combustion Science and Technology* **188** (6) pp 1011-1019

[3] Talukdar A, Das D and Saikia M 2014 Study of combustion characteristics of fuel briquettes *International Journal of Computational Engineering Research* **4** (3) pp 2250-3005

[4] Nikiforov A, Prikhodko E, Kinzhibekova A and Nurkina Sh 2020 Study of strength characteristics of fuel briquettes from organic waste *AIP Conference Proceedings* **2212**, 020044 https://doi.org/10.1063/5.0000951

[5] Lisý M, Lisá H, Jecha D, Baláš M, Križan P 2020 *Energies 13*, 1448

[6] Răzvan M, Ion V I, Florin P 2013 Testing of improved boiler for biomass briquettes, *Symposium Actual Tasks on Agricultural Engineering*, Opatija, Croatia pp 336-342

[7] Shuguang L, Jia Jiang, Xiaolong Ch, Bowen X, Min Deng. 2020 *IOP Conference Series: Earth and Environmental Science* **467** (1) 012153

[8] Chen R, Rao Y, Yue H, Ai Y, Zhang X, Zhang L and Zheng J 2020 Numerical simulation of bed combustion in biomass-briquette boiler *Journal of Energy Engineering* **146** (4) https://doi.org/10.1061/(ASCE)EY.1943-7897.0000653

[9] 'Aliman L and Pasek A 2018 The evaluation of experimental and numerical study of combustion process on mini traveling chain grate furnace (incinerator) by computational fluid dynamics method *AIP Conference Proceedings* **1984**, 030005 https://doi.org/10.1063/1.5046626

[10] Hartmann K, Letzki E and Schäfer W 1977 *Planning an experiment in the study of technological processes* (Moscow: Mir). p 552

[11] Schenck H 1968 *Theories of Engineering Experimentation* (Mcgraw-Hill Book Co) p 284

# Modeling the basis power in the systems for continuous transport of fluids and bulk materials

**O Dinolov[1] and K Ilieva[2]**

[1] Department of Electrical Power Engineering, "Angel Kanchev" University of Ruse, Ruse, Bulgaria
[2] Faculty of Electrical Engineering, Electronics and Automation, "Angel Kanchev" University of Ruse, Ruse, Bulgaria

E-mail: odinolov@uni-ruse.bg

**Abstract.** This paper presents a novel developed model for calculating the totally useful power in the systems for continuous transport of fluids and bulk cargoes, allowing deeper and more functional analysis of the energy efficiency of this systems by physically based decomposition of the electricity consumption of a useful and unhelpful ingredient. The model consists of nine mathematical expressions by which, after determining the type of system, the physically required minimum possible power consumption can be calculated and compared with the total consumption. A pumping system for the transport of condensate water from a deaerator to the drum of a steam generator of a thermal power plant has been selected as the subject of the model demonstration experiment. In the investigation carried out, the basis power of the pumping system was 606.6 kW and the total losses were about 302 kW. Other energy efficiency indices have also been identified. The researches should be continued in the direction of algorithmization of the model and its justification mainly for the Type II systems.

## 1. Introduction

Globally, there are efforts to increase the efficiency of the energy consumption, which predetermines the need for a methodology that accurately determine the useful part of the consumed energy, based on the fact that the rapid development of the standards for more efficient equipment and components are pushing manufacturers towards further improvements on a component level [1]. In the citied paper, more than 30 energy efficiency measures were identified as it promotes a model to generate improvement suggestions, which can be compared with the strategies for reducing the industrial power consumption [2].

The importance of the energy efficiency of the industrial systems is obvious and related to the economic factors [3]. However, with the increase of the requirements for electrical efficiency, rise the need for more accurate measurement instruments, which must be sold [4]. That is why new approaches are required.

The systems for continuous transportation of fluids and bulk materials (SCTFB) have a large distribution without being targeted in a specific industry. Currently, there are several models to determine the useful part of the energy consumed. In many cases, the efficient part is assessed through the no-load condition of the system, which leads to inaccuracies in the evaluation. Other approaches are focused of the energy performance indicators, which presents a ratio between energy spent per unit of product, area, volume, or other quantity directly related to production [5].

Some papers examine the efficiency of machines with water-lifting applications [6, 7], which can be considered as subsystems of those described in this article and a comparison can be made to examinate how much their efficiency can be improved. The topic is addressed in other works, such as the Journal of Cleaner Production article, which provides an overview of methods for developing better corrective measures for energy saving [8] or how losses in metallurgy can be reduced [9] or even how the estimated maintenance of facilities supports energy losses [10]. The energy efficiency not only in the cited, as well as other energy efficiency studies, is judged by the total power consumed by the system.

The aim of this study is to offer a model of the basis (fully useful) power in the SCTFB, allowing deeper and more functional analysis of the energy efficiency of this systems by physically based decomposition of the electricity consumption of a useful and unuseful component.

A pumping system for the transport of condensate water from a deaerator to the drum of a steam generator of a thermal power plant has been selected as the subject of the model demonstration experiment.

## 2. Theoretical model

### *2.1. Classification of the systems*

The SCTFB are found in a number of industrial facilities such as thermal power plants, warehouses, ports, chemical plants, mines, quarries, water supply, irrigation and other flow systems.

This paper proposes a new classification of the systems under consideration. The systems are divided into those of the first and second type (Type I and Type II). After determining the type, the system can be categorized according to the transported load and denivelation. According to the cargo that is carried, the systems are for the transport of fluids – gases and liquids, and for transport of bulk materials. According to the denivelation, rising, horizontal and descending systems can be distinguished. Type I systems are those where the cargo-transport parameters (for bulk cargo – the speed, and for fluids – the pressure) are irrelevant and Type II are required for a certain load speed or pressure (figure 1).

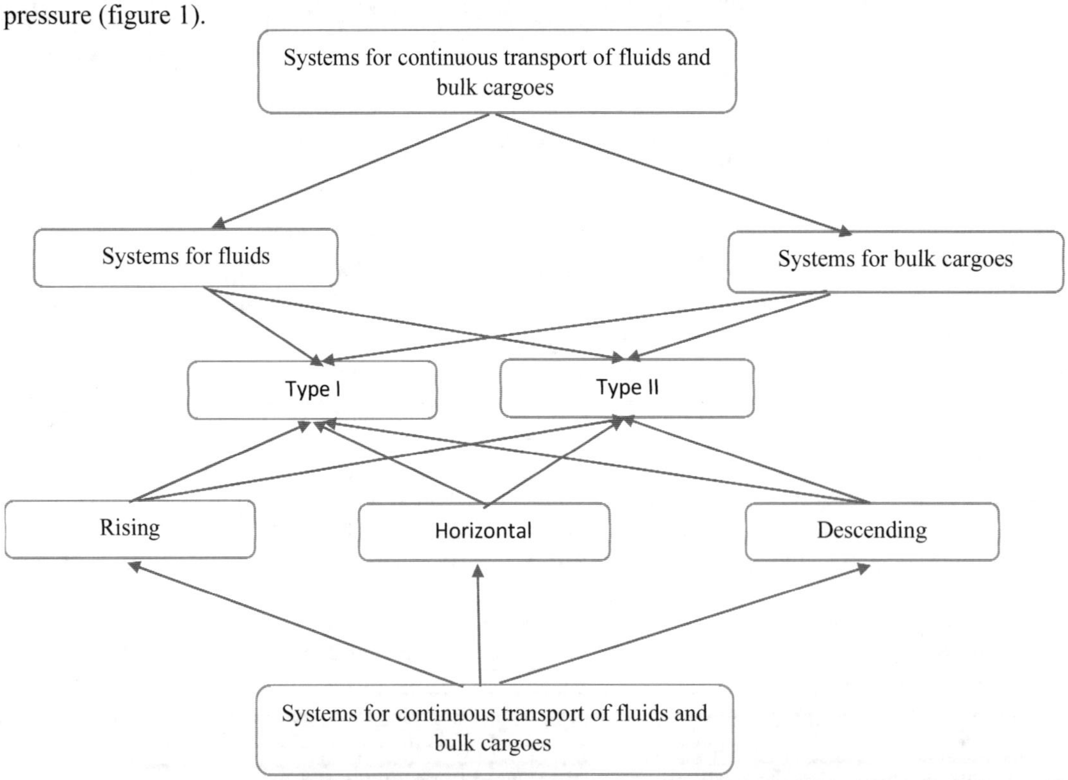


**Figure 1.** Parametric-based classification of SCTFB.

*2.2. Mathematical models*

In the case of continuous bulk transport systems, the basis power is determined by the gravitational constant and the variable parameters mass m, the difference in denivelation $\Delta h$, the time $t$. In the case of Type II systems, the desired load speed v at the output of the flow line is also considered. For Type I systems, the summary formula may be presented:

$$P_b = \frac{m.g.\Delta h}{t}, \tag{1}$$

and for those of Type II:

$$P_b = \frac{m.g.\Delta h}{t} + \frac{m.v^2}{2t} = \frac{m}{t}\left(g.\Delta h + \frac{v^2}{2}\right), \tag{2}$$

where m is the mass of the load, g – the gravitational constant, t – time of transportation, Δh - the difference between the altitude at the starting and the end point of the system and v – the desired load speed at the end point.

In the case of continuous fluid transport systems, the basis power shall be determined by the first according to the fluid type - liquid or gas. When we replace in equation (2):

$$Q = \frac{V}{t} = \frac{m}{\rho.t} \Rightarrow m = Q.\rho.t, \tag{3}$$

where V is the fluid volume, Q is the fluid flow rate, and ρ is the density of the load, the basis power of Type I fluid systems can be obtained:

$$P_b = \frac{Q.\rho.t.g.\Delta h}{t} = Q.\rho.g.\Delta h, \tag{4}$$

and for those of the Type II:

$$P_b = Q.\rho.t.g.\Delta h + Q.\Delta p_T = Q\left(\rho.t.g.\Delta h + \Delta p_T\right). \tag{5}$$

In the case of gas transport systems, the difference in pressure $\Delta p_T$ of the starting and the ending point of the system, as well as the load flow rate Q is essential:

$$P_b = Q.\Delta p_T. \tag{6}$$

According to the denivelation, the systems in question are horizontal, rising and descending. In the case of horizontal systems of Type I, the basis power is zero, since starting from equation (1) at $\Delta h = 0$, $P_b = 0$. In the case of Type II for bulk materials:

$$P_b = \frac{m.v^2}{2t}, \tag{7}$$

where v is the desired load speed at the end point of transportation, which is also a determining factor. For the fluids, the desired output pressure and flow rate are critical and equation (6) applies.

In the case of the ascending systems, the general formulae set out above apply, as there is no case where $\Delta h = 0$, $P_b = 0$ for the Type I systems, and the determining factor is the difference Δh in the height of the starting and ending point. For Type II systems, in addition to the difference in denivelation Δh, the desired initial parameters are also relevant. In this case for Type I systems, equation (1) applies for bulk cargoes, and equation (2) - to those of Type II. Equations (4) and (6) are in force for Type I lifting systems for liquids and fluids, and for Type II - equations (5) and (6), respectively.

Resulting from the gravitation, the basis power is zero in the case of descending systems of Type I. In the case of Type II systems, it is essential whether the necessary parameters are reached as a result of the mass of the load and the difference in the height of the starting and ending points of the system.

In the case of Type II bulk materials systems, based on the fact that the speed caused by the gravitation is:

$$v_g = \sqrt{2.g.\Delta h}\ , \tag{8}$$

if the desired output speed of the load is equal to that obtained from the physical parameters of the system, then the basis power is equal to zero ($P_b$=0). Otherwise, Δv is defined as a difference between the desired speed v and the $v_g$ obtained from the gravitational forces

$$\Delta v = v - v_g\ , \tag{9}$$

as the basis power is:

$$P_b = \frac{m.\Delta v^2}{2t}\ . \tag{10}$$

Similarly for lowering systems in the transport of fluids and gases, if the desired pressure is equal to that obtained by the physical parameters of the system, the basis power is equal to zero or in the other case we define $\Delta p = p - p_g$, as a difference between the desired pressure p and the pressure $p_g$ obtained by the gravity. The basis power is:

$$P_b = \Delta p.Q\ . \tag{11}$$

## 3. Case study

### *3.1. Object of investigation*

A characteristic pumping system for transporting condensate water from the deaerator to the drum of a steam generator in a thermal power plant has been selected as a case-study object (figure 2).

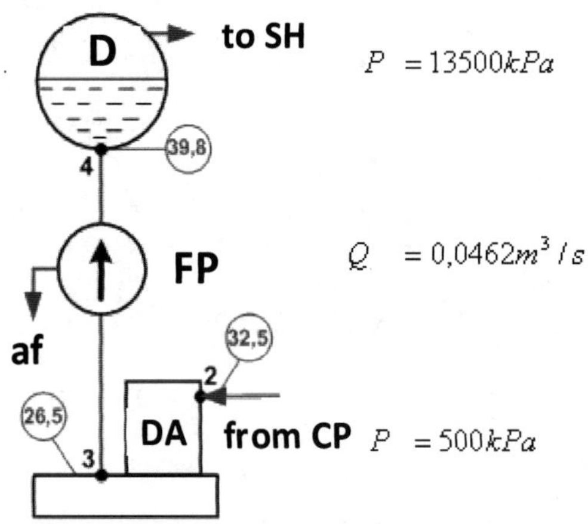


**Figure 2.** Diagram of a feeding pump system in a steam power plant: D – drum; SH – superheater; FP – feeding pump; af – auxiliary feeding; DA – deaerator; CP – condensate pump.

According to the proposed classification, the investigation object is a lifting system for continuous transport of fluids of Type II. The system consists of a deaerator, a power pump, after which two loops are distinguished – to the drum and to an auxiliary feeding, serving to provide different auxiliary processes from the water tract of the plant. A series of connected water heaters have significant influence on the hydraulic resistance of the track. On the other hand, the hydraulic resistance of the contour is dictated by the passage through the heated high pressure and the eco-oiler of the steam generator. The differences in the geodetic heights at which the water rises are determined by the elevations. The power pump is powered by an induction motor with a rated voltage of 6000 V. The rest parameters of the motor are: type 32V-188-02H; nominal power 2000 kW; nominal current 227 A; rotor speed 2975 $min^{-1}$; power factor 0.88; efficiency 96 %.

*3.2. Results*

The results are obtained using the methods described in [11]. The investigation methodology is complemented by the mathematical model in section 2.2.

The pump flow rate is taken from the monitoring system of the plant. The data is for a characteristic steam generator operation for a period of 1 hour and 7 minutes, with a sampling interval of 1 minute. The flow rate is almost constant. The average flow rate was 0.0461 $m^3/s$. The other parameters in the observations carried out are $\Delta p = 1.3.10^7$ Pa; $\Delta h = 13.3$ m, $\rho = 1000$ $kg/m^3$. Using these output data, the pump's basis power equals to 606.6 kW. The study also determined that the average measured power is 908.7 kW. Thus, the level of total losses can be easily specified as the difference between the power consumed and the basis power. In this case, the total losses are about 302 kW. The rest energy efficiency indices are presented in table 1.

**Table 1.** Energy-efficiency indices for a feeding pump system in a steam power plant.

| | | | | | | |
|---|---|---|---|---|---|---|
| A1 | A2 | A3 | A4 | A5 | A6 | B1 |
| kW | kVAr | min | kWh | kWh | - | kW |
| 1666.65 | 944.53 | 47.7 | 144.1 | 792.61 | 1.222 | 908.7 |
| B2 | B3 | B4 | B5 | B6 | B7 | C2 |
| kVAr | min. | kWh | kWh | kWh | - | kW |
| 501.3 | 67 | 649.06 | 323.25 | 972.31 | 1.498 | 302.1 |
| D1 | D2 | D3 | D4 | E1 | E2 | E3 |
| – | – | min. | kWh | % | % | % |
| 1.23 | 2.24 | 19.3 | 179.7 | 0 | 0.4 | 96 |
| E4 | E5 | E6 | F1 | F2 | F3 | F4 |
| - | - | - | kW | - | kVAr | - |
| 0 | 0.01 | 0.87 | 12.26 | 0.014 | 12.04 | 0.024 |

The indices used in table 1 are as it follows: A1 - active power in optimum operation; A2 - reactive power in optimum operation; A3 - optimum operating time; A4 - non-useful electric power consumption in optimum operation; A5 - total electric power consumption in optimum operation; A6 - relative electric power consumption in optimum operation; B1 - average active power in actual operation; B2 - average reactive power in actual operation; B3 - actual operating time; B4 - useful electric power consumption; B5 - non-useful electric power consumption in actual operation; B6 - total electric power consumption in actual operation; B7 - relative electric power consumption in actual operation; C2 - total power loss; D1 - coefficient of overrun of total energy consumed; D2 - coefficient of overrun of non-useful energy consumed; D3 - overrun of operating time; D4 - overrun of electric energy; E1 - deviation from the maximum electric motor efficiency in optimum operation; E2 - deviation from the maximum electric motor efficiency in actual operation; E3 - maximum electric motor efficiency; E4 - deviation of the power factor from the desired value in optimum operation; E5 - deviation of the power factor from the desired value in actual operation; E6 - desired power factor; F1 - standard deviation of the active power in actual operation; F2 - coefficient of variation of the active power in actual operation; F3 - standard deviation of the reactive power in actual operation; F4 - coefficient of variation of the reactive power in actual operation [11].

The primary energy-efficiency indices are B7, D1 and D2. They give information for essential analysis. B7 is always higher than 1 and it is equal to 1 for an ideal transport system. In this case, the index is equal to 1.498 relative units. This means that per each 1 kWh energy consumed for making useful work, the pump system consumes additional 0.498 kWh for covering the losses. On another hand, if D2 is higher than D1, the energy efficiency can be best increased by improving the working condition instead of making constructional changes. And vice versa.

From the value of D1 equal to 1.23 and the relatively small value of A6, a satisfactory structural perfection of the pump is shown. The values for actual mode of operation are statistically reliable. This is confirmed by the low coefficients of variation of about one and two percent.

Based on the coefficient D2 and on the difference between A6 and B7, which is 0.3 relative units, it can be concluded that there is an opportunity to improve the operating conditions of the pump by additional loading until it reaches the optimal level. This improvement can lead to savings of nearly 180 kWh of electricity (index D4). Given the continuous pump operation, the economy for a month reaches EUR 6 500 in average production price of EUR 50/MWh.

## 4. Discussion

This paper proposes a new way to determine the theoretically necessary, minimum possible electricity consumption by a basis power modelling method. In the case of the systems for continuous transport, the basis power modelling is an appropriate approach for calculating the minimum energy needed to operate and improve them. The developed method can also be applied to the construction of new systems with the best possible parameters, which would have both a greater environmental and economic effect.

The actual consumption is the total energy consumed and is determined by the power actually consumed from the network. The basis power level predetermines the amount of energy that has been spent solely to carry out a useful work. The difference between the actual electricity consumption and the energy determined by the basis power gives the total losses. The total losses level characterises the constructive perfection and the appropriate operating mode of the systems.

On the other hand, some pieces of research published at the time identified the efficiency of the systems as the ratio of the difference between the total energy consumed and the energy determined by the no-load power to the total electricity consumed. It should be considered that this approach does not take into account the variable losses component and the efficiency determined by the no-load level is not sufficiently reliable.

The method developed in this study allows to identify clearly and precisely this minimally needed energy and to compare it with the amount spent. This gains a better understanding of the total losses. The example with the pumping system above shows that approximately 1/3 of the energy is lost in the field. This makes clear that there are significant possibilities for optimizing the systems.

## 5. Conclusions

A new classification has been developed for bulk materials and fluid transport systems, through which the system type can be unambiguously determined.

An aggregate model including nine mathematical expressions is derived, by which, after determining the type of system, the required minimum possible power can be physically-based calculated and compared with the total consumption. The model can be used both in existing systems and when planning new ones.

The proposed methodology is demonstrated in a real working environment by setting the basis power for a rising system for continuous transport of fluids of Type II, part of a thermal power plant. On the basis of the proposed model, it is possible to plan optimizations of the systems environmentally and economically.

Future research covers: algorithmizing the model and developing relevant software products to computer-aided research; justification of the applicability of the model to the rising, horizontal and descending systems for the continuous transport of bulk materials of Type I and Type II, as well as to the Type I fluid systems.

## References

[1] Svensson A and Paramonova S 2017 An analytical model for identifying and addressing energy efficiency improvement opportunities in industrial production systems – Model development

and testing experiences from Sweden, *Journal of Cleaner Production* **142** (4) pp 2407-2422, https://doi.org/10.1016/j.jclepro.2016.11.034

[2] Grega G and Raab M 2004 *Tribology and Lubrication Technology* vol **60** issue 10 part 1 p 28-31

[3] Coroiu M and Chindris M 2014 Energy efficiency indicators and methodology for evaluation of energy performance and retained savings, *49th International Universities Power Engineering Conference (UPEC)*, Cluj-Napoca, pp. 1-6, doi: 10.1109/UPEC.2014.6934707

[4] Qun H and Wei S 2010 A Measurement Method about the Key Parameters of Nonlinear System Based on Generalized Predictive Control *International Conference on Challenges in Environmental Science and Computer Engineering*, Wuhan, China, pp 467-470, doi: 10.1109/CESCE.2010.188.

[5] Pitis C and Al-Chalabi Z 2016 Evaluation of energetic efficiency of the industrial systems by using benchmark energy factor, *IEEE Canadian Conference on Electrical and Computer Engineering (CCECE)*, Vancouver, BC, pp 1-5, doi: 10.1109/CCECE.2016.7726669.

[6] Asok Kumar A, Bindu GR, Cherian E and Parvathy M 2020 Energy saving and economic analysis of switched reluctance motor in agricultural applications. *Technol Econ Smart Grids Sustain Energy* **5** (2) https://doi.org/10.1007/s40866-019-0075-z

[7] Shankar A, Kalaiselvan V, Umashankar S, Paramasivam S and Hanigovszki N 2016 A comprehensive review on energy efficiency enhancement initiatives in centrifugal pumping system, *Applied Energy, Elsevier*, **181**(C), pp 495-513, doi: 10.1016/j.apenergy.2016.08.070

[8] Sihag N and Sangwan K 2020 A systematic literature review on machine tool energy consumption, *Journal of Cleaner Production* **275** 123125, https://doi.org/10.1016/j.jclepro.2020.123125

[9] Cheng Q, Gong J, Xiao G, Yang C, Liu Z and Qi B 2020 Research on energy-saving production planning of periodic forging resistance furnace, *Journal of Cleaner Production*, **275** 122897, https://doi.org/10.1016/j.jclepro.2020.122897.

[10] Abouelanouar B, Elkihel A, Khathyri F Gziri H 2021 Monitoring Energy Consumption Based on Predictive Maintenance Techniques. *Proceedings of the 2nd International Conference on Electronic Engineering and Renewable Energy Systems. ICEERE 2020. Lecture Notes in Electrical Engineering*, **681** Springer, Singapore. https://doi.org/10.1007/978-981-15-6259-4_6

[11] Andonov K, Dinolov O and Mihailov L 2012 Methodology for energy-efficiency investigation of electrical consumers, *Journal of Engineering Studies and Research* **18** pp 6-19

# Development of innovative methods of thermal waste processing in modular waste burners with a high degree capture of harmful emissions

**R K Orumbayev[1], A A Kibarin[1], I K Iliev[2], A A Torgayev[1] and M B Kumargazina[1]**

[1] Heat power units Department, Almaty University of Power Engineering and Telecommunications named after Gumarbek Daukeev, Almaty, 050013, Kazakhstan
[2] Heat, Hydraulics and Environmental Engineering Department, "Angel Kanchev" University of Ruse, Ruse, POB 7017, Bulgaria

E-mail: a.kybaryn@aues.kz

**Abstract.** Innovative methods have been developed for the thermal processing of solid waste in modular waste incineration plants with mechanical furnaces of boilers from 1.16 MW with gas afterburning over the layer and with a high degree of capture of harmful emissions in compact emulsifiers and activated carbon filters.

## 1. Introduction

More than 13 million m$^3$ of municipal solid waste (MSW) is accumulated in Kazakhstan annually, of which more than 2 million m$^3$ in the city of Almaty and its suburbs alone. The analysis carried out over the past 5 years indicates a tendency for an annual increase of solid waste at disposal facilities by an average of 10 - 12%. Solid waste management (SWM) is one of the main points in the concept of transition to a green economy [1]. In particular, the government has been tasked with increasing the share of solid waste processing to 40% by 2030.

Each year significant areas of land, often valuable farmland, are allocated to landfills for solid waste-as of 2019 the allotted land for landfills in the country exceeds 1200 hectares. At the same time more than 90% of landfills do not have a hydraulic protection system, and less than half of the landfills have a sanitary protection zone. In general, the analysis carried out in the regions of the Republic of Kazakhstan showed that the system for sorting and processing of secondary resources and solid waste works inefficiently. Technical support and preparatory work among the population for the collection of solid waste is low, so up to 50% of all annual paper resources and up to 20% of ferrous and non-ferrous metals and other valuable components go to landfills. The norms for solid waste accumulation from housing, public facilities, trade (wholesale markets) and cultural institutions, educational institutions and schools have been exceeded. A separate problem is the transportation of solid waste to waste transfer stations and beyond.

In most developed countries, environmental issues relating to the protection of the environment from garbage are high priority. Garbage is a source of various kinds of pollution and offers an environment for bacteria to thrive, resulting in diseases among the population. Meanwhile transportation using garbage trucks with large capacities, using railway platforms, river and sea barges, is developing. Sanitary, environmental, and techno-economic effective methods for the disposal of solid waste are also being developed [2, 3, 4].

To a greater extent utilization methods should be applied with the maximum use of all the useful properties of the waste by processing parts of solid waste into organic fertilizer and biofuel. To a lesser extent liquidation methods should be used that do not provide for the use of the beneficial properties of waste: landfill, dumping into mines, quarries, removal to the sea and burning of waste without using the produced heat [4].

According to the technological essence, the methods of garbage neutralization can be divided into the following types:

- biological - liquidation and disposal with the processing of solid waste by microorganisms [5,6];
- thermal - burning without and with using heat, pyrolysis to produce combustible gas [7-12, 21-24];
- chemical - hydrolysis [13];
- mechanical - pressing and manufacturing of various blocks using binders and additives [14, 15, 16].

## 2. Analysis of recent research, publications and unsolved parts of the issue

Currently, there are more than 2500 waste-to-energy plants in the world, most of them in developed countries across Europe, as well as the US and Australia.

Examples are the most efficient and environmentally friendly waste-to-energy plants in Germany, Italy, Austria, and France (one of the plants operates within the city centre of Vienna) [10, 17, 18]. In France, the utilization of household waste at 34 enterprises with the generation of heat and electricity on their basis allows for energy savings in terms of the oil equivalent of more than 300 thousand tons per year. In Germany, more than 33% of the more than 25 million tons per year of household waste incinerated at 42 plants is used to generate electricity and steam. More than 70% of household waste is used in Denmark, more than 80% in Sweden and half of this is used as fuel. According to the estimates of Swedish scientists, burning 5 tons of household waste gives the same amount of energy as burning 1 ton of fuel oil or 2 tons of coal. At the same time, 1 ton of fuel oil costs 1,400 SEK and preparation for burning 1 ton of household waste costs 160 SEK, i.e. almost nine times cheaper. Recalculation of the cost in preparation for the incineration of solid waste in Kazakhstan and comparison with the cost of fuel oil in tenge will be almost the same ratio.

In Italy, waste-to-energy plants successfully operate in Modena with an electric capacity of 19 MW, in Bologna 37 MW, in Ferrara 70 MW, in Ravenna 28 MW, in Forlì and in Rimini Coriano of up to 80 MW. At the same time, the cleaning and capture systems are under constant and continuous control of Italian supervisory authorities. Moreover, from the chimneys of the indicated plants there are practically no emissions visible from the unaided eye or smoke of a bluish and dark color.

Starting from 2017, about 200 waste-to-energy plants are commissioned in the Russian Federation. The expanded production of powerful boilers at the Machine-Building Factory of Podolsk with a coating of steel all-welded screens with a special composition to increase the service life during the combustion of solid waste was operationally launched [11, 19]. The widespread and successful use of furnaces for the incineration of household waste and efficient generation of electricity in the Scandinavian countries, the USA and the UK should also be noted. In New Jersey (USA) a plant was put into operation back in 1988, designed to recycle 106 thousand tons per year of waste with a design heat capacity of combustion of 12.3 MJ / kg (with a base of 10.5 MJ / kg). The installation provided district heating for a population of 550 thousand people. The area occupied by the station was 10 ha. It should be noted that the power plant consists of two independent technological units, each of which is designed to burn 180 tons of waste per day. There is experience in the successful operation (since 1976) of the Bernard Road power plant in Sheffield (England). At this station, 102 thousand tons of household waste are burned annually. A boiler with a thermal capacity of 24.4 MW, a steam capacity of 32 t / h with a pressure of 1 MPa (10 kg / $cm^2$) provides heat to the area with a population of 100 thousand people. Significant successes in the design and development of solid waste incineration technology used to obtain heat and electricity were achieved in Switzerland [20].

Of certain interest is an incinerator steam boiler (Russian Federation) E-6.5-1.4-225 "O" (KE-6.5-14-225 SHW) with a steam capacity of 6.5 t / h designed for burning unsorted solid household waste (SHW) with a steam capacity of 3 tons / hour. The boiler produces superheated steam to meet the technological needs of the enterprise or for the purposes of heat supply, ventilation and hot water supply.

## 3. Purpose of the study and the main results of the development

An experimental verification of joint-combined burning of solid waste in a burning layer of coal was carried out on a KST-1 [21,22] (solid fuel steel boiler with heat capacity of 1 Gcal) boiler with a mechanical furnace with fire-bars for drop grate (figure 1) and showed low calorific value of unsorted solid waste (up to 1800 kcal / kg) and unpreparedness of solid waste for burning in such a wet form.

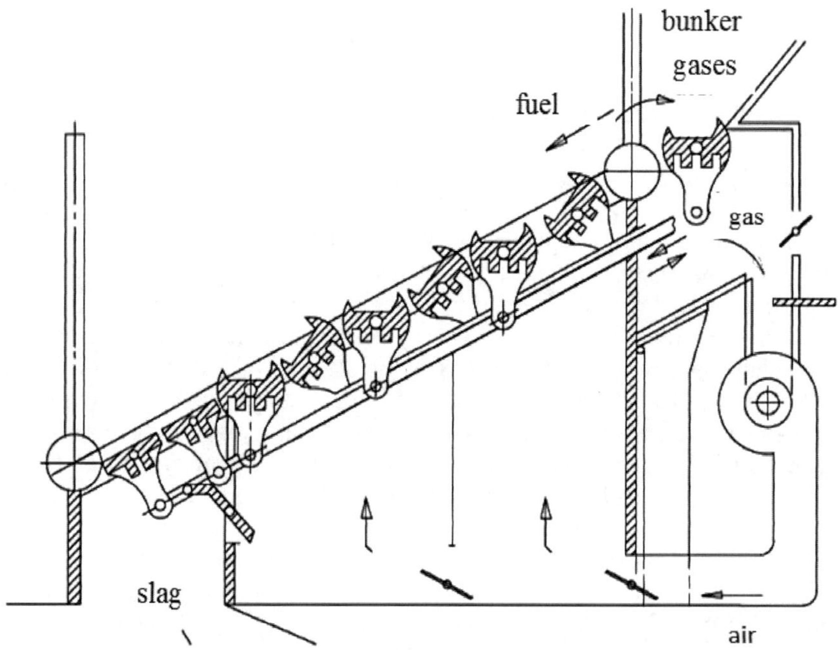


**Figure 1.** Mechanical furnace with fire-bars for drop grate.

The thermal power of the boiler decreased by 25–40% without the subsequent addition of coal. The duration of the experimental combustion was 4-5 hours. Also on the KST-1 boiler, substandard cardboard and paper were burned for 2.5–3.5 hours. The experiments on burning solid waste led to the conclusion that it is necessary to increase the combustion mirror and to partially modify the grate. Previous experiments and developments of the authors formed the basis for the developed working drawings of the KST-1.16 hot-water boiler with mechanical cast-iron fire-bars for drop grate, complete with the working documentation of a mobile modular boiler house with an automatic heating system up to 2.32 MW with a chimney. A draft design of a feeder hopper with a dispenser for feeding solid waste into the furnace of the KST-1.16 boiler was developed, with a preliminary design of an emulsifier with a capture coefficient of up to 99.7% of solid fractions when burning solid waste.

Initial data has been prepared for the development of technical documentation using modern 3D design software products to prepare the production of the first samples of modular waste-to-energy plants. The plants are equipped with new mechanical furnaces for hot water and steam boilers with a thermal capacity of 1.16 MW and up to 11.6 MW with afterburning over the bed complete with efficient purification of harmful emissions by compact emulsifiers and filters using activated carbon batteries. This is accomplished using automated combustion process control systems and shows potential for application in large cities and regional centers of the Republic of Kazakhstan.

The authors are preparing for experiments in semi-industrial conditions for the experimental burning of solid waste on a layered manual grate with the useful utilisation of heat. New dry systems for collecting solid fractions in the gas duct behind a steel boiler are being prepared for experimental verification.

## 4. Conclusion

Preliminary heat engineering tests and simulation in the special software have shown the technological possibility of burning low-calorie solid waste in hot-water solid fuel boilers. The developed working drawings for the KST-1,16 steel hot water boiler with a mechanical furnace and swinging grates for solid waste incineration in the layer can be an effective solution to the problem of waste disposal, provided that the equipment is highly environmentally friendly. Now the authors are working on the development and testing of a block-modular boiler house with a set of emulsifiers and an automated control system for the combustion process. New dry systems for capturing solid fractions in the flue behind a steel hot water boiler have been developed and are being prepared for experimental testing.

## References

[1] On approval of the action plan for the implementation of the concept for the transition of the Republic of Kazakhstan to a "green economy" for 2013 - 2020. *Resolution of the Government of the Republic of Kazakhstan*, 2013, p 750

[2] Kuznetsova N A, Trufanova E V and Makeev P V 2016 Features of the collection of solid waste in Russia and abroad. *New Science: Theoretical and practical view*. No. 8 (88), pp.183-184 (https://elibrary.ru/item.asp?id=26461579)

[3] Mohammadi M and Harjunkoski I 2020. Performance analysis of waste-to-energy technologies for sustainable energy generation in integrated supply chains. *Computers & Chemical Engineering*, **140**, 106905. https://doi.org/10.1016/j.compchemeng.2020.106905

[4] Halder P and Azad A K 2019 Chapter 7 - Recent trends and challenges of algal biofuel conversion technologies. In Woodhead Publishing Series in Energy, editor(s): Abul Kalam Azad, Mohammad Rasul, *Advanced Biofuel. Woodhead Publishing*, pp 167-179

[5] Usmani Z, Sharma M, Karpichev Y, Pandey A, Kuhad R Ch, Punia R, Aghbashlo M, Tabatabaei M and Gupta V 2020 Advancement in valorization technologies to improve utilization of bio-based waste in bioeconomy context. *Renewable and Sustainable Energy Reviews*, **131**, 109965, doi: https://doi.org/10.1016/j.rser.2020.109965

[6] Evdokimov S V, Orlova A A and Dubinina G F 2015 Provision of environmental safety during municipal solid waste recycling, *Ecology and Industry of Russia*, **19** (11), pp 36-40. (https://elibrary.ru/item.asp?id=24833270)

[7] Pham T, Kaushik R, Parshetti G K, Mahmood R and Balasubramanian R 2015. Food waste-to-energy conversion technologies: Current status and future directions. *Waste Management*, **38,** pp 399-408, https://doi.org/10.1016/j.wasman.2014.12.004

[8] Makarichi L, Jutidamrongphan W, Techato K 2018 The evolution of waste-to-energy incineration: A review. *Renewable and Sustainable Energy Reviews*, **91**, pp.812-821, https://doi.org/10.1016/j.rser.2018.04.088

[9] Uzoejinwa B, He X, Wang S, Abd El-Fatah Abomohra, Hu Y and Wang Q 2018 Co-pyrolysis of biomass and waste plastics as a thermochemical conversion technology for high-grade biofuel production: Recent progress and future directions elsewhere worldwide. *Energy Conversion and Management*, **163**, pp 468-492, doi: https://doi.org/10.1016/j.enconman.2018.02.004

[10] Ershov A G and Shubnikov V L 2014 Thermal waste disposal: theory and practice, myths and legends. *Solid household waste*, **5** (95), p. 46-53. (https://elibrary.ru/item.asp?id=21733306)

[11] Baskakov A P 2014 The prospects for incineration of municipal solid waste in Russia in order to produce heat and electric power. *Thermal Engineering (English translation of Teploenergetika)*, **61**, pp 265–273, doi:10.1134/S0040601514030033

[12] Sergeev V V, Aleshina A S 2010 The use of municipal solid waste as fuel. *Scientific and Technical Gazette of St. Petersburg State Polytechnic University*, **4** (110), pp 91-96, (https://elibrary.ru/item.asp?id=15561850)

[13] Mahmoodi P, Karimi K and Taherzadeh M 2018 Efficient conversion of municipal solid waste to biofuel by simultaneous dilute-acid hydrolysis of starch and pretreatment of lignocelluloses. *Energy Conversion and Management*, **166**, pp 569-578, doi: https://doi.org/10.1016/j.enconman.2018.04.067

[14] Zagorodniuk L Kh, Naumenko N A and Tutskaya I N 2019 Foreign experience in the use of household waste. *In the collection: High-tech technologies and innovations Electronic collection of reports of the International scientific and practical conference dedicated to the 65th anniversary of BSTU named after V.G. Shukhov*. pp 145-151. doi: 10.12737/conferencearticle_5cecedc2383321.46876442

[15] Tang Z, Li W, Vivian W Y Tam and Caihong Xue 2020 Advanced progress in recycling municipal and construction solid wastes for manufacturing sustainable construction materials. *Resources, Conservation & Recycling: X*, **6**, 100036. https://doi.org/10.1016/j.rcrx.2020.100036

[16] Nikishanin M, Zagrutdinov R, Senachin P 2015 Briquetting of solid household waste for gasification. *In the book: Combustion of fuel: theory, experiment, applications Abstracts of the IX All-Russian Conference with international participation*. p 104

[17] Weber K, Quicker P, Hanewinkel J and Flamme S 2020. Status of waste-to-energy in Germany, *Part I – Waste treatment facilities. Waste Management & Research*. **38**. 0734242X1989463. 10.1177/0734242X19894632

[18] The role of waste-to-energy in the circular economy 2017 *Communication from the Commission to the European Parliament, the Council, the European Economic and Social Committee and the Committee of the Regions*. Brussels, 26.1.2017 **(COM 2017) 34** final.

[19] Gunich S, Yanchukovskaya E and Dneprovskaya N 2015 Analysis of modern methods of processing municipal solid waste. *News of universities. Applied Chemistry and Biotechnology*, **2** (13), p. 110-115. (https://elibrary.ru/item.asp?id=23641688)

[20] Waste Management, 2011 *National Reporting to CSD* **18/19** by Switzerland. 19th session of the Commission on Sustainable Development.

[21] Orumbayev R, Smaga G, Mazur V, Aubekerov T, Kurmangaliev M and Nusupbekova D 1989 Inventors certificate No. 1508052 Mechanical stoking, *Publ. Bull. No. 34, 09/15/89*. (https://findpatent.ru/patent/150/1508052.html)

[22] Orumbayev R, Ibrashev M and Ponamarev O 1990. Inventors certificate No. 1592649 Grate for burning solid fuel, *Publ. Bull.№***34** *09/15/90*. (https://findpatent.ru/patent/175/1755010.html)

[23] Orumbayev R, Airich Y et al. 2007 Patent of the Republic of Kazakhstan No. 18797. Hot water boiler, *Publ. Bull.* **9**, *September 17, 2007*. https://gosreestr.kazpatent.kz/Invention/Details?docNumber=140168)

[24] Orumbayev R, Orumbaeva Sh, Kibarin A et al. 2020 Patent of the Republic of Kazakhstan. No. 34214. Hot water boiler. *publ. Bull.*, **9**, *Marth 5, 2020*. (https://gosreestr.kazpatent.kz/Invention/Details?docNumber=301088)

# Studies of integrated natural capillary-porous coatings

**A A Genbach[1], D Y Bondartsev[1] and H I Beloev [2]**

[1] Faculty of Heat & Power Units, Almaty University of Energy and Communications, Almaty, Republic of Kazakhstan
[2] Agricultural Machinery Department, University of Ruse, 8 Studentska Street, Ruse, 7017, Bulgaria

E-mail: d.bondartsev@inbox.ru

**Abstract.** There were developed devices with porous coatings that can eliminate the appearance of cracks in the chambers and nozzles due to the powerful thermodynamic screen of three heat sources, and to arrange cutting of reinforced concrete structures with detonation high-temperature flares flowing from combustion chambers and nozzles cooled by capillary-porous coatings. In studies, the holography method was used. The expected economic effect from the one burner is approximately 200-300 $. The environment needs to be reduced by the cooling system's energy consumption. The studied chambers and burners can be implemented in the thermal power plants, especially in the gas turbine equipment. Capillary-porous and flow-through cooling systems demonstrated high reliability, however, the first system can reduce the cooler consumption by tens of times (up to 80 times). The design and manufacturing of hybrid porous surfaces make them integrated and multifunctional, provides the possibility to work with higher critical loads, and enables a mobile process for monitoring the limiting condition of coatings.

## 1. Introduction

It is of interest in the design and manufacturing of hybrid porous surfaces with a hybrid micro-nano scale [1]. Such kinds of surfaces can be considered future technological products, they become more compact, integrated, and multifunctional, able to operate with higher thermal loads. Therefore, we are developing nanoscale and microscale porous structured surfaces in the form of integrated mesh structures with coatings from natural mineral media, which are examined in the present study. They have synergetic advantages of combining these two structures into an integrated technology for their manufacturing (structures and coatings) with a further increase of the relocate' limit of thermal loads and control of the boundary condition of porous coatings. Modern research continues to be ways of increasing the cooling efficiency of the combustion chambers and gas turbine nozzles with various coolers to increase the efficiency of the machines and machine cycles [2, 3, 4, 5]. Water, air, and steam coolers are considered. Therefore, we present new integrated natural capillary-porous coatings, which, unlike the existing ones, significantly increase the heat, power unit cycle efficiency [6, 7, 8, 9, 10].

## 2. Design and structure of the capillary-porous

The diverse tasks that were solved in the study for several devices of thermal power units TPU (choice of a heat carrier, consideration of its circulation type, choice of types of material of heat transfer apparatus and intensifiers, the system operating mode (SOM) under pressure (vacuum), energy supply and type, system orientation) [7, 10, 11].

The conducted studies allow to guiding principle of the choice of a heat carrier, take into account its circulation type, determine the composition and material of heat transfer apparatus and intensifiers, in regards to the SOM (under pressure or vacuum), type, energy supply, etc. To control energy processes, it is proposed to divide the energy into two components: the energy of a heatwave, an explosively generated steam nucleating seed, and the energy of compressed flow, which is also important for modeling and similarity of boiling in the pores of a structure (coating).

Figure 1 also shows the design technique of porous systems as applied to the designed equipment of TPU.

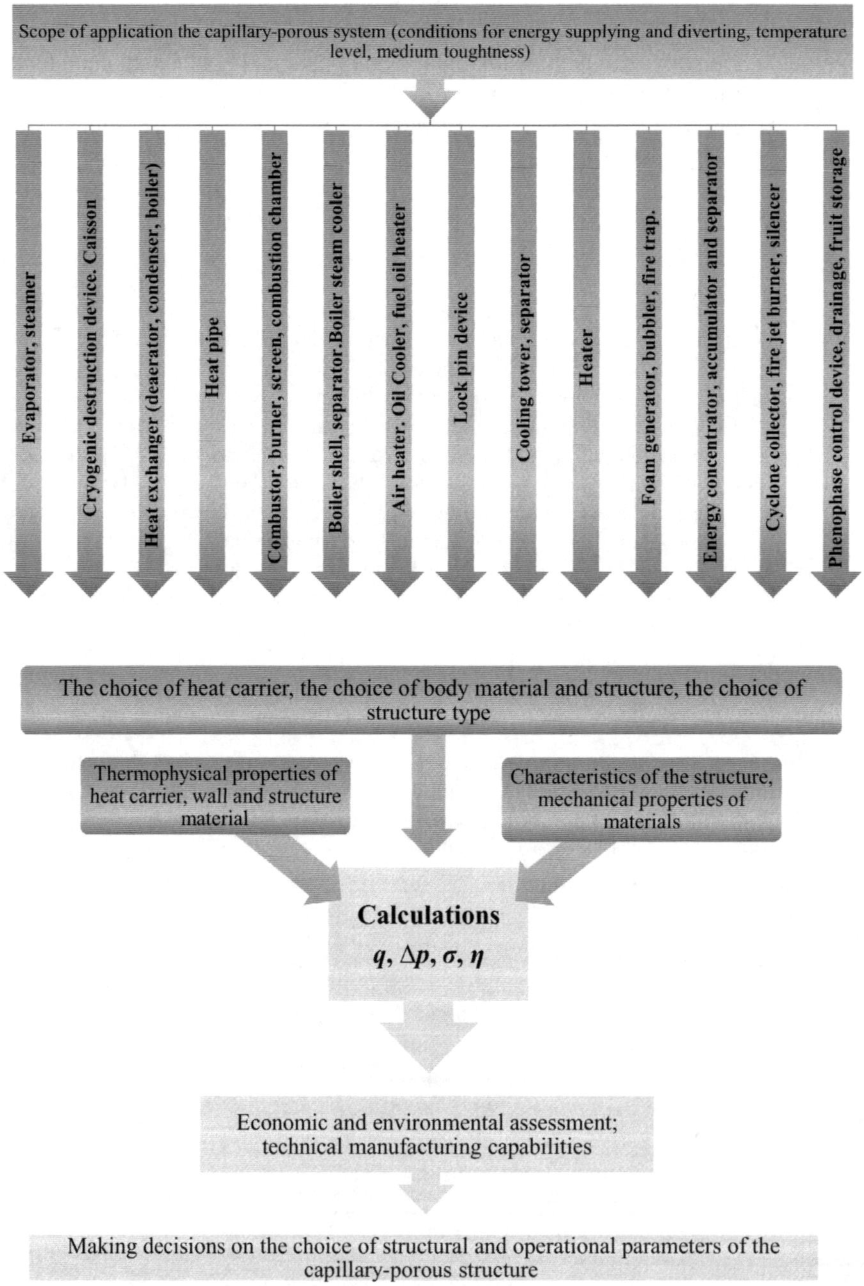


**Figure 1.** Block scheme of the suggested capillary-porous system.

The conducted studies [6, 7, 8, 9, 12] allow the supply of coolant liquid, housing material, and structure, the type of porous coating, the heat transfer calculation, resistance, thermal stresses, and provision of the economic and environmental assessment.

A promising thermal device for the reconstruction and modernization of turbo-foundations are considered the rocket-type burners with high-temperature supersonic pulsating detonation jets (flares). Further advancement of the cooling systems of the combustion chambers and nozzles is required to improve the reliability of the heat-loaded highly accelerated elements of their structures. The designed combustion chambers and nozzles are suitable for gas turbines (shown in figure 1).

Attempts of modern designers to design power units, including gas turbine units (GTU), are related to the advantages of the detonation method of fuel consumption.

By the way, scientific and design studies encounter two actual problems: how to implement a safe and reliable method of detonation combustion and how to provide highly accelerated operating modes of combustion chambers and supersonic nozzles using the most advanced high-intensity cooling systems.

The use of supersonic nozzles with shortened diffuse part enables the power units to work in detonation mode.

Furthermore, the cooling system is useful for tools designed for cutting concrete and reinforced concrete, and especially for fire-jet burners installed in whirling thermal drills for turbo - foundation works. In the case of using a capillary-porous cooling system in thermal drills (fire-jet burners designed for drilling blastholes and boreholes), the scheme changes slightly due to a different arrangement of arteries.

The cooling scheme is also applicable to air-cooled burners (benzo-air and kerosene-air). This makes it possible to reduce more than 1.5 times the consumption of compressed air, which is used only as an oxidant. The capillary-porous cooling system of such burners is highly effective and can significantly increase the time of operation on the combustion chambers and nozzle apparatus (from 100 – 150 hours to 500 – 600).

Due to a significant reduction in water consumption, refusal of pumps, high reliability of the system, design simplification, and low operating costs, the annual economic effect per burner equals at least 200 – 300 dollars. A significant drawback of the water system is that we must supply water under high pressure $(3.5 - 4)\cdot 10^5$ Pa. The nozzle and the combustion chamber of the burner will be destroyed immediately as soon as the water pressure in the line drops. The decrease in water pressure in the system leads up to service time reduction. To ensure high reliability, the water flow rate is expected to be high and, depending on the burner capacity, ranges from 0.2 to 3 and more kg/s.

With the purpose to eliminate the indicated drawbacks to ensure high efficiency and reliability, a capillary-porous cooling system is proposed on figure 2. It has high technical characteristics, reliable and simple to operate. Further, in the operating mode, the consumption of cooling water is sharply reduced (~ 10 times).

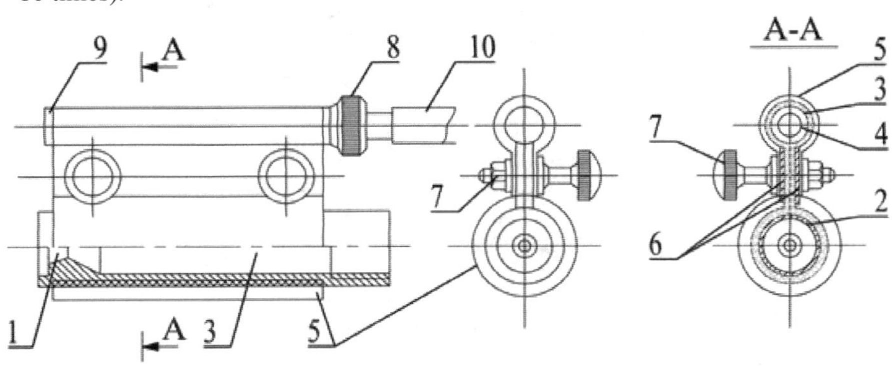


**Figure 2.** Cooling of the nozzle and combustion chamber of a fire-jet burner with a capillary-porous system: 1–nozzle; 2–combustion chamber; 3–capillary-porous structure; 4–artery; 5–casing; 6–pressure plate; 7–swivel screw; 8–sealing ring; 9–stopper plug; 10–rubber hose.

Figure 2 shows a device for cooling the nozzle 1 and the combustion chamber 2 of a fire-jet burner covered by a capillary-porous body 3 and finally, the ends are connected to the artery 4. The capillary-porous body 3 and the casing 5, which at the bottom with a gap cover the chamber and nozzle, protecting them from mechanical damage, and at the same time protecting the personnel from burns, are tightly against the perforated tube of the artery 4. The cooling agent is supplied from artery 5 through structure 3 to the cooled surfaces of chamber 2 and nozzle 1. The gap shaped by the casing 5 and the chamber provides the active removal of steam due to its ejection of the supersonic high-temperature gas jet flowing out of the nozzle. When the refrigerant is supplied at a pressure from 20 kPa to 25 kPa, the consumption is reduced by about (60-70) times, since heat is diverted by the steam generation heat.

The heat flow is transmitted through the chamber's walls and the nozzle. The heat flow rate varies in a wide range (from 1 to 2 orders of magnitude). It reaches a maximum in the critical section of the nozzle. However, the system's adaptive ability successfully copes with this work condition and ensure uniform distribution of the temperature field in the cooled walls.

Reduce the formation of a steam bubble is limited cooling channels, lead to burnout of the walls is eliminated. The design of the burner and its operation are simplified, which makes it possible to reduce the technical requirements for the seals of the combustion chamber and the burner head. The structure of the capillary-porous is easily removable - unscrew it from the distribution head together with the combustion chamber. When disconnected, the mechanism can be freed from mechanical and chemical deposits and again installed on the burner, so it can be renovated many times for a short time (a few minutes). The low rate of the water flow, as for a kerosene-oxygen burner (3 – 4) g/s instead of 200 g/s, permits the vapor emission and simplifies the design and work of the system.

## 3. Analysis and discussion of holographic interferograms of deformations and stresses in capillary-porous coatings.

For comparison, we present capillary-porous coatings interference patterns made by the toughest rock (granite) with porosity ε = 5% and viscous rock (tuff) with ε = 30%.

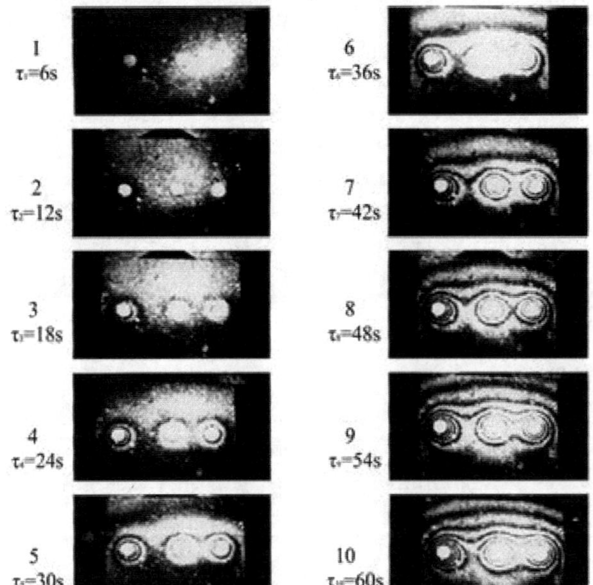


**Figure 3.** Holographic interferograms of tuff coating, displaying the state of the surface every 6 s, with a heat flux of 15 W. The sample loads were made from the condition fixed from time 126 s to condition 180 s (items 1-10) ($q = 2.1 \cdot 10^6$ W/m$^2$).

According to the properties of the material, both coatings can be used to create two environments with significantly different stiffness values, and thereby provide control of the heatwave penetrating the combustion chamber and nozzle of their cooling surface (figure 3, 4).

The analysis of interferograms of deformations and thermal stresses showed the following: the expansion of the coating sample occurs firstly around the heat source, and then throughout. The deformation of the sample volume depends on the distribution of the temperature field created by the heat sources and the resulting thermal stresses on the surface and in the body volume.

Figure 4 presents photographs of the granite coating every 6 seconds after the start of simultaneous heating from three sources with a total power of 7 W ($q = 1\cdot10^6$ W/m$^2$).

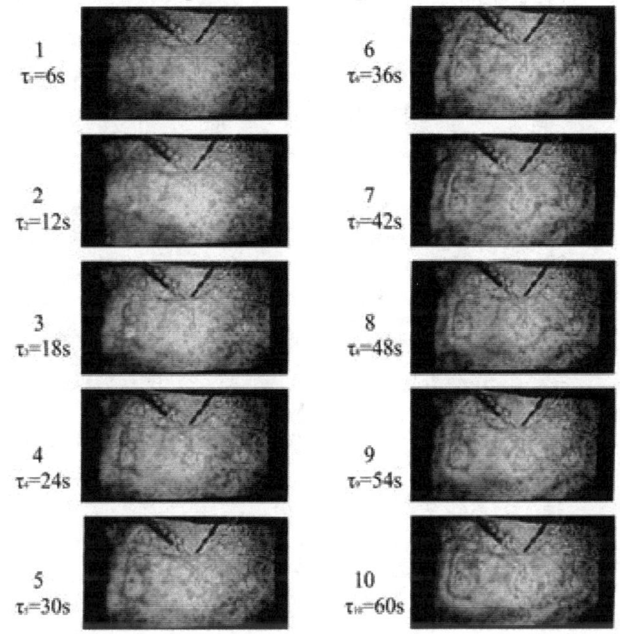


**Figure 4.** Holographic interferograms of the granite coating. The granite sample was loaded from the condition: $\tau_1$=6 s; $\tau_2$=12 s etc., to $\tau_{10}$=60 s (items 1-10), ($q = 1\cdot10^6$ W/m$^2$).

Figure 5 presents photographs of a marble coating under the same heating conditions as in figure 4. The comparison of the deformation patterns of holographic interferograms and thermal stresses on interferograms obtained by the optical polarization method revealed their analogy, which indicates the surface correlation and other processes [6, 7, 8, 9, 11]. It was revealed that samples with low porosity (granite) are in a more stressed condition compared to viscous and porous media (figure 3). A two-fold increase in the heat flux power led to an up to the three-fold increase in their surface deformation, while for samples with high porosity deformation increase was only by half. The number of interference bands per unit length is upper for coatings with low porosity values, which means the presence of large thermal stresses in them, the growth rate of which increases along with heat load increase (shown on figures 3÷5).

The process of coating destruction by thermal acoustic waves of the burner flare, amplified from the fronts of reflected waves from various surfaces of the chamber and the nozzle, as an outcome of thermal stresses arising from the nozzle sections from interference and diffraction of the waves (squash/stretch and displacement stress waves), will be neutralized by screens using the acoustic and thermodynamic effect principle (shown on figures 3, 4 and 5). Using the interferograms we can detect not visible defects and cracks, not only large inclusions in which area the equal deformation lines were winding (figure 5).

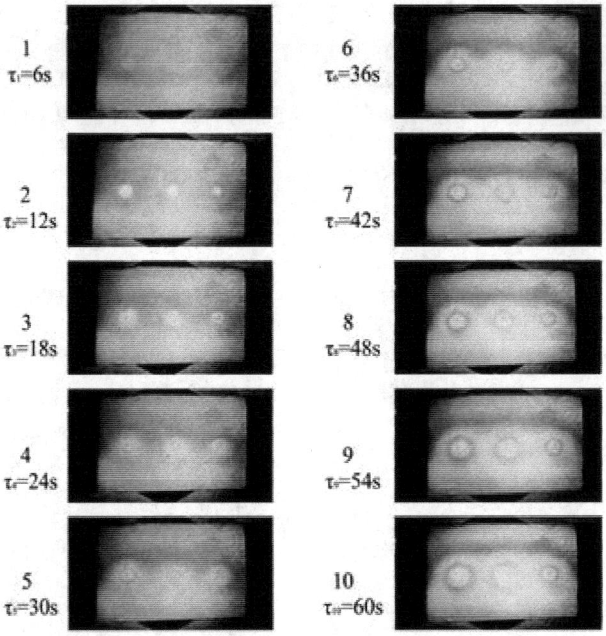


**Figure 5.** Holographic interferograms of the marble coating every 6 seconds after the start of simultaneous heating from three sources with a total power of 7 W. The marble sample was loaded from the condition: $\tau_1$=6 s; $\tau_2$=12 s etc. to $\tau_{10}$=60 s (items 1-10), ($q = 1\cdot 10^6$ W/m$^2$).

The type of thermodynamic screen is selected depending on the cooling surface material, taking into account the acoustic stiffness coefficient of the contacting media. The thermodynamic screen can be installed independently or in conjunction with screens containing a cooling tank with finning.

## 4. Conclusion

We have shown that due to new thermal devices with porous coatings, the likelihood of cracks development coming from the reflected wave in nozzles and combustion chambers, as well as during the modernization of turbo-foundations of TPU, can be reduced. These devices allow establishing powerful thermodynamic screens from detonation high-temperature jets preventing cracks development, and selected material properties of the fence surface. Holographic interferograms illustrate the concentration of deformations and thermal stresses in the solid under thermal loads up to $2.1\cdot 10^6$ W/m$^2$. Capillary-porous and flow-through burner cooling systems have shown high efficiency up to the limiting condition of the metal of combustion chambers and nozzles ($1\cdot 10^6$ W/m$^2$).

## References

[1] Khan S, Sezer N and Koç M 2019 Design, fabrication and nucleate pool-boiling heat transfer performance of hybrid micro-nano scale 2-D modulated porous surfaces. *Applied Thermal Engineering*. **153**, pp 168-180. https://doi.org/10.1016/j.applthermaleng.2019.02.133

[2] Wang W, Gao J, Shi X and Xu L 2013 Cooling performance analysis of steam cooled gas turbine nozzle guide vane. *International Journal of Heat and Mass Transfer*. **62**, pp 668-679. https://doi.org/10.1016/j.ijheatmasstransfer.2013.02.080

[3] Wang W. 2015 Efficiency study of a gas turbine guide vane with a newly designed combined cooling structure. *International Journal of Heat and Mass Transfer*. **80**, pp 217-226. https://doi.org/10.1016/j.ijheatmasstransfer.2014.09.024

[4] Yang X, Liu Z, Liu Z, Feng Z and Simon T 2019 Turbine platform phantom cooling from airfoil film coolant, with purge flow. *International Journal of Heat and Mass Transfer*. **140**, pp 25-40. https://doi.org/10.1016/j.ijheatmasstransfer.2019.05.109

[5] Moon S, Kwon H, Kim T, Kang D and Sohn J 2018 A novel coolant cooling method for enhancing the performance of the gas turbine combined cycle. *Energy*. **160**, pp 625-634. https://doi.org/10.1016/j.energy.2018.07.035

[6] Genbach A, Bondartsev D, Iliev I and Georgiev A 2020 Scientific method of creation of ecologically clean capillary-porous systems of cooling of power equipment elements of power plants on the example of gas turbines. *Energy*. **199**, 117458. https://doi.org/10.1016/j.energy.2020.117458

[7] Genbach A, Bondartsev D and Iliev I. 2019 Heat transfer crisis in the capillary-porous cooling system of elements of heat and power installations *Thermal Science*. **23**, **Issues 2**, pp 849 – 860. https://doi.org/10.2298/TSCI171016139G

[8] Genbach A and Bondartsev D 2020 Limiting thermal state of capillary-porous power-plant components. *Russian Engineering Research*. **40**, pp 384-389. https://doi.org/10.3103/S1068798X20050093

[9] Genbach A, Shokolakov K, Bondartsev D and Iliev I 2019 Research of the heat transfer crisis and limit state of the heat exchange surface covered by capillary porous medium, 2019, *IOP Conf. Ser.: Mater. Sci. Eng.* **595** 012007. https://doi.org/10.1088/1757-899X/595/1/012007

[10] Stoyanov I, Ivanov V, Iliev T and Ivanov H 2019 Yield and performance study of a 1MWP grid connected photovoltaic system in Bulgaria, *Journal of Engineering Science and Technology Review Special Issue on Conference in Telecommunications, Informatics, Energy and Management*, pp 206-209, ISSN: 1791-2377

[11] Genbach A, Bondartsev D, Iliev I and Terziev A 2019 Analogy in the processes of heat exchange of capillary-porous coatings in energy installations. *E3S Web Conf.*, **85**, 05003. DOI: https://doi.org/10.1051/e3sconf/20198505003

[12] Kamarova S, Abildinova S and Terziev A 2020 Study of the thermodynamic efficiency of solid fuel preparation systems of a thermal power plant based on an exergy analysis of the operation of subsystems, *E3S Web Conf* **207** *02005 (2020)*, DOI: 10.1051/e3sconf/202020702005

# A study of thermal devices with porous coatings

**A A Genbach[1], D Y Bondartsev[1] and H I Beloev [2]**

[1] Faculty of Heat & Power Units, Almaty University of Energy and Communications, Almaty, Republic of Kazakhstan
[2] Agricultural Machinery Department, University of Ruse, 8 Studentska Street, Ruse, 7017, Bulgaria

E-mail: d.bondartsev@inbox.ru

**Abstract.** Thermal devices (mechanisms) with porous coatings designed for combustion chambers of gas turbine units, as well as for cutting and drilling turbo foundations of power plants, which solve the problem affiliated with the appearance of cracks from the reflected wave, have been developed and studied. The porous coating provides a powerful thermodynamic screen in the similitude of three heat sources. The studies were unleashed on five samples with different porosity and hardness values (5 ÷ 30%). Thermal movements, deformations, and stresses were recorded on a holographic unit in real-time. A thermal device in the form of a rocket-type burner with a detonation jet has proved high efficiency for a capillary-porous and flow-through cooling system. Photographs of combustion chambers and nozzles studied depending on their geometry and thickening of the nozzle wall, an excess oxidizer (0.3 ÷ 0.8), and operating conditions until of the metal limiting condition (1106 W/m$^2$) are presented. The optimal geometry of the chambers and nozzles either the type of porous structure was specified. It has been found that the economic effect is 200-300 $ for 1 burner, and the consumption of coolant is reduced several times.

## 1. Introduction

The introduction of energy efficiency measures are among the priorities in all country's energy policy and correspond to the objectives in the new Energy Policy for Europe [1]. The paper aims to study the heat transfer during boiling of water in porous structures of thermal power units of thermal power plants (TPP) and devices for cutting natural and artificial mineral media. The experiments were carried out on a rocket-type burner. The combustion chambers and supersonic nozzles with different porous structures were cooled.

In the modern literature [2, 3] studies of heat transfer of various homogeneous and inhomogeneous porous surfaces of coatings (multilayer), as well as specially designed wicks able to increase their heat transfer abilities, are still of interest. Studies [4, 5] of the thermohydraulic parameters of boiling in porous media using thermocouple measurements and monitoring with a high frame rates camera (bubble sizes, discharge frequency, and density of nucleating seed) started to appear. According to the comparative analysis of different methods for calculating the heat transfer from boiling water with vertical subcooling is suggested [6].

The main task for the power units of TPPs is arranging a cooling system for highly heat-stressed parts and assemblies [7]. These include furnace walls of highly accelerated boiler units, combustion chambers, nozzles, and vanes of gas turbine units, thermal devices (tools) for the execution of construction and assembling works.

Created a powerful heat screen relative to the generation of deformation fields and thermal stresses is an obstacle to the propagation of the reflected blast wave, causing the appearance and growth of the cracks. The integration of mesh structures with capillary-porous coatings from natural mineral materials makes a synergistic effect of combining into the technology of their manufacture, expanding the removal of critical loads, monitoring and adjusting the coating limiting conditions.

Designed and studied combustion chambers and nozzles that can be used in gas turbines of TPPs and thermal devices for cutting natural stones and artificial mineral materials [8, 9, 10, 11, 12].

## 2. An experimental method for studying capillary-porous samples

The studies were conducted on five natural samples made from various rocks (figure 1). The shown investigated materials of capillary-porous samples are respectively: 1 – granite; 2 – marble; 3 – tuff; 4 – granite, 5 – tuff.

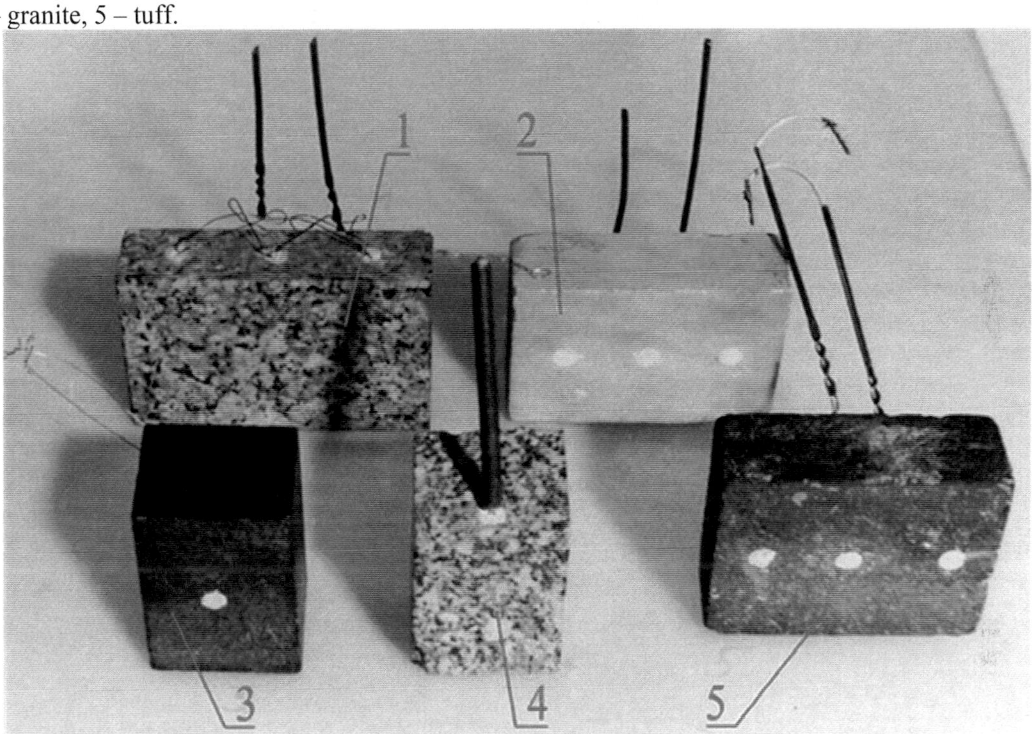


**Figure 1.** The studied samples of capillary-porous mineral media.

Three holes with dimensions $6 \cdot 10^{-3}$ m in diameter and depth of $12 \cdot 10^{-3}$ m were drilled perpendicular to the surface of the front side (samples №2 and №5), or perpendicular to the surface of the minor side (sample №1). Thus, a thermodynamic screen was created when three heat sources (electricity) were turned on. One heat source was also studied (samples №3 and №4).

On figure 2 is shown the operation of the evaporation-condensation circuit in a closed cycle. The operating principle of the installation is as follows: the electricity is supplied to the core heater from a welding transformer (TSD). The output voltage is: 2.5; 5; 7.5 and 10 V. For the current measurement of the heater was used a universal current transformer (UCT). The primary current is between 100 A and 2000 A, thus secondary - 5 A. The heater voltage drop was measured. The accuracy for the current is ±0.6%; voltage drop – ±1%; power – ±1.6%. The heater there is a backup power provided by the voltage regulator (VR).

The flow rating of the cooling and circulating liquids were measured with a rotameter. The flows of the draining liquid and condensate are stored in a measuring tank with a scale of $0.5 \cdot 10^{-3}$ l and the time for filling is measured with a chronometer, it has reached an accuracy of 0.1 s. The errors in measurement of the rotameter no more than ±3%, and for the volumetric method < ±2%.

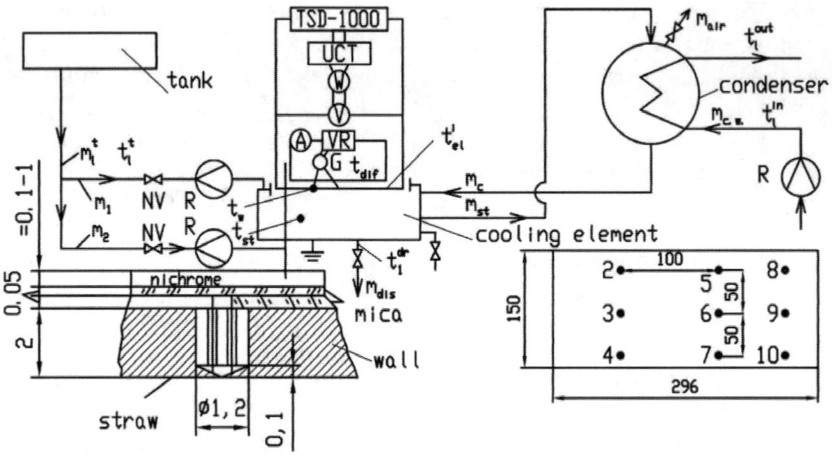


**Figure 2.** Experimental measurement of the flows on the heat-exchange surface and capillary-porous coating.

The input power Ne was equal to (7 ÷ 30) W. When the drilled diameter of the screw $d_s$=(3 ÷ 6).10$^{-3}$ m, the specific heat flux reach to following values:

$$q = \frac{N_e}{\pi\frac{d_s^2}{4}} = \frac{7 \div 30}{\pi\frac{(3\div6)^2\times10^{-6}}{4}} = (0.99 \div 1.06) \times 10^6, \text{ W/m}^2 \qquad (1)$$

To study porous thermodynamic screens, the method of holographic interferometry was used. The stressed and deformed state of the samples was studied. Deformations of the samples were recorded in real-time (figure 3).

The interferogram photographing frequency is 0.5 frames per second; 30 shots were taken from each sample. Photo prints were placed in two frames reflecting the surface state of the samples every 6 seconds.

The decoding of holographic interferograms was made according to the accepted technique. The displacement vector was determined $\bar{d}$ , according to:

$$/\bar{d}/ = \frac{N\lambda}{1+cos\varphi}, \qquad (2)$$

where $N$ – is the measured number of bands between the studied point and the zero-order band: λ – is the wavelength; φ – is the angle between the illumination of the studied $\bar{\tau_l}$ and the observation $\bar{\tau_s}$ point on the hologram, i.e. $\bar{\tau_s} \wedge \bar{\tau_l}$.

The deformation for one heat source can be determined by the displacement gradient which has the greatest value in the filled with the heater with a radius < 10 mm and increased with the extension of the heat exposure period.

Furthermore, three simultaneously acting heat sources with independent displacement field led to a simple superposition on the sample surface. This phenomenon took place both for displacement fields along a plane passing through the centers of sources and for planes spaced 6 mm apart from them.

**Figure 3.** Experimental stand for measurement of main processes of steam generation and stressing and deformation coatings by optical methods: 1 - laser; 2 - film camera SKS-1M; 3 - porous cooling element.

A holographic unit with samples from mesh structures and porous rock coatings allows to obtain real-time interferograms and to interpret them according to internal (thermohydraulic) boiling characteristics, thermal stresses, and deformations appearing in thermodynamic screens.

### 3. Results and discussion

The tested combustion chambers and nozzles having different lengths and wall thickenings are shown on figure 4, and some of the chambers were cooled with a capillary-porous system. The excess oxidizer coefficient α varied within 0.3 ÷ 0.8 [8].

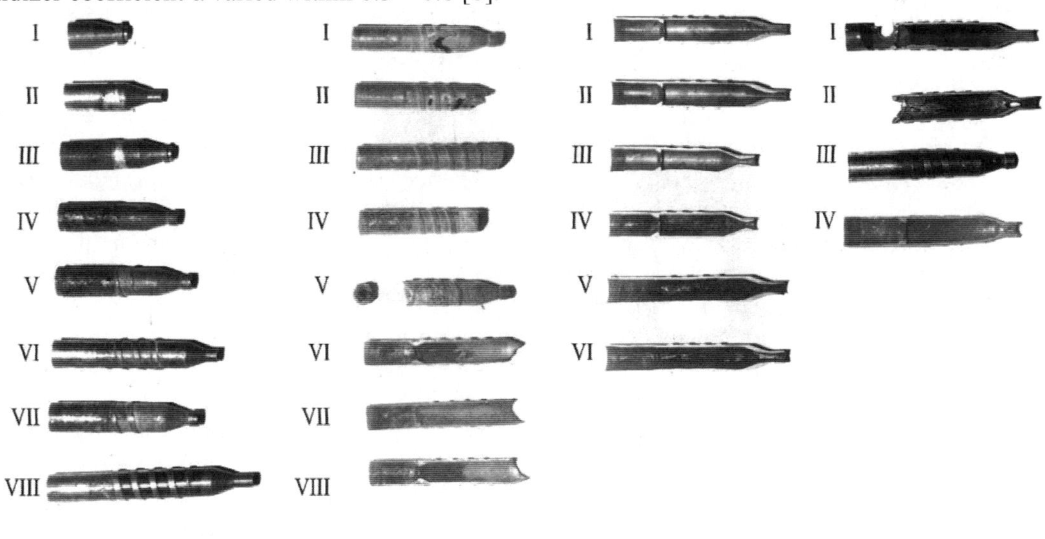


a) b) c) d)

**Figure 4.** Samples of combustion chambers and nozzles.

The structures were made of mesh with a hydraulic diameter of $b_h = 0.4 \cdot 10$ mm. We investigated some of the operational and technological capabilities of combustion chambers and nozzles up to their destruction. The following notation is accepted for figure 4:

a) samples of combustion chambers having different lengths and design on the nozzle exit: I – IV – oxidizer excess coefficient $\alpha = 0.3$; V – VII – $\alpha = 0.65 - 0.7$; VIII – $\alpha = 0.8$; I – IV – porous cooling is used at $q_{cr.c\text{-}s.} = 5 \cdot 10^6$ W/m$^2$; V – VIII – a water cooling system is used at $q_{cr.c\text{-}s} = 1 \cdot 10^6$ W/m$^2$, $\overline{w_c} = 10$ m/s;

b) the nozzles are made with a wall thickening: I - VIII – excess oxidizer coefficient $\alpha = 0.6...0.65$; the destruction occurred as a result of gases breakthrough into the water cooling system during seals depressurization; V – is an experimental sample of a combustion chamber with a melted swirl nozzle. The cooling system used is a capillary-porous cooling system at $q_{cr.c\text{-}s.} = 1 \cdot 10^6$ W/m$^2$;

c) nozzles are made without wall thickening: I - IV – before operation; V - VI – after 40 hours of operation (deflector rings are destroyed and nozzle cross-sections are widened); I - VI – $\alpha = 0.8$; III - VI – $\alpha = 0.6$; IV – a sample of the combustion chamber made with a shortened nozzle (provided the detonation mode of combustion). The cooling system used is a water cooling system at $q_{cr.c\text{-}s} = 1 \cdot 10^6$ W/m$^2$.

d) destroyed combustion chambers, nozzles, and "skirts" as a result of the penetration of bubbles due to poor brazing of the chamber afterbody (I - II); $\alpha = 0.65$; nozzles are made with a cylindrical outer surface [8]; III - IV – sample chamber with a shortened nozzle with wall thickening; destruction began to occur at the beginning of the operation on the external surface ahead of the critical section; the excess oxidizer coefficient $\alpha$ was equal to = 0.65; I - II – water flow cooling at $q_{cr.c\text{-}s.} = 1 \cdot 10^6$ W/m$^2$, $\overline{w_c} = 10$ m/s, III - IV – capillary-porous cooling at $q_{cr.c\text{-}s.} = 1 \cdot 10^6$ W/m$^2$, $\overline{w_c} = 10$ m/s.

## 4. Conclusion

The proposed investigated capillary-porous system has a better advantage than the pool boiling. Thus it's possible to regulate the heat flow of the processes. The main advantages of the proposed system are the unification of capillary forces and mass transfer. Thus resulting in an improvement of the thermal and strength characteristics of the system.

The conducted results confirmed the high efficiency of the proposed capillary-porous and flow-through burner cooling systems. The results obtained can be applied to the cooling towers in relation to the environment by a tenfold reduction in coolant consumption.

## References

[1] Stoyanov I, Ivanov V, Iliev T and Ivanov H 2019 Yield and Performance Study of a 1MWP Grid Connected Photovoltaic System in Bulgaria, *Journal of Engineering Science and Technology Review Special Issue on Conference in Telecommunications, Informatics, Energy and Management*, , pp 206-209, ISSN: 1791-2377

[2] Odagiri K and Nagano H 2019 Investigation on liquid-vapor interface behavior in capillary evaporator for high heat flux loop heat pipe. *International Journal of Thermal Sciences*. **140**, pp 530-538. https://doi.org/10.1016/j.ijthermalsci.2019.03.008

[3] Ji X, Xu J, Zhao Z and Yang W 2013 Pool boiling heat transfer on uniform and non-uniform porous coating surfaces. *Experimental Thermal and Fluid Science*. **48**, pp 198-212, https://doi.org/10.1016/j.expthermflusci.2013.03.002

[4] Chang Y and Ferng Y 2019 Experimental investigation on bubble dynamics and boiling heat transfer for saturated pool boiling and comparison data with previous works. *Applied Thermal Engineering*. **154**, pp 284-293. doi.org/10.1016/j.applthermaleng.2019.03.092

[5] Chuang T, Chang Y and Ferng Y 2019 Investigating effects of heating orientations on nucleate boiling heat transfer, bubble dynamics, and wall heat flux partition boiling model for pool boiling. *Applied Thermal Engineering*. **163**, 114358. https://doi.org/10.1016/j.applthermaleng.2019.114358

[6] Jamialahmadi M., et at. 2008 Experimental and theoretical studies on subcooled flow boiling of pure liquids and multicomponent mixtures. *International Journal of Heat and Mass Transfer*. **51, 9–10**, pp 2482-2493. https://doi.org/10.1016/j.ijheatmasstransfer.2007.07.052

[7] Kamarova S, Abildinova S and Terziev A 2020 Study of the thermodynamic efficiency of solid fuel preparation systems of a thermal power plant based on an exergy analysis of the operation of subsystems, *E3S Web Conf.* **207** *02005 (2020)*, doi: 10.1051/e3sconf/202020702005

[8] Genbach A, Bondartsev D, Iliev I and Georgiev A 2020 Scientific method of creation of ecologically clean capillary-porous systems of cooling of power equipment elements of power plants on the example of gas turbines. *Energy*. **199**, 117458. doi: https://doi.org/10.1016/j.energy.2020.117458

[9] Genbach A, Bondartsev D and Iliev I 2019 Heat transfer crisis in the capillary-porous cooling system of elements of heat and power installations *Thermal Science*. **23** (2), pp 849 – 860, doi: https://doi.org/10.2298/TSCI171016139G

[10] Genbach A and Bondartsev D 2020 Limiting Thermal State of Capillary-Porous Power-Plant Components. *Russian Engineering Research*. **40**, pp 384-389, doi: https://doi.org/10.3103/S1068798X20050093

[11] Genbach A, Shokolakov K, Bondartsev D and Iliev I 2019 Research of the heat transfer crisis and limit state of the heat exchange surface covered by capillary porous medium, *IOP Conf. Ser.: Mater. Sci. Eng*. 595 012007, doi: https://doi.org/10.1088/1757-899X/595/1/012007

[12] Genbach A, Bondartsev D, Iliev I and Terziev A 2019 Analogy in the processes of heat exchange of capillary-porous coatings in energy installations. *E3S Web Conf.*, **85**, 05003. doi: https://doi.org/10.1051/e3sconf/20198505003

# Solar energy potentials in different climatic zones of Nigeria

**S T Ogunjo[1], A A Obafaye[2] and A B Rabiu[2]**

[1] Physics Department, Federal University of Technology Akure, Ondo State, Nigeria
[2] Center for Atmospheric Research, National Space Research and Development Agency, Anyigba, Kogi State, Nigeria

E-mail: stogunjo@futa.edu.ng

**Abstract.** Solar energy is the cleanest and most abundant renewable energy available in Nigeria, a tropical nation. It is also a viable tool for solving most of our problems ranging from desertification to climate change. This study investigated the solar energy potentials at different climatic zones of Nigeria using four representative locations. Surface temperature and solar radiation measurements retrieved from the study locations were analyzed to obtain the potential power output from solar power systems. It was shown that the monthly mean output power from the solar power systems is affected by seasonal changes for the four locations under study. We also concluded that the increasing temperature is a critical factor limiting the PV production in Nigeria.

## 1. Introduction

Hitherto, fossil fuel has been the main source of energy for domestic and industrial power needs. However, fossil fuels have been found to cause air pollution through combustion, water pollution from transportation and accident spills, and soil pollution from residues. Apart from the limited reserves of different fossil fuels, the price of fossil fuels are easily affected by factors beyond demand and supply. The need for sustainable, environmentally friendly, cheap, and affordable power supply has created the urgent need to harness clean power from nature.

Solar is the most abundant source of energy in the world. The energy from the Sun can be harnessed through solar thermal systems, passive heating systems, and photovoltaic (PV) systems. The limitations of solar energy includes: efficiency of converters, affected by clouds and aerosols [1], and chaotic [2]. Solar energy is significant for tropical countries due to their population, better access to solar radiation, and impact on the economy through tourism. Ogunjo et al [3] reported that solar energy has more potential than wind energy in Akure, Ondo State of Nigeria.

Okoye et al [4] estimated the solar potential for three locations within Nigeria using simulated Typical Meterorological Year (TMY2) data. The authors reported solar radiation potential for Kano, Onitsha, and Lagos as 6.08, 4.43, and 4.42 $kWhm^{-2}$ respectively. The solar radiation potential in the northern and southern region of Nigeria has been reported as 5.62 up to 7.01 and 3.54 up to 5.43 $kWhm^{-2}$ respectively using artificial neural networks [5]. The need to study and investigate solar energy potential across different regions of the country at high temporal resolution using in-situ data has necessitated this work. We aim to study the variation of temperature and solar radiation, as well as, solar energy potential for selected locations in Nigeria.

**Table 1.** Geographical coordinates of the different locations engaged in this work.

| Location | Latitude (°N ) | Longitude (° E) | Altitude (m) | Climate [8] |
|---|---|---|---|---|
| **Abuja** | 9.067 | 7.483 | 536 | Midland |
| **Lagos** | 6.454 | 3.394 | 34 | Coastal |
| **Minna** | 9.615 | 6.546 | 298 | Midland |
| **Port Harcourt** | 4.463 | 7.004 | 508 | Coastal |

## 2. Methodology

Temperature and solar radiation data used in this study were obtained from the ongoing Tropospheric Data Acquisition Network (TRODAN) campaign hosted by the Centre for Atmospheric Research, National Space Research and Development Agency (CAR-NASRDA), Nigeria. The data were acquired using a Campbell Scientific Weather Station with an integration time of 5 minutes in different locations. The data for four locations (Abuja, Lagos, Minna, and Port Harcourt) during the year 2009 were used for this study. The geographical coordinates of the locations are presented in table 1.

### *2.1. Estimating PV power output*

The power output from a single photovoltaic panel was estimated using the expression from Jie et al [6] given as:

$$P = \eta_{T_{ref}} A G_T \left(1 - 0.0045\left(T_C - 298.15\right)\right), \tag{1}$$

where $\eta_{T_{ref}} = 0.14$ is the cell electrical efficiency at reference temperature, $T_C$ is the air temperature in Kelvin, $G_T$ is the solar radiation flux on the module, and $A$=0.5m$^2$ is the area of the PV module. Parameters used were obtained from [7].

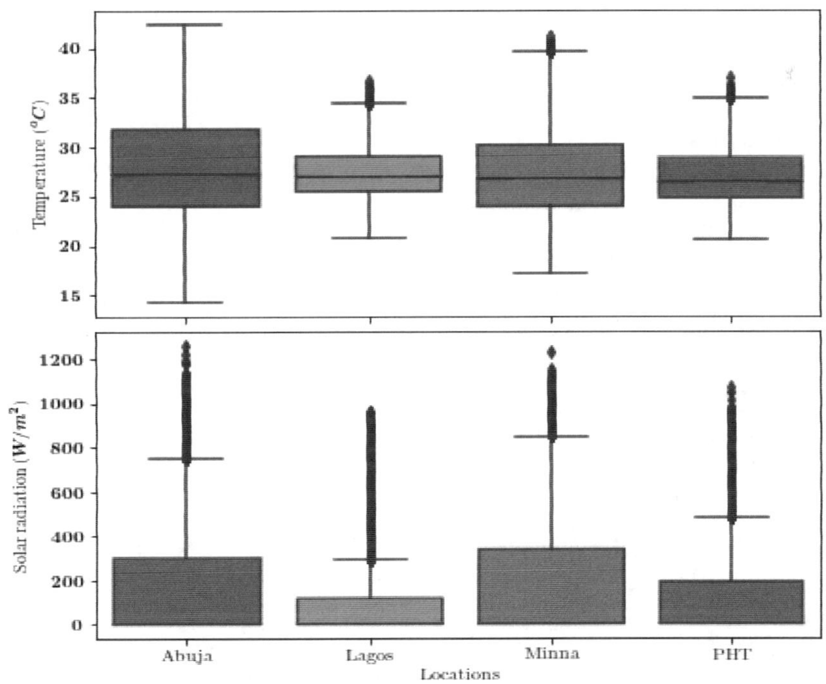


**Figure 1.** Boxplot showing the statistical distribution of 5 minutes temperature (top plot) and solar radiation (bottom plot) in the four locations under consideration.

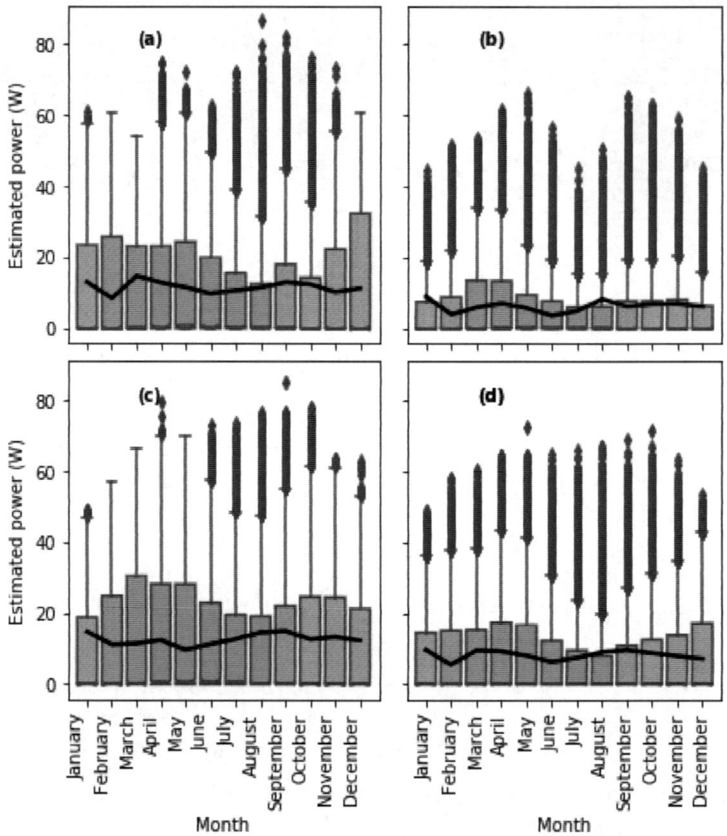


**Figure 2.** Monthly variation of estimated PV output power for (a) Abuja (b) Lagos (c) Minna and (d) Port Harcourt. Mean values are shown in black lines.

## 3. Results and discussion

The statistics of temperature and solar radiation obtained at 5 minutes interval for the four locations are shown in figure 1. Northern locations (Abuja and Minna) have a wider spread (interquartile range) than the southern locations. The mean solar radiation values were observed to be 172.31, 91.94, 185.20, and 120. 51 W/m² for Abuja, Lagos, Minna, and Port Harcourt respectively. The low values in both Lagos and Port Harcourt are due to the closeness to the coastal areas. This reflects in the maximum values of 1263, 963, 1230, and 1074 W/m² for Abuja, Lagos, Minna, and Port Harcourt respectively. Port Harcourt has the minimum temperature (26.98 °C) while Abuja has a maximum temperature of 27.95 °C.

Using the temperature and solar radiation data, the potential solar energy from each of the locations was estimated using equation 1. The monthly variation of the potential solar energy is shown in figure 2 while the annual statistics are presented in table 2. From figure 1, the lowest potential solar energy in all locations was found to occur during July and August, except for Minna with a minimum in January. This can be attributed to the Intertropical convergence zone (ITCZ) [9] and climatic conditions that characterize the locations during those months [10]. The maximum potential energy was obtained during May for Minna and Port Harcourt; April for Lagos, and December for Abuja; these were the months with the greatest clearness index [10]. Fewer or no outliers were found in the northern locations during February and March compared with the southern locations. This suggests larger fluctuations in solar energy potential in the southern locations compared to the northern locations. The monthly and daily variations in air temperature are shown in figures 3 and 4 respectively.

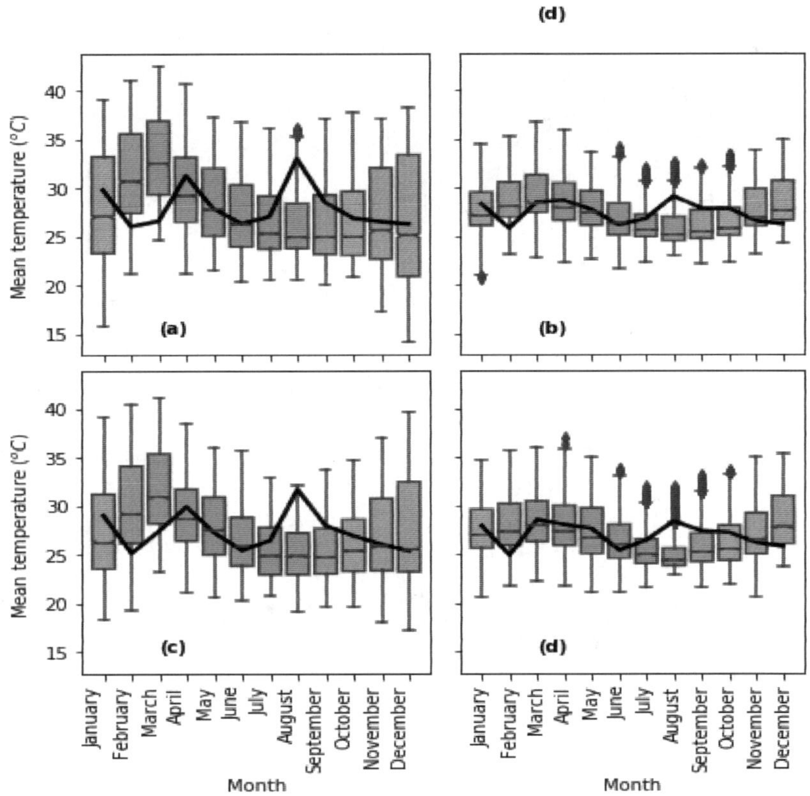


**Figure 3.** Monthly variation of mean temperature for (a) Abuja (b) Lagos (c) Minna and (d) Port Harcourt. Mean values are shown in black lines.

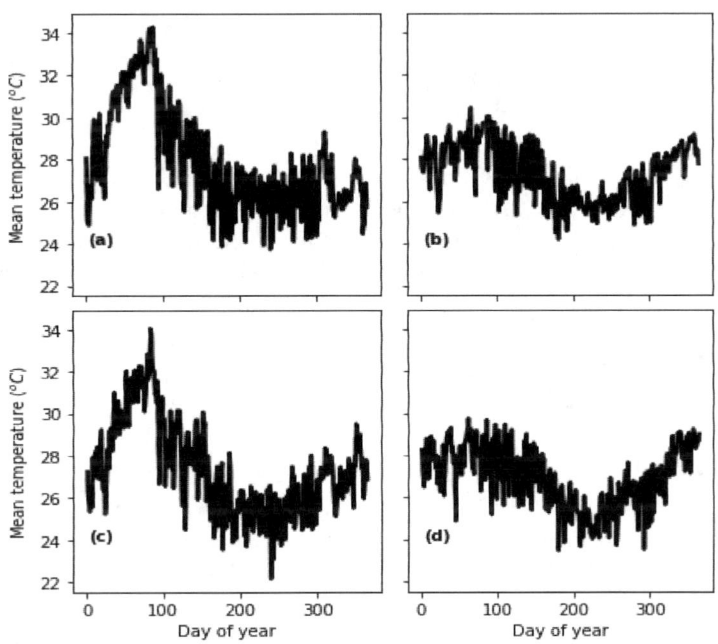


**Figure 4.** Mean daily variation of temperature for (a) Abuja (b) Lagos (c) Minna and (d) Port Harcourt.

## 4. Conclusion

In this study, we have considered the variation of temperature and solar radiation within four locations in different climatic zones of Nigeria. Furthermore, we evaluated the potential solar energy output from the locations. Our results suggest that the northern locations have greater potential and less variability compared to the southern locations investigated in this study. Our approach uses data obtained in a 5 minutes interval over a period of 1 year. Results obtained and presented in this study will be useful in the achievement of the sustainable development goals within the country.

## References

[1] Son J, Jeong S, Park H, Park CE. The effect of particulate matter on solar photovoltaic power generation over the Republic of Korea. *Environ. Res. Letters.* 2020; **15**(8):084004

[2] Ogunjo ST, Adediji AT, Dada JB. Investigating chaotic features in solar radiation over a tropical station using recurrence quantification analysis. *Theoretical and Applied Climatology*. 2017;**127**(1-2):421-7

[3] Ogunjo ST, Adedayo KD, Ashidi AG, Oloniyo MI. Investigating wind-solar hybrid power potential over Akure, southwestern Nigeria. *J Niger Assoc Math Phys*. 2013;**23**:511-6

[4] Okoye CO, Taylan O, Baker DK. Solar energy potentials in strategically located cities in Nigeria: Review, resource assessment and PV system design. *Renewable and Sustainable Energy Reviews*. 2016; **55**:550-66

[5] Fadare DA. Modelling of solar energy potential in Nigeria using an artificial neural network model. *Applied energy*. 2009; **86**(9):1410-22

[6] Jie J, Hua Y, Gang P, Bin J, Wei H. Study of PV-Trombe wall assisted with DC fan. *Building and Environment*. 2007; **42**(10):3529-39

[7] Skoplaki E, Palyvos JA. On the temperature dependence of photovoltaic module electrical performance: A review of efficiency/power correlations. *Solar energy*. 2009; **83**(5):614-24

[8] Ogunjo S, Ife-Adediran O, Owoola E, Fuwape I. Quantification of historical drought conditions over different climatic zones of Nigeria. *Acta Geophysica.* 2019; **67**(3):879-89

[9] Dommo A, Philippon N, Vondou DA, Sèze G, Eastman R. The June–September low cloud cover in western central Africa: mean spatial distribution and diurnal evolution, and associated atmospheric dynamics. *Journal of Climate*. 2018; **31**(23):9585-603

[10] Falayi EO, Rabiu A, Prediction of Clearness Index For Some Nigerian Stations Using Temperature Data. *Journal of Science and Technology*, 2008; **28**(2);94 – 101

## Acknowledgments

The results presented in this paper rely on TRODAN data collected and managed by the Centre for Atmospheric Research, National Space Research and Development Agency, Federal Ministry of Science and Technology, Anyigba, Nigeria. We thank the Centre for Atmospheric Research and their partners for promoting high standards of atmospheric observatory practice as well as the Federal Government of Nigeria for the continuous funding of the Nigerian Space programme (www. carnasrda.com).

IOP Conf. Series: Materials Science and Engineering 1032 (2020) 012041 doi:10.1088/1757-899X/1032/1/012041

# Wind energy assessment for low complex terrain using reference meteorological mast

**A K Terziev[1], Y Panteleev[1], and M Ivanov[1]**

[1] Technical University of Sofia, Dept. of Power Engineering and Power Machines,8 Kliment Ohridski blvd., 1000 Sofia, Bulgaria

E-mail: aterziev@tu-sofia.bg

**Abstract.** The wind flow distribution over semi-complex terrains is evaluated by the roughness factor of the topography and the terrain peculiarities. The energy yield from the turbines in the park is based on the wind parameters extrapolated from the on-site measurement point to wind turbine hug height. For the lower zones of the terrain, the wind shear is influenced by the boundary layer, which can hardly be modeled using only tall tower data and a linear numerical model. The implementation of a second mast (reference), however, gives enough reliable data about the wind parameters at the lower part of the location and is used to adjust the flow and directional behavior from the main mast. The current study shows that for the low complex terrains using the proposed methodology an improvement of the assessed energy yield from the farm with 3% might be anticipated. Furthermore, for more complex terrains where the effect of the orography on the wind shear is more significant the expected improvement should be higher.

## 1. Introduction

The wind is a renewable energy source, utilization systems of which are developing rapidly in the last decades. Thus, performing an accurate analysis of the site wind potential is essential for the profitability of the wind farm. Generally, wind parameters are registered with specialized hardware, installed on tall towers. When performing measurements with meteorological masts, it is necessary to install the measuring equipment far enough, so that it is not exaggerated by the disturbing wind flow around the mast. Therefore, it is essential to follow the relevant standards, applicable to wind measurements [1-3]. Also, when conducting wind measurements, the measuring equipment needs to be calibrated and allowing proper accuracy. In order to estimate the wind velocity distribution in height, it is needed to measure the wind parameters for at least three points above the earth surface. In flat and low complex terrain surfaces, the use of a 50 m mast is enough to determine the wind potential in the area of the site. However, in complex topographies it is required to use higher weather masts.

Once the wind parameters at the measurement point are determined, they should be extrapolated to the hub elevation of the wind turbine, which is at a higher height and can be spaced up to several kilometers. The land topographies as well as the site roughness factor [4-7] have the highest impact on the wind behavior over a given terrain.

In [8] the behavior of the wind shear is analyzed, when reaching a hill, and the results are accomplished experimentally. The behavior of the wind velocity distribution in height between two hills is even more complicated, when the developed flow between them is turbulent [9]. But, not only the topology of the terrain influences the behavior of the velocity profile. The presence of orographic elements on the terrain, such as tall grasses, shrubs or even forest belts and tall trees, have a substantial influence on the wind shear distribution [10].

The possibility of using empirical correlation for the foreseen of the data from the measuring point to the turbine tower height, is presented in [11]. The wind shear factors impact over the energy production from such turbines is presented in [12-13]. Recently, numerical modeling techniques for wind flow analysis over terrains are used [14-16]. Several studies indicate the specifics in the modeling of wind turbulent flows over terrains with varying complexity and further assessment of wind energy yield are presented in [17-22].

The performed analysis demonstrates that the effect of the ground boundary layer on wind shear distribution has a strong relation with the terrain roughness. Thus, the aim of the presented work is to reveal how the use of a second meteorological mast would improve the prediction of the distribution of the wind shear over the terrain, resp. refining the estimated energy production.

## 2. On-site wind data measurements

A semi-complex terrain is chosen in the presented study, and the elevation alteration among the higher and lower part of the site is 150 m. Two masts with a height of 20 and 60 meters (figure 1) are installed. The 60-meter high meteorological mast (Mast 1) is installed on the high part of the site, and the measurements of the wind-flow parameters are at three heights - 40, 50 and 60 m. The low mast (20 m) (Mast 2) is prepared with two measuring devices for wind speed, mounted at 10 and 20 m height. The low-lying sensor is used to determine the effect of the terrain boundary layer on the wind behavior in height. Photography of the two measuring masts is shown on figure 2.

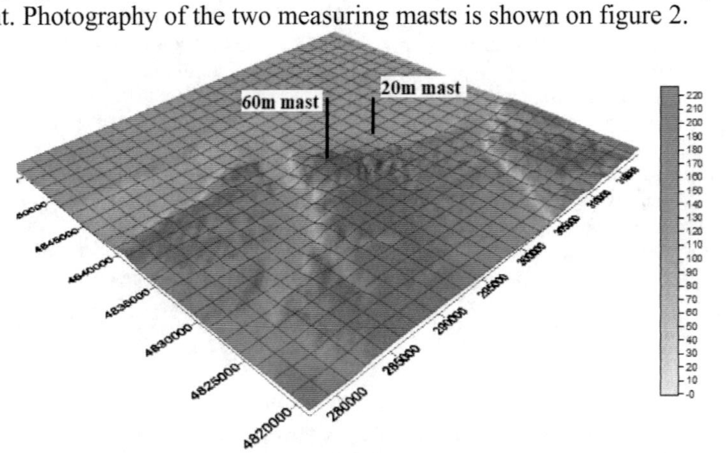


**Figure 1.** 3D orography of the terrain with the installation points of 60 and 20 masts.

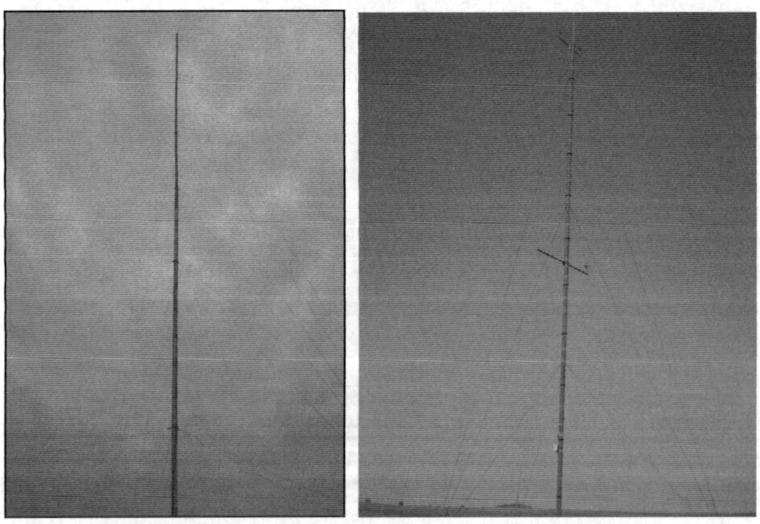

**Figure 2.** View of 60 m and 20 m measuring masts.

The measuring equipment is calibrated, and the operating range and measurement error are presented in table. 1.

**Table 1.** Summarized technical information for the equipment, used for the on-site study.

| Sensor | Model | Range | Accuracy |
|---|---|---|---|
| Cup anemometer | NRG #40C | 1 up to 96 m/s | ±0.14m/s |
| Wind vane | NRG 200P | 0 up to 360° | ± 1° |
| Temperature sensor | TS 21 | -40 up to +70 °C | ± 0.2 °C |
| Pressure sensor | BP 20 | 15 up to 115 kPa | ± 1.5 Pa |

## 3. Wind data analysis

Mast 1 was used to monitor the wind potential for one year, and mast 2 was installed three months later. The average velocity at 60 m is 4.99 m/s, at 50 m - 4.80 m/s and at 40 m - 4.68 m/s. Based on this data, the wind shear profile is built for the high mast, using the two laws - logarithmic and power law:

- **Logarithmic law**

$$U(z)=\begin{cases}\frac{u_*}{k}\ln\left(\frac{z}{z_0}\right), & if\, z>z_0\\ 0, & if\, z<z_0\end{cases} \tag{1}$$

where:

$U(z)$ - wind speed at certain height above ground $z$, $m/s$;

$z_0$ - surface roughness, $m$;

$k$ - von Karman's constant (0.4);

$u_*$ - friction velocity, $m/s$.

- **Power law**

$$U(z)=u(z_1)\left(\frac{z}{z_1}\right)^{\alpha} \tag{2}$$

where:

$U(z)$ - wind speed at some height above ground $z$, $m/s$;

$z_1$ - height of the known wind speed, $m$;

$\alpha$ - power law exponent.

After constructing the wind shear, it is seen that the site roughness factor is 0.014, and the power law exponent, according to the above equation is 0.123.

The measured average wind speed for the considered period with the low mast shows an average wind speed at an elevation of 20 m to be 2.89 m/s, and that at 10 m - 2.15 m/s. The lower velocity is due to the less altitude at which mast 2 is installed. Using the same approach to calculate the wind shear as for mast 1, for the roughness factor is obtained 1.47 m, and for the power coefficient - 0.445. The variance in the two profiles in height is significant. This is mostly because the velocity sensor, mounted at a height of 10 m, is in the zone affected by the ground boundary layer and thus registers significantly lower velocity, which affects the profile behavior in height. The aim of the current study is to assess the error in determining the energy production only when using a low mast on weakly complex terrains, as well as the enhancement when using a reference high mast.

Data for the change of the average wind speed by months, for the studied period, for the two masts at the indicated heights, is presented on figure 3. It is seen a good correlation between the average velocity, both between the individual sensors on one measuring mast and between the two masts. The relation coefficient between the two masts is 89%, which is an indicator of a relatively stable height profile of the wind for the site.

The wind frequency distribution in several directions is shown on figure 4. It is seen that the predominant wind direction is from East and West. The correlation coefficient between the two masts with respect to the wind rose is over 80%, which evidently is acceptable.

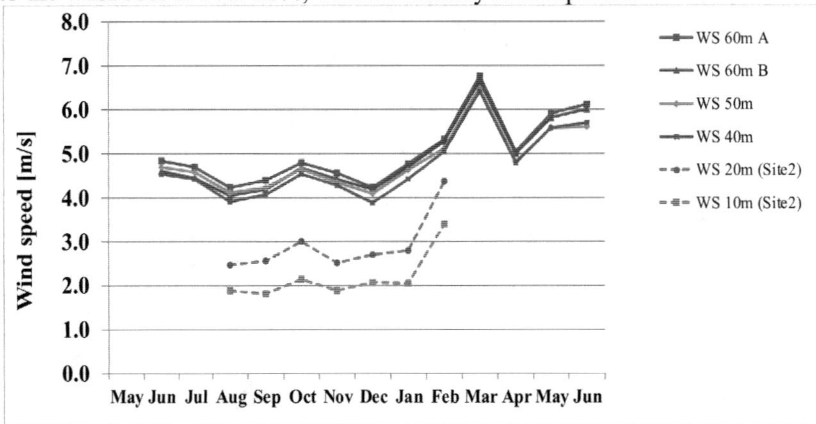


**Figure 3.** Average wind velocity from the measuring masts.

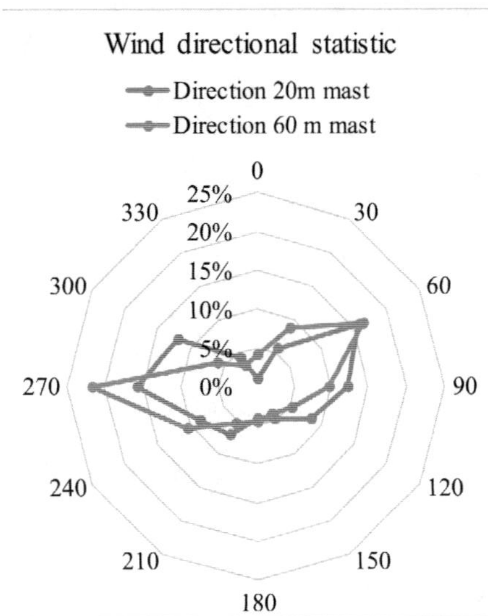


**Figure 4.** Wind rose for the sites of the two masts.

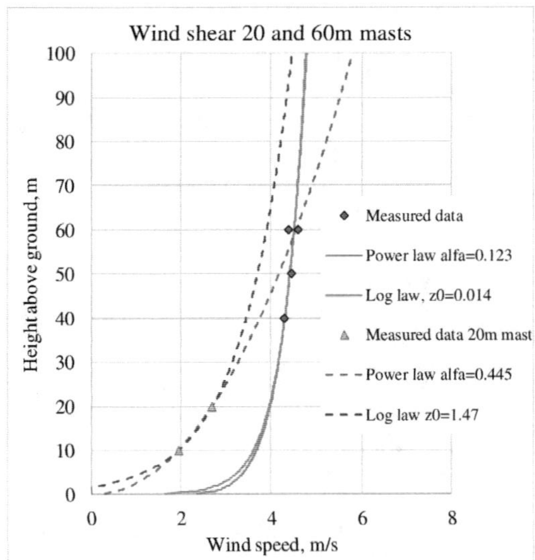


**Figure 5.** Velocity profile in height.

On figure 5 are presented the average measured wind shear for both masts, as well as foreseen data in height up to 100 m, using the logarithmic and power laws. It must be noted that when the wind velocity is measured at several heights above the earth's surface, and when at least two measuring instruments are outside the shadow of the boundary layer, the two laws give similar values for an elevation of 100 m. Nevertheless, when the measurements are up to 20 m high and one sensor, as well as higher locations, falls partially or completely in the shadow of the boundary layer, the difference in extrapolated speeds at 100 m is over 30%, which would significantly compromise a prefeasibility study.

The average turbulence intensity for the corresponding mast at wind velocity of 15 m/s is shown on figure 6. For Mast 1 it amounts to 0.090, and for Mast 2 - 0.109. Both masts fall into class C, which determines the turbulence as low.

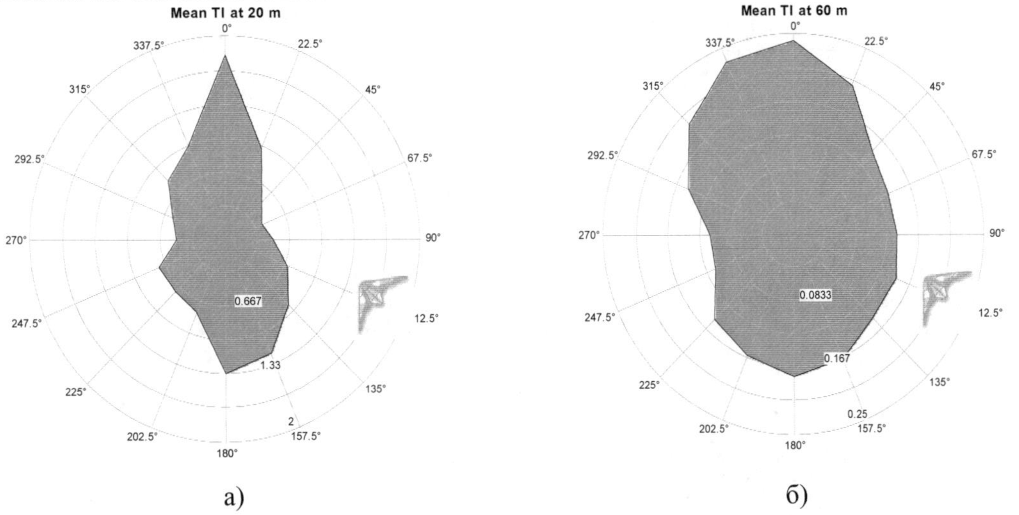


**Figure 6.** Average turbulent intensity as velocity of 15 m/s for: a) 20 m meteorological mast; b) 60 m meteorological mast.

## 4. Numerical study

Measurements from a high meteorological mast, for not less than 1 year, are usually used as an input data for calculating the energy production from a wind farm. The measurements usually last at least two years, and it is obligatory to correlate the data with long-term ones, either from other masts or from the local meteorological stations of BAS (Bulgarian Academy of Sciences).

In the presented numerical analysis, two cases will be assessed - determining the production from a given site when using each of the masts separately, and when using mast 1 as the main and mast 2 as the reference.

The numerical modeling is performed with the WAsP software package. This is a specialized software, used to predict wind behavior on the terrain using a linear model. The aim of the study is to determine how the use of reference measurements can more accurately predict the production from a given terrain with semi-complex geometry.

A three-dimensional surface of the site has been built by digitizing a topographic map. The wind masts and wind generators are situated on it. The numerical modeling used a wind generator type Vestas V90 with an installed capacity of 2.0 MW. The height of the wind turbine hub is 100 m.

### *4.1. Analysis of the energy production using the high meteorological mast (mast 1)*

The data from the field measurements with the high mast (mast 1) were used as input data for the numerical study. The conducted numerical analyses shows that the wind velocity in the vicinity of the high mast is 5.66 m/s (Location 1), and for the location of a wind generator in the vicinity of the low mast is 5.50 m/s (Location 2). Summarized data about the wind generators for the two locations is presented in table. 2.

**Table 2.** Summarized data from numerical study

| Wind Generator | Velocity, m/s | Altitude, m | Energy production, GWG |
|---|---|---|---|
| Location 1 | 5.74 | 182 | 4.407 |
| Location 2 | 5.50 | 30 | 4.048 |

### *4.2. Analysis of the energy production using the low meteorological mast (mast 2)*

The data from the field measurements with the low mast (mast 2) were used as input data for the numerical study. The conducted numerical analyses shows that the wind velocity in the vicinity of the high mast is 5.79 m/s (Location 1), and for the location of a wind generator in the vicinity of the low mast is 5.59 m/s (Location 2). Summary information about the wind generators for the two locations is presented in table. 3.

**Table 3.** Summarized data from numerical study

| Wind Generator | Velocity, m/s | Altitude, m | Energy production, GWG |
|---|---|---|---|
| Location 1 | 5.79 | 182 | 4.302 |
| Location 2 | 5.59 | 30 | 3.980 |

When comparing the results of the numerical studies, it is seen that when using input data in the studies from a high mast, the annual production is 2.3% lower compared to the use of the data from the low meteorological mast (Fig. 7).

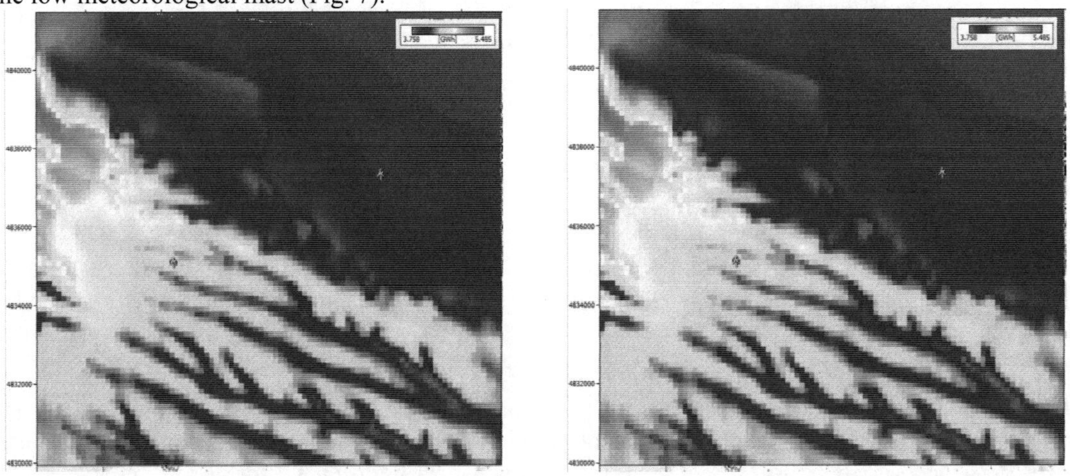

a) б)

**Figure 7.** Results from the numerical solutions with the use of input data from a) 20 m meteorological mast; b) 60 m meteorological mast.

## 5. Conclusions

The presented study reveals the opportunity of using reference wind measurements in the assessment of energy production from a wind farm installed on low complex terrain surfaces. Two masts with a height of 60 m and 20 m, equipped with calibrated equipment, were used. The distance between the masts is about 4 km. A correlation was made with respect to the average velocity registered by the two masts, as the correlation coefficient is over 90%. In addition, a correlation was made with respect to the frequency distribution of the wind in the directions. A high correlation coefficient was also found - over 80%.

The numerical results show that the use of data only from the low meteorological mast leads to some inaccuracy due to the low-lying sensor in the shadow of the boundary layer. With low complex terrain surfaces, this can increase the expected energy production by about 3%. In more complex terrains, however, where the wind shear is strongly influenced the terrain roughness, this increase will be significantly higher. The proposed methodology can be successfully used for flat and low complex terrains. For very complex terrains methods other than linear extrapolation need to be used.

## References

[1] Standard for the preparation of wind energy assessments; German wind energy association; Advisory board for wind resource; 2. Edition 2005-02

[2] Recommendations for the documentation of wind measurements in the context of assessment of energy resource for wind energy projects; German wind energy association; Advisory board for wind resource; I edition 2003-09-25.

[3] VDI 3786 Environmental Meteorology – Meteorological Measurements, Part 1: Fundamentals, edition, November 1995.

[4] Santos M and González M 2019 Factors that influence the performance of wind farms, *Renewable Energy*, **135**, pp 643-51
[5] Gualtieri G and Secci S 2011 Wind shear coefficients, roughness length and energy yield over coastal locations in Southern Italy, *Renewable Energy*, **36** (3), pp 1081-94
[6] Rehman S and Al-Abbadi N M 2005 Wind shear coefficients and their effect on energy production, Energy Conversion and Management, **46** (15–16), pp 2578-91
[7] Fırtın E, Güler Ö and Akdağ S A 2011 Investigation of wind shear coefficients and their effect on electrical energy generation, *Applied Energy*, **88** (11), pp 4097-105
[8] Bradley E F 1980 An experimental study of the profile of wind speed, shearing stress and turbulence at the crest of a large hill, *Q. J. R. Meteorol. Soc.* **106**, pp 101–124
[9] Cao S and Tamura T 2006 Experimental study on roughness effects on turbulent boundary layer flow over a two dimensional stee phill, *J. Wind Eng. Ind. Aerodyn.*, **94**, pp 1–19
[10] Taylor P A, Mason P J and Bradley E F 1987 Boundary-layer flow over low hills, *Bound. Layer Meteorol.*, **39**, pp 107–132
[11] Rehman S, Al-Hadhrami L M, Mahbub Alam Md and Meyer J P 2013 Empirical correlation between hub height and local wind shear exponent for different sizes of wind turbines, *Sustainable Energy Technologies and Assessments*, **4**, pp 45-51
[12] Firtin E, Guler O and Akdag S A 2011 Investigation of wind shear coefficients and their effect on electrical energy generation, *Appl. Energy*; **88**, pp 4097–105
[13] Rehman S, Al-Abbadi N M 2008 Wind shear coefficient, turbulence intensity and wind power potential assessment for Dhulom, Saudi Arabia, *Renew Energy,* **33**, pp 2653–60.
[14] Antonini E G A, Romero D A and Amon C H 2019 Improving CFD wind farm simulations incorporating wind direction uncertainty, *Renewable Energy*, **133**, pp 1011-23
[15] Makridis A and Chick J 2013 Validation of a CFD model wind turbine wakes with terrain effects, *Journal of wind engineering and industrial aerodynamics*, **123**, pp 12-29
[16] Castellani F, Astolfi D, Burlando M and Terzi L 2015 Numerical modelling for wind farm operational assessment in complex terrain, *Journal of Wind Engineering and Industrial Aerodynamics*, **147**, pp 320-29
[17] Vuorinen V, Chaudhari A and Keskinen J 2015 Large-eddy simulation in a complex hill terrain enabled by a compact fractional step OpenFOAM solver, *Adv. Eng. Softw*, **79**, pp 70–80
[18] Kenjereš S and Hanjali´c K 2009Tackling complex turbulent flows with transient RANS, *Fluid Dyn. Res*, **41**, pp 1–32
[19] Kenjereš S and Hanjali´c K 2006 LES, T-RANS and hybrid simulations of thermal convection at high Ra numbers, *Int. J. Heat Fluid Flow*, **27** (5), pp 800–810
[20] De Mello P and Yanagihara J 2010 Numerical prediction of gas concentrations and fluctuations above a triangular hill within a turbulent boundary layer, *J. Wind Eng. Ind. Aerodyn*, **98**, pp 113–119
[21] Cansev G, Sakalli A, Stoyanov I, Iliev T, Mihaylov G and Beloev I 2020 Development of Wind Energy and the Installed Wind Power Plants in Turkey, E3S Web of Conferences **207** (2020), 02013, doi: https://doi.org/10.1051/e3sconf/202020702013
[22] Cansev G, Sakalli A, Stoyanov I, Iliev T, Mihaylov G and Beloev I 2020 Wind Energy Potential for Electricity Production in Hatay, Turkey, E3S Web of Conferences **207** (2020), 02014 doi: https://doi.org/10.1051/e3sconf/202020702014

**Acknowledgments**
The paper was supported under the contract No 202ПД0015-02 with the Research and Development Sector at the Technical University of Sofia.

IOP Conf. Series: Materials Science and Engineering 1032 (2020) 012042 doi:10.1088/1757-899X/1032/1/012042

# Estimation of wind speed and energy potential by atmospheric model for day-ahead market and wind power plants in Turkey

**M A Devrim**[1], **A Sakalli**[2]

[1] Department of Energy Systems Engineering, Energy Institute, Iskenderun Technical University, Hatay, 31200, Turkey
[2] Department of Industrial Engineering, Science Institute, Iskenderun Technical University, Hatay, 31200, Turkey

E-mail: mdevrim.ee18@iste.edu.tr

**Abstract.** The requirement for energy has expanded day by day due to excessive consumption, developing technology, and population growth, especially after the Industrial Revolution. Since the investments for promoting renewable energy sources, The Turkish energy sector has been progressing in the way of renewable energy. This inevitable change in the energy sector and variable power generation values in wind energy directly affect the energy markets. Accompanying with the structural transformation of the energy sector, Turkey had launched the Day-Ahead electricity marketing system in December 2011. The price per megawatt-hour is determined on the open market by the supply and demand. Hatay province was selected in this study because of its highest wind potential. Temperature and wind speed values, observed in Hatay Samandağı for October 2019, were determined. The data was configured with The Weather Research and Forecasting Model. The results indicate that the wind speed at high altitudes could reach above 7 meters per second on Turkey's eastern Mediterranean coast, Hatay. The results were compared with the real-time data, and the error rates were determined. In the selected region, the electrical energy values generated using the model results were calculated the day before.

## 1. Introduction

Since the highly increasing consumption of energy and population growth rate, in other words, raised demand on the global scale and the decline of fossil-based fuels, the motive for alternative energy is advancing day by day. Obstacles caused by climate change and the social sensitivity about the environmental policies force the energy sector into ongoing reform. Due to the international sanctions, most countries have been tasked with reducing CO2 emissions, so that the incentive programs for renewable energy have widespread internationally. In 2015, the Paris Agreement was ratified along with international accession, including Turkey. In the agreement, Turkey has set its target limits for 2030 and has made commitments for its realization. The commitment is endorsed to reduce the anticipated greenhouse gas emissions by 21% within the next 10 years and increase solar energy power usage to 10 GW and wind energy power to 16 GW [1]. Because of the policies such as credit facilities, full purchase of guaranteed renewable energy, investment incentives; wind power generation has been advancing over the last decades. Managing wind energy potential not only gear up economic development and reduces carbon emission but also contributes to environmental quality, energy security and independence, and conditions to restore life quality [2]. In 2019, wind power generation expanded by 3.34% compared to the previous year and reached an energy capacity of 7615 MWh [3].

The electric power generated from wind energy depends directly on wind speed parameters, and deviations in wind speed are at a level that will affect energy markets [4]. Zhao et al had the idea that fluctuations in wind speed, stochasticity, and wind outages could challenge the stabilization of electrical networks in high penetration level situations [5].

Turkish Energy Market, as of December 2011, market swap prices and purchase and sale amounts are announced in line with the offers offered by market participants in the Day Ahead Markets. After the new system, stock market prices are created with bids submitted by market participants. The price of electricity per MWh, together with the pre-day markets, is determined according to supply and demand conditions and is intended to balance the energy markets [6]. In deregulated markets, daily prices are considered reference prices. Unpredicted situations in the operation systems, such as deviations in demand, generation unit failures, and other supply issues, can lead to price fluctuations in the real-time market, referred to as price imbalances [7]. In order to eliminate these problems, high wind potential regions of Turkey have been identified, and studies have been started.

Turkey's average wind speed is 2.45 m/s over a year period. Wind energy sources are commonly located in the coastal areas of Turkey.[8]. Turkey's top-level wind energy potential areas are the Marmara and Aegean coasts and higher altitude soils with an average annual wind speed of 6.9 m / s at 50 m altitudes. East parts of Mediterranean coastal areas also have high potential wind energy. [9]. Analysis of statistical data about wind distribution, average wind density, and speed for the Hatay region demonstrate that wind speeds above 4 m/s are observed at an altitude of 10 m above the surface for more than 80% of the year [10].

## 2. Materials and Methods

*2.1 Study Area and Data*

Hatay province, one of the highest wind potential areas of Turkey, is considered as this study area. In the first stage, the fields to run the model have been determined. Turkey was chosen as the main region. The area covering the Samandag coast, a part of Hatay province, was then defined as the sub-region. The main domain of the work area was also demonstrated in figure 1.

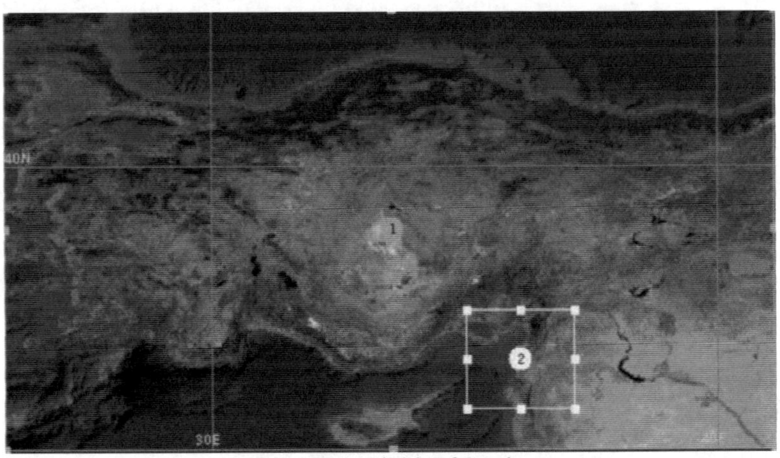


**Figure 1.** Field Of Study.

2.2 *Methods*

The data analyzed in this research were obtained by Weather Research and Forecasting Model (WRF). A basic set of physics components is included in WRF to determine and calculate variables and to get data outputs at high accuracy. The model was configured with the physical conditions as in table 1 [11].

**Table 1.** WRF physics options.

| Function | Parameter | Description |
|---|---|---|
| ra_lw_physics | longwave radiation | RRTM scheme: Rapid Radiative Transfer Model. An accurate scheme using look-up tables for efficiency. |
| ra_sw_physics | shortwave radiation | Dudhia scheme: Simple downward integration allowing for efficient cloud and clear-sky absorption and scattering. |
| bl_physics | planetary boundary layer | YSU scheme |
| cu_physics | cumulus | Kain-Fritsch (new Eta) scheme: deep and shallow sub-grid scheme using a mass flux approach with downdrafts and CAPE removal time scale and no cumulus |
| mp_physics | microphysics | WRF Single-Moment (WSM) 3-class simple ice scheme: A simple, efficient scheme with ice and snow processes suitable for mesoscale grid sizes. |
| sfclay_physics | surface layer | Monin-Obukhov Similarity scheme: Based on Monin-Obukhov with Carslon-Boland viscous sub-layer and standard similarity functions from look-up tables |

Study areas with defined limits are structured to receive 25 days of data. After the configuration was completed, first static geographic data and then the global prediction system (GFS) data were defined in the study areas. The defined GFS data were applied in 3-hour periods with a resolution of 0.25 degrees. The model was run through October 2019. The data taken from the model is configured to predict 24 hours before each working day. Version 1.9.3 of the Climate Data Operators (CDO) software is used to visualize the data. Measured data were observed from 25 locations for the working area. The average wind speed of 25 locations has been measured, and the wind energy power that can be generated from these locations has been calculated.

## 3. Results

As a result of the data received from the atmospheric model of October 2019 in Turkey, the average temperature and wind speed values are visualized. As a result of the data outputs, it was determined that high wind speed values were reached in the locations where the temperature increased and in the coastal regions where the temperature difference was intense. Average temperature and wind speed maps are shown in figure 2 and figure 3.

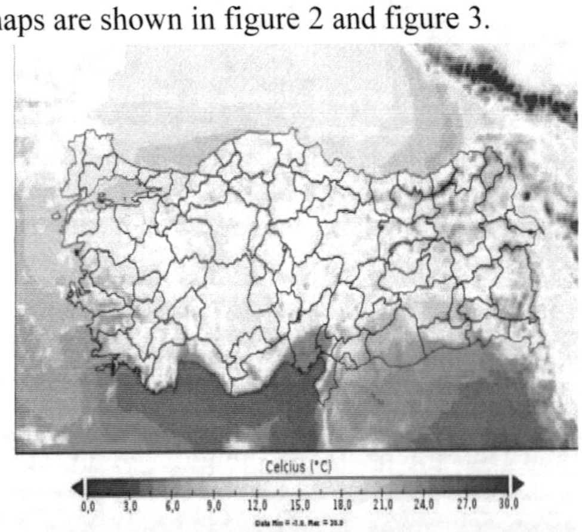


**Figure 2.** October 2019 Temperature Map.

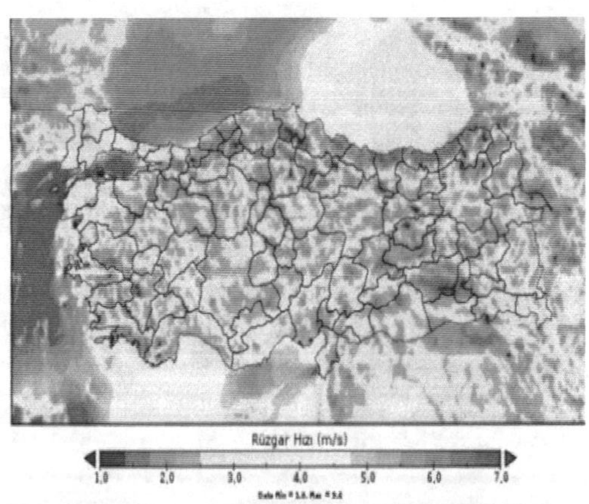


**Figure 3.** October 2019 Wind Speed Map.

Wind speed values were taken from the measurement centers in the Samandag region during October 2019 at eighty meters. The average of the measurement data taken in hourly periods is shown in figure 4.

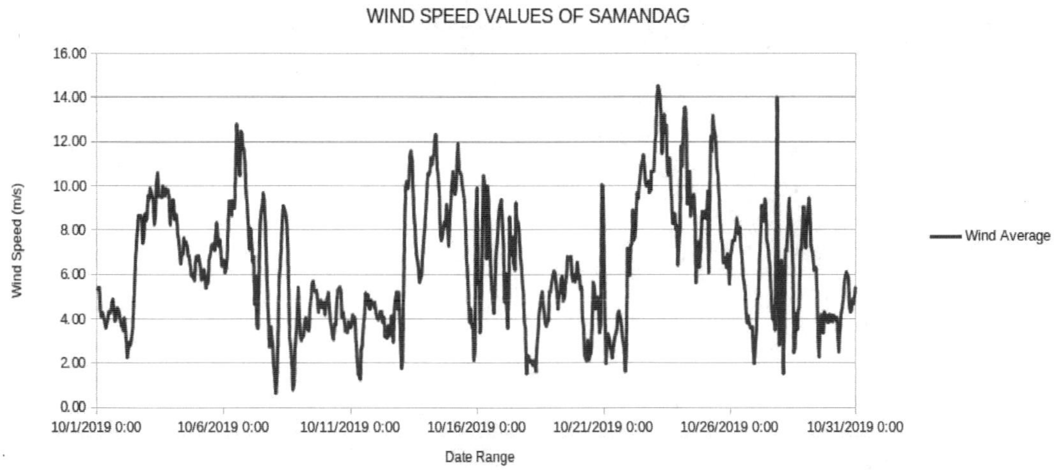


**Figure 4.** Wind velocity graph at measured locations.

The wind speed fluctuated at a low level between the 9th and 13th of October in the Samandag region. Between 23th and 25th of October, the wind speed fluctuated at a high level. While the lowest wind speed values were encountered between the 9th and 13th days of October, the highest wind speed values were reached on the 22nd and 25th days. The amount of electrical power that can be produced with the wind velocity values obtained from the measurement and atmospheric model data is shown in figure 5.

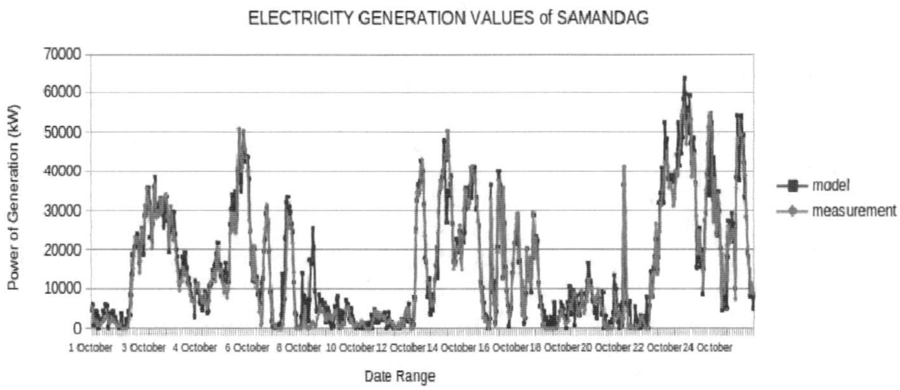


**Figure 5.** Electric power graph that can be generated from model and measurement data.

As a result of the measurement data and the wind parameter values obtained from the atmospheric model, both production values were examined by regression analysis. Analysis results are shown in figure 6.

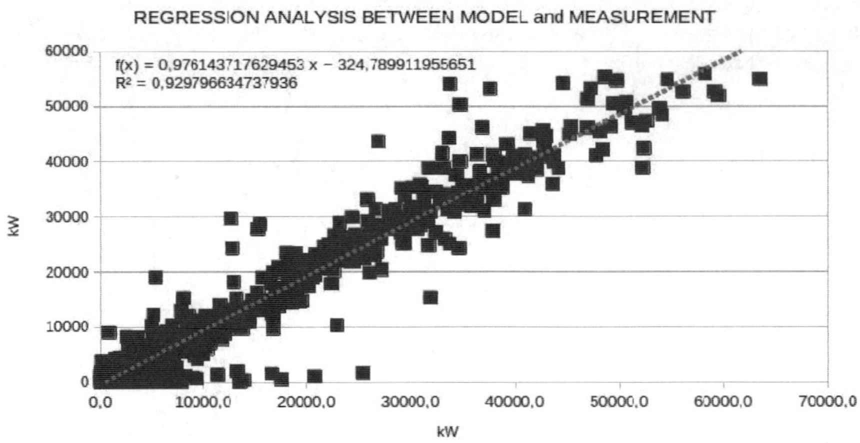


**Figure 6.** Regression analysis graph of model and measurement data.

As a result of our analysis, the consistency between our model and the measurement values was observed at a rate of 92%.

## 4. Discussion

In our study, atmospheric wind velocity parameter estimates with the high resolution WRF model were determined with pre-day reports. The shares of the obtained data outputs in electricity generation were compared with real-time data, and the consistency between them was observed at a rate of 92%. As a result of the analysis, wind production values can be estimated one day earlier, and the fluctuations in the Turkish Energy Market can be surpassed. In their study, Sahin et al. (2005) found that October wind velocity values of the Samandag region between 1997 and 2001 were between 3 and 4 m / s at an altitude of 10 meters [12]. According to the previous study, in the WRF model we used, the wind speed values at 10 meters height were higher than 1 m / s. The two studies' value changes may be caused by differences in data set and positions in the study area. Also, the physics packages we used in our WRF model and the degree of resolution affected the data's accuracy. Temporal differences and atmospheric changes caused by climate change can also affect the wind speed and electricity generation power. In the coming years, changes in the temperature difference and unplanned urbanization may cause a sudden decrease and increase in wind speed and electrical energy generated from wind energy.

## References

[1] Kat B, Paltsev S, & Yuan M 2018 Turkish energy sector development and the Paris Agreement goals: A CGE. *Energy Policy* **122**, p 84 – 96
[2] Genc C, Sakalli A, Stoyanov I, Iliev T, Mihaylov G and Beloev I 2020 Development of wind energy and the installed wind power plants in Turkey, *E3S Web Conf.*, **207** (2020) 02013, doi: https://doi.org/10.1051/e3sconf/202020702013
[3] Turkish Wind Energy Statistic Report July. Ankara*: Turkish Wind Energy Association.* 2019
[4] Lei M, Shiyan L, Chuanwen J, Hongling L and Yan Z 2009 A review on the forecasting of wind speed and generated power. *Renewable and Sustainable Energy Reviews* **13**, p 915-920
[5] Zhao X, Liu J, Yu D and Chang J 2018 One-day-ahead probabilistic wind speed forecast based on optimized numerical weather prediction data. *Energy Conversion and Management* **164**, p 560-569
[6] Sahin C 2018 Consideration of Network Constraints in the Turkish Day Ahead Electricity Market. *Electrical Power and Energy Systems* **102**, p 245-253

[7] Gayretli G, Yücekaya A and Bilge A H 2019 An analysis of price spikes and deviations in the deregulated Turkish power market. *Energy Strategy Reviews* **26**, 100376

[8] Ilkiliç C and Aydin H 2015 Wind power potential and usage in the coastal regions of TurkeyWind power potential and usage in the coastal regions of Turkey. *Renewable and Sustainable Energy Reviews* **44**, p 78-86

[9] Emeksiz C and Demirci B 2019 The determination of offshore wind energy potential of Turkey by using novelty hybrid site selection method. *Sustainable Energy Technologies and Assessments* **36**, 100562

[10] Genc C, Sakalli A, Stoyanov I, Iliev T, Mihaylov G, Beloev I 2020 Wind energy potential for electricity production in Hatay, Turkey. *E3S Web Conf.,* **207** (2020) 02014 doi: https://doi.org/10.1051/e3sconf/202020702014

[11] Carvalho D, Rocha A, Gomez-Gesteira M, and Santos C 2012 A sensitivity study of the WRF model in wind simulation for an area of high wind. *Environmental Modelling & Software* **33**, p 23-34

[12] Sahin B, Bilgili M and Akilli H 2005 The wind power potential of the eastern Mediterranean region of Turkey. *Journal of Wind Engineering and Industrial Aerodynamics* **93**, p 171-183

# Impacts of climate change and distribution of precipitation on hydroelectric power generation in Turkey

**U Bulut**[1], **A Sakalli**[2]

[1] Department of Energy Systems Engineering, Energy Institute, Iskenderun Technical University, Hatay, 31200, Turkey
[2] Department of Industrial Engineering, Science Institute, Iskenderun Technical University, Hatay, 31200, Turkey

E-mail: umutbulut.ee18@iste.edu.tr

**Abstract.** The growing energy demand besides the rapidly developing consumer sector is of great importance on a global scale. In this study, hydroelectric energy recovery in Turkey, which is affected by the negative consequences of climate change, were taken into account. Increased irregular rainfall patterns and drought events as a consequence of global climate change pose a significant danger to hydroelectric power potential. Kızılırmak River has an important position on account of hydroelectric energy. In this study, the yearly flow rates of 2 stations determined by months in the Kızılırmak basin were supplied from the State Water Works (DSI) current and observation annuals. Analysis was conducted for two stations between 2005 and 2015. In addition, using ERA5 Ensemble satellite data, the group was analyzed using " climate data operators" (CDO), a map of Turkey's total annual average rainfall and annual average temperature with a resolution of 12.5 km x 12.5 km between 1979 and 2019. According to our analysis, Turkey's average annual precipitation between 1979-2019 was 527.61 mm, and the highest annual average temperature in the 30-year data set between 1979-2019 was 15.31°C in 2010 and 2018. The average temperature in 2019 is 14.96°C.

## 1. Introduction

Climate change, by bringing along its negative consequences, has become one of the main problems in the world and has attracted the interest of Turkey and the energy sector and has brought about the need to work on this issue.

The IPCC report found that climate change and global climate change causes are among the causes of anthropogenic causes. Estimated anthropogenic global warming increases by 0.2°C every ten years owing to past and ongoing emissions. Human activity is predicted to be in the range of about 0.8°C to 1.2°C, resulting in warming of about 1.0°C from pre-industrial levels. If global warming continues to increase at the current pace, it seems like to reach 1.5°C by 2030-2052 and a peak temperature of 2°C by 2100 [1]. Given this risk because of climate change, the final share of renewable energy should reach 65% by 2050, up from 19% in 2017 [2]. Turkey has an important position in connection with the potential of hydroelectric energy and other renewable energy sources [3,4].

In our study, the effects of climate change and rainfall distribution on hydroelectric energy in Turkey were examined. Turkey's average annual temperature between 1981-2010 increased by 1.9°C, with 15.4°C in 2018 being the second warmest year since 1971. Turkey's average rainfall in 2018 (658.7 mm) was observed to increase by 30% compared to the average rainfall in 2017 (506.6 mm) and an increase of 14.8% compared to the country's normal [5].

In Turkey and other countries of the world, there are studies on the change of temperature and precipitation parameters on hydroelectric production, which are among the important elements of climate and climate change. A significant increase in drought was observed in the Kızılırmak Basin during 12 and 60 months using the SYI (Standard Precipitation Index) method [6]. In the same basin, according to the trend analysis approach, it is seen an increase in temperature and a decrease in rainfall [7]. An upward trend in average annual temperature and a downward trend in total annual precipitation may be the menace of drought [8]. It has also been emphasized that the water flow is directly mean to the rainfall, and therefore the hydroelectric output power will fall in proportion to the precipitation [9].

## 2. Materials and methods

The Kızılırmak Basin (1355 km) covers an area of 78.180 $km^2$, a significant portion of Turkey's water potential (3.5%) with an annual flow volume of 6.48 billion $m^3$ [6], [7], [10]. In this study, annual flow rates of 2 stations determined in Kızılırmak Basin according to the months obtained from State Water Works (DSI) current and observation annals. 2 stations (Sarımsaklı Baraj Girişi, Hüseyinli) indicated on the Google Earth map are shown in figure 1. Analysis was conducted for the two stations between 2005-2015. Furthermore, using ERA5 Ensemble satellite data, a map of the total annual average rainfall of Turkey with a resolution of 12.5 km x 12.5 km between 1979 and 2019 and the average annual temperature was analyzed using "Climate Data Operator" (CDO) [11]. In addition, adjacent grids in the study area were identified and regional precipitation amounts were analyzed.

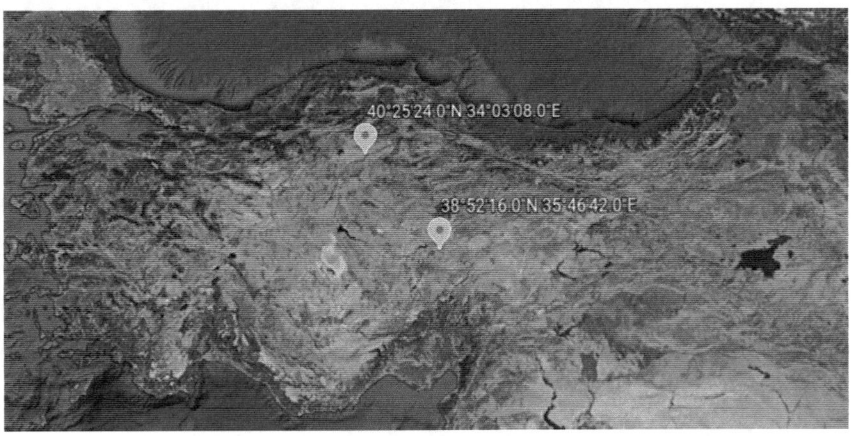


**Figure 1.** Study area.

## 3. Result

The average precipitation of Turkey between 1979-2019 is shown in figure 2. According to this map, the region with the most precipitation is the Eastern Black Sea region.

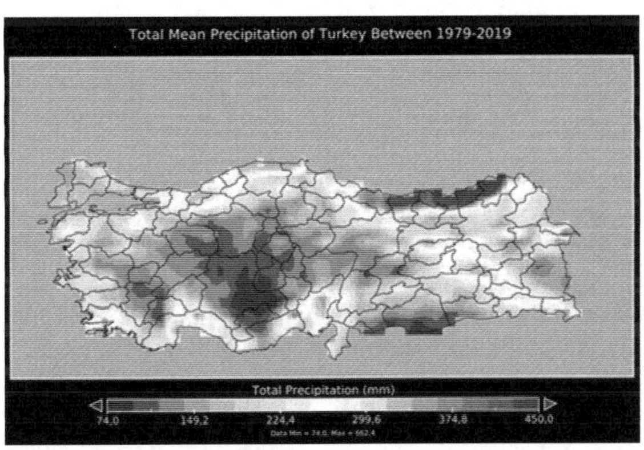


**Figure 2.** Map of total average precipitation in Turkey between 1979-2019.

Turkey's total annual average precipitation between 1979-2019 was 527.61 mm. In the 30-year data set, the highest precipitation was 614.9 mm in 1988. The lowest precipitation was observed in 2008 with 418.2 mm. The average rainfall in 2019 was 578.5 mm. The average annual rainfall of Turkey is shown in figure 3.

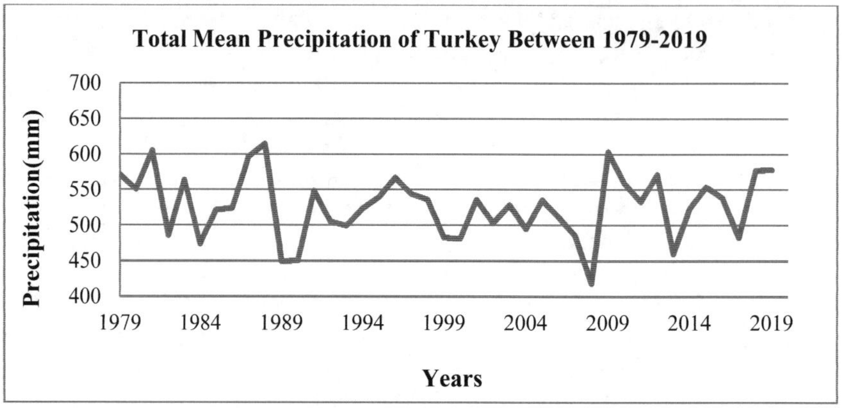


**Figure 3.** Total mean precipitation of Turkey between 1979-2019.

The highest average annual temperature in the 30-year data set from 1979-2019 is 15.31°C in 2010 and 2018. The average temperature in 2019 is 14.96°C. Average annual precipitation amounts between 1979-2019 and temperature values between the same years are shown in figure 4.

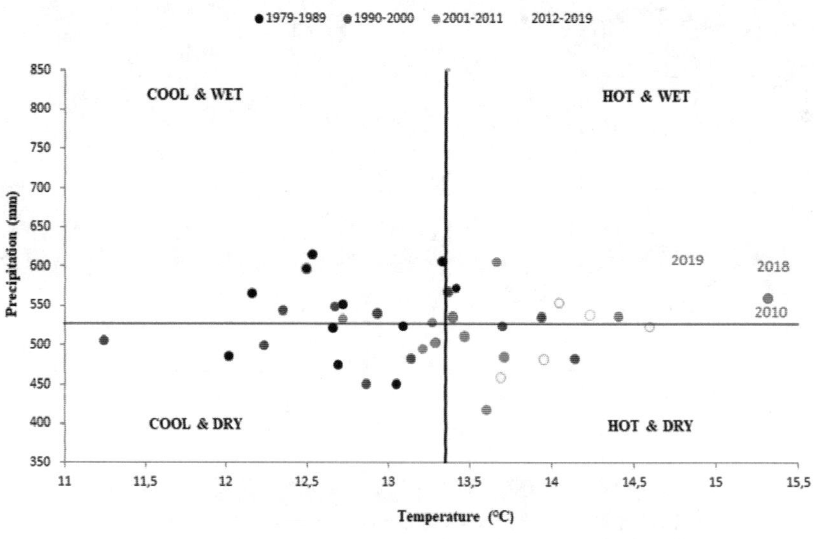


**Figure 4.** Average annual rainfall and temperature graph.

| | | |
|---|---|---|
| 1. Grid | 2. Grid | 3. Grid |
| 118.34 mm | 128.48 mm | 143.33 mm |
| 4. Grid | Sarımsaklı Baraj Girişi* | 6. Grid |
| 125.41 mm | 136.96 mm | 136.96 mm |
| 7. Grid | 8. Grid | 9. Grid |
| 136.02 mm | 160.30 mm | 157.00 mm |

**Figure 5.** The adjacent grids of the Sarımsaklı Baraj Girişi and average annual precipitation amounts.

Reference points are also marked in figure 5 and figure 6 to determine the regional precipitation amount of the 2 stations selected in this study. Accordingly, 8 different neighboring grids close to the reference points (1.Grid, 2.Grid, 3.Grid, 4.Grid, 6.Grid, 7.Grid, 8.Grid, 9.Grid). The amount of precipitation at these points was determined and studied. Accordingly, the total average rainfall of Sarımsaklı Baraj Girişi station between 1979-2019 is 136.96 mm, and for Hüseyinli station is 124.47 mm. (It is accepted that the grids show a homogeneous distribution). Precipitation in regions where the stations are located in close value with average precipitation of Turkey. This analysis shows that the results are consistent.

| | | |
|---|---|---|
| 1. Grid | 2. Grid | 3. Grid |
| 124.00 mm | 111.74 mm | 120.70 mm |
| 4. Grid | Hüseyinli* | 6. Grid |
| 121.64 mm | 124.47 mm | 133.90 mm |
| 7. Grid | 8. Grid | 9. Grid |
| 136.73 mm | 140.97 mm | 134.61 mm |

**Figure 6.** The adjacent grids of the Hüseyinli and average annual precipitation amounts.

The average monthly flow rate of the stations examined between 2005-2015 is shown by year in figure 7 and figure 8. The maximum flow rate for Hüseyinli station in February 2012 was 185 $m^3/s$. It had a dry season of two months in 2007 and four months in 2010. Also, no data entry was found in 2009. The maximum flow rate for Sarımsaklı Baraj Girişi station in April 2015 was 4.29 $m^3/s$. In July 2014, a minimum flow rate of 0.61 $m^3/s$ was observed.

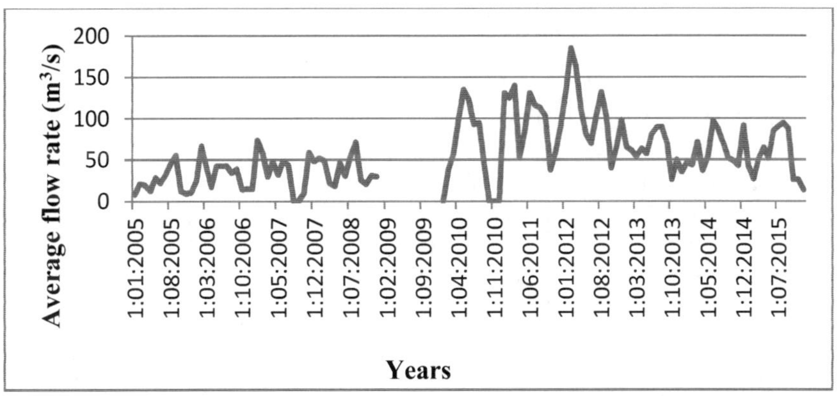


**Figure 7.** Hüseyinli station average annual flow rate graph from 2005 to 2015.

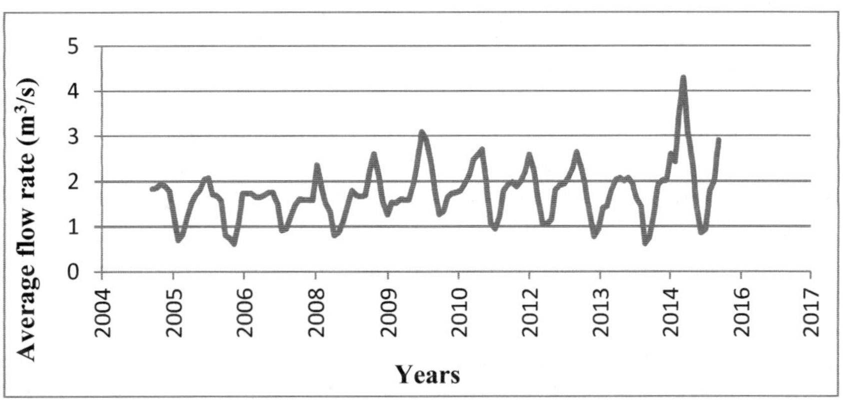


**Figure 8.** Sarımsaklı baraj girişi station average annual flow rate graph from 2005 to 2015.

## 4. Discussion

In our study, the total annual average precipitation in Turkey with a resolution of 12.5 km x 12.5 km between 1979 and 2019 using ERA5 Ensemble data and the average annual temperature amount were analyzed using CDO. An increase of 4.07°C was found when looking at the lowest and highest annual average temperatures between 1979-2019. Between 1979-2019, Turkey's average annual precipitation (527.61 mm) was not significantly reduced after 2008.

Average annual flow rates of 2 stations determined for our study were obtained from the State Water Works (DSI) current observation annuals. The average monthly flow amounts compared to the years shown in figure 7 and figure 8 were compared with the average precipitation and temperature data compared to the years shown in figure 4. A decline in rainfall and an increase in temperature tended to decline in flow rates at 2 stations. A decline in flow rate means a decline in hydroelectric output power.

In 2019, Turkey had an average annual Area precipitation of 585.1 mm. This value was 1.9% more than the 1981-2010 average precipitation (574 mm). Monthly precipitation in February, March, May, September, October, November of 2019 was below the average amount of rainfall in 1981-2010, and in the other months of 2019 has been above average in the work [12]. The value differences between the two studies can be caused by data set differences in the workspace or by the methods used.

## References

[1] Zahai P *et al* 2018 Intergovernmental Panel on Climate Change Global Warming of 1.5°C. *An IPCC Special Report on the impacts of global warming of 1.5°C above pre-industrial levels and related global greenhouse gas emission pathways, in the context of strengthening the global response to the threat of climate change, sustainable development, and efforts to eradicate poverty* chapter 4,5

[2] Solaun K and Cerda E 2019 Climate change impacts on renewable energy generation. A review of quantitative projections *Renewable and Sustainable Energy Reviews,* **116**, 109415, https://doi.org/10.1016/j.rser.2019.109415

[3] Genc C, Sakalli A, Stoyanov I, Iliev T, Mihaylov G and Beloev I 2020 Development of wind energy and the installed wind power plants in Turkey, *E3S Web Conf.*, **207** (2020) 02013, doi: https://doi.org/10.1051/e3sconf/202020702013

[4] Genc C, Sakalli A, Stoyanov I, Iliev T, Mihaylov G and Beloev I 2020 Wind energy potential for electricity production in Hatay, Turkey. *E3S Web Conf.*, **207** (2020) 02014 doi: https://doi.org/10.1051/e3sconf/202020702014

[5] Turkish State Meteorological Service *2018 Climate Assessment* Jan. 2019 Available at: https://www.mgm.gov.tr/FILES/iklim/yillikiklim/2018-iklim-raporu.pdf

[6] Arslan O, Bilgil A and Veske O 2016 Meteorological drought analysis in Kızılırmak Basin using standardized precipitation index method *Nigde University Journal of Engineering Sciences* **5**(2) pp 188-194, https://doi.org/10.28948/ngumuh.295572

[7] Bahadır M 2011 A statıstical analysis of the flow changes of Kızılırmak River, *Turkish Studies - International Periodical For The Languages, Literature and History of Turkish or Turkic* vol **6**(3) pp 1339-1356, http://dx.doi.org/10.7827/TurkishStudies.2120

[8] Ercan B and Yüce M I 2017 Trend analysis of hydro-meteorological variables of Kızılırmak Basin, *Nevsehir Journal of Science and Technology* **6** pp 333-340, doi:10.17100/nevbiltek.323640

[9] de Jong P, Tanajura CAS, Sánchez AS, Dargaville R, Kiperstok A and Torres EA 2018 Hydroelectric production from Brazil's São Francisco River could cease due to climate change and inter-annual variability, *Science of the Total Environment* **634** pp 1540–1553, https://doi.org/10.1016/j.scitotenv.2018.03.256

[10] Kisi O and Ay M 2014 Comparison of Mann-Kendall and innovative trend method for water quality parameters of the Kızılırmak River, Turkey *Journal of Hydrology* **513** pp 362–375

[11] CDO 2015 Climate Data Operators. Available at: http://www.mpimet.mpg.de/cdo

[12] Turkish State Meteorological Service *2019 Climate Assessment* Jan. 2020 Available at: https://www.mgm.gov.tr/FILES/iklim/yillikiklim/2019-iklim-raporu.pdf

# Utilization of the energy potential of waste meal and creating of "green" workplaces

**M K Mladenov, S A Yaneva**

University of Chemical Technology and Metallurgy, Sofia, Bulgaria, 1756, 8 St. Kliment Ohridski bul.

E-mail: mladenov@uctm.edu

**Abstract.** The issue of finding a suitable eco-technological approach for the utilization of waste meals generated in the production of vegetable oil is extremely relevant in terms of achieving a high "non-waste" degree in modern oil production. In practice, the direction related to the application of the meal as a material resource in animal husbandry is widely known – mainly for feeding of farm animals. At the same time, there are alternative areas related to other economic sectors, such as composting, biodiesel production, heat treatment and others, whose economic efficiency is still not sufficiently substantiated. Along with the topic of waste recovery, the measures for prevention of workers and ensuring safe and healthy working conditions are considered. The creation of "green" jobs, which directly or indirectly contribute significantly to the protection and restoration of the environment and to the reduction of the consumption of energy and raw materials, is extremely relevant in this direction.

## 1. Introduction

World olive oil production is concentrated in the Mediterranean region, with Spain, Greece and Italy being the three main producers, amounting to about 80% of world production. Olive oil production is one of the major sectors of agriculture, whose growth is expected to continue in the future. This industry generates significant amounts of waste and by-products (olive meal wastes) during the olive processing. This type of waste is a serious environmental problem for olives and olive oil producing countries, mainly due to emerging soil and water pollution [1-5]. According to [6], 35-40% of the processed olives are turned into waste meal (cake). We can point out that depending on the type of production process used, the generated quantities of the olive mill wastewater can be achieved up to 1.8 $m^3$/ton olives at continuous process and up to 0.5 $m^3$/ ton olives at discontinuous process [7].

The use of the generated meal as a raw material or energy resource will help to achieve a positive economic effect for processing enterprises in two directions: reducing the price of produceced olive oil and reducing the cost of treatement of the generated meal. The use of the meal as a material resource is realized mainly for feeding farm animals [8, 9], but it is possible to apply other methods such as composting, biodiesel production, direct application in soils to restore organic carbon stocks, etc. [10, 11]. Opportunities for the use of this type of waste as an energy resource are the biogasification, incineration, etc. [1, 4, 5, 12, 13].

Meal is characterized by high residual oil content and high humidity. For this reason, its utilization for energy production must be carried out after pre-treatment in order to reduce the moisture content or whith mixed with additional mass of combustible materials with low humidity. In [14] was reported that for this purpose, the combination with wood particles would be extremely appropriate.

Balancing economic growth and the need to protect the environment are challenging goals to increase energy efficiency and to promote renewable energy, as well as to reduce waste. In the occupational safety sector, this has led to the emergence of a wide range of "green" workplaces – jobs that contribute to protecting the environment or restoring it to its original state. "Green" workplaces cover a wide range of different jobs in different sectors and include a diverse workforce. These may include workplaces that protect ecosystems and biodiversity or reduce energy and raw material consumption, or reduce waste and pollution [15]. However, for "green" workplaces to be sustainable, they must provide safe, healthy and worthy working conditions.

Work in agriculture and manufacturing is among the most dangerous occupations, and as a type of work, agriculture and forestry rank third or fourth among the most dangerous occupations in the European Union [16]. According to the International Labor Organization (ILO), about two million people die each year as a result of work-related accidents and occupational illnesses. It is estimated that 160 million people suffer from work-related illnesses, and 270 million fatal and non-fatal work-related accidents occur each year [17, 18]. This requires an in-depth workplace risk analysis to be considered at each stage of a work cycle, which in turn will lead to higher efficiency and better economic performance.

In the present study, emphasis is placed on the possibility of using the generated waste meal from olive production in energy production and the possible dangers for workers, which need to be taken into account for the creation of "green" workplaces.

## 2. Materials and methods

The production of olive oil is directly related to the production and cultivation of olive plants. A diagram illustrating this connection, summarized with the hazards for workers at each stage of the process, is shown in figure 1.

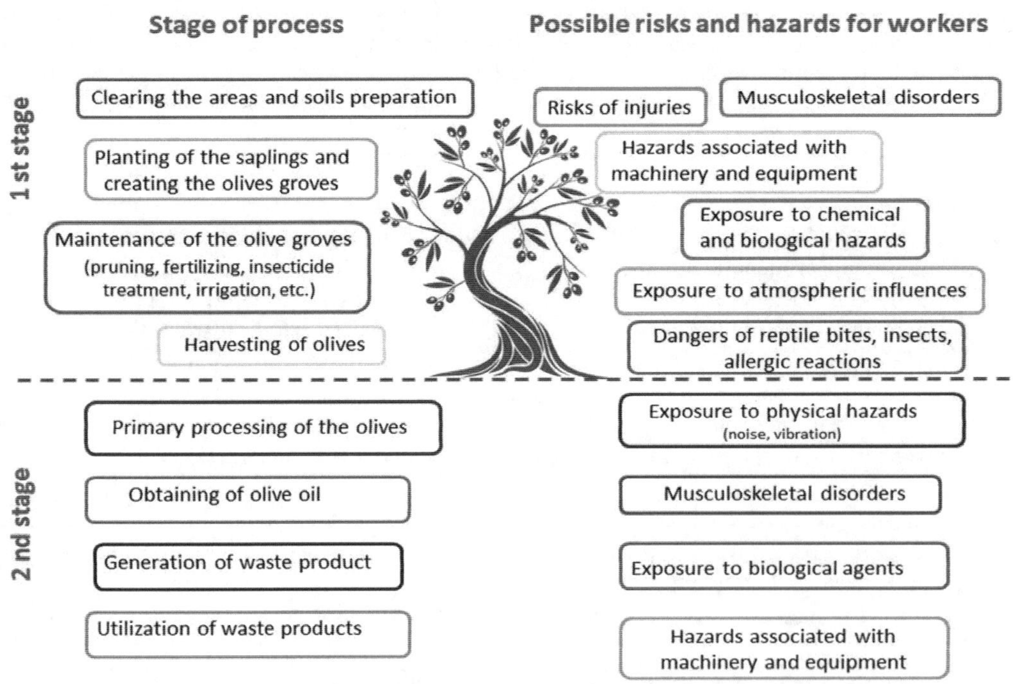


**Figure 1.** Scheme of production of olives and olive oil and possible dangers for workers.

*Calculation of the combustion process.* When a material is considered a fuel, only some of the main elements that play a role in the combustion process are taken into account. Carbon (C), "free" hydrogen (H-free) and volatile sulfur (Sl) (contained mainly in the organic compounds of the fuel) are considered as "valuable" elements for the process, and for the purposes of calculating the combustion the so-called "working mass" (WM), are taking into account [19, 20]. This mass includes the contents of carbon (C), hydrogen (H), oxygen (O), nitrogen (N), sulfur (S), ash (A) and moisture (W), the sum of which must be equal to 100 %:

$$X^W \stackrel{\text{def}}{=} 100\% = C^W + H^W + O^W + N^W + S^W + A^w + W^W \quad (1)$$

For the purposes of the calculations presented in this article, data on the thermal characteristics of two types of olive meal are taken into account, according to the results presented by Mladenov et all. [21] and Miranda et all. [14] – see table 1.

**Table 1.** Thermal properties of olive meal, in the work mass.

| Material | $W^w$ % | $A^w$ % | $O^w$ % | $C^w$ % | $H^w$ % | $N^w$ % | $S^w$ % |
|---|---|---|---|---|---|---|---|
| *Olive meal [21]* | *71.80* | *1.67* | *15.72* | *6.29* | *4.28* | *0.21* | *0.03* |
| *Olive meal (20 % moisture)[1] [21]* | *20.00* | *4.72* | *44.6* | *17.86* | *12.14* | *0.60* | *0.09* |
| *Olive meal (pomace) [14]* | *6.86* | *5.55* | *27.53* | *51.42* | *6.56* | *1.98* | *0.1* |

[1] *The parameters are calculated theoretically on the basis of tolerance for reduced moisture content of 20% and no losses with the emitted gases from the other elements.*

The calculations are made for one kilogram of the working mass of the meal, under the following prerequisites:

- *dry atmospheric air in proportion k* (k = $\%N_2/\%O_2$ = 3.76), *are use*
- *stoichiometric calculations are converted to normal conditions* (P=1 atm, t=0°C, V=22.4 m$^3$)
- *the absolute humidity value is assumed to be 13* $[20] - g_{H2O} = 13\ [g.water/m^3 dry\ air]$
- *the combustion reactions that going at actual combusting process*, are:

$C + O_2 = CO_2$; $2H_2 + O_2m = 2H_2O\ (vapour)$; $S + O_2 = SO_2$

The calculating of theoretical (stoichiometric) combustion, before calculating the actual combustion, must be done. That stage is intermediate. Actual combustion is a process that requires an amount of oxygen that is actually necessary for the chemical reactions to take place to produce complete combustion products. It is always higher than the theoretical one and for its calculation an air flow rate value (α) is used, which depends on various factors. In the present study, in order to ensure comparability between the obtained results, a value for this coefficient equal to 1.40 is accepted.

In the present study, the formula (2) proposed by M. Ioelovich [22] for the calculation of heat of combustion (heating value) was used.

$$q\ (MJ/kg) = 0{,}344\,.\,C + H + 0{,}105\,.\,S - 0{,}110\,.\,O - 0{,}015\,.\,N \quad (2)$$

where C, H, S, O and N is percentage of the corresponding elements in the raw material.

The calculation of the saved energy for households was made on the basis of an assumption for the average monthly electricity consumption for an average family of four, equal to 350 kWh and assuming that the generated meal is burned in small combustion plants up to 30 tons per year.

## 3. Results and discussion

The description of figure 1 gives an idea of the two main stages for the production of olive oil and its main waste product – olive meal. The first stage includes the agricultural activities for the establishment and maintenance of the olive groves and the dangers to which the workers in the sector are exposed. The main dangers are the when working with machines and equipment, musculoskeletal disorders, exposure to dangerous chemicals – fertilizers, pesticides and insecticides. Last but not least, the exposure to atmospheric influences - mainly solar radiation - must be taken into account. There is also a risk of exposure to biological agents when working in agriculture, and the various skin infections are very common. Prolonged contact with biological agents, soil, plants, pesticides and fertilizers, animals, manure and timber, can lead to a number of infections, tumors, scars and the development of fungal diseases. The second stage, related to the processing of the harvest and the production of the final product, in addition to the above-mentioned hazards for workers, also includes the exposure to physical hazards - such as noise and vibration. All this requires an analysis of the risk in the workplace for each stage, to take into account the dangers and to take the necessary measures to ensure safe and healthy working conditions for workers.

The life cycle assessment (LCA) of olive oil production "from cradle to grave" includes consideration of all processes and operations from the arrival of the harvested olives in the factory in beginning of the olive processing process to the extraction and use of the olive oil, and also treatment of the packaging waste in which it is stored and distributed, which are generated after consumption. Schematically such production is presented in figure 2.

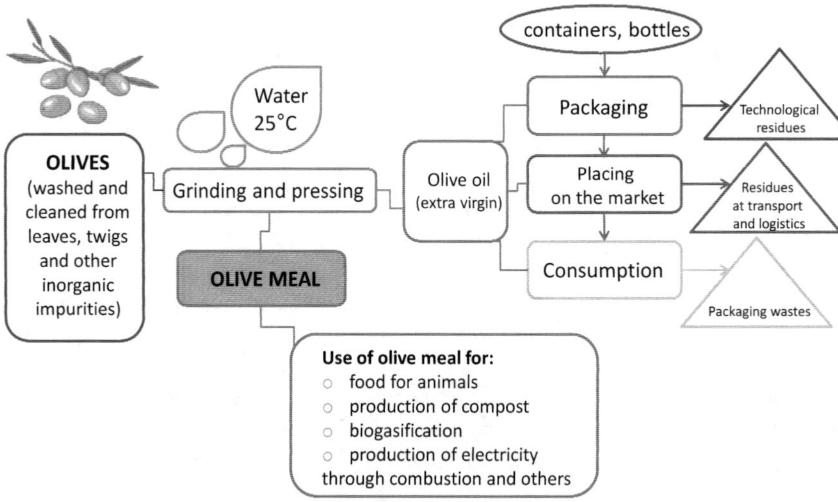


**Figure 2.** LCA of olive oil production process.

In order to prepare "extra virgin" olive oil, the harvested olives are subjected to a preliminary treatment, during which the fallen branches, leaves and inorganic impurities are removed, after which they are washed. In the next stage, the cleaned olives are ground and pressed together with the stones, at which stage the grounded mass (pulp) are produced. To this mass is added warm water (up to 25 degrees), which is mixed with the pulp for 30-60 minutes. This is followed by a step in which the olive oil is separated (usually by centrifugation) from the rest of the mass, which is the waste olive meal, and then packed. Packaged olive oil is distributed in the commercial network and after its consumption, waste is generated from packaging in which it was stored and transported. The following tables present the results of the performed calculations.

The results for the calculated amounts of oxygen required for the actual combustion process and the amounts of excess air required for combustion show that these amounts are directly dependent on the composition of the feedstock and especially on the moisture content of the olive meal (table 2).

**Table 2.** Results of calculations for the amounts of oxygen and excess air required for combustion process.

| Material | Oxygen required for actual combustion, $m^3$ $O_2$/kg. meal | Excess air, $m^3$ EA/kg. meal |
|---|---|---|
| *Olive meal [21]* | *0.247* | *0.478* |
| *Olive meal (20 % moisture)[1] [21]* | *0.701* | *1.358* |
| *Olive meal (pomace) [14]* | *1.135* | *2.196* |

*EA – excess air,*

For example, in the case of raw meal, the required amounts of oxygen and excess air are 0.247 $m^3$ $O_2$/kg meal and 0.478 $m^3$ EA/kg meal, respectively, at a moisture content of 71.8%. When the moisture is partially removed and reaches a value of 20.0%, the required amounts of oxygen and excess air increase almost three times to 0.701 $m^3$ $O_2$/kg meal and 1.358 $m^3$ EA/kg meal.

The calculations for the composition of the generated flue gases are given in the table 3.

**Table 3.** Results of flue gas composition calculations, in %.

| Material | $CO_2$ | $H_2O$ | $SO_2$ | $N_2$ | Excess oxygen from combustion products |
|---|---|---|---|---|---|
| *Olive meal [21]* | *3.71* | *44.21* | *0.0066* | *48.95* | *3.13* |
| *Olive meal (20 % moisture) [21]* | *4.98* | *25.16* | *0.0094* | *65.66* | *4.19* |
| *Olive meal (pomace) [14]* | *10.13* | *9.95* | *0.0074* | *75.11* | *4.80* |

The results of table 3, as expected, show that the share of generated water vapour in the flue gases decreases with decreasing humidity of the meal – the lowest is the content of generated water vapour in the meal with the lowest humidity. Simultaneously with the decrease in humidity, there is an increase in the share of the other components – mainly nitrogen gases and $CO_2$, which is why during the construction of the facilities cleaning facilities for these gases must be provided. The $SO_2$ content is directly related to the sulphur content of the original olive meal.

The calculations for the energy value, the amount of energy produced and the possible amount of saved energy produced from conventional fuels given in table 4 are performed without taking into account the heat losses of the heat generator due to incomplete combustion of the fuel due to chemical or mechanical reasons [14, 21]. These calculations give a sufficient idea of the potential of meal to be used for energy production.

**Table 4.** Results for heating value, electricity end economized energy from conventional sources.

| Material | Heating value, MJ/kg | Electricity, kWh/ton | Electricity at combusting of 30 t, MW/year | Economizes electricity from conventional sources for 10 household, about months* |
|---|---|---|---|---|
| Olive meal | 4.7145 | 1309.6 | 32.3 | 11 |
| Olive meal (20 % moisture) | 13.378 | 3716.2 | 111.5 | 32 |
| Olive meal (pomace) | 21.20 | 5889.2 | 176.7 | 50 |

The results of the calculations of heating value confirm the requirement for reduced humidity of the raw material intended for combustion – e.g. olive meal [21] with a moisture content of only 20% generates 2.8 times more heat, and meal with a moisture content of 6.86 (according to [14]) has the potential to generate 4.5 times more heat than wet.

The obtained results show significant differences between the values obtained for heating value and the electricity obtained from the combustion of a ton of meal, with the highest values expected to be obtained from the meal with the lowest humidity – 21.2 MJ/kg and 5 889.2 kWh/t, respectively.

Burning low-moisture meal can save electricity produced by conventional fuels to cover the needs of 10 families in 4 years and 2 months. However, achieving such low humidity requires the use of additional energy, which can be partially or completely provided by the combustion of meal.

On the other hand, burning high-moisture meal can lead to energy savings from fossil fuels to cover the needs of 10 households with an average monthly electricity consumption of 350 kWh for 11 months, without the need to invest additional energy to reduce humidity. However, this is accompanied by additional technological complications related to difficulties in the combustion process, the need for the addition of materials for mechanical reduction of humidity, specific requirements for burners and many others.

## 4. Conclusions

The results of the calculations show that the use of waste meal for energy production is not only possible but also economically justified in order to save energy extracted from fossil fuels. The highest values for heating value and electricity were obtained when calculating the combustion of tonnes of meal with low humidity (6.86%) – 21.2 MJ/kg and 5 889.2 kWh/t, respectively.

It is recommended that the combustion of the meal be carried out in specialized combustion plants in combination with combustible materials with low humidity and equipped with specialized equipment for capturing and cleaning the generated waste gases.

The description of the production process and the identified hazards for the workers show that the "green" from an ecological point of view manufactures are not always "safe" for the workers.

## References

[1] Arvanitoyannis I and Kassaveti A 2008 Chapter: Olive Oil Waste Management: Treatment Methods and Potential Uses of Treated Waste, *in Book: Waste Management for the Food Industries*
[2] Nunes M et all, 2018 *Sci Tot Env*, 644, pp 229-236
[3] Stillitano T et all. 2016 *Bulg J Agric Sci*, **22** (4), pp 517-526
[4] Christoforou E and Fokaides P 2016 *Waste Manag*, **49**, pp 346-363
[5] Ruggeri B et all 2015 *Chem Eng J*, **259**, pp 630-639
[6] Yanik D 2017 *LWT-Food Sci Tech*, **77**, pp 45-51
[7] Al-Khatib et al. 2009 *Bulg J Agric Sci*, **15** (6), pp 544-551
[8] Molina-Alcaide E and Yanez-Ruiz D 2008 *Animal Feed Science and Technology*, **147**, 247-264
[9] Wedyan M at all 2017 *Bulg J Agric Sci*, **23** (5), pp 866-872
[10] Tekin A and Dalgic A 2000 *Resources, Conservation and Recycling*, **30**, pp 301-313
[11] Al-Hamamre Z 2015 *Energy Sources, Part A: Recovery, Utilization, and Environmental Effects*, **37**, pp 2609-2615
[12] Ion V and Popescu F 2017 *Scientific Works of University of Food Technologies*, **64** (1)
[13] Saifuddin M et all 2014 *Bulg J Agric Sci*, **20** (1), pp 186-192
[14] Miranda T et all 2012 *Fuel Processing Technology*, **103**, 91-96
[15] Wandzich D and Plaza G 2017 *New and Emerging Risks Associated With "Green" Workplaces*, Workplace Health & Safety, **65**(10), 493-500
[16] *A non–binding guide to best practice with a view to improving the application of related directives on protecting health and safety of workers in agriculture, livestock farming, horticulture and forestry*, Publications Office of the European Union, 2012
[17] Priorities *for occupational safety and health research in Europe: 2013-2020*, Luxembourg: Publications Office of the European Union, 2013
[18] *Biological agents and work-related diseases: results of a literature review, expert survey and analysis of monitoring systems European Risk Observatory Literature Review*, Luxembourg: Publications Office of the European Union, 2019
[19] Totev I 1998 *Solid municipal wastes treatement*, UASG, Sofia
[20] Paraschiva L et all 2020 *Energy Reports*, **6**, pp 36-45
[21] Mladenov M et all 2020 *J Chem Techn Met*, **55**(4), pp 797-800
[22] Ioelovich M 2020 Short Overview of Methods for Calculation of Combustion Heat

**Acknowledgement**
The authors would like to thank the Science and Research Sector of the UCTM–Sofia for the financial support (Contract № 12019/2020).

# Change point detection in recent hydropower generation in Romania

**A Neagoe[1], E-I Tică[1], F Popa[2], B Popa[1]**

[1] Department of Hydraulic, Hydraulic Machineries and Environmental Engineering, University POLITEHNICA of Bucharest, Romania
[1] Department of Electronics, Telecommunications and Energy Engineering, University VALAHIA of Targoviste, Romania

E-mail: bogdan.popa@upb.ro

**Abstract.** In Romania, about thirty percent of the installed capacity in National Power Grid and a third of the electricity generation is in hydropower plants. Thus, any aspect related to this form of renewable energy source is very valuable. Chronological records of yearly hydropower generation in Romania represent a time series that can be statistically analysed. The aim of this work is to determine the trend of this time series and, also, to locate the change points, where the time series has significant change. The period analysed is from 1980 to the present. To determine the change point of this time series, the Pettitt method was applied.

## 1. Introduction

Often, in the field of meteorology and hydrology the issue arises of identifying points where significant changes occurs in series of data over long periods for which their justification is sought. A major change in the natural regime of flows occurs when a dam is built on a river and, also, climate change is often the reason for changes in the series of meteorological parameters such as rainfall [1], [2], humidity [3], air temperature [4], [5], wind speed [6], etc. To determine the moments in which this change occur the statistical analysis of the recorded data set is used.

For each country, the trend for generation and consumption of electricity is an interesting subject. In Romania, about a third of the total energy generation is from hydropower plants [7], and therefore it is important to follow the evolution over time and trend of this variable.

In [8] was presented a review of evolution in time – between 1884 and 2015 – of hydropower plants. The aim of this paper is to determine the evolution of hydropower generation as time series from a statistical approach and to determine the change point of this time series.

## 2. Time series data

The analysed time series are composed by annual values of recorded data of energy generation in hydropower plants in Romania between 1980 and 2019, data published in annual report of Hidroelectrica [9] and Transelectrica website [10] regarding the National Power Grid (NPG).

In Romania, during 1950-1995, most of the important hydropower plants were built, which contributed to a major increase of energy production. For example, between 1980 and 1990, 47 new hydropower plants were put into operation, including Iron Gates II, Retezat, Şugag, Petreşti, Tismana, Bradişor, etc. This evolution easily indicates a continuous increase for hydropower production over this period.

The development over time of annual hydropower generation for the period 1990 – 2019 decrease, when the development of new hydropower plants was slowed down. The recorded data used for the hydropower generation, *G*, for the analysis period are shown in figure 1.

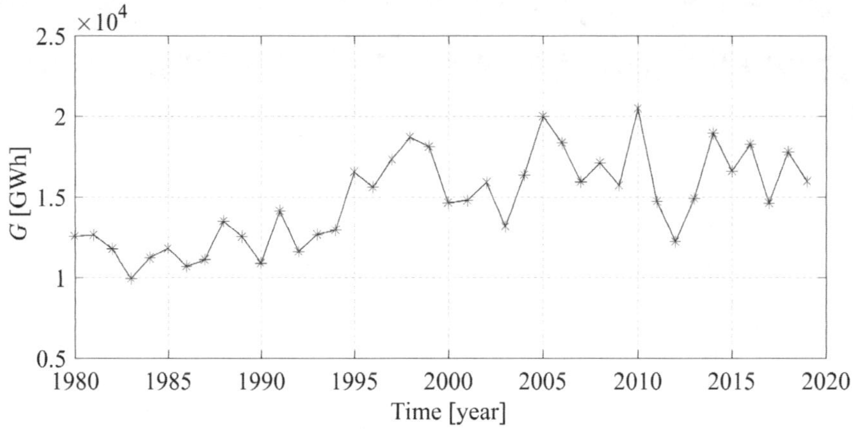


**Figure 1.** Time evolution of the annual hydropower generation in Romanian.

From figure 1 it can be noticed that for 2 years, 2005 and 2010, the hydropower generation is very high. It is noted that the extreme values observed on the graph are not change points for the time series and the average of all 40 values is 14,804 GWh.

## 3. Pettitt's method

The Pettitt method is commonly used to detect change point in the mean value of observed series [9].

If the series $x_1, x_2, ..., x_N$ has a change at time *t,* then $x_1, x_2, ..., x_t$ and $x_{t+1}, x_{t+2}, ..., x_N$ are two homogeneous subsets with different distribution function. The $U_t$ statistics is:

$$U_t = \sum_{i=1}^{t} \sum_{j=t+1}^{N} sign(x_i - x_j), \tag{1}$$

$$sign(x_i - x_j) = \begin{cases} 1, \text{if } (x_i - x_j) > 0 \\ 0, \text{if } (x_i - x_j) = 0 \\ -1, \text{if } (x_i - x_j) < 0 \end{cases}. \tag{2}$$

Its statistic $K_t$ and the associated probabilities used in the significance testing are:

$$K_t = \max|U_t|, \tag{3}$$

$$p = \exp\{-K_t / (N^3 + N^2)\}. \tag{4}$$

It can be considered that where the function $U_t$ has a peak of, there is the moment when in the time series occurs a change.

The Pettitt method can be used if one change point is present, i.e., one abrupt change of the mean. To find if another change point exists, the original time series can be split into two sub-series and repeat the analysis for each sub-series [10].

## 4. Results

After applying Pettitt method for series between 1980-2019, it can be observed that the maximum of function $U_t$ is obtained for the year 1994. The change in trend is obvious at that time and is since new hydropower plants were commissioned in Romania between 1980 and 1994 and it was slowed down afterwards.

For recorded values between 1994 and 2019 the Pettitt method was applied again and the function associated with this method is shown in figure 2.b. In this situation, because the length of the string is shorter, the main maximum is not so obvious that in figure 2.a, however, from this period the probability that one year would represent a change point is 2004.

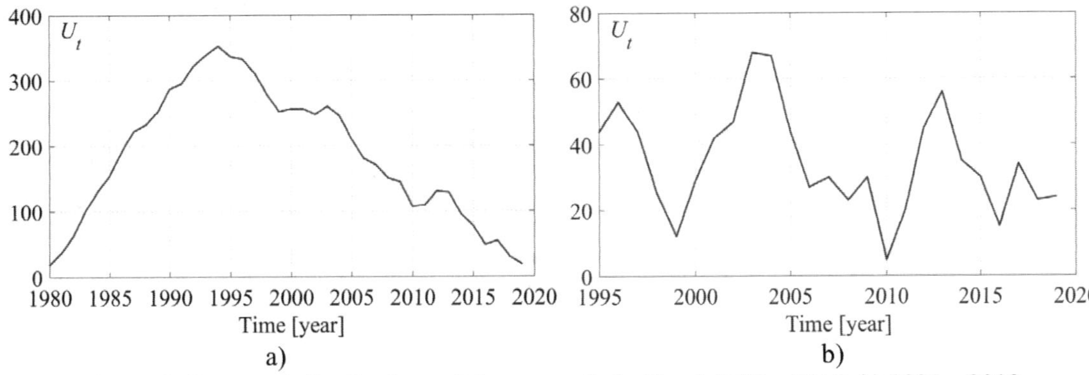


**Figure 2.** Temporal distributions of the test statistic $U_t$ : a) 1980 – 2019; b) 1995 – 2019.

In table 1 are presented two values of $K_t$ for first and second change points.

**Table 1.** The $K_t$ statistic of Pettitt method.

| | $K_t$ | Year |
|---|---|---|
| First change point | 353 | 1994 |
| Second change point | 68 | 2004 |

For analysis, the time series will be split into three subsets: $S_1$ contains $N_1$ values for the period 1980-1994, $S_2$contains $N_2$ values for the period 1995-2004 and $S_3$ contains $N_3$ values recorded in the period 2005-2019. The main statistical characteristics of these three subsets are given in table 2, with the specification of the year in which the minimum and maximum values of each period were recorded.

**Table 2.** The main statistical characteristics of the subsets of time series.

| | Average | Standard deviation | Min (year) | Max (year) |
|---|---|---|---|---|
| ***E* [GWh] – $S_1$ (1980-1994)** | 11975 | 1128 | 9930 (1983) | 14110 (1991) |
| ***E* [GWh] – $S_2$ (1995-2004)** | 16103 | 1685 | 13130 (2003) | 18690 (1998) |
| ***E* [GWh] – $S_3$ (2005-2019)** | 16767 | 2246 | 12200 (2012) | 20470 (2010) |

The three subsets will be analysed separately because they have different statistical characteristics and averages. Also, for the period 2005-2019 it is noted that the limits within which the recorded values are situated are higher, implying a higher average standard deviation. For each of the three subsets, a linear regression is found which best approximates the discrete values recorded.

For the period 1980-1994 shown in figure 3, the recorded data are well approximated by a regression equation, in the form of:

$$y_1 = 11340 + 79 \cdot t\,,\; t = 1,2,\ldots,N_1\,. \tag{5}$$

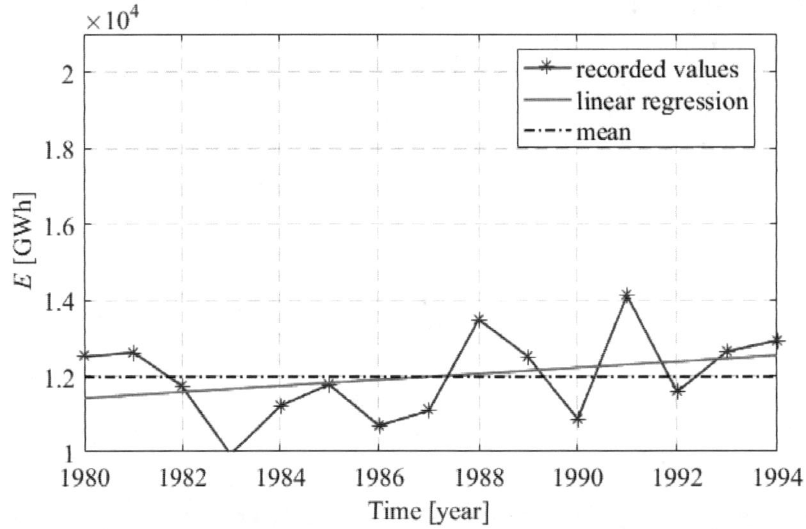


**Figure 3.** The linear regression of hydropower generation in 1980-1994.

The average trend for this period is upward.

For the period shown in figure 4, 1995-2004, a decrease in the energy produced in hydropower plants can be observed.

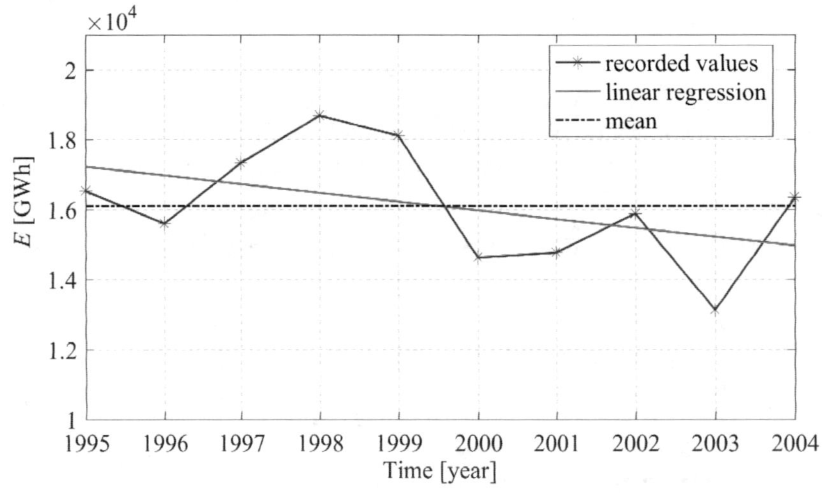


**Figure 4.** The linear regression of hydropower generation for the period 1995-2004.

The slope of the regression line is strongly downward as shown by the regression equation found for the variation of hydropower generation over the period 1995-2004:

$$y_2 = 17481 - 251 \cdot t\,,\; t = 1,2,\ldots,N_2\,. \tag{6}$$

From figure 5, it can be observed that the trend of hydropower generation in the time period 2004-2019 is decreasing, but the slope of the regression equation is low, i.e. the downward trend is smaller than for the period 1994-2004:

$$y_3 = 17734 - 121 \cdot t,\ t = 1,2,...,N_3. \quad (7)$$

The average values of the last analysed period are also the highest, around 16.7 TWh annually. The increase is most likely due to the upgrading of hydropower plants during this period.

The Pettitt method finds the moment where the statistical characteristics before and after this differ significantly. It can be observed that for the 3 subsets in which the original series was divided, both the averages and the variants are very different. If it relates to 1980, the time series can be considered to have an upward trend and an average of 14804 GWh. This value is well above the average of the last analysis period 2005 to 2019, which is 16767 GWh. When considering the series from 2004 onwards it can be concluded that although there are annual values with high standard deviation, they fluctuate around the average, with a slightly decreasing trend.

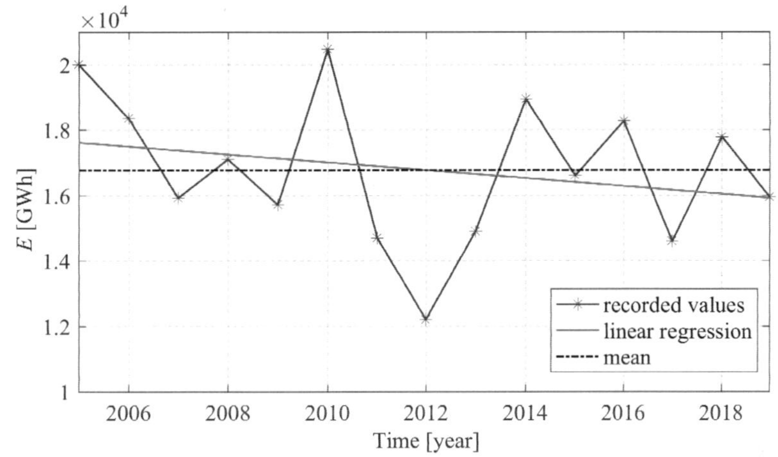


**Figure 5.** The linear regression of hydropower generation for the period 1995-2004.

## 5. Conclusions

After applying Pettitt method for time series represented by hydropower generation recorded between 1980-2019, it can be observed that the maximum of the function $U_t$ is obtained for the year 1994. The change in trend at that time is justified since new hydropower plants were commissioned in Romania between 1980 and 1994 and the development of the sector was almost stopped afterwards.

Using the Pettitt method, the time series was fragmented, so the years of changes in the time series of the annual energy generation in hydropower plants in Romania over the last 40 years were determined. Using all the values from 1980 to 2019 it was obtained that 1994 represents the year when the change point is abrupt. Instead, if only the range of values from 1995 to 2019 are analysed, in 2004 the change was not so radical, however, the analysis can conclude that for the period 2004-2019 it was recorded in average an increase in hydropower generation compared to previous years.

## References

[1] Liu Q, Yang Z F and Cui B S 2008 Spatial and temporal variability of annual precipitation during 1961-2006 in Yellow River Basin, China, *J. of Hydrology* **361** (3-4) pp 330–338

[2] Swanson K L and Tsonis A A 2009 Has the climate recently shifted? *Geophysical Research Letters* **36** (6) 06711

[3] Bryson C B Richard E C and Adrian W B 2012 Trend estimation and change point detection in individual climatic series using flexible regression methods, *J of Geophysical research* **117** D16106

[4] Zhang P, Ren G, Xu Y, Wang X L, Qin Y, Sun X and Ren Y 2019 Observed changes in extreme temperature over the global land based on a newly developed station daily dataset, *J of Climate* **32** (24) pp 8489-8509

[5] Jaiswal R K, Lohani A K and Tiwari H L 2015 Statistical analysis for change detection and trend assessment in climatological parameters, *Environ Process* **2**, pp 729–749

[6] Birsan M, Marin L and Dumitrescu A 2013 Seasonal changes in wind speed in Romania, *Romanian Reports in Physics* **65** (4) pp 1479–1484

[7] Neagoe A, Tică E-I, Diminescu M A, Popa B and Cismaş S C 2019 The analysis of hydropower ratio from total energy production in Romania *E3S Web Conf.,* **85** 06009

[8] Năstase G, Şerban A, Năstase A F, Dragomir G, Brezeanu A I and Iordan N F 2017 Hydropower development in Romania. A review from its beginnings to the present, *Renewable and Sustainable Energy Reviews,* **80** pp 297-312

[9] Pettitt A N 1979 A non-parametric approach to the change-point problem, *Journal of the Royal Statistical Society. Series C (Applied Statistics)* **28** (2) pp 126–135

[10] Busuioc A and Von Storch H 1996 Changes in the winter precipitation in Romania and its relation to the large-scale circulation *Tellus A* **48(4)** pp 538–552

# The comparison of five proton linac beams in order to set up a thermal neutron radiography installation

**J G Fantidis[1], G E Nicolaou[2], G Tsakalos[1], A Thysiadou[3]**

[1] Department of Electrical Engineering-Department of Physics, International Hellenic University, Agios Loukas, 65404 Kavala, Greece
Email: (Corresponding author)
[2] Laboratory of Nuclear Technology, Department of Electrical and Computer Engineering, 'Democritus' University of Thrace, Xanthi, Greece
[3] Department of Chemistry, International Hellenic University, Agios Loukas, 65404 Kavala, Greece

Email: fantidis@teiemt.gr

**Abstract.** Radiography without doubt is maybe the most famous non-destructive technique with many applications in a number of scientific fields. In the present study, the performance of five photo-neutrons which delivered by proton linear accelerators with energy equal with 8, 12, 16, 20 and 24 MeV for a thermal neutron radiography unit has been designed and simulated with the help of the MCNPX 2.5.0 software which based on the Monte Carlo statistical method. The geometrical arrangement of the unit has been simulated with primary goal to provide the highest thermal neutron flux in the location of the examined item. For every energy of the proton beam all the significant parameters with the thermal neutron radiography theory like the flux of the thermals neutrons, the (collimator length)/(aperture diameter) ratio and the percentage of the thermal neutron within the neutron beam simulated for a wide series of values. Owing to the fact that the there are many fast neutrons inside the beam, it is obligatory to decrease the quantity of the fast neutrons. For this reason, the addition of a single sapphire for fast neutron filtering has as a consequence a neutron beam with better quality. According to the simulations, the presence of the filter improves the % presence of the thermal neutrons with no substantial abatement in the flux of the slow neutrons.

## 1. Introduction

Thermal neutron radiography is a complementary method with the classical radiography method which uses γ and X-rays, and this characteristic gives new potential in Non-Destructive Evaluation (NDE) area. Neutron radiography (NR) is a very useful method with intention to study many heavy materials like copper, steel or Nickel. Additionally, it is also appropriate and for numerous light isotopes with high cross-section values like the hydrogen, the gadolinium, the lithium or the carbon. Thermal NR today has many applications in a really huge number of scientific files such as the medicine, the archaeology and almost in all engineering departments [1-6].

Research reactors are the most common neutron sources for thermal NR facilities however have significant drawbacks such as extremely high cost of construction, colossal bureaucracy to manage, and often have low public acceptability. Spallation Neutron Sources even though are powerful neutron sources require billion € for their development. Isotopic neutron sources (eg $^{252}$Cf, $^{241}$Am/Be) suffer from low intensity [7-10]. A similar problem exists for deuterium-deuterium neutron generators which provide fast neutrons with energy ≈2.5 MeV but provide fluxes no more than $10^{12}$ n $s^{-1}$ [11-13]. Deuterium-Tritium neutron generators, even though they have harder spectrum, are capable fluxes about two orders of magnitudes higher than that of deuterium-deuterium neutron generators [14].

Proton accelerators are cheaper than power research reactors but can provide relatively high neutron fluxes and seem like an interesting compromise between neutron flux and initial cost. Beryllium and lithium are the most common material as targets for the proton beams because it provides softer spectra. Beryllium target does not require any unique design compared to lithium which has poor mechanical properties and low melting temperature [15-18].

The present article has threefold goal: First to lay out a unit appropriate for thermal NR intentions, 2) Second to simulate the deigned facility and last but not least to estimate the performance of 5 proton beams with energies 8, 12, 16, 20 and 24 MeV on beryllium target which uses as a photo-neutron source. All these aims have been realized with the use of the MCNPX 2.5.0 software [19].

The needed neutrons provided when protons with energy between 8 - 24 MeV strike on a thick beryllium target. The beryllium target has a disk shape of 6 mm thickness and 16 cm diameter. Figure 1 illustrates the produced neutron spectra for five discrete energies namely 8, 12, 16, 20 and 24 MeV protons beams. These spectra were provided after experimental works by Paul et al. [20]. For proton beam current equal with 1 mA the full neutron fluxes are $1.10\times10^{13}$, $4.23\times10^{13}$, $7.23\times10^{13}$, $1.02\times10^{14}$ and $1.42\times10^{14}$ $ncm^{-2}s^{-1}$ for protons with energy 8, 12, 16, 20 and 24 MeV respectively. The gamma-ray spectrum received from works by Cortes-Giraldo et al. [21].

## 2. Materials and Methods

### *2.1. Thermal neutron radiography parameters*

The most important factor which describes the quality of the radiographic images is the ratio of the collimator length (L) versus the inlet aperture diameter D. The neutron flux which can measured at the collimator output given by the formula

$$\phi_o = \left(\frac{D}{4L}\right)^2 \cdot \phi_i \tag{1}$$

where $\phi_i$ indicates the flux of the neutrons in the entrance of the collimator while the geometric unsharpness can be calculated using the formula

$$u_g = \frac{L_f \cdot D}{L_s} \tag{2}$$

where $L_s$ is the distance between the inspected matter and the neutron source and $L_f$ is the distance between the inspected matter and image plane (usual 0.5 cm). The beam divergence (θ) is the half-angle of the beam and gives the highest dimension of the image plane. This described mathematically by the equation [22-25]:

$$\theta = \tan^{-1}\left(\frac{0.5 * I}{L}\right) \tag{3}$$

Thermal neutron content (TNC) express the percentage (%) of the thermal neutrons within the neutron beam. Finally, the ratio of the neutron flux vs. to the γ component (n/γ ratio), indicates a parameter which produces noise in the radiographies. The suggested value for this ratio is > $10^4$ n/($cm^2mSv^1$) [26].

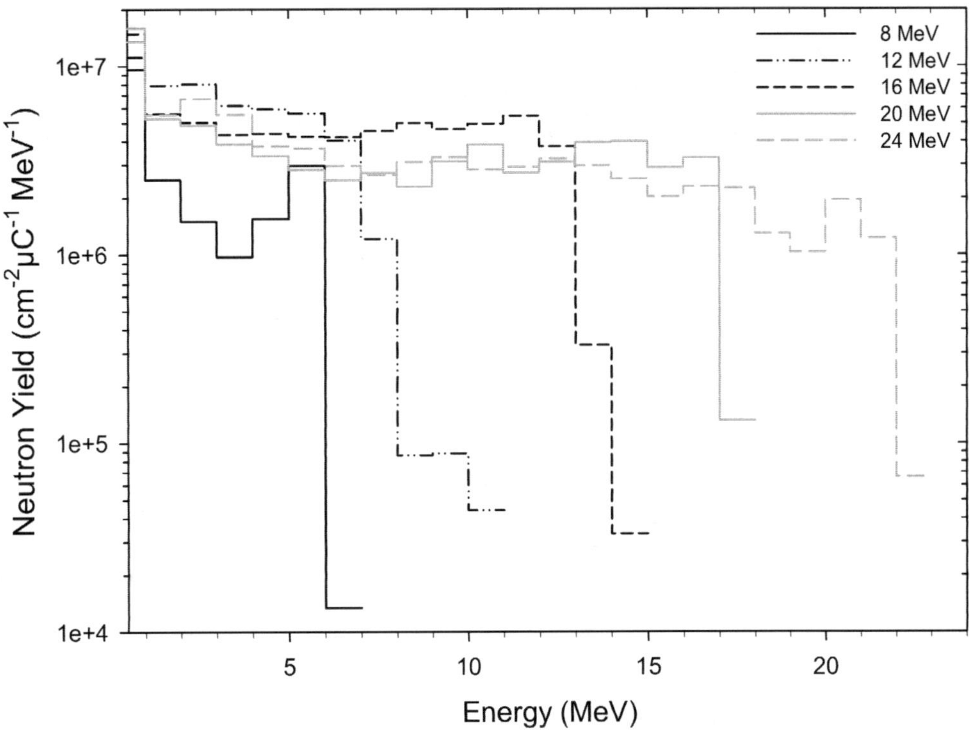


**Figure 1.** The provided neutron spectra for proton beams with current equal to 1mA.

*2.2. Design of the facility*

The design of the simulated geometry is depicted in figure 2. The main goal for this design was to maximize the thermal neutron flux ($f_{th}$) at the position of the investigated matter. In accordance with the spectra from figure 1, it is obvious that all the proton beams provides hard spectra for radiography which needs thermal neutrons hence is crucial to moderate these spectra using a moderator material. Based on the data from previous work by Fantidis et al. [15] the high density polyethylene (HD-PE) is the most suitable material in order to minimize the neutron energy offering the maximum $f_{th}$. In figure 2 a is the distance connecting to the beryllium target and the convergent collimator. This distance according to the simulated results is 2.5 cm in the case of 8 MeV proton beam and 2.7 cm for all the others cases. Convergent collimator has a conical shape with big radius 5 cm and small equal to 1 cm. The length b of this collimator is 9 cm in the cases of 8, 12 and 16 MeV protons beams, 11 cm in the case of 20 MeV proton beam and 12 cm using the 24 MeV proton beam. This collimator in the first set up was void, and in order to enhance the values of the TNC ratio in the second configuration was filled from with a single sapphire which is the best fast neutron filter [26]. Close to this collimator is a second collimator with changeable length (L = 100–300 cm) and inlet aperture (D) equal to 2 cm. The divergent collimator has boral as lining material and two layers one from borated polyethylene (PE-B) and one other from bismuth (bi). The thicknesses of these materials three were 0.8, 3.2 and 1 cm, correspondingly. The exit aperture ($D_0$) has a dimension from 14 up to 18 cm. In order to minimize the unwanted (scattered) γ rays and neutrons to exit from the divergent collimator were used bi and Boral with thicknesses of 1.2 and 0.8 cm respectively.

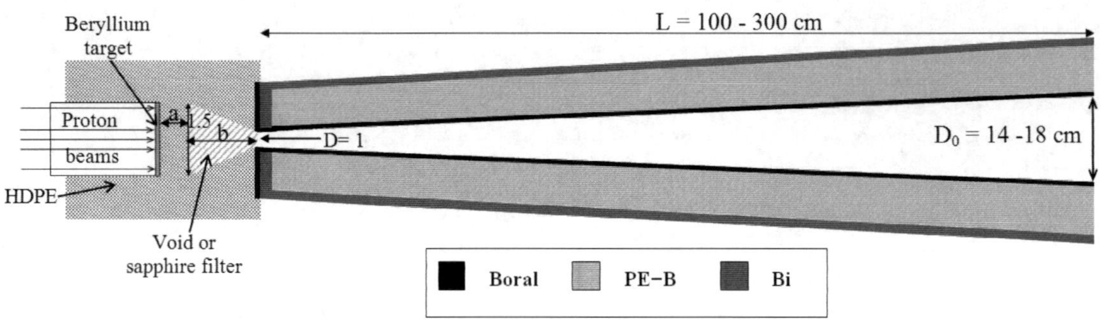


**Figure 2.** Configuration of the unit (dimensions are not in scale).

## 3. Results and discussion

All the beams were simulated for an extensive set of simulations for the most crucial parameter of every thermal NR unit the L/D ratio. The L/D has 3 values (50, 100 and 150) whilst the θ fluctuates from 0.86–2° and $u_g$ has values from $3.33\times10^{-3}$ to $1\times10^{-2}$. In this article the simulation of the n/γ ratio do not presented because have values significantly higher from the suggested limits. Both $f_{th}$ and TNC were estimated for every source using the F2 tally cards (which calculates surface flux). The thermal neutron energy band was 0.01–0.3 eV. The results are given in Table 1 and indicates that the 24 MeV proton beam offers both the maximum $f_{th}$ values and the minimum TNC. In contrast, protons with energy 8 MeV provides both the minimum values for the $f_{th}$ and the maximum for the TNC.

**Table 1.** The simulated values for three different L/D ratio values

| | | | | | 8 MeV proton | | 12 MeV proton | | 16 MeV proton | | 20 MeV proton | | 24 MeV proton | |
|---|---|---|---|---|---|---|---|---|---|---|---|---|---|---|
| L/D | L (cm) | D0 (cm) | θ (deg) | $U_g$ (cm) | $f_{th}$ ($n/cm^2s^1$) | TNC (%) | $f_{th}$ ($n/cm^2s^1$) | TNC (%) | $f_{th}$ ($n/cm^2s^1$) | TNC (%) | $f_{th}$ ($n/cm^2s^1$) | TNC (%) | $f_{th}$ ($n/cm^2s^1$) | TNC (%) |
| 50 | 100 | 14 | 2.00 | 1.00E-2 | 1.12E+6 | 2.06 | 3.07E+6 | 1.44 | 4.23E+6 | 1.08 | 5.62E+6 | 1.05 | 7.55E+6 | 1.02 |
| 100 | 200 | 16 | 1.14 | 5.00E-3 | 2.76E+5 | 1.92 | 7.51E+5 | 1.33 | 1.01E+6 | 0.99 | 1.26E+6 | 0.87 | 1.87E+6 | 0.94 |
| 150 | 300 | 18 | 0.86 | 3.33E-3 | 1.26E+5 | 1.95 | 3.33E+5 | 1.30 | 4.43E+5 | 0.96 | 5.62E+5 | 0.86 | 8.51E+5 | 0.94 |

According to the results from the Table 1 in every simulation the TNC has rather poor values. In order to enhance these values the presence of the single sapphire as fast neutron filter is really useful. Table 2 demonstrates that using the single sapphire inside of the convergent collimator in the case of the 8 MeV proton beams improves drastically the TNC values from about 2% to 16% with only an insignificant reduction to $f_{th}$. Similar results occur and for the 12 MeV proton beams as shown in the same table, with TNC around to 8%. In accordance with the data analogous is the trend of the TNC values and for the other neutron beams which produced from protons with energy 16, 20 and 24 MeV. The TNC values in these cases were between 3.5 and 4% approximately. Again the reduction of the $f_{th}$ in each source is not considerably. Figure 3 shows the relevant TNC values with and without single sapphire filter.

**Table 2** The simulated values for three different L/D ratios with the use of single sapphire filter.

| | 8 MeV proton | | 12 MeV proton | | 16 MeV proton | | 20 MeV proton | | 24 MeV proton | |
|---|---|---|---|---|---|---|---|---|---|---|
| Filter depth (cm) | 9 | | 9 | | 9 | | 11 | | 12 | |
| L/D | $f_{th}$ ($ncm^{-2}s^{-1}$) | TNC (%) | $f_{th}$ ($ncm^{-2}s^{-1}$) | TNC (%) | $f_{th}$ ($ncm^{-2}s^{-1}$) | TNC (%) | $f_{th}$ ($ncm^{-2}s^{-1}$) | TNC (%) | $f_{th}$ ($ncm^{-2}s^{-1}$) | TNC (%) |
| 50 | 8.09E+6 | 16.94 | 2.21E+6 | 8.33 | 3.05E+6 | 3.76 | 3.80E+6 | 3.96 | 4.93E+6 | 3.70 |
| 100 | 1.99E+5 | 16.48 | 5.41E+5 | 8.37 | 7.30E+5 | 3.59 | 8.49E+5 | 3.45 | 1.22E+6 | 3.62 |
| 150 | 9.10E+4 | 16.51 | 2.40E+5 | 8.24 | 3.19E+5 | 3.47 | 3.80E+5 | 3.43 | 5.56E+5 | 3.64 |

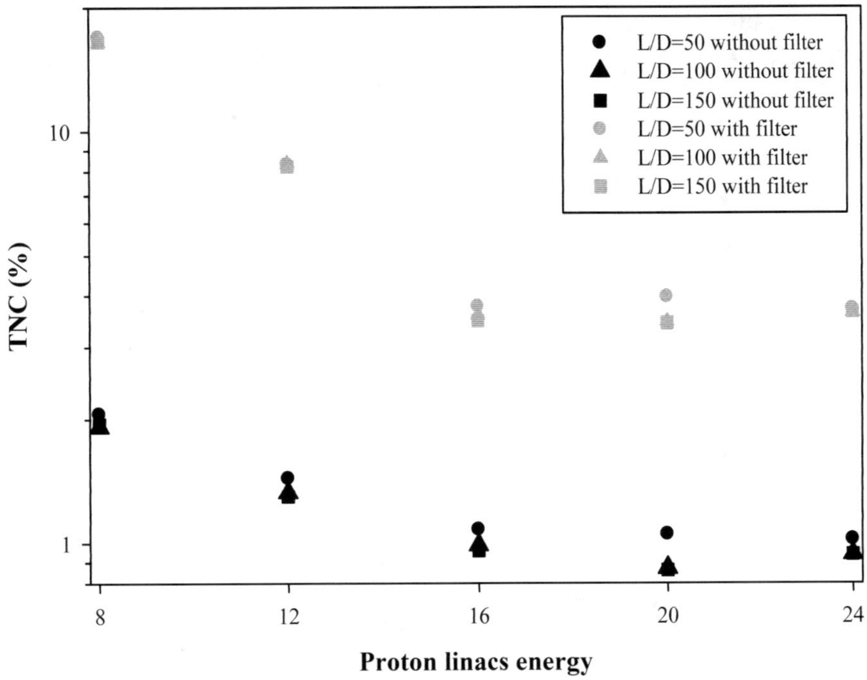


**Figure 3.** The TNC values without and with single Sapphire fast neutron filter.

## 4. Conclusions

In this article five proton beams which bombard a beryllium target were used as photo-neutron sources in order to set up a facility for a thermal neutron radiography purposes. Both the design and the simulations of the presented unit has been realised using the MCNPX 2.5.0 software. According to the results neutron spectrum from 8 MeV protons is capable to offer the highest values for the TNC ratio for each L/D ratio. On the other hand, owing to the smaller total neutron yield the thermal neutron fluxes are the lowest in each simulation. On the contrary, neutrons spectrum derived by the 24 MeV protons is really harder. This photo-neutron source gives the maximum neutron flux but simultaneously has the minimum values for the thermal neutron content. Last but not least the use of the single sapphire for filtering the fast neutrons ameliorates the thermal neutron content with only a non importance decrement in the flux of the thermal neutrons.

## References

[1] Chen Z and Wang X 2006, *Port Technol. Int*, **30**, pp 163-165
[2] Fantidis J G and G E Nicolaou 2011, *Nuclear Instruments and Methods in Physics Research Section A: Accelerators, Spectrometers, Detectors and Associated Equipment*, **648**(1), pp 275-284
[3] Takahashi Y, Misawa T, Pyeon C H, Shiroya S, and Yoshikawa K, *Applied Radiation and Isotopes*, vol. 69 (7), pp. 1027-1032, 2011
[4] Elsheikh N, Viesti G, El Agib I, and Habbani F 2012, *Applied Radiation and Isotopes*, **70** (4), pp 643-649
[5] Fantidis J G and Nicolaou G E 2013, *Journal of Radioanalytical and Nuclear Chemistry*, **295**(2), pp. 973-977
[6] Moradllo M K, Reese S R, and Weiss W J 2019, *Advances in Civil Engineering Materials* **8.**1, pp 71-87
[7] Fantidis J G , Nicolaou G E, Tsagas N F 2010, *Nuclear Instruments and Methods in Physics Research Section A: Accelerators, Spectrometers, Detectors and Associated Equipment*, **618**(1-3), pp 331-335

[8] Bergaoui K, Reguigui N, Gary C K, Cremer J T, Vainionpaa J H, and Piestrup M A 2014, *Journal of Radioanalytical and Nuclear Chemistry*, **299**(1), pp 41-51
[9] Taylor M, Sengbusch E, Seyfert C, Moll E, Radel R 2017, *Physics Proce*dia, **88**, pp 175-183
[10] Lehmann E, Frei G, Nordlund A, Dahl B 2005, *IEEE transactions on nuclear science*, **52**(1), pp 389-393
[11] Da Silva A X and Crispim V R 2001, *Radiation Physics and Chemistry*, **61**(3-6), pp. 515-517.
[12] Fantidis J, Nicolaou G, & Tsagas N 2010, *Journal of radioanalytical and nuclear chemistry*, **284**(2), pp 479-484
[13] Jafari H, & Feghhi S A H 2012, *Radiation Physics and Chemistry*, **81**(5), pp 506-511
[14] Fantidis J G 2017, *Journal of Taibah University for Science*, **11**(6), pp 1214-1220
[15] Fantidis J G 2018, *Nuclear Instruments and Methods in Physics Research Section A: Accelerators, Spectrometers, Detectors and Associated Equipment*, **908**, pp 361-366
[16] Zou Y, Wen W, Guo Z, Lu Y, Peng S, Zhu K and Zhou Q 2011, *Nuclear Instruments and Methods in Physics Research Section A: Accelerators, Spectrometers, Detectors and Associated Equipmen*t, **651**(1), pp 62-66
[17] Shaaban I 2017, *Indian Journal of Pure & Applied Physics (IJPAP*), **55**(2), pp 135-144
[18] Fantidis J 2012 *Journal of Radioanalytical and Nuclear Chemistry*, **293**(1), pp 95-101
[19] Hendricks J S 2003 *MCNPX version 2.5. c (No. LA-UR-03-2202*), Los Alamos National Laboratory
[20] Sabyasachi P, Sahoo G S, Tripathy S P, Sharma Ramjilal S C, Ninawe N G, Sunil C, Gupta A K, Bandyopadhyay T et al. 2014 *Review of Scientific Instruments* **85.6**: 063501
[21] Cortes-Giraldo, M A, Quesada J M, Palomo F R, & Garcia-Sanchez, E 2011 *Progress in Nuclear Science and Technology*, **2**, 568-575
[22] Domanus J C, 1987 *Collimators for thermal neutron radiography: An overview*, Markgraf J.F.W. (Ed.)
[23] Domanus J C and Matifield R S 1981, *Neutron Radiography Handbook*, D. Reidel Publ. Co., Dordrecht
[24] MacGillivray G M 2000 Imaging with neutrons: the other penetrating radiation. In Penetrating Radiation Systems and Applications II, *International Society for Optics and Photonics*, **4142**, pp 48-58
[25] Hawkesworth M R 1977 Atomic Energy Review, **15**(2), pp 169-220
[26] Mildner D F R and Lamaze G P 1998 *Journal of applied crystallography*, **31**(6), pp 835-840

# Automatic detection of characteristic segments of a recorded ECG signal with noise handling methods

**B Glushica**[1], **B Aleksovski and A Kuhar**[1]

[1] Faculty of Electrical Engineering and Information Technologies, Ss. Cyril and Methodius University in Skopje, North Macedonia

E-mail: kuhar@feit.ukim.edu.mk

**Abstract.** Depolarization and repolarization of the myocard results in specific wave shapes that are recognizable in an ECG signal and are characterized by their length, rise time and amplitude. Changes in these characteristics usually indicate anomalies in the function of the heart. Automatic detection of characteristic segments of the ECG signal is of crucial importance for fast and reliable recognition of artefacts that are further analyzed as a means for setting a diagnosis. In this paper a contribution has been made towards efficient automatic segment detection of real ECG signals recorded in arbitrary conditions in the presence of noise from various sources. The method of accumulated differential and a technique for tracing the wave shapes of the ECG signal using their local extrema have been implemented in the developed algorithm. The implementation of these methods yields an advanced way of handling the noise present in the ECG signal resulting with high precision segment detection and low computational power requirement. The high accuracy of the algorithm has been verified using real signals recorded on 2 lead ambulatory electrocardiograph. Another benefit of the developed algorithm is its speed – it is able to process a 5 minute long recorded ECG signal in 9 seconds.

## 1. Introduction

Electrocardiography is a standard method for recording electrical pulses generated by the heart muscle activity. As a result of this procedure an electrocardiogram is obtained – graphical recording of the changes in the bioelectric potential of the heart. The activity of the heart is controlled by so-called action potentials generated by specialized cells located at the right atrium. Depolarization (contraction) and repolarization (relaxation) of the heart muscle are detected as changes in the ECG signal amplitude in the range of millivolts [1], [2], which can be recognized as particular segments. Detection of these segments is of crucial importance since their presence/absence as well as their shape, amplitude and duration reflect the health condition of the patient's heart.

While being recorded the ECG signal is polluted by noise from various sources. In addition to this, ECG signals are sometimes recorded for several days which results in large quantities of data. Reviewing noisy and large ECG recordings by professionals is a time consuming and ineffective process.

All of the above leads to the conclusion that automatic detection is of essential importance for fast and reliable segment detection in the ECG signal, as well as extraction of problematic artefacts for further analysis. In previous research authors have used methods such as discrete wavelet transform, Hilbert transform, adaptive filtering etc. for handling the noise within ECG signals [3], [4], [5], [6]. Some of the approaches incorporate machine learning algorithms [6], which requires certain computational requirements and in some cases foreknowledge of the recordings. However, some of the recordings do not have typical waveforms of the segments as a result of the different equipment used for recording. Our contribution is an efficient algorithm which can be used on ECG signals recorded in arbitrary conditions implementing only techniques with low computational requirements.

## 2. Features of the ECG signal

The ECG signal is formed by quasi-periodic repetition of the typical waveform shown in figure 1. There are three main characteristic components (segments) of the ECG signal waveform [7]:

- P wave – segment characterized with a slight increase in amplitude followed by a decrease. The P wave has relatively low amplitude and is registered during the atrial depolarization;
- QRS complex - segment characterized with steep increase in amplitude followed by steep decrease. The QRS complex has the highest amplitude and is detectable during the depolarization of the ventricles;
- T wave – segment characterized with a mild increase in amplitude followed by another decrease. The T wave has slightly higher amplitude that the P wave and represents the repolarization of the ventricles.

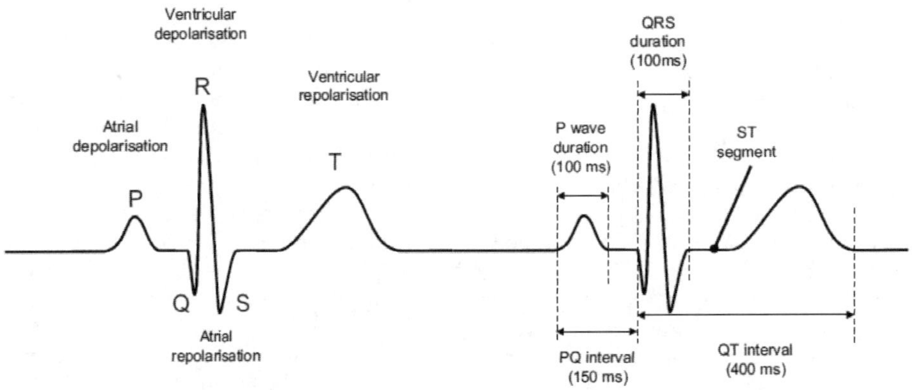


**Figure 1.** Characteristic segments of an ECG signal [7].

Figure ,1 also shows the following segments of the ECG signal: the PR segment which marks the time from the beginning of the contraction of the atria to the beginning of the contraction of the ventricles and the QT interval - a segment of time between the beginning of the contraction of the ventricles and the end of their relaxation. The QRS complex has the highest amplitude and a short duration being less prone to noise interference, with the exception of the noise created by small movements of the electrodes on the patient's skin. The P wave typically has a lower amplitude and shorter duration than the T wave while the PR segment has a shorter duration than the QT segment. The discussed segments of the ECG signal are most notable and have a characteristic duration, amplitude and waveform all depending on the health status of the particular heart that generated them. For example, with proper detection of the QRS complexes the average heart rate, i.e. the pulse is determined. The length of the QR interval provides information regarding the potential risk of developing lethal arrhythmia. Registering a change in the position of the PQ segment is the most reliable indication of atrial infarction [6].

The analysis of the discussed parameters of the segments is the main function of the algorithm developed within this research.

## 3. Noise sources and its influence

Low-pass and band-pass filters are used during the pre-processing of ECG signals due to the numerous sources of noise that influence the quality of the recorded ECG signal. Some significant factors that produce noise within the ECG signal are [8]: movement of skin under the electrodes; interference caused by the power grid; the electrical activity of the skeletal muscles; breathing etc.

The algorithm developed within this research paper includes sections for reducing the influence of the mentioned types of noise. In the pre-processing section a band-pass filter with lower limit of 0.667 Hz and upper limit of 50 Hz is implemented in order to filter out of the types of noise mentioned without losing useful information. Further in the process, the filtered ECG signal undergoes additional filtration with a low-pass filter with a limiting frequency of 30 Hz. Since the QRS complex also contains higher frequencies - up to 50 Hz, the removal of those between 30 and 50 Hz influences a certain decrease in its amplitude. The tests show that the influence of the loss of amplitude of the QRS complex is negligible with regard to the benefits that this kind of filtering introduces in the future usage of the segmentation methods.

## 4. An overview of the developed algorithm

The algorithm for automatic segment detection is written in the free programming language Python and consists of two parts: detection and labelling of the QRS complexes with the method of accumulated differential, and detection of P and T waves by tracing their local extrema.

### *4.1. Finding the R peaks and extraction of QRS complexes with the method of accumulated differential*

The first task of the algorithm is to locate the amplitudes of the QRS complexes, i.e. the points marked with R (figure 1). Because of the steep increase and high amplitude of the QRS complex, differential and accumulated differential is used according to these formulas:

$$d(i) = ecg(i+1) - ecg(i-1) \tag{1}$$

where $d(i)$ is the differential of the signal, $ecg(i)$ are the values of the ECG signal and:

$$a(i) = \sum_{j=i-m}^{i+m} d(j), \qquad (m < n \in N) \tag{2}$$

where $a(i)$ is the accumulated differential at point $i$.

The precise value of the differential is not of interest so the equation (1) is not scaled. The accumulated differential represents the sum of the differential in a sufficiently large number of points so that approximately half of the duration of the QRS complex is covered. In figure 2 the calculated accumulated differential according to (2) is shown. This approach enables the recognition of shapes that have steep changes, but with duration sufficiently long so that the steep changes as a result of noise are ignored.

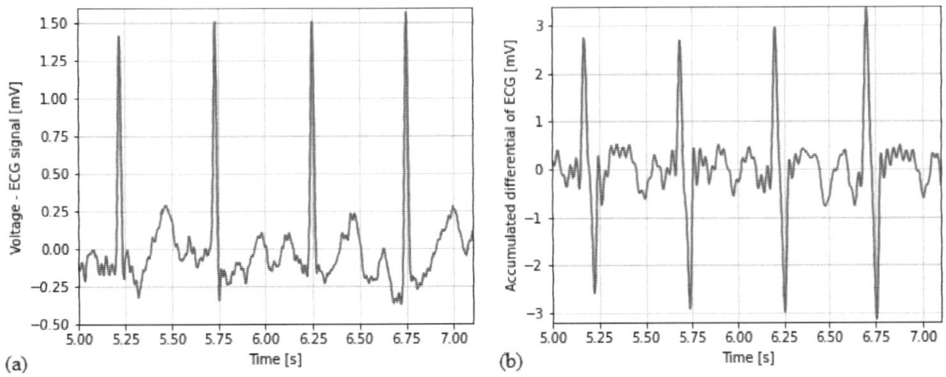


**Figure 2.** Part of an ECG signal (a) and the resulting accumulated differential (b).

The procedure starts by calculating the greatest difference between two consecutive maximum and minimum (in this order) of the whole $a(i)$ and then that value is scaled and it represents a minimum boundary value. The points candidates which are detected by the algorithm must satisfy this boundary value in order to be detected as R peaks of the QRS complex. Because the noise can cause a significant decrease in amplitude of the QRS complex, we continue with local calculation of the greatest difference between two consecutive maximum and minimum points. This difference is scaled down and represents the boundary threshold for the separation of the QRS complex. The R peak can be located by finding the local maximum after which we proceed to determine the boundaries of Q and S. To be able to do this, we take into consideration the fact that the values of the QRS complex start with a decrease in value which is followed by a sharp increase in value up to the R point for then to sharply decrease down to a certain minimum and finish with another slight increase. Looking at this from the location of the peak point R, we can find the coordinates of the S point by finding the local minimum of the original signal in a small area after the point R after which we locate a small peak (if it exists) in the next few points. Similar operations are performed for finding the Q point. The results of this section of the algorithm are presented in figure 3.

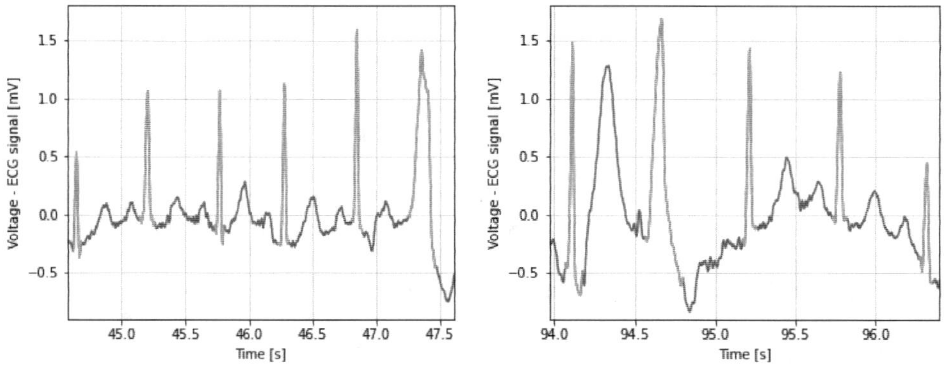


**Figure 3.** Correctly detected QRS complexes with different amplitudes from two parts of the ECG signal.

### *4.2. Detection of P and T waves using their local extrema*

In some parts of the ECG signal the P and T waves may be camouflaged because of the fast and drastic oscillations caused by noise interference. However, we can still recognize a group displacement in the oscillations of the noise in relation to the shapes of P and T waves. This indicates that the P and T waves can be traced using their local extrema. In figure 4 we observe the oscillations caused by noise influence and its bypassing with this method. The tracing of the envelope of the noisy signal greatly reduces the number of checks and calculations that follow in the next phase.

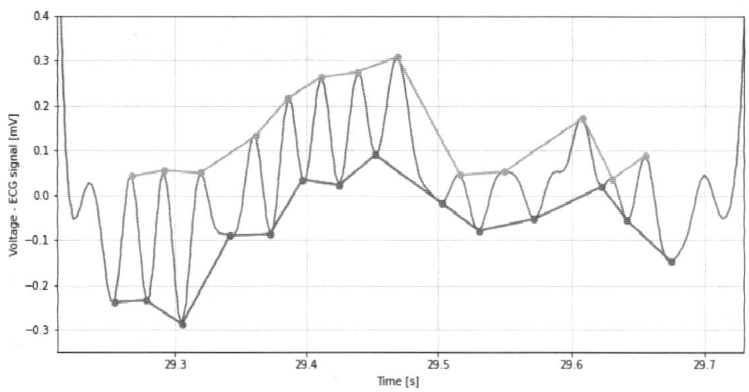


**Figure 4.** Tracing of P and T waves by analyzing their local extrema.

Since there is a possibility that the waves are shifting on the time axis or changing their duration as a result of noise, all checks for P and T waves aren't done in the area around one QRS complex, instead they are done between two consecutive QRS complexes. By analyzing information for the rising and falling of the values from the local extrema, we can detect the occurrence of the T wave. The existence of the P wave is examined in a similar manner.

IOP Conf. Series: Materials Science and Engineering 1032 (2020) 012047 doi:10.1088/1757-899X/1032/1/012047

## 5. Obtained results

The algorithm is tested on ECG signals with high and medium level of noise from the MIT-BIH Arrhythmia Database [9] and MIT-BIH ST Change Database [10]. The detection of P and T waves (marked with green and red colour respectively) is shown in figure 5 which have different shapes as a result of the noise influence.

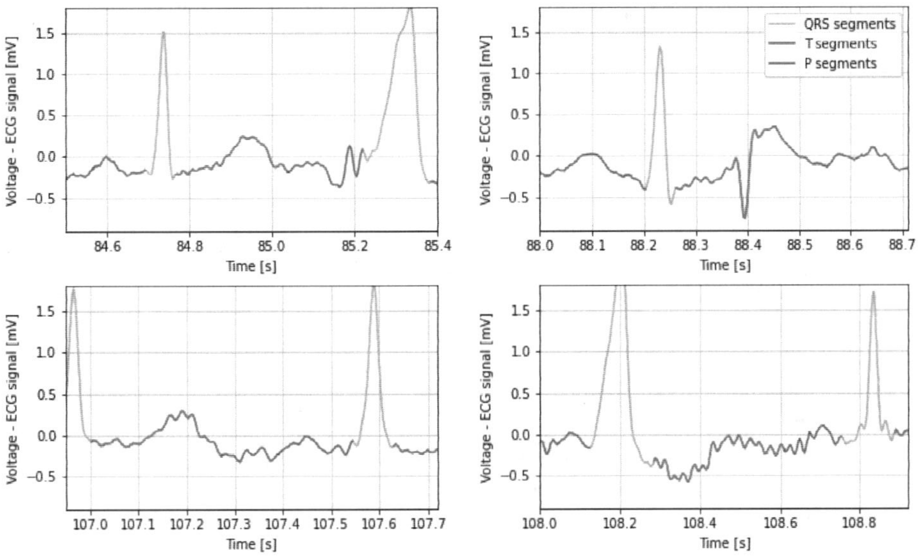


**Figure 5.** Detection of different P and T waves.

There is a possibility that some of these waves are not detected as a result of a heart anomaly that causes drastic signal deformation or as a result of high noise level. In that case, the examined section between the two QRS complexes is classified as a problematic segment and is marked as black on the graphs. A small percentage of the analyzed signals results in falsely classified segments. The reason for this is the detection of a third wave which is the result of the lenient criteria for detection set in order to process signals polluted by noise. As a part of this criteria, additional processing is done to avoid consecutive sign changes in the differential in a short time interval or to locate a transitioning point. The dominant factor is to determine which extrema to be ignored according to the boundary thresholds for the difference between the time points and amplitudes. By setting up stricter boundaries, it is possible to classify a noisy segment as problematic even in the case of clear recognition of the waves. That is why a compromise is made when determining the values of the boundaries.

The verification of the developed algorithm is done by comparing the accuracy of segment detection and classification on representative samples of the ECG signals in which the segments have been labelled by a cardiologist. According to this, table 1 presents the average results of the accuracy of segment detection in signals without heart anomalies in (%).

**Table 1.** Verification of the accuracy of the algorithm.

| Signal | Correctly detected P waves (%) | Correctly detected QRS complexes (%) | Correctly detected T waves (%) | Accuracy of borders between segments (%) |
|---|---|---|---|---|
| (I) #208 | 96.64 | 99.40 | 96.16 | 95.57 |
| (I) #209 | 95.49 | 99.59 | 95.52 | 98.07 |
| (II) #116 | 98.84 | 100.0 | 98.18 | 99.74 |
| (II) #124 | 97.61 | 99.61 | 97.21 | 97.21 |
| (II) #205 | 98.87 | 99.57 | 98.87 | 100.0 |
| (II) #300 | 98.27 | 99.81 | 98.46 | 99.02 |
| (II) #304 | 99.61 | 100.0 | 99.61 | 100.0 |

The algorithm has been implemented on 5 minute samples of ECG signals with high noise level (section I) and ECG signals with medium noise level (section II). The signals #300 and #304 are part of MIT-BIH ST Change Database and were recorded during exercise stress test. The rest of the signals are part of MIT-BIH Arrhythmia Database which contains ambulatory ECG signals. Each signal contains noise of various sources that interferes with the signal. In addition, some of the recordings do not have typical waveforms of the segments as a result of the different equipment used for recording.

The time duration for processing a 5-minute ECG signal on a computer with 2.5GHz CPU and 1.6GHz RAM is 9 seconds. Hence, we can conclude that the developed algorithm is optimized to work relatively fast while achieving a high level of accuracy (table 1).

## 6. Conclusions and future work

In this research we developed an algorithm in the free programming language Python for automatic detection of P, QRS and T segments in real ECG signals, employing the method of accumulated differential and a technique for tracing the specific wave shapes of the ECG signal using their local extrema. It is shown here that by implementing the two techniques we are capable of handling the significant amount of noise present in real i.e. measured ECG signals. The contribution of the developed algorithm is implementation of techniques with low requirements for computational resources while ensuring fast execution – a 5 minute long recorded ECG signal is processed in 9 seconds. In addition, the algorithm can be used on ECG signals recorded in arbitrary conditions with no foreknowledge of the recordings. The accuracy of the algorithm has been verified by testing on real ECG signals recorded in ambulatory settings.

Future research in this field would be focused on upgrading the procedures for noise classification. Specifically, post-processing techniques would be developed to prevent interpretation of noise components as signal artefacts. The main benefit of the implementation of such procedures would be to further improve the algorithm's precision.

## References

[1] Peterson D Bronzino J 2006 *The Biomedical Engineering Handbook* (Taylor & Francis Group), pp 1-4

[2] Sornmo L Laguna P 2005 *Bioelectrical signal processing in cardiac and neurological applications* (Elsevier Academic Press) p 360

[3] Zhu, Huaiyu *et al* 2018 A lightweight piecewise linear synthesis method for standard 12-lead ECG signals based on adaptive region segmentation *PloS one* vol. **13**,10 e0206170. 19, doi:10.1371/journal.pone.0206170

[4] Lyon A Mincholé A Martínez JP Laguna P Rodriguez B 2018 Computational techniques for ECG analysis and interpretation in light of their contribution to medical advances. *J R Soc Interface* , doi:10.1098/rsif.2017.0821

[5] Aspuru J Ochoa-Brust A Félix RA *et al* 2019 Segmentation of the ECG Signal by Means of a Linear Regression Algorithm. *Sensors* (Basel), doi:10.3390/s19040775

[6] Li Hong Zu Boulanger P 2020 A Survey of Heart Anomaly Detection Using Ambulatory Electrocardiogram (ECG) *Sensors* (Basel, Switzerland) vol. **20**,5 1461, doi:10.3390/s20051461

[7] P R Hoskins P V Lawford B J Doyle Editors 2017 *Cardiovascular Biomechanics* (Springer International Publishing Switzerland) p 115

[8] Кухар А 2019 *Предавања од предметот Биомедицинско инженерство* (ФЕИТ Скопје)

[9] Moody GB Mark RG 2001 The impact of the MIT-BIH Arrhythmia Database *IEEE Eng in Med and Biol* 20(3), pp 45-50, available at: http://www.physionet.org/physiobank/database/mitdb/

[10] Albrecht 1983 *P S-T segment characterization for long-term automated ECG analysis,* M.S. thesis (MIT Dept. of Electrical Engineering and Computer Science) , available at: https://www.physionet.org/content/stdb/1.0.0/

# Plexiglass glove box for organic solar cells

**L A Szolga[1] and C A Stan[1]**

[1] Faculty of Electronics, Telecommunications and Information Technology, UTCN, Basis of Electronics Department, Optoelectronics Group, Cluj-Napoca, Romania

E-mail: Lorant.Szolga@bel.utcluj.ro

**Abstract.** The term Glove Box probably creates an image of a glass box with black gloves, but Glove Boxes are as varied as the applications in which they are used. It is a hermetically sealed, watertight box, built for the purpose of sensitive applications. In this paper we present the design, the implementation and the testing of a Glove Box using plexiglass of 6mm thickness. The entire control of the system is flashed on an ATmega328 microcontroller. The sensors inside the box include oxygen sensor, gas sensor, humidity and temperature sensor. The plexiglass used provides robustness for the system but being lightweight it can be easily moved in different places in a research laboratory, compared with the steel Glove Boxes which are difficult to handle. The Glove Box was tested, by filling it up with Argon (inert gas) and introducing bulk heterojunction organic solar cells (BHJ OSC) and leaving them inside for 12hours. The measurement of the OSCs after 12h revealed that their efficiency decreased with only 20% compared to those left outside of the Glove Box which had an efficiency decrease of more than 70% from their initial value.

## 1. Introduction

A Glove Box is a sealed room that aims to allow a person to manipulate different objects in an environment with a desired atmosphere. On the sides of the Glove Box the gloves are arranged in such a way that the user can put his hands in the gloves and perform tasks inside the box, without contaminating the interior. The entire box is transparent to allow the user to see what is inside. There are two types of Glove Boxes: the first allows a person to work with hazardous substances [1-2], such as radioactive materials or infectious disease agents, and the second allows the handling of substances that must be contained in an inert atmosphere of very high purity, such as argon or nitrogen [3]. It is also possible to use the Glove Box to handle objects or substances in a vacuum chamber [4].

The applications of this product are extremely varied being used in biochemistry, metallurgy, electronics, chemistry, geology, minerals, pharmaceutical industry [5] and in other sectors. They are also used in other fields like electronic materials, magnetic materials, battery materials, organic cultivation, vacuum food packaging and many more.

In work environments where we have an inert atmosphere, the gas pumped into a Glove Box through a series of devices offer a treatment to remove solvents, water and oxygen from the inside medium. A heated metal, for example copper or another finely divided metal, is commonly used to remove oxygen. While heated it is passed through a mixture of hydrogen or nitrogen and through the column formed the oxygen normally evaporates. The formed water is removed from the box with an excess of hydrogen and nitrogen. In order to remove water, a series of molecular sieves are used that absorb it into the pores of the sieves. Such a Glove Box is often used by organometallic chemists to transfer dry solids from one container to another [6]. Glove Box improvements are still under research [7-8].

## 2. Design and implementation

Having as starting point the high price of a Glove Box made-up by steal and the necessity of a Glove Box for the analysis of the organic solar cells (OSC) we wanted to design and implement one for the researchers in this field.

The first stage in the implementation of the Glove Box was the choice of a more affordable material from which the box will be assembled. One of the most suitable materials for creating this chamber are safety glass, double glazing and plexiglass. We chose to use the plexiglass because it offers a very good resistance and it is much lighter than the other options. The dimensions of the designed box are 100 x 62 x 65 cm (length x width x height) and were chosen based on other Glove Boxes that already exist on the market. The box is provided with a door on the side and two holes through which the hands will be inserted into the box using a pair of gloves.

The second stage was to choose the proper sensors and build the electronic measurement and control system knowing that we must monitor four important medium parameters inside the box: temperature, humidity, gas and oxygen level. After a large market research, the four chosen sensors were: DHT22 temperature and humidity sensor, MQ-2 gas sensor and ME2-O2-D20 oxygen sensor. Among these sensors a Reed magnetic sensor was placed to close/open the door.

In the last stage we looked for two suitable solenoid valves. Because the designed box has a small volume of $0.4m^2$ it was not necessary to use some big ones.

The block diagram of our design is presented in figure 1.

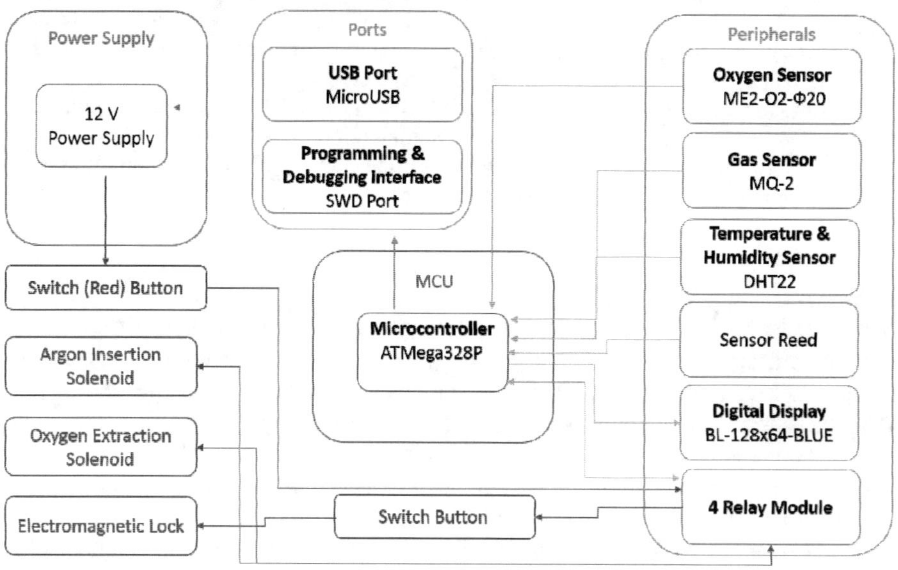


**Figure 1.** Block diagram of the system.

The schematic of the entire electronic system was implemented in Eagle CAD and presented in figure 2. It consists of several blocks (B1 to B5) responsible for various actions in the circuit.

The control centre is an ATmega328 microcontroller (B2) to which all other components are connected. To power up the system a 5V stabilizer must be used (B1). In our case we used the Arduino Uno development board which incorporates the microcontroller and the voltage stabilizer.

The third block (B3) consists of: temperature and humidity sensor, oxygen sensor, Reed sensor and gas sensor. In addition to these sensors there are also the 4 LEDs with the 4 resistors related to them. Two of the LED pairs are connected to the gas sensor, and when the threshold value is exceeded, a switch will be made from the green LED to the red LED. The second pair of LEDs will make the same switch when the box door closes at the Reed magnetic sensor action.

The fourth block (B4) is an LCD screen on which it is displayed the information related to temperature, humidity, oxygen and gas level inside the box.

The last block (B5) consists in a row of four relays connected as follows: one to the electromagnetic sensor responsible for closing and opening the door actioned by a switch button; one for the solenoid responsible for inserting the gas into the box; one for the solenoid valve which has the role of extracting the oxygen from the box; one extra relay for further connections if necessary. A main switch button was placed to power up the entire system.

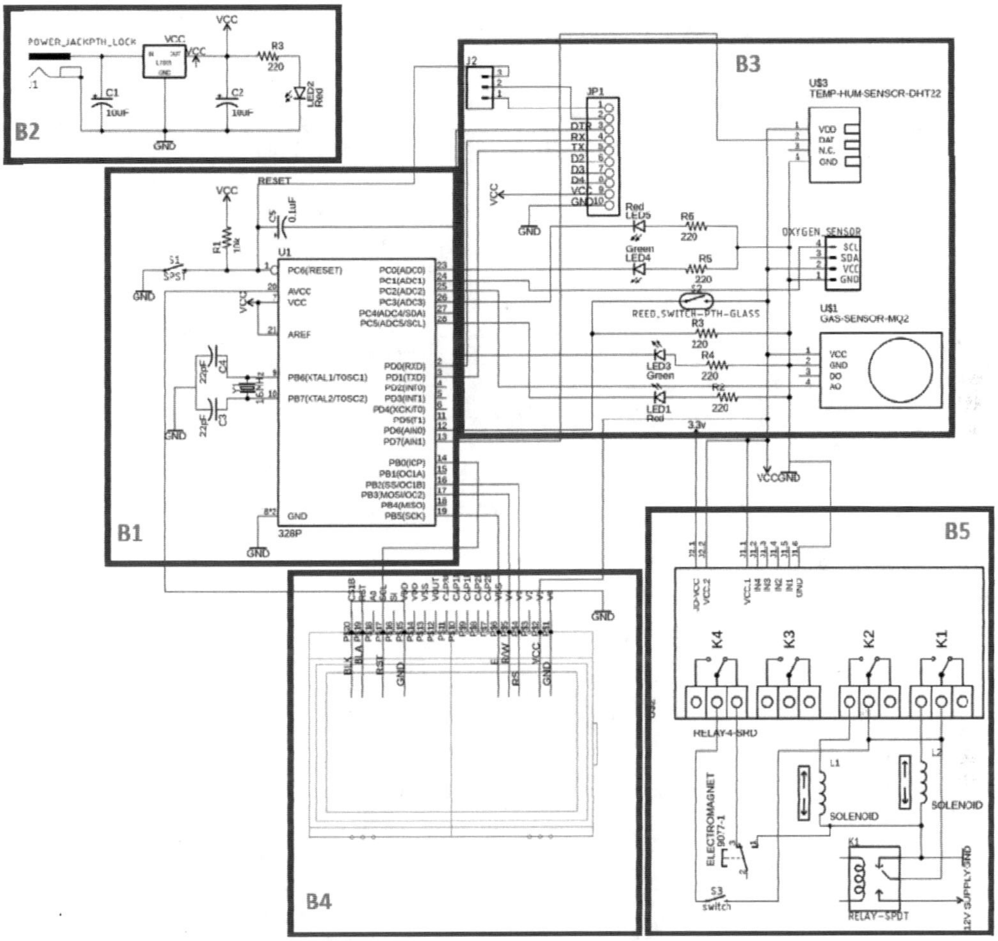


**Figure 2.** Eagle schematic of the electronic system.

The system was tested initially on a breadboard prototype and after that transferred into a PCB level (figure 3 – top side) implementation which contains two PCBs: one for the sensors unit which must pe placed inside the box to monitor the inside medium and one to command and control the unit mounted outside the box. The PCBs were encapsulated in custom designed and 3D printed boxes (figure 3 – bottom side). For the design we used the Autodesk platform: Eagle and Fusion360.

The electro-valve for the insertion of the inert gas into the box was mounted on the right-bottom side of the box and the electro-valve for the extraction of the oxygen from the box in the left-top side of the box as presented in figure 4. Near the oxygen extraction valve the sensor unit was mounted. We used this setting knowing that argon has a higher density than oxygen, so the argon will fill the bottom of the box and push the oxygen to the top. In this way we will have an accurate measurement for the oxygen level inside the box.

IOP Conf. Series: Materials Science and Engineering 1032 (2020) 012048 doi:10.1088/1757-899X/1032/1/012048

Among the holes for the glove mounting an extra sealed hole was placed to connect the sensor unit to the control unit. This hole permits the insertion of a power cord inside the box for powering up different equipment intended to be used under inert gas.

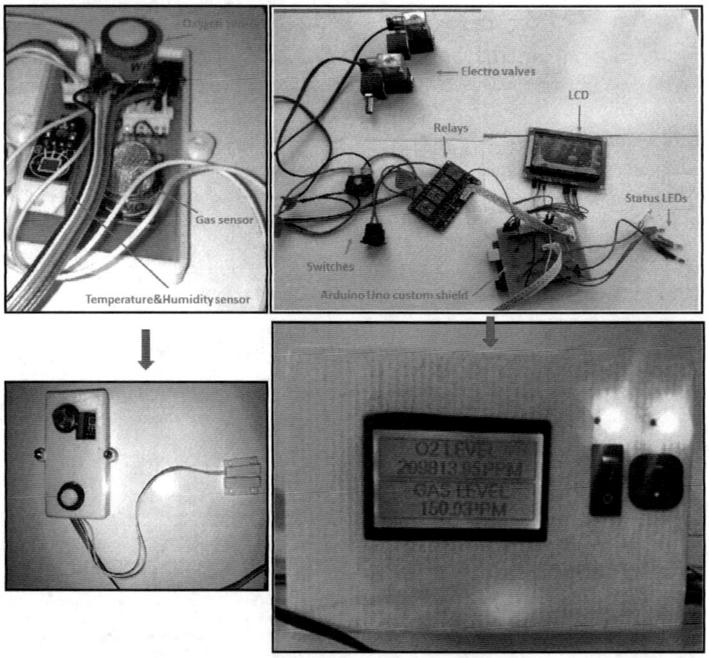


**Figure 3.** PCB level implementation: detailed (top), mounted in enclosures (bottom), sensor unit (left), command and control unit (right).

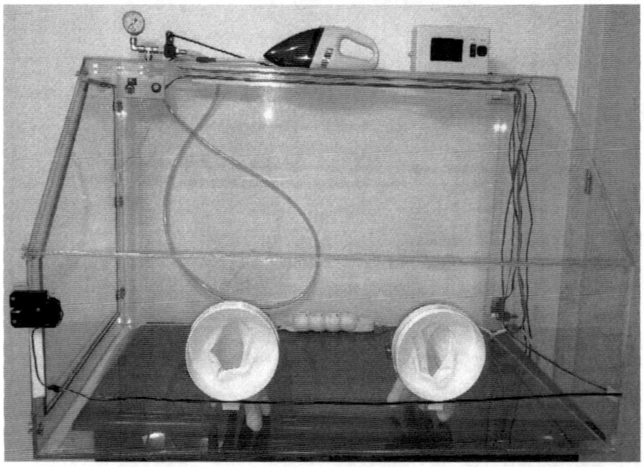

**Figure 4.** Final assembly of the Glove Box.

The microcontroller receives all the information from the sensors and redirects it to the display. Also, it controls the relays which in turn control the valves and the electromagnetic lock. The whole operation algorithm is presented in figure 5 and explained beneath.

It all starts with the system initialization then it continues with the activation of the interruption and initialization of the LCD. Both solenoids and the door lock are closed. The LEDs related to the Reed sensor and the gas sensor are ON depending on which condition is accomplished. The sensors start the monitorization and the data is read and transferred to the microcontroller and then displayed on the LCD. If the oxygen value is higher than 50.000 PPM, then the valves will open and start the process of argon insertion and oxygen extraction. To speed up the extraction process a vacuum pump can be mounted. When the condition is accomplished the valves will close automatically and inside the chamber, we will have a hermetic environment ready for the experiments.

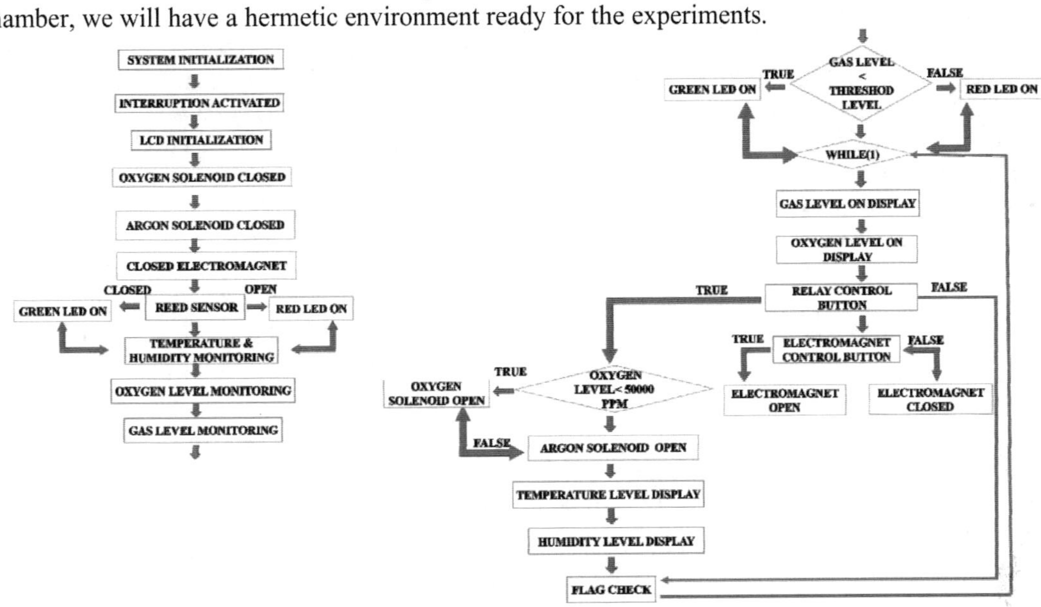


**Figure 5.** Logic diagram of operation.

## 3. Experimental results

The values read by the sensor at the beginning of the first test were: O2: 209808.69PPM; Gas: 30.54PPM; Temperature: 35.2°C; Humidity: 44.9%.

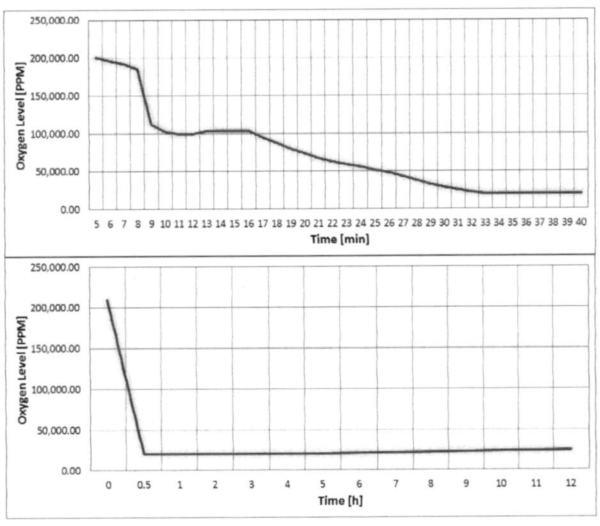


**Figure 6.** Glove Box operation during Argon injection.

The entire evolution of oxygen level during this first test is presented in figure 6 (top graph). We started the insertion of argon with a pressure of 1.25Bar for a time period of seven minutes and after that we doubled the insertion for two minutes. We could see that the decrease of the oxygen level is much faster in the case of the 2.5Bar than in the one with 1.25Bar insertion pressure. The 2.5Bar pressure is the maximum pressure allowed in the laboratory where we took the testing. Keeping the valves sealed for 5 minutes the oxygen level remained constant around 100.000PPM. After this, we inserted more argon at 2.5Bar and we reached in sixteen minutes the minimum oxygen level of

20.000PPM that could read by the sensor. At this point we closed the valves for eight minutes and observed that the oxygen level remained unchanged at 20.000PPM. At the end of this test the other sensors read: Gas: 3.66PPM; Temperature: 36.2°C; Humidity: 22.4%.

For the second test (figure 6 - bottom), we start the whole process from the ambient medium oxygen level, around 210.000PPM. After closing the door, we inserted argon at 2.5Bar for almost 30min. when the read oxygen level dropped to 20.000PPM. At this point we closed the valves and monitored the oxygen level for 12 hours. Analysing the data saved for this time period we found that the oxygen level was kept at 20.000PPM for five hours and after that we had small changes every hour. After 12 hours the oxygen level was around 24.000PPM.

The last test was taken with eight organic solar cells and after their initial measurement, four of them were introduced in the Glove Box and kept at 20.000PPM oxygen level medium and four of them outside the Glove Box for 12 hours. After this storage time the OSCs were measured again and the results revealed a decrease of the efficiency (PCE) with 20% for the cells kept in the inert medium and with more than 70% for the cells kept in the high oxygen level medium. The measurement results are presented in table 1.

**Table 1.** BHJ OSC testing results with and without Glove Box.

| Cell Name | Initial PCE [%] | PCE [%] after 12h | PCE decrease percent [%] | Storage condition |
|---|---|---|---|---|
| **P1_E1** | 2.33 | 1.86 | 20 | Inside Glove Box |
| **P1_E2** | 2.01 | 1.67 | 17 | Inside Glove Box |
| **P2_E1** | 2.76 | 2.29 | 17 | Inside Glove Box |
| **P2_E2** | 2.62 | 2.14 | 18 | Inside Glove Box |
| **P3_E1** | 2.08 | 0.56 | 73 | Outside Glove Box |
| **P3_E2** | 2.14 | 0.51 | 76 | Outside Glove Box |
| **P4_E1** | 2.57 | 0.57 | 78 | Outside Glove Box |
| **P4_E2** | 2.31 | 0.67 | 71 | Outside Glove Box |

## 4. Conclusions

The Glove Box design and the implementation of it proved to work properly for a time period of at least 12 hours where the oxygen level remains almost constant at 20.000PPM. The test made on the organic solar cells shows that the Glove Box offers a good environment for them. From the price point of view the Glove Box materials (plexiglass and electronics) are less than 1000EUR, which is a big plus compared to the more expensive Glove Boxes available on the market. Future work can be considered by replacing the oxygen sensor with one with a higher sensibility.

## References

[1] Oka J 2012 *A human factors approach towards the design of a new glovebox glove for Los Alamos National Laboratory* (Master Thesis The university of New Mexico)
[2] Cournoyer M E *et al* WM'04 Conference, WM-4027
[3] Kohli R 2012 Elsevier *Methods for Monitoring and Measuring Cleanliness of Surfaces,* pp 107-178
[4] Olander L *et al* 2001 Elsevier *Local Ventilation from Industrial Ventilation Design Guidebook*, pp 807-1023
[5] Huynh J *et al* 2015 *Evaluation and Improvement of Medical GloveBox Holder*, (Master of Science Thesis, Stockholm, Sweden)
[6] Velaga S K *et al* 2009 *Twentieth annual conference of Indian nuclear society – INSAC*
[7] Choppin G R *et al* 2002 Elsevier *Radiation Biology and Radiation Protection* pp 474-513
[8] Steckle W P *et al* 2002 *American Glovebox Society Annual Conference*

## Acknowledgements

The present work was accomplished by the support of the grant "REGRENPOS" PN-III-P2-2.1-PED-2019-2601, founded by the Romanian Ministry of Education and Research.

IOP Conf. Series: Materials Science and Engineering 1032 (2020) 012049 doi:10.1088/1757-899X/1032/1/012049

# Robotic arm controlled by Android app through Bluetooth connection for organic solarcell manipulation

**L A Szolga and G Flueras**

Faculty of Electronics, Telecommunications and Information Technology, UTCN, Basis of Electronics Department, Optoelectronics Group, Cluj-Napoca, Romania

E-mail: Lorant.Szolga@bel.utcluj.ro

**Abstract.** In this paper we present the design of a three joint robotic arm printed out in polyethylene terephthalate glycol material which provide durability and has good chemical resistance. To move the arm, we used four servomotors for a complex orientation in horizontal and vertical plane. At the tip of the robotic arm a 5V vacuum pump was mounted to handle organic solar cells printed out on indium tin oxide treated glass sheets with a dimension of 25/24/1mm. For an easy and precise control of the robotic arm an Android application was developed with an intuitive user-friendly interface. The communication between the robotic arm and the Android App is established through Bluetooth protocol. The control of the robotic arm is executed by a custom program on an ATmega328 microcontroller.

## 1. Introduction

Organic photovoltaic technology has received increased attention due to its promising qualities, such as the processability of solutions, their adjustable electronic properties, the manufacture of low temperatures and the use of cheap and light materials. While many other photovoltaic technologies are more efficient, organic photovoltaic cells remain advantageous because the toxicity of the materials used, the production costs and their impact on the environment is low. So far, organic photovoltaic cells and maintenance are set in terms of efficiency, which is 13%, reaching almost the efficiency value obtained with low-cost silicon solar cells [1].

The first step towards the mass production of organic solar cells is the choice of the right technique for depositing the organic compound on the desired surface that offers a high yield and low production cost. The main factor that makes photovoltaic technology so attractive is the potential for Roll-to-Roll processing on low-cost substrates, through standard coating and printing processes. Several techniques for printing or coating organic solar cells were established and demonstrated on a laboratory scale. These include inkjet, offset, engraving, mold, rotary coating and more. Each technique has certain advantages and disadvantages, which makes a technique more attractive in certain conditions than others. The next step is to transfer the knowledge developed in the laboratories to an industrial scale of Roll-to-Roll production [2].

The most used deposition technique for manufacturing laboratory scale devices is rotary coating, which is not compatible with the Roll-to-Roll type of production. However, rotating coating is still widely used to study and understand the fundamental principles of organic solar cells [3].

The spin coating process is done by placing ITO treated glass sheets inside a spin-coater device. While the device is rotating the glass sheet at different speeds and time periods, an organic compound is released on top of it which is spread in a very uniform, thin and pattern less manner. After completion of the rotation process, the cell is extracted from the device and moved to an area where the excess of the organic compound is wiped. Subsequently an aluminum or silver electrode is deposited by PVD (physical vapor deposition) technique onto the organic compound [4].

In all these processes the glass sheets must be handled with extreme care because the organic compound is extremely fragile. Since the organic compound and the electrodes are deposited on their surface, the probes can only be trapped in certain areas so as not to damage the deposited layers. Currently, in the process of manufacturing the organic solar cells, these are manipulated using metal tweezers. Given the level of fragility of cells and any human errors that may occur during their movement, there is a very good chance that mistakes will occur that can lead to premature cell destruction.

To combat this, we proposed for their manipulation a robotic arm [5-9] controlled remotely by a custom developed Android App through a Bluetooth communication [10]. This method avoids the use of tweezers, and thus considerably reduces the risk of cell damage during the manufacturing process.

## 2. Design and implementation

The block diagram of the proposed robotic arm system is presented in figure 1, which expose its operation. The user can enter the manipulation data in the mobile application, which is transmitted via Bluetooth to the ATmega328 microcontroller. The microcontroller is incorporated on an Arduino Nano development board. The electricity needed to operate the entire system is provided by an external 5V power supply, which connects directly to the Arduino board, the Bluetooth module, the servo motors and the vacuum pump. With the facilities given by the microcontroller, the data transmitted by the user through the application is interpreted and transmitted to the robotic arm.

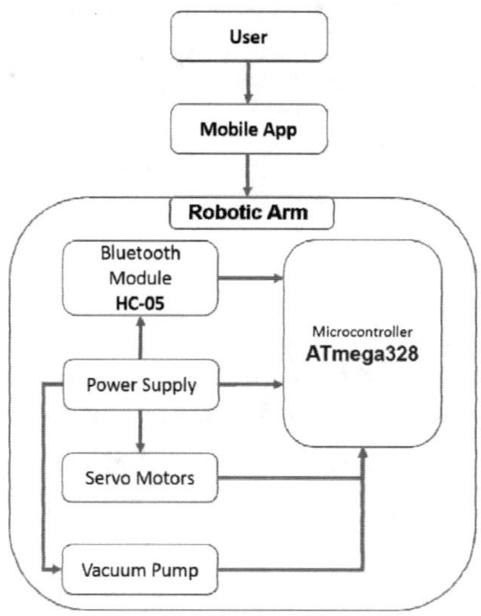


**Figure 1.** Block diagram of the system.

Based on the block diagram we started out the design of the electronics needed to control the robotic arm. The schematic of the entire electronic system was implemented in Eagle CAD and presented in figure 2 (top). To have a complex control on the robotic arm we considered four servo motors: one for the horizontal movement (waist servo) and three for the vertical movements to mimic a human hand (shoulder/elbow/wrist servo). Digital pins 5 to 8 of the Arduino board were used to transmit signals to the four servo motors. For a finger style action of the robotic arm a vacuum pump was considered, that is controlled through digital pin 9 on the Arduino.

Knowing that the glass probes are weighting a few grams we choose a small vacuum pump, which works with a voltage of 5V. Since it has only two connectors, we decided to control it with a BD243 NPN transistor. The transistor is fed into the collector, the signal is connected to the base of the transistor and the positive terminal of the vacuum pump is connected to the emitter.

The entire circuit is powered using a jack connector and a power supply that is connected to a wall outlet. The power supply provides a voltage of 5V and a current of 4A.
The PCB level implementation was conducted as presented in figure 2 (bottom).

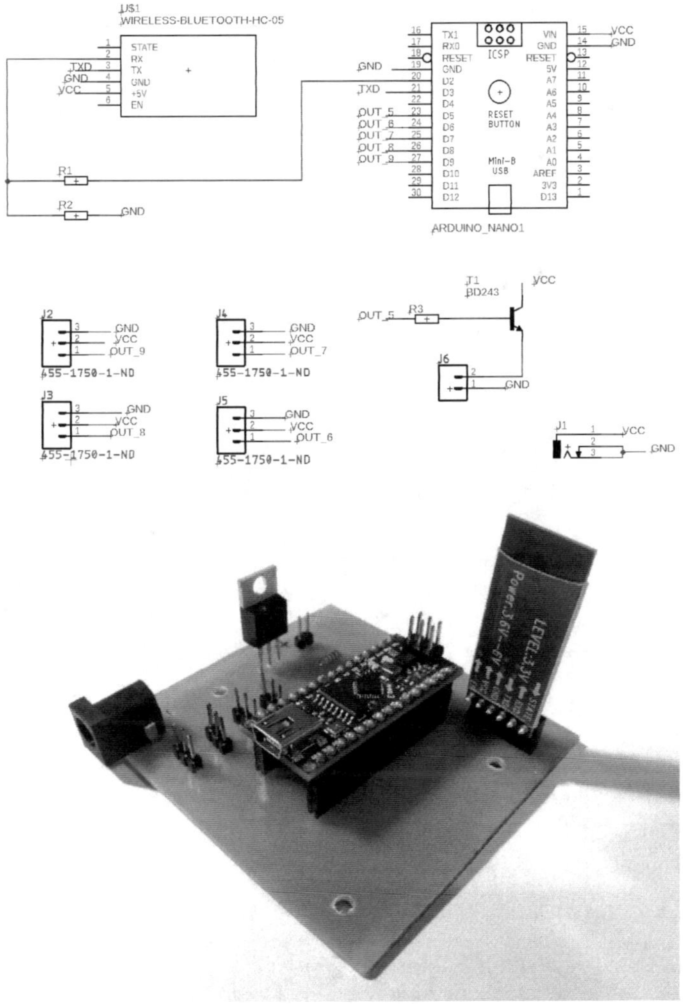


**Figure 2.** Robotic arm: Eagle schematic (top), final PCB (bottom).

The design of the robotic arm was done in SolidWorks CAD. The robot has been designed to be easy to implement, use and maintain. Inside the base of the robot (figure 3-left), which also acts as an enclosure, there is a place of approximately 11cm x 8.5cm in which all the electronic components will be positioned. Along the two arms of the robot, a channel was created through which the wires from the servo motors and the tube for the suction cup can be pulled. This channel starts from the vacuum pump. The use of this channel reduces the risk of damage of the wires while using the robot. At the same time, it gives a more compact and pleasant appearance. The robot has six main parts (figure 3-middle):

- The electronics enclosure, which will be attached to the work surface, to provide stability and contains a servo motor, the vacuum pump and the PCB;
- The lid, which can be easily removed to access the electronic components inside the enclosure;

- The rotating base, on which the rest of the assembly is attached. This rotating base is mounted on a servo motor that could rotate between 0 and 180 degrees.
- Two almost identical parts that have the role of the arm and the forearm;
- The tip, to attach the suction cup with which the organic solar cells will be manipulated.

All the parts were 3D printed in PETG (polyethylene terephthalate glycol) material which gives robustness to the robot.

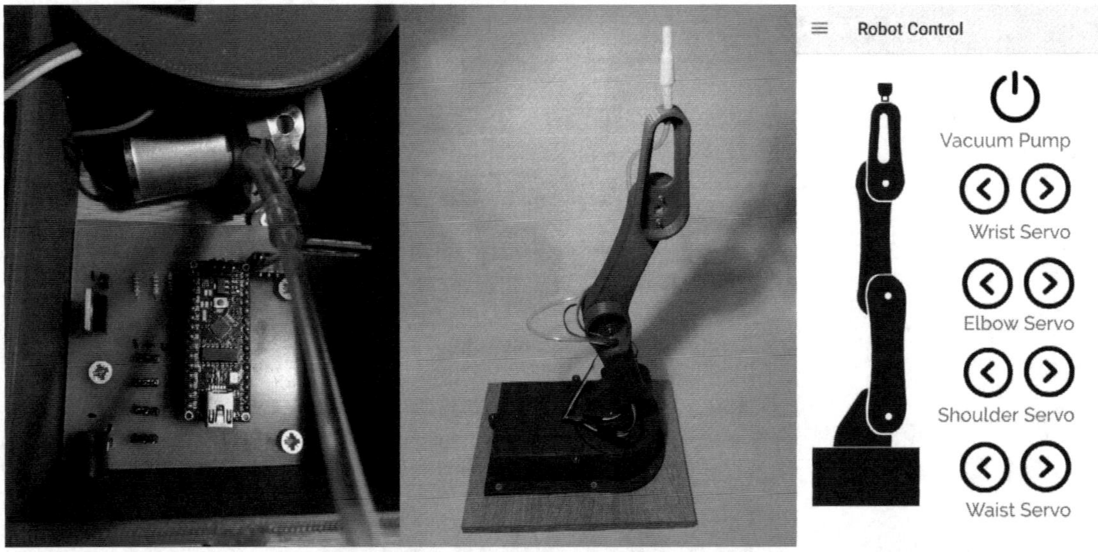


**Figure 3.** Robotic arm: electronics (left), in homing position (middle), mobile app control interface (right).

The logic diagram in figure 4 represents the operation of the Android application developed especially for this project. The application was created using the Android Studio development environment together with the Kotlin programming language.

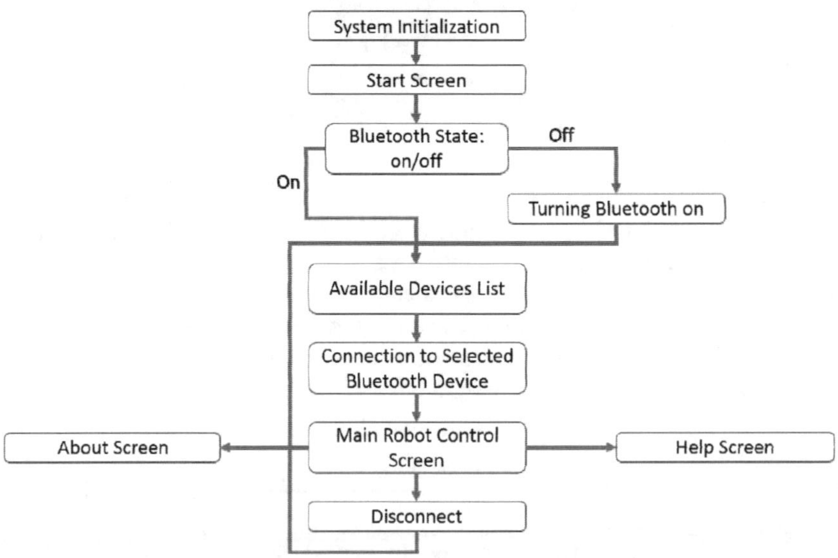


**Figure 4.** Logic diagram of operation for the mobile app.

In the first phase, the user reaches a home page, where he sees a brief description of the functionality of the application and has the option to go further. This first page can be seen in figure 5 (left). Subsequently, the application checks whether the Bluetooth is activated on the device that is in use. If it is turned off, the user is asked if he wants to turn it on.

After activating the Bluetooth, the user is presented with a list of all Bluetooth devices that were previously connected to that smartphone. It chooses the desired device, and the application goes to the next step, namely establishing the connection between the smartphone and the robotic arm. When the connection between the two devices is successful, the application reaches the main screen. From here, the robotic arm can be controlled using the buttons on this page (figure 3-right). The buttons are placed next to each robot joint so that the application can be used intuitively. Under each set of buttons there is also a label with the name of the servo motor operated by the respective buttons.

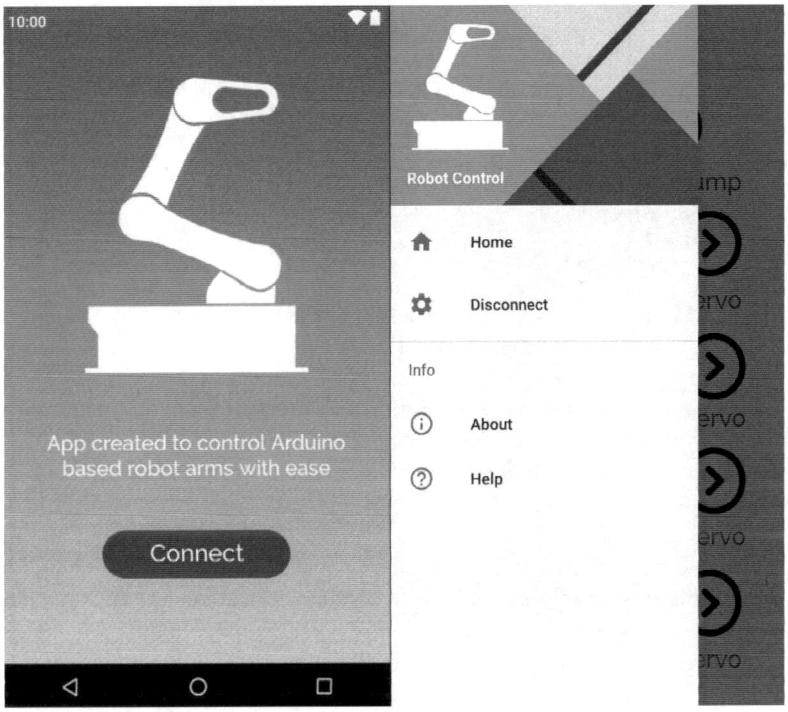


**Figure 5.** Android App: home page (left), side menu (right).

A side menu was implemented on the main screen of the application (figure 5-right), which is accessed using the button next to the page name. Once it is pressed, the user can see additional functions of the application:

- The "Disconnect" button is used when the user of the robotic arm wants to end its activity. Pressing this button will move the robotic arm to a neutral position and the connection will be broken.
- The "About" and "Help" buttons in the "Info" section give the user the opportunity to learn more about the application, either about its role and functionality, or about how to use it.

## 3. Experimental results

The robotic arm controlled by the Android App was tested in OSC manipulations for proper placing of the probes into the matrix substrate, which was introduced after that in the PVD chamber. A view of the robotic arm operation for this purpose is presented in figure 6.

**Figure 6.** Robotic arm in action with OSC manipulation.

## 4. Conclusions

We presented the design and the implementation of a robotic arm for OSC manipulation. The design incorporates three main parts: electronic design, mechanical design and dedicated software application design for mobile devices to control the robot. The robot mimics a human hand movement. The implementation of the robot was a real success, proven by the laboratory testing for OSC handling where the probes were moved from one place to another well-defined spot. The Bluetooth control permits the usage of the robot in different areas where it would be very hard to handle small elements with the human hand, like inside of a glove box. Soon the application will be extended to automatize the artificial solar measurement of the OSCs.

## References

[1] Zhao W 2017 *J.Am.Chem.Soc*, **139** p 7148−7151
[2] Galagan Y *et al* 2012 *Third Generation Photovoltaics from Organic Photovoltaics: Technologies and Manufacturing* (Intechopen)
[3] Rafique S *et al* 2018 *Renewable and Sustainable Energy* **84** p 43-53
[4] Anagnostou K. *et al* 2019 *Energies* **12**(11) p 2188
[5] Henderson H 2006 *Modern Robotics: Building Versatile Machines* (Infobase Publishing)
[6] Lynch K M 2017 *Modern Robotics: Mechanics, Planning, and Control* (Cambridge University Press)
[7] Saeed B N 2020 *Introduction to Robotics: Analysis, Control, Applications* (Whiley)
[8] Mckinnon P 2016 *Robotics: Everything You Need to Know About Robotics from Beginner to Expert* (CreateSpace Independent Publishing Platform)
[9] Siciliano B and Khatib O 2017 *Springer Handbook of Robotics* (Springer)
[10] Khan A A 2016 *DIU journal of Science and Technology*, **11**(1)

## Acknowledgements

The present work was accomplished by the support of the grant "REGRENPOS" PN-III-P2-2.1-PED-2019-2601, founded by the Romanian Ministry of Education and Research.

# High-level model based calibration technique design for SAR ADC

**M N Skripnichenko**[1, 2]

[1] Department of Design and Engineering of Integrated Circuits, Moscow Institute of Electronic Technology, 124498 Moscow, Zelenograd, Russia
[2] Mixed-signal IC Department, ELVEES RnD center, 124498 Moscow, Zelenograd, Russia

E-mail: mskripnichenko@elvees.com

**Abstract.** This article presents a method of developing and debugging algorithms for mixed-signal schemes using mathematical model. SAR ADC is used as an example to present the method. A divided capacitor array with is used in the mathematical model. The Laplase transform is used to replace differential integral equations with algebraic ones. The input dependence of the comparator on the switches in the capacitor array is the basis of the mathematical model. There are two additional models – capacitance mismatch model, which was performed by randomly adding the value of the uniform distribution from -0.01 to 0.01 unit of the smallest capacitance, and noise model, which was performed as randomly changing comparator offset. As a result, there is a mathematical model of SAR ADC with event-driven architecture. Switching of K0 sample switch and bit switches leads to changing of circuit state. Mismatch in the capacitor array during chip manufacturing leads to deterioration of key characteristics of ADC – signal-to-noise ratio and spurious-free dynamic range. There is an algorithm, that can calculate the capacitors mismatch and correct the output code. Such an algorithm shows the efficiency under the conditions of capacitor mismatch and comparator input noise. With this algorithm, the signal-to-noise ratio has been increased by 7.3 dB, which is equivalent to 1.2 bit ENOB. The spurious-free dynamic range has been increased by 16.8 dB. Using the mathematical model when designing and debugging calibration algorithms speeds up these processes.

## 1. Introduction

As the ADC bit rate increases, the influence of negative effects in the fabrication and operation of the chip increases. Variation of capacitors during fabrication [1] and offset of differential pairs [2] are the main undesirable effects.

The following types of calibration are used to compensate the effects mentioned above:

- preliminary, taking into account the effects of production;
- background, taking into account the effects of operation - temperature rise, voltage drop in certain places of the chip;
- external, taking into account both the effects of operation and the effects of production.

The use of calibration involves the development of a complex algorithm that requires long-term modeling. Modern mixed signal design tools [3] can execute such modeling to verify the algorithm implementation using the analog part. Designer, that uses this flow, has to implement the algorithm in HDL and debug it directly during mixed modeling. Such design flow requires a lot of time.

As an alternative to the traditional mixed modeling flow, it is proposed to use the behavioral model of the analog part. Developing a behavioral model on Verilog-A, which implies using the digital part on HDL [4-6], is a common practice. In this paper, it's proposed to develop a model and corresponding algorithms for digitization and calibration.

The mathematical model should have the same functions as scheme in an analog simulator, but it can only take into account certain effects.

## 2. SAR ADC mathematical model

### *2.1. Principle of ADC operation*

Figure 1 shows a block diagram of a SAR ADC. The key elements of the converter are the successive approximation register as part of the digital part of the ADC, the digital-to-analog converter (DAC) with capacitor array, the comparator, and the sample/hold circuits (s/h). In modern converters [7], s/h circuits are combined with the C2C matrix in the DAC.

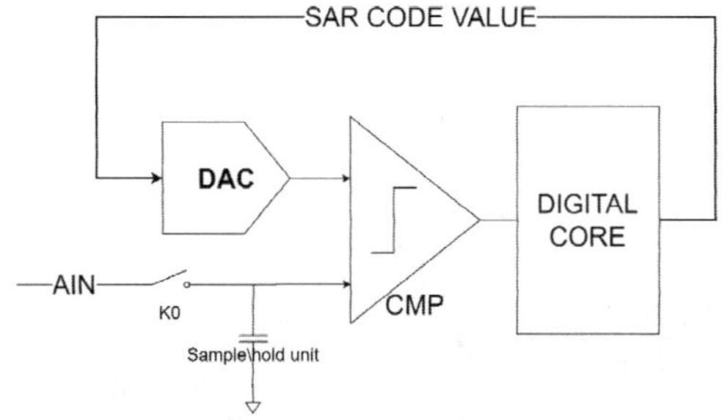


**Figure 1.** SAR ADC scheme.

### *2.2. Capacitor array structure*

A split capacitor matrix is used as an intermediate option [5] between array with binary weighted capacitors and C2C array. This approach combines the advantages of both versions - reasonable size of capacitors and a small number of sampling switches. An example of a circuit with split capacitor array is shown in figure 2.

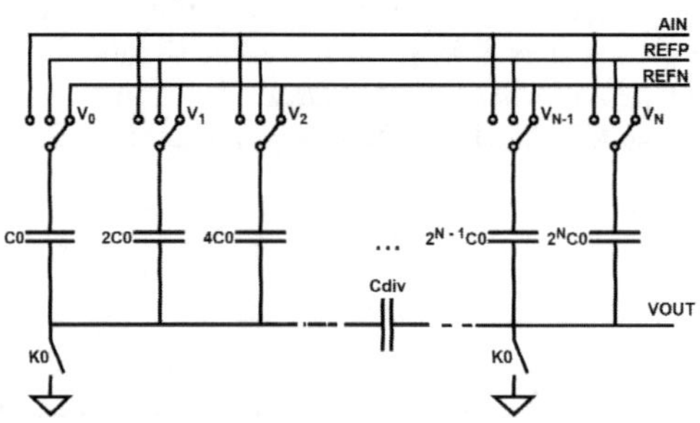


**Figure 2.** Splitted capacitor array.

### *2.3. Mathematical model developing*

The operator method [8] that replaces differential-integral expressions with algebraic ones to simplify the calculation process is used for mathematical model.

Let's take a look at the block diagram of the capacitor array. The basis of the mathematical model is the dependence of the voltage at the comparator input from the array depending on the key switching. This can be described by creating a system of equations based on the Kirchhoff Current Law for the common output of capacitors in each section.

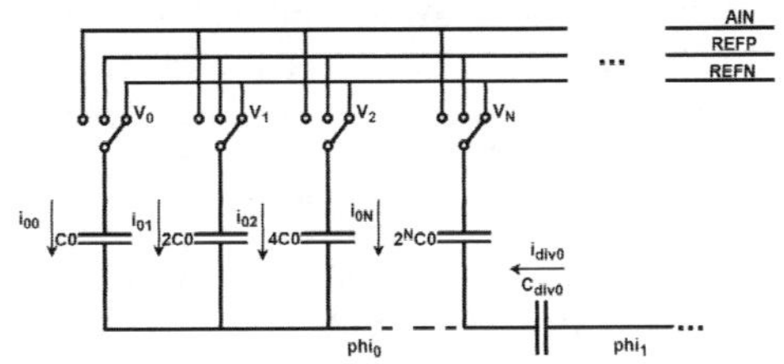


**Figure 3.** Current through capacitors.

Let's take a look at a capacitor array with 3 sections, each with N capacitors. In the case of an array with a large number of sections, the number of equations for intermediate sections increases, based on the same principle as the equation for the middle section in the example below.

Let's make the equations of currents in the lower section of the capacitor matrix in the state when the key that attracts the total potential to the earth is opened. According to Kirchhoff current law, the sum of currents converging at a single node is zero. In this case, we assume that at the moment when voltage Vi is changing, the current flows through the bit capacitors and the divider capacitor to the common potential phi0, as indicated in figure 3. Thus, the following equations are obtained:

$$\sum_{k=0}^{N-1} i_{0k} + i_{div0} = 0 \tag{1}$$

$$\sum_{k=0}^{N-1} C_{0k} \frac{\mathrm{d}(U_{0k}(t) - phi_o(t))}{\mathrm{d}t} + C_{div0} \frac{\mathrm{d}(pho_1(t) - phi_o(t))}{\mathrm{d}t} = 0 \tag{2}$$

After performing a direct Laplace transform we will get the following expression:

$$\sum_{k=0}^{N-1} C_{0k} \frac{U_{0k}(p) - phi_o(p)}{p} + C_{div0} \frac{pho_1(p) - phi_o(p)}{p} = 0 \tag{3}$$

From this expression we can deduce the dependence of the voltage at phi0 on the Vx and phi1 voltages and perform the inverse Laplace transform:

$$phi_o(t) = \frac{\sum_{k=0}^{N-1} C_{0k} * U_{0k}(t) + C_{div0} * phi_1(t)}{\sum_{k=0}^{N-1} C_{0k} + C_{div0}} \tag{4}$$

After similar calculations for the middle and high sections, the dependence of the voltage changes in the nodes $phi_1$ and $phi_2$ when the voltage $V_x$ changes is obtained. After substituting formula (4) in the formula for the middle section, and then expressing phi1, the variable phi0 disappears from the expression. After similar operations with the expression for the higher sections, the expression of the dependence of the node $phi_2$ on the values of $V_x$ is obtained, i.e. the formula for the dependence of the voltage at the comparator input on the voltages on the capacitors. The resulting formula is the basis for a mathematical model of the capacitor array.

For example, the formula for an array of two sections with one capacitor in each will take the following form:

$$V_{OUT} = \frac{V_0 * C_0 * C_{div0} + V_1 * C_1 * (C_0 + C_{div0})}{C_0 * C_1 + C_0 * C_{div0} + C_1 * C_{div0}} \tag{5}$$

### *2.4. Noise model and capacitor mismatch model*

To simulate the capacitors mismatch, the value of the uniform distribution function from zero to 0.01 of the weight of the less significant capacitor in the section was randomly added or subtracted from each capacitor. The noise model is an additive source at the comparator input [9].

### *2.5. Usage of mathematical model*
Based on the above formulas, a mathematical model of a successive approximation register ADC with an event-oriented architecture is described [10]. Events that change the state of the schema are changing the state of sampling switches and bit switches. Mathematical model is written in Python3 with NumPy library. Difference between capacitor array output voltage in math model and output voltage in Cadence Virtuoso XL with ideal elements is smaller than 1e-9 V.

## 3. Preliminary calibration algorithm of capacitor array
One of the reasons for the degradation of dynamic characteristics of the ADC is the capacitors mismatch during the chip fabrication. Algorithms that take into account the mismatch when generating the ADC output code are used to improve the characteristics. There will be a description of the algorithm for pre-calibration of capacitors and the results of modeling the algorithm using a mathematical model.

The calibration algorithm is based on the ability to represent a capacitor as a linear combination of smaller capacitors:

$$C_n = \sum_{i=0}^{n-1} C_i * K_{ni} \tag{6}$$

where $C_n$ is the capacity of the n-th capacitor,

$K_{ni}$ is the coefficient of representation of the i-th capacitor in the linear expansion of the n-th capacitor, belongs to the set {0, 1}.

It is difficult to achieve acceptable accuracy due to the small number of capacitors represented in the linear expansion when measuring the ratio of smallest capacitors. To calculate the inter–bit ratios, capacitors are divided into two groups - smallest non-calibrated capacitors and major calibrated. The concept of "weight" of a capacitor is introduced – a dimensionless quantity reflecting its capacity. The ratio of the weights of the two capacitors is equal to the ratio of the capacitances with accuracy equal the weight of the lowest capacitor used for calibration. The weights of the lower capacitors are known constants.

Weight of each calibrated capacitor $W_n$ can be calculated as the sum of the weights of the capacitors included in the decomposition $W_i$ (6):

$$W_n = \sum_{i=0}^{n-1} W_i * K_{ni} \tag{7}$$

The final step is the full scale calibration. The purpose of full scale calibration is to establish a correspondence between the weight of the capacitor and its contribution to the ADC output code.

To get an output code with an OW (Output Width) bit width, each capacitor's weight should be recalculated as:

$$Wfs_n = \frac{\sum_{i}^{n-1} W_i * K_{ni}}{\sum_{l} W_l} \tag{8}$$

where $W_l$ is the weight of the capacitors that are part of the sample/hold circuit.

Thus, the resulting code will be a scalar product of the $W_{fs}$ capacitor weight vector and the code on the capacitor array obtained as a result of digitization:

$$\mathrm{Re}\,sult = \sum_{i=0}^{n-1} Wfs_i * CODE_i \quad (9)$$

## 4. Simulation results

Simulation of SAR ADC preliminary calibration of capacitors and digitization of 2000 samples of sinusoidal signal was done. Calibration and digitization took place under different conditions – without the physical noise model and the capacitor mismatch model, taking into account both models and taking into account each model separately.

When modeling the algorithms, a split capacitor array with 3 sections and 7 capacitors in each one was used. 4 most significant capacitors of were used as sample/hold circuit. In each section the binary ratio of the capacitances of two adjacent capacitors is set before using mismatch model.

The bridge capacitor is selected in such a way that the voltage difference at the comparator input when switching the smallest capacitor of the high section is twice as large as when switching the biggest discharge of the low section. The noise value is set by a normal distribution function with an expected value of 0 V and a standard deviation of 32 mV for a full scale ADC value of 3 V. The values are obtained from the results of circuit modeling of the comparator block using 180 nm technology.

**Table 1.** Simulation results

| **Simulation parameters** | Pre-calibration | *SINAD, db* | ENOB, bits |
|---|---|---|---|
| **Without capacitors mismatch and noise** | Yes/No | 113.54/113.57 | 18.56/18.57 |
| **With noise, without capacitors mismatch** | Yes/No | 89.3/89.9 | 14.55/14.56 |
| **Without noise, with capacitors mismatch** | Yes/No | 96.5/86.46 | 15.7/14.1 |
| **With capacitors mismatch and noise** | Yes/No | 89.7/82.7 | 14.6/13.4 |

As we can see from the simulation results shown in Table 1, without noise and capacitors mismatch, pre-calibration does not produce any effect, since the binary inter-bit ratio is preserved.

If noise model is used, performed pre-calibration even worsens the results, since the calibration is performed with noise. To improve results, each calibration is performed 4 times to reduce the impact of noise. As a result, the signal-to-noise ratio deteriorated by only 0.6 dB relative to the SNR of the signal measured without calibration.

When the capacitors are mismatched without noise, an improvement can be seen during calibration – more than 10 dB SNR. Meanwhile, the number of effective bits has decreased, since the effective capacity of the matrix has also decreased due to changes in the capacity of the separation capacitors.

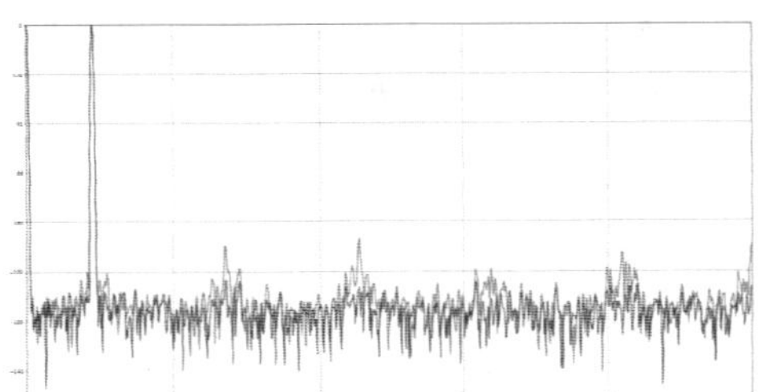

**Figure 4.** Spectrum of the digitized signal when simulating model with noise and capacitor mismatch without pre-calibration (red graph) and with one (blue graph).

The spectrum of signals with and without correction is shown in figure 4. The signal to noise ratio increased by 7.3 dB or 1.2 effective bits. In addition, the spurious-free dynamic range increased by 16.8 dB – from 86.7 to 103.5 dB.

## 5. Conclusion

A method for using the model in the development and debugging of calibration algorithms that accelerates these processes is proposed in article.

A mathematical model of a successive approximation register analog-to-digital converter based on a split capacitor array that is both a digital-to-analog converter and a sample-hold circuit has been developed. With the developed model, the effect of capacitor mismatch on reducing the number of effective bits was studied.

The algorithm that takes into account capacitors mismatch has shown efficiency in the conditions of the capacitors mismatch and noise at the comparator input. Using the algorithm, the signal-to-noise ratio was increased by 7.3 dB, which is equivalent to 1.2 effective bits. The spurious-free dynamic range was increased by 16.8 dB.

## References

[1] Lei S, Qinyuan D, Chuangchuan L and Gaoshuai Q 2009 Analysis on Capacitor Mismatch and Parasitic Capacitors Effect of Improved Segmented-Capacitor Array in SAR ADC *2009 3rd Int. Symp. on Intelligent Information Technology Application*

[2] Shen J, Shikata A, Fernando L D, Guthrie N, Chen B, Maddox M, Mascarenhas N, Kapusta R and Coln M C W 2018 A 16-bit 16-MS/s SAR ADC With On-Chip Calibration in 55-nm CMOS *IEEE Journal of Solid-State Circuits* **53** 1149–60

[3] Santhanalakshmi M and Yasoda K 2015 Verilog-A implementation of energy-efficient SAR ADCs for biomedical application *2015 19th Int. Symp. on VLSI Design and Test*

[4] CADENCE 2013, Cadence SIMPLI Mixed-Signal/MEMS Co-design Methodology, Available at: https://www.cadence.com/content/dam/cadence-www/global/en_US/documents/services/cadence-vcad-simpli-ds.pdf

[5] Li Y and Lian Y 2014 Improved binary-weighted split-capacitive-array DAC for high resolution SAR *ADCs Electronics Letters* **50** 1194–5

[6] Osipov D 2012 Capacitor split array matrix based DAC for SAR ADC *Problems of perspective micro-and nanoelectronic systems design*

[7] Chang K-H and Hsieh C-C 2019 A Calibration-Free 12-bit 50-MS/s Full-Analog SAR ADC With Feedback Zero-Crossing Detectors *IEEE Journal of Solid-State Circuits* **54** 1624–35

[8] Marin M and Öchsner A 2018 *Essentials of partial differential equations* (Springer International Publishing) 137–68

[9] Luo Y, Jain A, Wagner J and Ortmanns M 2019 Input Referred Comparator Noise in SAR ADCs *IEEE Transactions on Circuits and Systems II: Express Briefs* **66** 718–22

[10] Clark T and Barn B S 2011 Event driven architecture modelling and simulation *Proceedings of 2011 IEEE 6th Int. Symp. on Service Oriented System (SOSE)*

## Acknowledgments

The reported study was funded by RFBR, project number 19-31-90081.
I would like to pay my special regards to my supervisors Andrey Aleksandrovich Belayev and Alexey Viktorovich Zaytsev.

# Information management and monitoring system for a grapes harvesting robot

**K Tziridis, A Nikolaou, T Kalampokas, E Vrochidou, T Pachidis, G A Papakostas and V G Kaburlasos**

Human-Machines Interaction Laboratory (HUMAIN-Lab), Department of Computer Science, International Hellenic University, Agios Loukas, 65404 Kavala, Greece

E-mail: gpapak@teiemt.gr

**Abstract.** In recent years, there has been an increasing integration of high technology in the execution of agricultural tasks towards improving the quality of the produced products, while reducing production costs. In this context, there is a special interest in the development of automatic agricultural robots (Agrobots) with advanced techniques of collecting, processing data, decision making and applying actions. A key element of the successful integration of such autonomous robots is the information management and monitoring system (IM2S) that accompanies it. This work presents the IM2S developed for an autonomous grapes harvesting robot, as part of a national project. Details regarding the system's design are presented, covering system requirements, functional specifications, Graphical User Interface (GUI) design, database design and system-robot communication. Along with the presentation of the IM2S design, the special functional and technological requirements that must be met, are also discussed. Thus, the system could be a useful tool for the human operator. Moreover, the designed system can be easily adapted to collaborate with any robot meeting specific communication requirements. This makes the proposed IM2S a universal tool for any grape harvesting Agrobot already deploy or will be developed in the future.

## 1. Introduction

The technological advancement of automation and robotics, consequently resulted to the advent of robots in agriculture, able to drastically increase the productivity, reduce the operating costs and the lead time of agriculture [1]. Agricultural robots, namely *Agrobots*, are currently used towards the automation of agricultural tasks. Recently, several research attempts propose vineyard agrobots for different purposes; pruning [2], watering [3], harvesting [4], spraying [5]. The new generation of agrobots demonstrates a higher level of sophistication using artificial intelligence for decision-making on the spot. The agrobot of our interest is an autonomous grape harvesting robot (AGR), as part of a co-funded national project. ARG is developed to perform harvest, green harvest and defoliation, based on commercial hardware, machine vision and innovative computational intelligence algorithms. The developed algorithms are intended for real-time processing of the sensory data, and decision making under the robot's sensory awareness [6,7]. The robot consists of a mobile vehicle, a robotic arm with adequate end-effector and mounted sensors such as light detection and ranging (LiDAR), a global positioning system (GPS) and cameras.

The focus of the proposed work is on information management and visual monitoring for ARG. The motivation behind the proposed work lies in reducing human labor and bringing efficiency in dexterous grape-related agricultural operations through the use of an autonomous agrobot. The novel contributions of this work are: 1) a solid information management system that can be generalized and adopted by different projects that meet the same requirements, 2) a vineyard monitoring system based on inexpensive sensors and 3) a friendly interactive software, specially developed to monitor the vineyard addressed to people with minimum knowledge of technology.

The rest of the paper is organized as follows. System requirements are presented in section 2 and the functional specifications in section 3. Database design and GUI are shown in sections 4 and 5, respectively. Section 6 summarizes the system-robot communication. Finally, section 7 concludes the paper.

## 2. System requirements

The IM2S application that communicates and interacts with the agrobot is an application designed to run on a portable computer. Thus, the central unit of the system will be a laptop, placed in the operating range of the robot, in order to send and receive real-time information to/from the robot. The system requirements for the hosting machine so as to run smoothly and interact loosely with the robot is a Windows 10 64bit operating system with at least 1.5GB hard disk space for the application installation and 4GB disk space for the vineyard maps. The orthomosaic vineyard maps are extracted from the acquired images of an unmanned aerial vehicle (UAV) by using the Agisoft software. The application uses the vineyard maps to navigate the robot between the vineyard rows [8]. Additionally, it should have at least 8GB RAM memory and a CPU of at least i5-2540m. There are no GPU requirements for the laptop, due to the fact that the application's main purpose is to preview and organize the received information from the robot and to send back commands to the robot through the database. Moreover, the robot is supported by Robot Operating System (ROS), uses MongoDB database and needs a WiFi connection.

## 3. Functional specifications

The users need to Login and have access to vineyard maps of their interest. All actions of the users need to be stored to an individual history file. The administrator needs to register new users and update the application. The administrator can access all maps and history files. The user can see all real-time measurements of the robot and live stream from its cameras, knowing exactly from which part of the vineyard these videos come from. All measurements and video sequences need to be able to get saved in files. Measurements need to grouped and visualized on the maps in an understandable way. Measurements regarding the status of the robot, such as battery level, the current position of the robot, working time, also need to be displayed. The users need to be able to select the agricultural task for the robot and customize the selected task from a setting panel based on their personal practices. The IM2S needs to keep a history of the tasks and choices made by the users. Statistics regarding the selected tasks, such as the number of bunches removed, how long the work took to complete, etc., should also be recorded. Since the measurements are numerous, the users need to have the possibility to choose which of the information will be displayed at the screen at any time. It should be possible for the users to check that all the individual components that comprise the robot work properly and are connected to each other. The users need to be able to alternate between their maps and see all possible paths on every stored vineyard map. Moreover, the users need to be able to select one or more paths on each map, where the robot should navigate and perform the selected task. If the users do not select a path or specific settings for an agricultural task, then the system will use default settings and perform the tasks on the closest path from its current location. However, the users need to select the agricultural task to further proceed, so that the system to send their options to the robot and start executing the task. The users need to be able to zoom in/out on the displayed maps so as to clearly inspect the paths and measurements shown on the maps. The users need to be able to give the command to the robot to start a task but also to stop working at any time.

## 4. Database design

The database of the IM2S is the most important part of the application since it binds the communication of the application with the robot. It is a MongoDB non-relational database and it is selected due to its ability to store JavaScript Object Notation (JSON) documents instead of data in tables, like SQL. This architecture permits the storing of different data categories to different collections. In figure 1 it is illustrated a part of the database diagram which shows extensively the type of data exchanged between the database and the application, and between the database and the robot.

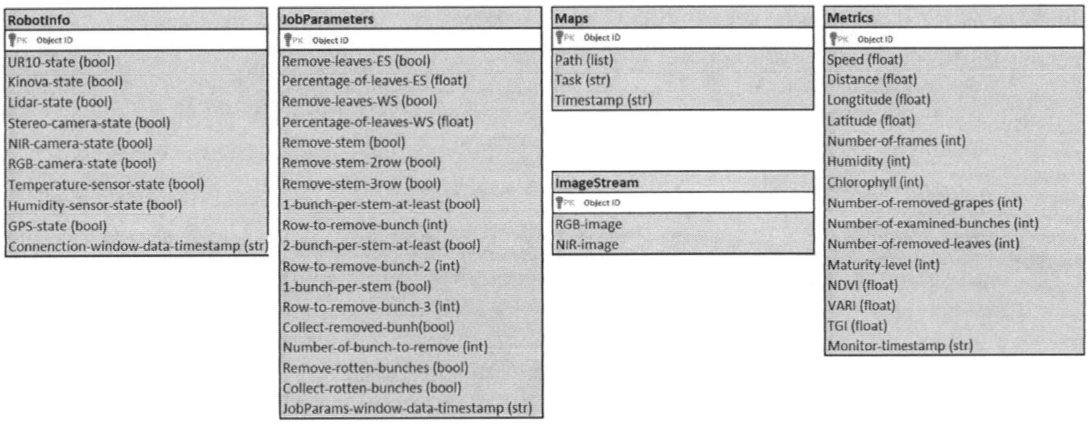


**Figure 1.** Database diagram.

The *User collection* stores all the information about the accounts that have access to the application. More analytically, the data of this collection are primarily the usernames and passwords of the users. Additionally, the history of every user is stored along with the vineyard maps of users' interest. The *Maps collection* stores the vineyard maps that are provided to the user. In these maps, the crop rows are identified and the navigation route mapping of the robot is designed, i.e. the paths between the crop rows where the robot should navigate and execute the agricultural tasks. In this collection, the geographic coordinates of the paths (the start and end points) are also stored. The *Spatial maps collection* stores 3 different maps. These maps are formulated on the main vine map, based on the collected data of the robot during its navigation in the vineyard paths. The three spatial maps are depicting *temperature*, *moisture* and *vegetation*. Thus, the user can have a complete image of the aforementioned metrics distribution in the vine, in real-time. The *Path Planning collection* is created based on the orthomosaic maps and it stores the path coordinates of the maps that the user selects. More specifically, the user selects the exact paths on the vineyard map of his interest, where the robot should navigate to and perform agricultural operations. The *Robot Info collection* stores information about the robot parts (manipulator, end-effector, encoders etc.) so as the user to be able to monitor the system's hardware and to detect in time any malfunction, knowing from which part it came from. Moreover, this collection includes information regarding the robot's speed, battery level, the traveled distance and its current longitude and latitude. The *Metrics collection* includes measurements of humidity, temperature, chlorophyll, grapes ripeness level, number of grape clusters and number of leaves that the robot removed. The *Job Parameters collection* stores the settings for the three agricultural tasks; harvest, green harvest and defoliation. These settings concern the leaves and grapes to be removed and are defined by the user. Finally, the *Image Stream collection* stores the images from the robot's cameras. All the information is stored and displayed at the GUI for real-time monitoring.

## 5. Graphical User Interface (GUI) design

The GUI of the application is implemented with Qt5. This minimal user interface is meant to be as simple as possible, addressed to the average user. The GUI displays all relevant information regarding the agricultural tasks that the robot can execute and regarding the robot itself. The GUI comprises the *Login Option* window and the *Main GUI* window which consists of 7 tabs.

In the *Login Option window*, one can select to login either as a user or as an administrator. In the user login window, the user needs to enter his account credentials, i.e. username and password that are stored in the database. If the inserted credentials exist in the database, then the user can continue to the

*Main GUI* window. In the administrator login window, the same process takes place for the administrator to enter the GUI. When entered, the administrator can create accounts for new users. The administrator creates the credentials for the new users, which are stored in the database at the *User collection*, giving them access to the GUI.

The *Main GUI* window consists of the following seven tabs:

- **Maps**: this is the main tab. The user can load the desired orthomosaic map, can choose the paths on the map where the robot will navigate to and can choose the agricultural operation that the robot will perform. The *Start* button is inactive until the Job parameters, which are found in the next tab, are defined. Then, the Start button becomes available for sending all the parameters to the robot. The user can press at any time the *Stop* button to interrupt the robot's operation.
- **Job parameters**: the user defines the parameters of the each agriculture task. For the defoliation, the user can choose the number of leaves to be removed and from which side of the vineyard (east or west). For the green harvest, the user can choose how many grape bunches to remove per tree and the way to remove them; starting from both edges or alternately per shoot. For the harvest, there are no parameters to be defined. The robot removes and collects all remaining grape bunches on the vines. After the user had selected the task parameters and had stored them, for the application is ready to send them to the robot. All information from this tab is stored in the *Job Parameters collection.*
- **Connections**: the user can monitor the robot components' status. With a *green* indicator next to them are marked the components that work properly while with a *red* indicator those in which a problem is detected. The information about this tab are retrieved from the *Robot Info collection* of the database.
- **Monitor**: the user can access useful information about the vineyard, the selected robot task and the robot components. This information is retrieved from the *Metrics collection*, the *Robot Info collection* and the *Image Stream collection* of the database. The actions that a user can do in this window are to hide a selected group of the information that is not of his interest, to save all displayed information in a .csv file and to save images from the stream.
- **Spatial maps**: the user can see *Spatial maps collection* information displayed on the orthomosaic map of the vineyard. Three different maps can be displayed and alternate. The selected metric to be displayed is presented as a dot on the orthomosaic map, located at the exact point on the map where the robot took the corresponding measurement. All spatial maps can also be stored by the user.
- **History**: the user has access to all previous actions. A list of all up-to-date selected tasks, settings that have been sent to the robot along with timestamps, is provided. The user can export this data in a JSON file and preview it at any time.
- **Help**: at this tab the user can find the manual of the GUI.

The main tab (Maps) of the GUI, is illustrated in figure 2, and the "Spatial maps" tab in figure 3.

## 6. System-Robot communication

The robot while functioning, it senses the environment and gathers data. This data has to be sent back to the main computer, to the system's database in real-time to update the IM2S and the displayed information on the GUI for correct decision making by the user. On the other hand, the user needs to send data about the tasks and settings to the robot, stored through the GUI, in the database and transmitted to the robot so as to start or/and stop executing. Therefore, the robot and the system write and read data to and from the database. In order to set up a real-time communication, between the robot and the main computer system, a Wi-Fi based communication network is established. Therefore, the used mobile robot vehicle needs to have Wi-Fi communication capabilities. Otherwise, the mobile robot needs to be controlled via a Wi-Fi wireless network with the help of simple software and hardware, such as chip-based modules that are well-known cost-effective tools of performing Wi-Fi wireless control functions.

**Figure 2.** Main tab of the GUI.

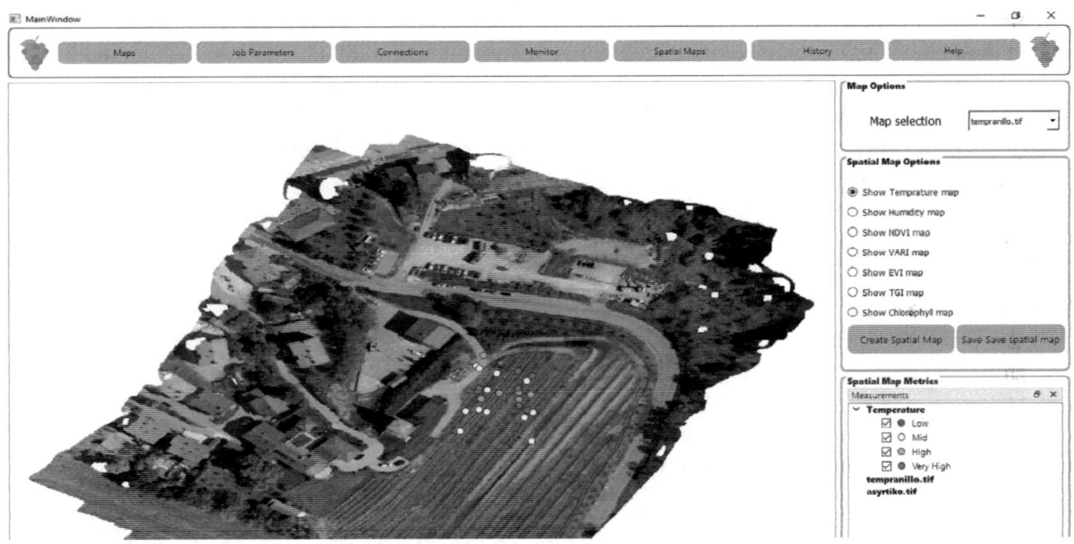


**Figure 3.** Spatial maps tab of the GUI.

Figure 4 illustrates the communication between the application and the robot. As it can be seen, the IM2S –robot communication depends on Wi-Fi. The database runs on the host computer, where the desktop application is installed, in order to avoid overloading the robot's internal disk drive. The application writes a json file into the database, containing details regarding the desired working area for the robot and all parameters of the agricultural task that will be performed. When the robot receives all details, it starts executing. While operating, it collects data from the environment which sends back to the database. Simultaneously, the collected data are shown in real-time on the monitor window of the application.

## 7. Conclusions

The application of agricultural robots in viticulture has experienced an increase in investment and research. In this work, an information management and monitoring system for an autonomous grapes harvesting agrobot is presented. System requirements, functional specifications, GUI design, database design and system-robot communication are some of the aspects that are discussed. The proposed system is meant to be a useful tool for the robot's operators. Moreover, the designed system can be easily adapted to collaborate with any robot that meets the communication requirements. This is due to the fact that the system depends on a robotic platform running on ROS. All robots that run on top of ROS come with their vendor packages preinstalled. Thus, the proposed IM2S is generic and can be considered as a universal tool that can be generalized for any grape harvesting agrobot.

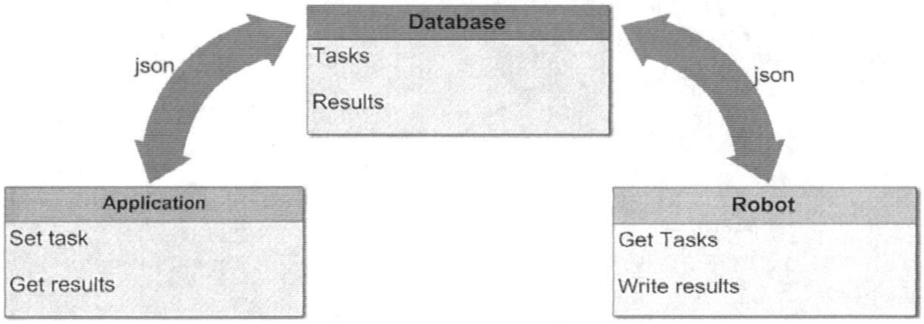


**Figure 4.** IM2S-robot communication.

## References

[1] Mavridou E, Vrochidou E, Papakostas G A, Pachidis T and Kaburlasos V G 2019 Machine vision systems in precision agriculture for crop farming, *J. Imaging* **5** 89

[2] Botterill T, Paulin S, Green R, Williams S, Lin J, Saxton V, Mills S, Chen X and Corbett-Davies S 2017 A Robot System for Pruning Grape Vines, *J. F. Robot*, **34** 1100–22

[3] Thayer T C, Vougioukas S, Goldberg K and Carpin S 2018 Routing algorithms for robot assisted precision irrigation, *IEEE International Conference on Robotics and Automation (ICRA)*, Brisbane, QLD, pp. 2221-2228, doi: 10.1109/ICRA.2018.8461242

[4] Luo L, Tang Y, Lu Q, Chen X, Zhang P and Zou X 2018 A vision methodology for harvesting robot to detect cutting points on peduncles of double overlapping grape clusters in a vineyard, *Comput. Ind.* **99**, pp. 130–9

[5] Adamides G, Katsanos C, Constantinou I, Christou G, Xenos M, Hadzilacos T and Edan Y 2017 Design and development of a semi-autonomous agricultural vineyard sprayer: Human-robot interaction aspects, *J. F. Robot*, **34**, pp 1407–26

[6] Kalampokas T, Tziridis K, Nikolaou A, Vrochidou E, Papakostas G A, Pachidis T and Kaburlasos V G 2020 Semantic Segmentation of Vineyard Images Using Convolutional Neural Networks, *21st International Conference on Engineering Applications of Neural Networks (EANN 2020)*, pp 292–303

[7] Badeka E, Kalabokas T, Tziridis K, Nicolaou A, Vrochidou E, Mavridou E, Papakostas G A and Pachidis T 2019 Grapes Visual Segmentation for Harvesting Robots Using Local Texture Descriptors, *12th International Conference on Computer Vision Systems (ICVS 2019) (Thessaloniki)*, pp 98–109

[8] Badeka E, Vrochidou E, Tziridis K, Nicolaou A, Papakostas G A, Pachidis T and Kaburlasos V G 2020 Navigation Route Mapping for Harvesting Robots in Vineyards Using UAV-based Remote Sensing, *IEEE 10th International Conference on Intelligent Systems (IS)*, pp 171–7

## Acknowledgments

This research has been co-financed by the European Union and Greek national funds through the Operational Program Competitiveness, Entrepreneurship and Innovation, under the call RESEARCH – CREATE – INNOVATE (T1EDK-00300).

# Kinematic model of a cow milking manipulator with simultaneous positioning at four given points

**S S Ruzin[1], D V Shilin[1], A S Dorokhov[1], D Yu Pavkin[1] and E L Chepurina[2]**

[1] Federal Scientific Agroengineering Center VIM, 5, 1st Institutskiy drive, Moscow, 109428, Russia
[2] Russian State Agrarian University - Moscow Timiryazev Agricultural Academy, 49, Timiryazevskaya st., Moscow, 127550, Russia

E-mail: ruzin.s.s@yandex.ru

**Abstract.** During the study, the authors have solved the problems associated with the kinematics of a manipulator intended for the automatic care of cattle. The working unit of the manipulator is a system positioning all four milking cups in the individual location of the udder teats. A significant advantage of this manipulator kinematics is the ability to simultaneously connect milking cups to the teats of a cow's udder to provide physiologically more correct milking and reduce the total milking time.

## 1. Introduction

Industrial robots are widely used in manufacturing and in science. As a rule, industrial robots are represented by automatic programmable manipulators that perform complex operations of spatial moving [1-4]. These robots are used in hazardous or high-precision manufacturing. They are used in light and heavy industry, and the list of robotic systems used in agriculture has been recently expanding. For example, large dairy farming companies are widely using milking robots, self-propelled robots for distributing and transporting feeds [5-10]. The most technologically advanced solutions are milking robots, as they perform a number of operations, such as locating the teats, their pre-cleaning with brushes or washing in the milking cup, feeding, weighing an animal, one-by-one fitting teat cups over cow's teats, individual milking each quarter of the udder, assessing qualitative indicators of milk, and successive disengaging teat cups. This robotic system contributes to complete replacement of human labor and improved product quality, as well as ensures preventive protection against teat diseases resulting from dry milking [11-14]. The most expensive and complex part of the milking robot is the system for manipulating the working units, and a feedback system for spatial orientation required for manipulations [15]. But even if this part is the most complex and expensive, it does not provide the maximum completeness of milking, since the cups are connected to the teats in turn, which weakens the response to the irritation phase of the teat receptors to stimulate the maximum return of alveolar milk, which reduces complete emptying of the alveoli. Delays during the attachment of teat cups also reduce the ability to fully utilize the neurohumoral phase of milk transfer, since oxytocin is secreted only once and enters the udder 35-50 seconds after the first teat has been stimulated, by this time existing milking robots are only finishing fitting the last teat cup over the teat still containing a portion of cistern milk [16,17].

IOP Conf. Series: Materials Science and Engineering 1032 (2020) 012052 doi:10.1088/1757-899X/1032/1/012052

## 2. Development of robotic manipulator

To solve the problem of simultaneous fitting teat cups over a cow's teats, the authors have developed a robotic manipulator shown in figure 1.

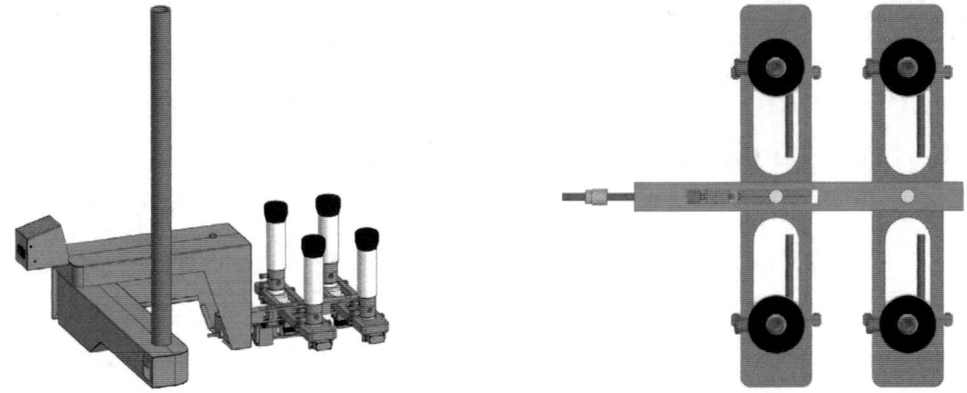


**Figure 1.** View of the robotic manipulator of a milking unit.

The design presented above was developed based on the kinematics of the previously designed manipulator [18], supplemented with a module that provides the ability to control milking cups in a horizontal plane shown in figure 1 B. And then the authors solved the direct and inverse kinematics problems and tested them in the boundary conditions.

## 3. Direct kinematic problem

The core difficulty of finding a solution to kinematic problems lies in the manipulator links, which are located in the 0xy plane, the kinematic diagram of which is shown in figure 2. Solving the direct kinematic problem involves finding the positioning coordinates of the working units. In the kinematic diagram, the working units are indicated by the points $K(x_1,y_1)$, $L(x_2,y_2)$, $M(x_3,y_3)$ and $N(x_4,y_4)$.

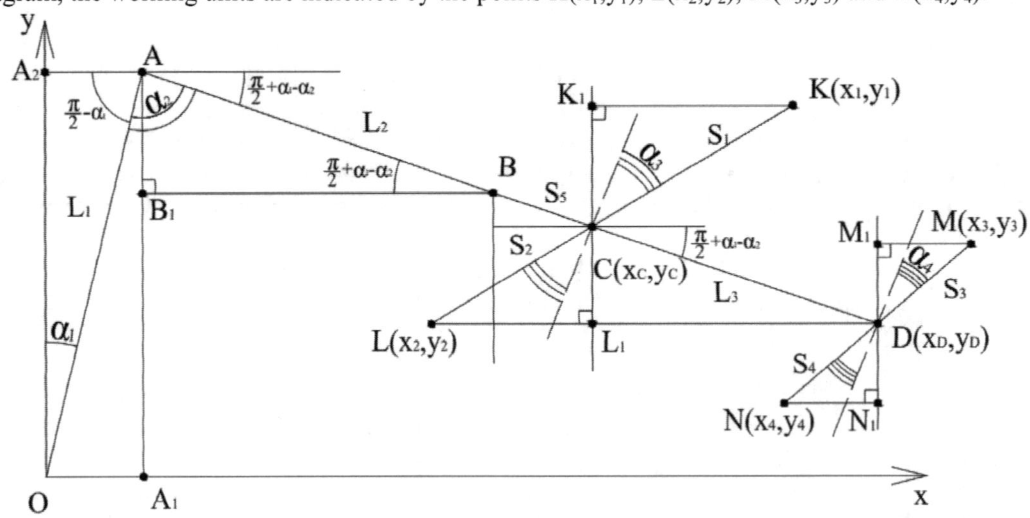


**Figure 2.** Kinematic diagram of the manipulator robot of a milking machine, the links of which are located in the 0XY plane.

A system of equations was compiled to find a solution to the direct kinematic problem, presented below:

$$\begin{cases} x_1 = |AA_1| + |BB_1| + |CC_1| + |KK_1|, \\ y_1 = |AA_2| + |AB_1| + |BC_1| + |CK_1|, \\ x_2 = |AA_1| + |BB_1| + |CC_1| + |LL_1|, \\ y_2 = |AA_2| + |AB_1| + |BC_1| + |CL_1|, \\ x_3 = |AA_1| + |BB_1| + |DD_1| + |MM_1|, \\ y_3 = |AA_2| + |AB_1| + |BD_1| + |DM_1|, \\ x_4 = |AA_1| + |BB_1| + |DD_1| + |NN_1|, \\ y_4 = |AA_2| + |AB_1| + |BD_1| + |DN_1|, \end{cases} \tag{1}$$

The system of equations (1), consisting of 8 equations and 8 unknowns, describes the direct kinematic problem.

## 4. Inverse kinematic problem

The inverse kinematic problem involves finding the angles $\alpha_1 - \alpha_4$ and distances $S_1 - S_5$ from known positioning points – $K(x_1,y_1)$, $L(x_2,y_2)$, $M(x_3,y_3)$ and $N(x_4,y_4)$ (see figure 2).

There can be numerous solutions to the inverse kinematic problem, which must be limited by boundary conditions. Below is the derivation of a system of equations that analytically describes the kinematic diagram shown in figure 2 and is intended to search for a unique solution according to given boundary conditions.

## 5. Boundary condition for solving the inverse kinematic problem

Figure 3 shows the working unit of the manipulator.

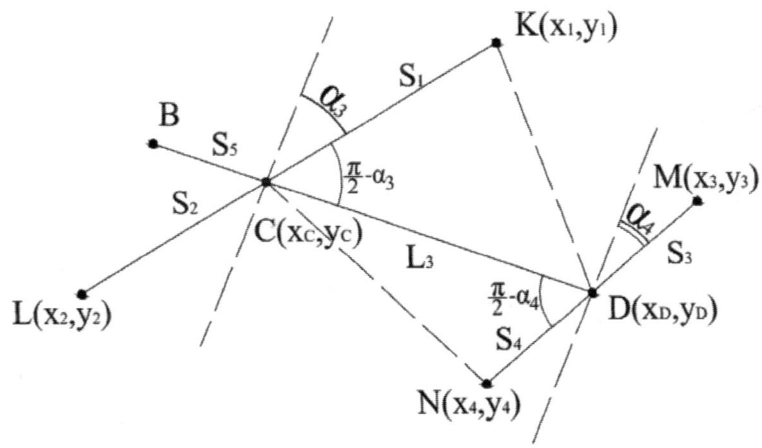


**Figure 3.** Kinematic diagram of the working unit of the manipulator.

The boundary condition for finding a solution to the inverse kinematic problem is that the distance $S_1 = S_2$ and $S_3 = S_4$. In other words, the point C will be equidistant from the points $K(x_1,y_1)$ and $L(x_2,y_2)$ in the same way as the point D will be equidistant from $M(x_3,y_3)$ and $N(x_4,y_4)$. Based on the above, the following system of equations is valid:

$$\begin{cases} \alpha_1 = \dfrac{\pi}{2} - \arccos\left[\dfrac{L_1^2 + x_C^2 + y_C^2 - (L_2 + S_5)^2}{2 \cdot L_1 \cdot (x_C^2 + y_C^2)^{1/2}}\right] - \arccos\left[\dfrac{x_C}{(x_C^2 + y_C^2)^{1/2}}\right], \\ \alpha_2 = \arccos\left[\dfrac{L_1^2 + (L_2 + S_5)^2 - x_C^2 - y_C^2}{2 \cdot L_1 \cdot (L_2 + S_5)}\right], \\ \alpha_3 = \dfrac{\pi}{2} - \arccos\left(\dfrac{S_1^2 + (L_3 - S_5)^2 - (x_D - x_1)^2 - (y_D - y_1)^2}{2 \cdot S_1 \cdot (L_3 - S_5)}\right), \\ \alpha_4 = \dfrac{\pi}{2} - \arccos\left[\dfrac{S_4^2 + (L_3 - S_5)^2 - (x_C - x_4)^2 - (y_C - y_4)^2}{2 \cdot S_4 \cdot (L_3 - S_5)}\right], \\ (x_2 - x_1)^2 + (y_2 - y_1)^2 = 4 \cdot S_1^2, \\ (x_1 - x_2)^2 + (y_1 - y_2)^2 = 4 \cdot S_2^2, \\ (x_4 - x_3)^2 + (y_4 - y_3)^2 = 4 \cdot S_3^2, \\ (x_3 - x_4)^2 + (y_3 - y_4)^2 = 4 \cdot S_4^2, \\ 4 \cdot (L_3 - S_5)^2 = (x_1 + x_2 - x_3 - x_4)^2 + (y_1 + y_2 - y_3 - y_4)^2. \end{cases} \quad (2)$$

To determine the desired quantities $S_1 - S_5$, it is necessary to solve the system of equations:

$$\begin{cases} (x_C - x_1)^2 + (y_C - y_1)^2 = S_1^2, \\ (x_C - x_2)^2 + (y_C - y_2)^2 = S_2^2, \\ (x_D - x_4)^2 + (y_D - y_4)^2 = S_3^2, \\ (x_D - x_5)^2 + (y_D - y_5)^2 = S_4^2, \\ 4 \cdot (L_3 - S_5)^2 = (x_1 + x_2 - x_3 - x_4)^2 + (y_1 + y_2 - y_3 - y_4)^2. \end{cases} \quad (3)$$

To solve the system of nonlinear equations (2) we can use the approximate Newton method.

## 6. Checking kinematics by boundary conditions

To verify the direct and inverse kinematic problems, studies on boundary conditions were carried out:

A. $\alpha_1 = \pi/2$, $\alpha_2 = \pi$, $\alpha_3 = \alpha_4 = 0$, $L_1 = L_2 = 0{,}5$ m., $L_3 = 0{,}4$ m., $S_1 = S_2 = S_3 = S_4 = S_5 = 0{,}1$ m. In this case, the manipulator will be operating along the abscissa axis over its full length equal to the total length of all the links $L_1 + L_2 + L_3$. Figure 4A shows a graphical representation of the manipulator in the 0XY plane.

B. $\alpha_1 = 0$, $\alpha_2 = \pi$, $\alpha_3 = \alpha_4 = 0$, $L_1 = L_2 = 0{,}5$ m., $L_3 = 0{,}4$ m., $S_1 = S_2 = S_3 = S_4 = S_5 = 0{,}1$ m. In this case, the manipulator will be operating along the ordinate axis over its full length equal to the total length of all the links $L_1 + L_2 + L_3$. Figure 4B shows a graphical representation of the manipulator in the 0XY plane.

C. $\alpha_1 = 0$, $\alpha_2 = \pi/2$, $\alpha_3 = \alpha_4 = 0$, $L_1 = L_2 = 0{,}5$ m., $L_3 = 0{,}4$ m., $S_1 = S_2 = S_3 = S_4 = S_5 = 0{,}1$ m. In this case, the manipulator will be operating along the ordinate axis by the length $L_1$ and along the abscissa axis by the length $L_2 + L_3$. Figure 4C shows a graphical representation of the manipulator in the 0XY plane.

D. $\alpha_1 = \pi/4$, $\alpha_2 = \pi$, $\alpha_3 = \alpha_4 = 0$, $L_1 = L_2 = 0{,}5$ m., $L_3 = 0{,}4$ m., $S_1 = S_2 = S_3 = S_4 = S_5 = 0{,}1$ m. In this case, the manipulator will be operating at an angle of 45 degrees to the abscissa axis and ordinate axis to its entire length equal to the total length of all the links $L_1 + L_2 + L_3$. Figure 4D shows a graphical representation of the manipulator in the 0XY plane.

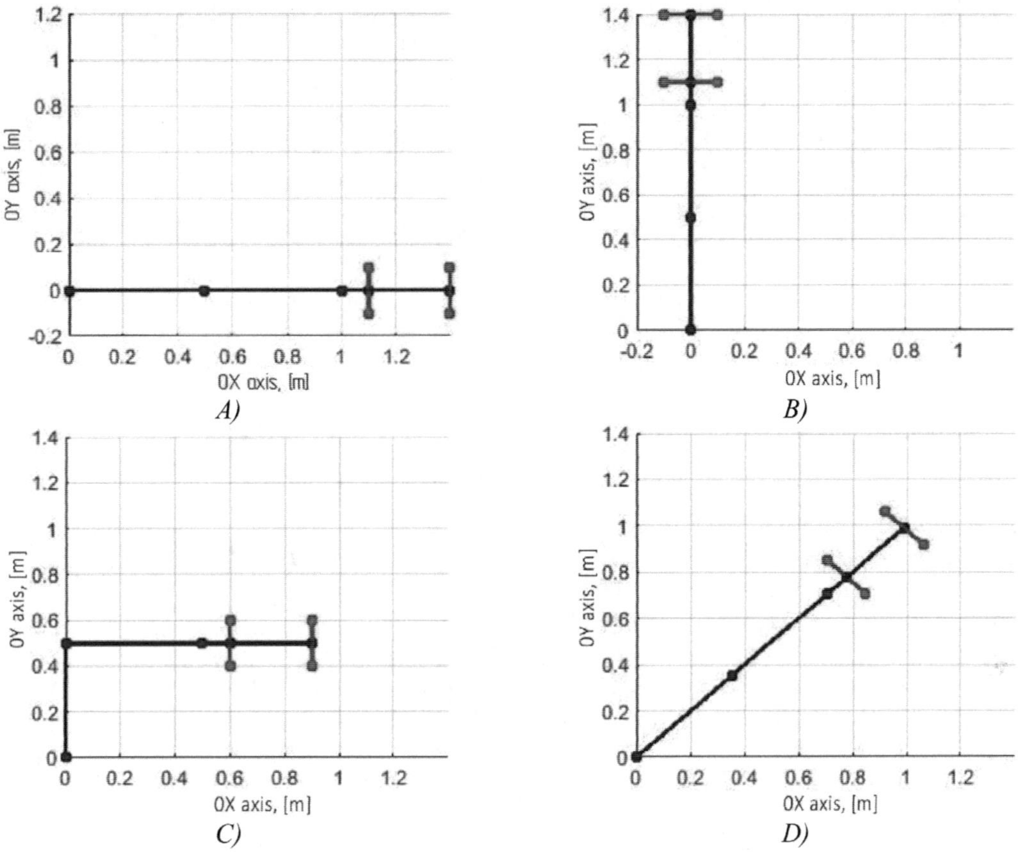


**Figure 4.** Graphical display of the manipulator in the 0XY plane for performing a direct task.

## 7. Conclusions

The study has offered solutions to the direct and inverse kinematic problems. When solving the inverse kinematic problem by the Newton method to achieve a correction of 0.001 m, the number of approximations did not exceed 10. The obtained reachability zone of the manipulator is symmetrical with respect to the second link.

The issues discussed in this paper will significantly simplify the search for control algorithms for the manipulator of the milking unit. With the help of the developed program, it is possible to study the kinematics of similar design schemes with various dimensions.

## References

[1] Tsai Y C and Soni A H 1984 The effect of link parameter on the working space of the general 3R robot arms. Mechanisms and Machine Theory, chapter 19 pp 9-16. DOI: 10.1016/0094-114X(84)90004-1

[2] Raghavan M and Roth B 1993 Inverse kinematics of the general 6R manipulator and related linkages. Trans. ASME, Journal of Mechanical Design, chapter 115 pp 502-508. DOI: 10.1115/1.2919218

[3] Gogu G 2005 Mobility of mechanisms: a critical review. Mechanism and Machine Theory 40:1068-1097.DOI: 10.1016/j.mechmachtheory.2004.12.014

[4] Ganin P, Kobrin A, Shilin D, Moskvin V, Shestov D 2019 Synthesis of real-time control systems for multilink industrial robots based on hybrid neural network approach of solution inverse kinematics problem, DAAAM International Week - 2019 The 30th DAAAM International Symposium Intelligent Manufacturing & Automation, 23-26th, Hotel Kolovare, Zadar, Croatia

[5] Lassen B 2011 Relationsship betweens farm structure, milking technology and productivity - survey results from European dairy farmers. Berichte uber landwirtschaft, chapter 89 pp 376-399

[6] Chepurina E L 2014 Conditions and factors of selection of machines for the mechanization processes in animal husbandry. Mekhanizatsiya i elektrifikatsiya sel'skogo khozyaystva, 2: pp. 17-22.

[7] Kormanovskiy L P, Tsoy Yu A, Kirsanov V V, Nikitin E A, Ruzin S S 2017. Priority Areas of Robotization Processes on Dairy Farms. Tekhnika i oborudovaniye dlya sela, chapter 12 pp 24-27

[8] Hogenboom J A, Pellegrino L, Sandrucci A, Rosi V, D'Incecco P 2019 Invited review: Hygienic quality, composition, and technological performance of raw milk obtained by robotic milking of cows. Journal of dairy science, chapter 102 pp 7640-7654. DOI: 10.3168/jds.2018-16013

[9] Switlyk M, Koloszycz E, Sompolska-Rzechula A 2019 Selected technical and organizational problems in various milk production systems. Bulgarian journal of agricultural science, chapter 25 pp 1069-1075

[10] Wildridge A M, Thomson P C, Garcia S C, Jongman E C, Kerrisk K L 2020 Transitioning from conventional to automatic milking: Effects on the human-animal relationship. Journal of dairy science, chapter 103 pp 1608-1619. DOI: 10.3168/jds.2019-16658

[11] Bruckmaier R M, Schams D, Blum J W 1994. Continuously elevated concentrations of oxytocin during milking are necessary for complete milk removal in dairy-cows. Journal of dairy research, chapter 61 pp 323-334. DOI: 10.1017/S0022029900030740.

[12] Bruckmaier R M, Blum J W 1996. Simultaneous recording of oxytocin release, milk ejection and milk flow during milking of dairy cows with and without prestimulation. Journal of dairy research, chapter 63 pp 201-208. DOI: 10.1017/S0022029900031708.

[13] Sandrucci A, Tamburini A, Bava L, Zucali M 2007 Factors affecting milk flow traits in dairy cows: Results of a field study. Journal of dairy science, chapter 90 pp 1159-1167. DOI: 10.3168/jds.S0022-0302(07)71602-8

[14] Besier J, Schupbach-Regula G, Wellnitz O, Bruckmaier R M 2017 Technical note: Effects of attachment of hind teats before cleaning and attachment of front teats on milking characteristics in automatic milking systems. Journal of dairy science, chapter 100 pp 3091-3095. DOI: 10.3168/jds.2016-11712

[15] Dorokhov A, Kirsanov V, Pavkin D, Yurochka S, Vladimirov F 2019 Recognition of Cow Teats Using the 3D-ToF Camera When Milking in the “Herringbone” Milking Parlor. In Proc. 2nd International Conference on Intelligent Computing and Optimization, vol 1072, Koh Samui, Thailand, pp 128-137. DOI: 10.1007/978-3-030-33585-4_13

[16] Pfeilsticker H U, Bruckmaier R M, Blum J W 1996 Cisternal milk in the dairy cow during lactation and after preceding teat stimulation. Journal of dairy research, chapter 63 pp 509-515. DOI: 10.1017/S0022029900032040

[17] Bruckmaier R M, Macuhova J, Meyer H H D 2001 Specific aspects of milk ejection in robotic milking: a review. Livestock production science. chapter 72 pp 169-176. DOI: 10.1016/S0301-6226(01)00277-9

[18] Dorokhov A, Kirsanov V, Pavkin D, Shilin D, Shestov D, Ruzin S 2019. Calculation of the Manipulator’s Kinematic Model and Mounting Points of the Drive Equipment. In Proc. 2nd International Conference on Intelligent Computing and Optimization, vol 1072, Koh Samui, Thailand, pp 339–348. DOI: 10.1007/978-3-030-33585-4_34

# A review on the methods for big data analysis in agriculture

**B I Evstatiev[1] and K G Gabrovska-Evstatieva[2]**

[1] Department of Electronics, University of Ruse Angel Kanchev, 8 Studentska Str., Ruse 7017, Bulgaria
[2] Department of Computer Science, University of Ruse Angel Kanchev, 8 Studentska Str., Ruse 7017, Bulgaria

E-mail: bevstatiev@uni-ruse.bg

**Abstract.** The integration of information and communication technologies into agriculture lead to the development of precision agriculture. Nowadays it relies on Internet of Things devices, geospatial data, historical and real-time information, which has the potential to transform farming into smart farming. However the use of Big Data requires significantly different skills and knowledge, compared to what many farmers and agronomists possess, which is an obstacle for their effective use. This study aims to summarize and provide insight into the common methods used for data analysis in a wide variety of agricultural applications. Initially, the basic characteristics and sources of agricultural data are explained. Next, a review of the common data analysis methods (classification, clustering and regressions) is provided, containing information about the data sources used as well as the desired goal of the analysis. At the end of the paper, a summary is given on the applicability of data analysis methods depending on the desired goals.

## 1. Introduction

The rapid development of information and communication technologies in the last decades has had a tremendous impact on agriculture. The use of Internet of Things (IoT) sensors, online datasets and navigation systems opened many new opportunities for automation of agricultural processes, which have the potential to solve the problems with food shortage [1]. On the other hand, the availability of such enormous amount of data became a major challenge for people involved in agricultural activities for numerous reasons. It is necessary to analyse the available information in either real-time or near real-time if it is to be useful [1,2].

Numerous methods and tools for Big Data analysis were developed in the last decades, which are aimed at facilitating the process. However, their application requires significantly different knowledge and expertise, compared to what agronomists and other people involved in agriculture possess. The problem is exaggerated by the fact that many of them are not engineers, which limits the possibilities for effective application of precision agriculture [3]. Therefore, it is important that agricultural specialists are provided with adequate guidelines and information about the different methods for data analysis.

This study is aimed at performing a review on the common methods for analysis of agricultural data, as well as their area of application. It should analyse and summarize their characteristics and requirements in order to ease agricultural specialists when selecting the appropriate method based on the desired goals.

## 2. Big data in the agriculture

### *2.1. Big data characteristics*

According to [2,4], big data can be characterized with the following dimensions:

- Volume: The size of data collected for analysis. For example, there are numerous open databases with remote sensing data with volume varying from terabytes to petabytes.
- Velocity: The time window in which data is useful and relevant – a certain data should be analysed in a reasonable amount of time in order to be useful. For example, the identification of pests and diseases would only be useful if identified in its early stages.
- Variety: The numerous data sources lead to significant difference in the properties of the collected data. It could vary in type (image, sensor data, etc.), time interval, resolution, etc.

Furthermore, other characterizations of big data also exist [1,6]:

- Veracity: This property refers to how accurate or truthful the data is. It could be considered from different points of view, such as is the data accurate, complete, consistent, objective, credible, etc.
- Valorization: Among the enormous amount of data, it is important to identify that one which is useful and has the potential to add value. Furthermore, it is important to determine how to make use of it.

The above dimensions and characteristics generally describe the big data; however, it is not mandatory that data analysis satisfy all of them as the analysis might get too complicated.

### *2.2. Sources of data in the agriculture*

Several key sources of data used for analysis of agricultural process could be identified, which are analysed below.

**Ground sensor data**

Ground sensors are sensors that are installed literally on the ground. In the context of BIG data, they are considered as IoT devices. The main modules of an agricultural IoT system are responsible for data acquisition, data transmission and monitoring [6]. Different architectures exist for the data transmission, in terms of both communication medium as well as security [7,8].

**Remote sensing data**

Remote sensing is based mainly on the processing of images, obtained using satellites, airplanes, drones and radars. Different types of cameras are used, providing colour (RGB) images, infrared images, etc. [9].

**Weather stations**

Real-time information from weather stations is an important source for monitoring the environmental conditions in agricultural areas [1,5]. They could be considered as a set of ground sensors (temperature, humidity, rainfall, air velocity, pressure, radiation, etc.), providing remote data. Furthermore, weather station systems commonly provide forecasting services, which is crucial for optimization and management of the agricultural activities.

**Historical information and datasets**

Historical information and datasets can have many dimensions and numerous sources, varying from archive weather information, different datasets with images and characteristics of plants, crops, diseases, agricultural areas, variation of prices and many more [1].

In theory everyone could make their own data analysis using the available historical data, however that requires a significant qualification. Therefore, for many agricultural specialists involved into decision making, it is easier to rely on statistical data, provided by the government, public or local authorities.

**Geographic Information System (GIS)**

The GIS technology commonly enriches the data obtained from other sources, by providing precise information about their location, thus identifying the agricultural objects and their parameters. This allows summarizing the big data on geographical maps [1,5].

**Data obtained from humans**

The human opinion is also an important source of data. People act as non-technical sensors relying on intelligence and comprehensive knowledge. The human "measurements" are created through personal and subjective observations, such as air quality, weather observations, statements on public safety, etc. This information could be collected using different tools, such as mobile or web applications, surveys, feeds from social media, etc. When the human opinion is analysed historically, it can provide valuable knowledge about the trends and expectations [1].

## 3. Data analysis in the agriculture

Considering the enormous amount and variety of agricultural data, it is very important to analyse and present it in an appropriate form that is easy for understanding even by non-engineer specialist. Several data analysis methods exist, which are explained below.

### *3.1. Regression analysis*

Regression analysis is used to estimate the relationship between one or more independent variable(s) and a single dependent variable. It provides statistical means to:

- Investigate the significance of the relation between the dependent and independent variables;
- Assess the strength of impact of multiple independent variables on a single dependent one.

Furthermore, regression allows to compare variables on different scales with different units. Different regressions exist, depending on the number of independent variables, the shape of the regression line (linear, nonlinear, etc.), the type of the dependent variables (numeric, Boolean, etc.). A summary of the common applications of regression for analysis of agricultural data is presented in table 1.

**Table 1.** Application of Regression for analysis of agricultural data.

| Sources of data | Type of regression | Goal of the analysis |
|---|---|---|
| Temperature; pH of soil; Nitrogen content in soil; Phosphorus content in soil; Potassium; Rainfall; Water required; Yield. | Multiple regression | Yield forecasting [10] |
| Milk yield; Temperature; Temperature-humidity index (THI); Delay impact | Nonlinear/Multiple regression | Dependency of milk yield on THI [11] |
| Irrigation; Fertilization | Linear regression | Dependency of crop yield on fertilization and irrigation [12] |
| Satellite images; Crop yield | Linear regression | Dependency of average crop yield on average NDVI [13] |
| Farm physical characteristics; Survey results; Management practices | Multiple regression | Influence of management practices [14] |

### *3.2. Clustering*

Clustering is a set of methods that allow grouping together different agricultural objects. The clustering could be done for different reasons, such as optimization, identification of management zones, differentiating between object properties and quality. According to [15], a common clustering approach includes (figure 1):

- Step 1. Identification of the clustering parameters of the different agricultural objects;
- Step 2. Obtaining the optimal number of clusters;
- Step 3. Obtaining the centre of each cluster;
- Step 4. Obtaining the main characteristics of each cluster (profile, size, etc.).

An overview of common applications of cluster analysis in agriculture is presented in table 2.

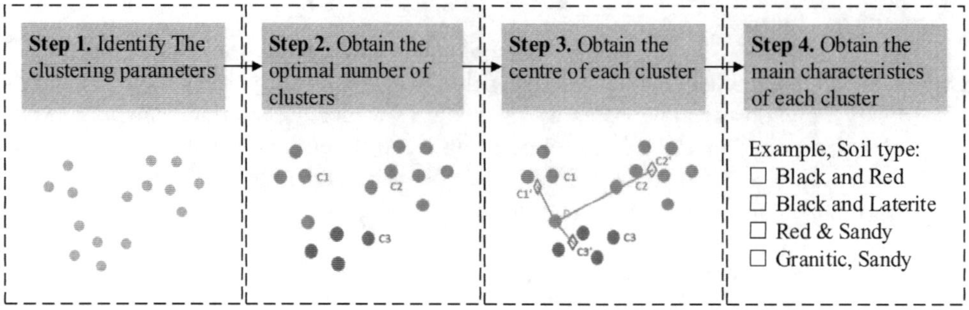


**Figure 1**. Flowchart diagram of a common clustering approach.

**Table 2.** Application of Clustering for analysis of agricultural data.

| Sources of data | Methods used | Goal of the clustering |
|---|---|---|
| Road distance between farms | Hierarchical cluster analysis | Optimization of supply chains by distance [15] |
| Map coordinates, NDVI | Clustering method | Simplification of data [5] |
| Datasets with average temperatures, average rainfall, soil properties | Density based clustering (DBSCAN algorithm) | Clustering of regions based on similarities in temperature/ rainfall/ soil type [10] |
| Number if water tanks; Number of bore wells; Number of open wells | K-means algorithm | Clustering of regions based on water level (to be used for yield forecasting) [16] |
| Dataset of images with plants and pests | K-means algorithm | Identification of plant pests [17] |
| Soil physical properties | Fuzzy c-means clustering | Defining irrigation management zones [18] |

### *3.3. Classification*

Classification is a different method, which is aimed at categorizing objects based on their properties, which are called predictors. It starts with finding an appropriate model and training it. Once the model is trained, it can be used to predict the classes of new objects, i.e. it allows their categorization. An overview of common applications of different classification methods with agricultural data is presented in table 3.

**Table 3.** Application of Classification for analysis of agricultural data.

| Sources of data used | Methods used | Goal of the classification |
|---|---|---|
| Data from multispectral, RGB and thermal sensors; Coordinates; Enhanced vegetation index | Decision trees | Monitoring fertilizer efficiency [19] |
| Images of plants with weed and soil on the background | Support Vector Machine (SVM) | Weed detection [20] |
| Images of plants under different outdoor conditions | k-Nearest Neighbour | Detection of plants [21] |
| Different vegetation indices (NDI47, NDI45, EVI, MCARI and GNDVI) | SVM | Detection of land and crop type [22] |
| Numerous crop images classified as either healthy or with a certain disease | Convolutional Neural Networks (CNN) | Crop disease detection [23] |
| Images with daily development of diseases; Different indices (NDVI; Simple ratio; Structure insensitive vegetation index; etc.) | SVM; | Disease early detection [24] |
| C-band radar backscatter data (satellite) | Image processing; Deep learning | Crop type detection [25] |

## 4. Conclusions

The present study was aimed at performing a review of the common data analysis methods used for agricultural applications. Initially, the key characteristics of Big Data are explained and the main sources of agricultural data are overviewed (ground sensor data, remote sensing data, weather stations, historical information and datasets, GIS geospatial data and human data).

Next, the key data analysis methods are analysed. The performed review showed that regression analysis is the obvious solution when the goal is to investigate the dependencies between agricultural data presented in numerical form. Image sources are also applicable though they are also converted to numerical arrays, such as vegetation indices and coordinates. Examples for successful application of regression analysis is yield forecasting, assessment of management routines, etc.

The clustering approach can be used with any type of input data when the goal is to define zones by some of their properties or by their mutual distances. Examples for such applications are optimization of supply chains, simplification of classified data, categorization of zones (management, yield, rainfall, etc.), identification of pests and other.

The third group of methods is classification analysis, and can be used with any type of input data. Its key agricultural applications include object detection and categorization by one or several properties. It is commonly used for detection of weeds, plants, crops, diseases and pests as well as for categorization of agricultural objects for easier understanding and analysis.

The review conducted in this work could be useful for agronomists and people involved into agricultural activities, who wish to get initial guidelines on the methods for analysis of farming data. The presented results and conclusions provides information about the recommended data analysis methods based on the goals of the investigation.

## References

[1] Kamilaris A, Kartakoullis A, Prenafeta-Boldú F 2017 A Review on the practice of big data analysis in agriculture *Computers and Electronics in Agriculture* **143**

[2] Chi M, Plaza A, Benediktsson J, Sun Zh, Shen J and Zhu Y 2016 Big data for remote sensing: challenges and opportunities *Proceedings of the IEEE* **104** (11) pp 2207-2219

[3] Madhuri J, Indiramma M 2019 Role of big data in agriculture *International Journal of Innovative Technology and Exploring Engineering (IJITEE)* **9** (2) pp 3811-3021

[4] Ivanova E, Iliev T, Stoyanov I 2020 Modelling and simulation of big data networks *2020 7th International Conference on Energy Efficiency and Agricultural Engineering (EE&AE)*, 12-14 November 2020, Ruse, Bulgaria, pp 1-4

[5] Saiz-Rubio V and Rovira-Más F 2020 From smart farming towards agriculture 5.0: a review on crop data management *Agronomy* **10** 207 doi:10.3390/agronomy10020207

[6] Zhang Ch and Liu Z 2019 Application of big data technology in agricultural Internet of Things *Advanced Big Data and Artificial Intelligence Technologies for IoT Systems* **15** (10)

[7] Alkhatib A, Baicher G 2012 Wireless sensor network architecture *2012 International Conference on Computer Networks and Communication Systems(CNCS 2012) IPCSIT* **12** IACSIT Press, Singapore

[8] Hristova T 2020 Blockchain application in mining management *E3S Web of Conferences* **168** 00009.

[9] Montes de Oca A, Arreola L, Flores A, Sanchez J and Flores G 2018 Low-cost multispectral imaging system for crop monitoring *International Conference on Unmanned Aircraft Systems (ICUAS)*

[10] Majumdar J, Naraseeyappa S and Ankalaki S 2017 Analysis of agriculture data using data mining techniques: application of big data *J Big Data* **4** (20)

[11] Türkmen E, Yaslioğlu E 2018 The effect of heat stress on milk yield in a Ipard supported dairy cattle enterprise *Proceedings of the XIX World Congress of CIGR* Turkey pp 51-61

[12] Maharjan B, Liang W Z, Panday D and Qiao X 2019 Evaluating crop sensor in maize grown in semi-arid condition under varying irrigation and nitrogen levels *The 12$^{th}$ European Conference on Precision Agriculture*, France

[13] Paixão L, Marques da Silva J R, Terron J M, Ramiro A and Ordóñez F 2019 Pisum sativum L. (Pea) yield modelling using Sentinel-2 NDVI maps *The 12th European Conference on Precision Agriculture*, France

[14] Townsley R and Parker W 1987 Regression analysis of farm management survey data *New Zealand Journal of Experimental Agriculture* **15** (2) pp 155-162

[15] Bălan A-V, Toma E, Dobre C and Soare E 2015 Organic farming patterns analysis based on clustering methods *Agriculture and Agricultural Science Procedia* **6** pp 639-646

[16] Manjula E and Djodiltachoumy S 2017 A model for prediction of crop yield *International Journal of Computational Intelligence and Informatics* **6** (4)

[17] Fina F, Birch F, Young R, Obu J, Faithpraise B and Chatwin C 2013 Automatic plant pest detection and recognition using k-MEANS clustering algorithm and correspondence filters *International Journal of Advanced Biotechnology and Research* **4** (2) pp 189-199

[18] Jiang Q, Fu Q and Wang Z 2011 Study on delineation of irrigation management zones based on management zone analyst software *Computer and Computing Technologies in Agriculture IV. CCTA 2010. IFIP Advances in Information and Communication Technology* **346** Springer

[19] Florence A, Revill A, Gibson-Poole S, Vigors B, Rees RM, MacArthur A, Barnes AP, Hoad SP and Williams M 2019 Advanced technologies for efficient crop management (ATEC) *The 12th European Conference on Precision Agriculture*, France

[20] Wellhausen C, Pflanz M, Pohl J-P and Nordmeyer H 2019 Automatic weed recognition for site-specific herbicide application *The 12th European Conference on Precision Agriculture*, France

[21] Arroyo J, Guijarro M and Pajares G 2016 An instance-based learning approach for thresholding in crop images under different outdoor conditions *Computers and Electronics in Agriculture* **127** pp 669–679

[22] Devadas R, Denham R J and Pringle M 2012 Support vector machine classification of object-based data for crop mapping, using multi-temporal landsat imagery *International Archives of the Photogrammetry, Remote Sensing and Spatial Information Sciences, ISPRS Congress*, Australia

[23] Sambasivam G and Opiyo G 2020 A predictive machine learning application in agriculture: Cassava disease detection and classification with imbalanced dataset using convolutional neural networks *Egyptian Informatics Journal* https://doi.org/10.1016/j.eij.2020.02.007

[24] Rumpf T, Mahlein A-K, Steiner U, Oerke E-C, Dehne H-W and Plümer L 2010 Early detection and classification of plant diseases with Support Vector Machines based on hyperspectral reflectance *Computers and Electronics in Agriculture* **74** pp 91-99.

[25] Teimouri N, Christiansen M P, Sørensen C A G and Jørgensen R N 2019 A deep learning-based approach for crop classification using dual-polarimetric C-band radar data *The 12th European Conference on Precision Agriculture*, France.

**Acknowledgments**

This work was supported by the Erasmus+ Program of the European Union under contract no 609544-EPP-1-2019-1-PS-EPPKA2-CBHE-JP "Boosting Innovation in Education aNd REsearch oF Precision AgriculTure in Palestine (BENEFIT)".

# Modeling of three-phase inverter power supply system in Simulink

**V B Ivanov**[1]

[1] Department of Physics, University "Prof. D-r Asen Zlatarov" – Burgas, Bulgaria,

E-mail: vasil_bi@abv.bg

**Abstract.** A model of a three-phase electronic converter of electric energy for an autonomous power supply system has been studied. The objective is measuring the percentage of the higher harmonics in the output of the three-phase inverter voltage and the effect of adding a passive selective Notch filters. Their inclusion in the output circuit creates a model of the system with improved electromagnetic compatibility, and hence an increased power factor and efficiency. A virtual architecture of the three-phase inverter power supply system has been developed in Matlab - Simulink software environment. The invertor is built of separate functional blocks simulating the components that can be embedded in a real project. The schematic diagram includes three-phase symmetric loads with three-phase filters models. The main operation modes of the invertor are modeled and the effects of adding three-phase filters to the output circuit after the voltage inverter are analyzed. From the pulse with modulation simulation, the frequency spectrum for the output current $I_L$ and voltage $U_L$ has been synthesized.

## 1. Introduction

A specific component of the construction of a three-phase inverter power supply system (TPSS) is the inverter. Many research results are published in this field. Deadbeat control of output voltage and inductor or capacitor current for three-phase pulse with modulation (PWM) inverters has been proposed in different papers [1, 2].

The control of electronic converters for supplying power is an important part of the technological process in electrochemical devices. It is expected to deliver optimal mode of the converters, keeping the electrical regime of the electronic devices optimal and achieve the best electrical properties [3].
In another research is shown some possibility for control. Some of them use a voltage oriented control (VOC) which originates from field-oriented control (FOC). All of this is to reduce the current harmonics and the total harmonic distortion (THD). This will improve the power factor of the invertor. Nowadays, most rectifier devices usually adopt non-controlled or semi-controlled rectification technology [1, 2, 4].

The dynamic response of DC bus voltage for the PWM rectifier is very important because a higher response means a little DC bus capacitor. The obtained results from simulations are shown that the suggested system there is a better dynamic performance than that of a traditional PI controller [5, 6].
The specific unit in the construction of a TPSS is the DC / AC voltage inverter. It converts the DC voltage into a variable with a frequency of the first harmonic of the amplitude-frequency characteristic (AFC) of the current I and the voltage U in the output circuit with appropriate parameters. The shape of the voltage depends on the control algorithm of the power devices in the circuit, and that of the current I - on the type of load.

It is known that nonlinear power components generate harmonics multiples of the main harmonic. They are characterized by the coefficient of THD for both voltage and current, respectively $THD_U$ and $THD_I$ [7, 8]:

The power part of the inverters are powerful semiconductor devices (IGBT, power MOSFET, GTO - thyristors), controlled by switching method [1, 2, 5, 7]. Switching is performed by pulse-width modulation (PWM), i.e. a section of the power elements is applied, obtaining the highest efficiency. The main disadvantage of the switching mode of operation of inverters is the deteriorated electromagnetic compatibility, which has a negative impact on the consumers [7].

The power circuit of the inverter contains transistors with anti-parallel diodes, which is used for PWM generation as well as the power bidirectional conversion, DC source ($U_d$) and symmetric three-phase consumer ($Z_A$, $Z_B$, $Z_C$).

The aim of this paper is to study the level of high harmonics at the output of the three-phase voltage inverter and to evaluate the effect of adding passive selective notch filters.

## 2. Methodology of the investigation of three-phase inverter power supply in Simulink

Compilation of the architecture of TPSS is in the programming environment "Simulink" [9]. The first step is simulating the model with three different loads. Virtual measurement of the spectral composition of the output current and voltage with and without filters in the output load electrical circuit [10]. Comparative analysis of the results of the measurement of the spectral composition of AFC with and without filters has been accomplished. The synthesis of the frequency spectrum is observed by a block analyser "powergui" in "Simulink".

TPSS simulation without filters in the output circuit has been realised. A three-phase consumer model has been included and the spectral composition of the current and voltage has been measured at different load regimes. After determining the size of the influential harmonics in the spectrum, notch filters and their adjustment by the cutoff frequency for each harmonic multiple of the main one ($h_1$ = 50Hz) have been selected.

Simulation of TPSS with filters in the output circuit: models of notch filters are added and the spectral composition of the frequency response of AFC for current and voltage is measured. The filters are adjusted to suppress the influential high harmonics in the spectrum of the output voltage and current, which is the reason for the decrease in THD. The frequency response reports are modeled by the virtual spectrum analyser from the "powergui" block.

After reporting the results of the two options, a comparative analysis of the frequency response is performed to determine the degree of reduced harmonic composition when operating the TPSS with the added filters in the output circuit and overall improved electromagnetic compatibility. The study uses models of passive frequency filters operating on the resonance principle. Finally, the obtained results are analyzed.

### *2.1. Investigation of a TPSS in SIMULINK*

The simulation study has been carried out in MATLAB / Simulink environment. According to the methodology, a TPSS has been built and tested (figure 1). It is grid-connected. The obtained results are for the output current $I_{outA}$ and the output voltage $U_{outA}$ for one phase.

In the diagram of figure 1, notch filters (green blocks) are connected to the output circuit and are marked with the numbers of the harmonics $h_2$, $h_3$, $h_5$, $h_7$.

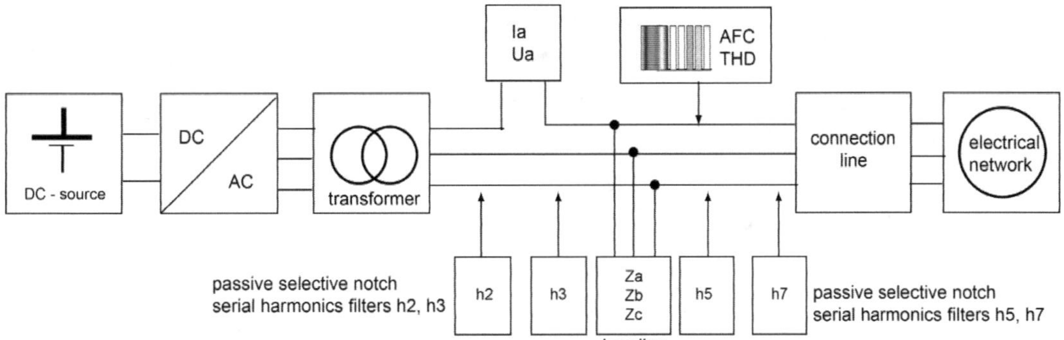


**Figure 1.** Structure of the TPSS model with added notch filters for $h_2$, $h_3$, $h_5$, $h_7$.

The additional use of notch filters improves the amplitude-frequency characteristic (AFC) and electromagnetic compatibility (EMC) of the inverter.

The modelling of the measurement processes is performed with virtual components S_1 and S_2. They are used to measure the $U_L$ and $I_L$,.

The spectrum analyzer's model calculates the individual report of the harmonic in the Fourier series, presenting graphically and tabularly the frequency response for the $THD_U$ and the $THD_I$.

*2.2. Model of TPSS in Simulink.*

The modelling of TPSS in figure 2 is built on a set of rated power 2000W, a voltage source ($R_b$, $V_{DC}$), DC voltage generator with simulation model (photovoltaic modules, battery or DC generator). The "PWM Generator" block on Figure 4 is with a PWM control algorithm.

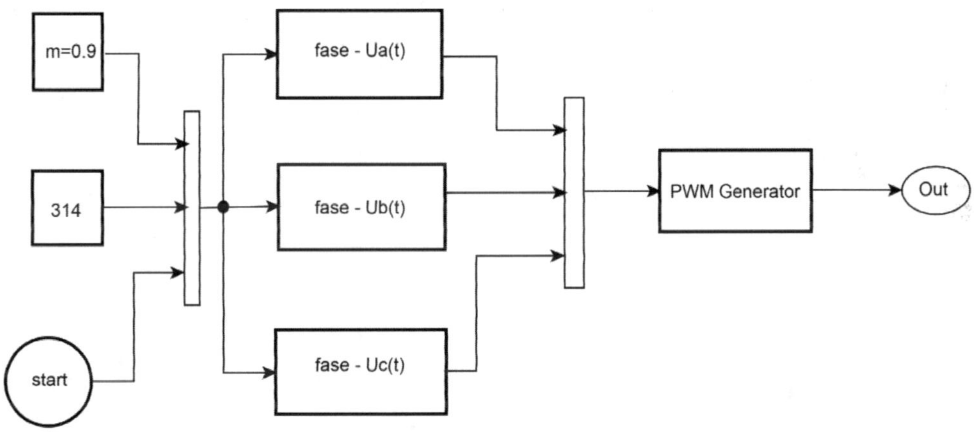


**Figure 2.** TPSS model in Simulink.

The used models for the currents and voltages set in the $F_{cn}$ function blocks are respectively:

$$U_A(t) = U_{Am}\sin(\omega.t + \psi_U),\text{ V} \quad i_A(t) = i_{Am}\sin(\omega.t + \psi_I), A \tag{1}$$

$$U_B(t) = U_{Bm}\sin(\omega.t - \tfrac{2.\pi}{3}),\text{ V} \quad i_B(t) = i_{Bm}\sin(\omega.t - \tfrac{2.\pi}{3}) \tag{2}$$

$$U_C(t) = U_{Cm}\sin(\omega.t + \tfrac{2.\pi}{3}),\text{ V} \quad i_C(t) = i_{Cm}\sin(\omega.t + \tfrac{2.\pi}{3}) \tag{3}$$

Where $\omega = 314, s^{-1}$ is the angular frequency, at $f = 50,\ Hz$.

The plots from dependences (1), (2) and (3) are sine waves dephased at $120^0$ and a model of the LV public electricity distribution network. The phase and voltage dephasing is determined by the difference between the initial phase of the voltage $\psi_U$ and the initial phase of the current $\psi_I$ :

IOP Conf. Series: Materials Science and Engineering 1032 (2020) 012054 doi:10.1088/1757-899X/1032/1/012054

$$\varphi = \psi_U - \psi_I \tag{4}$$

There are settings for parameters of the power devices from the Universal Bridge 3 arms unit and realized are the virtual connection of a real TPSS to the public LV electrical network. A complex symmetrical load with the option of adjusting the capacitive load is added.

The virtual model of the "power line" with the set parameters simulates real losses in the wires (impedance included). As the line length increases, the active and inductive resistances (R, L) increase, too.

The connection group (star / triangle) is also programmed, as well as the quality factor Q = 2. The notch filters $h_2$, $h_3$, $h_5$, $h_7$ work on the resonant principle, with a cut-off frequency determined by the resonant frequency $f_0$.

## 3. Results and discussion

The analysis of the obtained simulation results shows that in operating mode at minimum symmetrical load with minimum power (P =200W, $Q_C$ =10VAR, $Q_L$=50VAR) the $THD_I$=48,67% without filters is with high value, which will lead to losses and poor performance of consumers. When adding filters, the $THD_I$ values decrease to 20,15%. The obtained results shows that the $U_L$ voltage on the load without the filter group is with a value $THD_U$=7,18%, but $U_L$ with notch filters has $THD_U$ = 2,73%, which is under 3% and meets the requirements for the EMC.

Simulations in symmetrical load mode with average power (P = 800W, $Q_C$ = 10VAR, $Q_L$ = 50VAR ).

The analysis of the obtained results shows that for the output current $I_L$ without added notch filters $THD_I$ is 18,14%. It was found that increasing the active power P also improves the quality of $I_L$, as the best indicators are obtained at nominal operating mode. The reduction of $THD_I$ to 14,23% has been obtained by connecting notch filters. They suppress the harmonic components that have the greatest influence on the shape of the current $I_{h2}$, $I_{h3}$, $I_{h5}$, $I_{h7}$. The $U_L$ has $THD_U$=4,22% (without added notch filters), which is over 3% and does not meet the requirements for EMC. After adding the notch filters, the $U_L$ results show an improved harmonic composition up to $THD_U$=2,55%.

Simulation in mode of operation with maximum symmetrical load with set powers reaching the size set in the designed model of TPSS: P=2000W, Qc=10VAR, $Q_L$=50VAR. On Figure 3 are presented the results of the simulations performed for $I_L$. (figure 3).

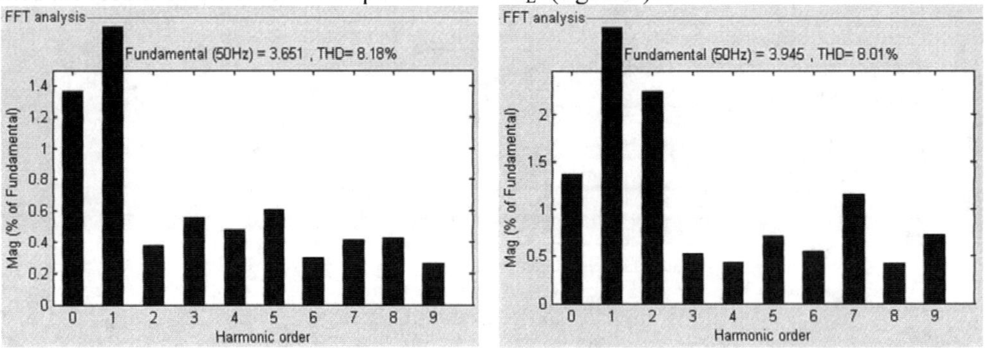


A B

**Figure 3.** Results for output $I_L$ without notch filters (A) and with added notch filters (B).

The measured values in $I_L$ have the percentage of harmonics $THD_I$=8,18% without notch filters. The data on the total content of harmonics with added filters, improve insignificantly the coefficient $THD_I$=8%, because the operating mode goes to nominal load. This is an indicator for effective use of the installed power, which leads to high efficiency.

The analysis of the obtained results for the $U_L$, are with allowable values for form of the wave and total harmonic coefficient $THD_U$=2,55% and without the filters $h_2$, $h_3$, $h_5$, $h_7$. The added group of notch filters for load $U_L$, with the reports in figure 4,B (on the right), reflects their effect on $THD_U$=2,25% (figure 4).

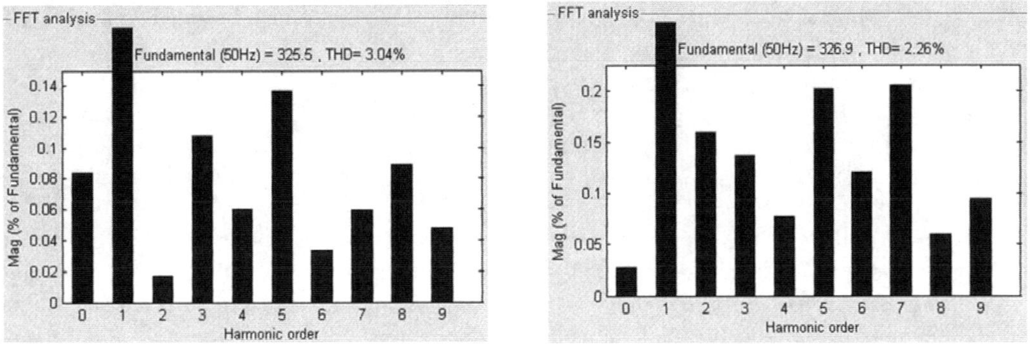

A B

**Figure 4.** Results for $U_L$ output without notch filters (A) and with notch filters (B).

In table 1 are shown the summarized results from made simulation.

**Table 1.** The summarized results of the output parameters for TPSS.

| Load of TPSS | | I, A | $THD_I$, % | $U_L$, V | $THD_U$, % |
|---|---|---|---|---|---|
| **Minimum power** | | | **P=200W** | | |
| | without filters | 0,48 | 48,67 | 231,6 | 7,18 |
| | with filters for $h_2$, $h_3$, $h_5$, $h_7$ | 1,1 | 20,15 | 233,1 | 2,73 |
| **Average power** | | | **P=800W** | | |
| | without filters | 1,2 | 18,14 | 231 | 4,22 |
| | with filters for $h_2$, $h_3$, $h_5$, $h_7$ | 1,5 | 14,2 | 232 | 2,55 |
| **Maximum power** | | | **P=2000W** | | |
| | without filters | 2,6 | 8,18 | 230 | 3,04 |
| | with filters for за $h_2$, $h_3$, $h_5$, $h_7$ | 2,7 | 8,01 | 231 | 2,26 |

The analysis of the data leads to the conclusion that the added group of filters at the point of connection, successfully suppresses the influencing harmonics with sequence numbers 2, 3, 5, and 7 and generally reduces the total coefficient of harmonic components THD. On the other hand, it can be said that the filters work more efficiently at low load.

## 4. Conclusions

The main modes of operation of TPSS have been modelled, and the effects of adding three-phase filters in the output circuit after the voltage inverter have been analysed. From the PWM simulation, the frequency spectrum for the output current $I_L$ and voltage $U_L$ has been synthesized.

The analysis of the data shows that the added group of filters, at the point of attachment, successfully suppresses the influential harmonics with sequence numbers 2, 3, 5 and 7 and generally reduces the total coefficient of harmonic components THD. At lower load capacities, the filters work more efficiently. It is proved that according to the designed model a real three-phase inverter power supply system can be created, which can be connected to the public power supply network. The reduction of $THD_I$ to 14,23% has been obtained by connecting notch filters. The notch filters can be effectively used in working inverter systems as their inclusion in the output circuit creates a model of the system with improved electromagnetic compatibility, and hence an increased power factor and efficiency.

## References

[1] Fekik A, Denoun H, Benamrouche N, Benyahia N and Zaouia M 2015 A fuzzy – logic based controller for three phase PWM rectifier with voltage oriented control strategy, *International Journal of Circuits, Systems and Signal Processing*, **9** pp 412–419

[2] Juncan C, Minhua Y, Liping J, Yingchao Z and Long L 2011 Voltage oriented control for three-phase PWM rectifier based on fuzzy PI control, *Int. Conference on Electric Information and Control Engineering*, Wuhan, China, pp 3784-3787

[3] Kadirova S, Evstatiev I, Kajtsanov D and Nenov T 2019 Microprocessor based electronic system for control of induction heating in machine industry, *16th Conference on Electrical Machines, Drives and Power Systems (ELMA)*, Varna, Bulgaria, pp 1-5, doi: 10.1109/ELMA.2019.8771487

[4] Stoyanov I, Iliev T, Mihaylov G and Ivanova E P 2017 Synthesis and design of virtual measurement system for investigation on photovoltaic modules, *10$^{th}$ International Symposium on Advanced Topics in Electrical Engineering ATEE*, Bucharest, Romania, pp 915-918, doi: 10.1109/ATEE.2017.7905070.

[5] Jawad F and Shahgholian G 2009 Modeling and simulation of a three-phase inverter with rectifier-type nonlinear loads, *Armenian Journal of Physics*, **2** (4) pp 307-316 Available at: https://arar.sci.am/dlibra/publication/25965/edition/23253

[6] Shanshan L, Chen B and Shi M 2020 *IOP Conf. Ser.: Earth Environ. Sci.* **467** 012091, https://doi.org/10.1088/1755-1315/467/1/012091

[7] Stoyanov I, Ivanov V, Iliev T and Ivanov H 2019 Yield and performance study of a 1MWp grid connected photovoltaic system in Bulgaria, *Journal of Engineering Science and Technology Review, Special Issue on Telecommunications, Informatics, Energy and Management*, ISSN: 1791-2377 pp 206-209

[8] B Gavaskar Reddy et al 2017 *IOP Conf. Ser.: Mater. Sci. Eng.* **225** 012071. https://doi.org/10.1088/1757-899X/225/1/012071

[9] Batard C, Poitiers F, Millet C and Ginot N 2012 *Simulation of Power Converters Using Matlab-Simulink*, ed V Katsikis (IntechOpen, United Kingdom) doi: 10.5772/46419

[10] Evstatiev B, Doychinov Y and Gabrovska K 2018 Method for creating virtual copies of real labs in a 2D environment, representing a 3D virtual reality, *IEEE 24$^{th}$ International Symposium for Design and Technology in Electronic Packaging SIITME*, Iasi, Romania, pp 138-141, doi: 10.1109/SIITME.2018.8599266

# AUTHOR INDEX